With the recognition of the importance of cell reproduction for the creation and maintenance of animal body form came the realization that cell division is a central event in organismal, as well as cell, biology. Division of living cells could be observed with relative ease long before it could be satisfactorily analyzed by experimentation. Its regularity and apparent simplicity prompted much speculation at first, but little concerted effort at objective analysis. This book attempts to trace the long history of some of the major ideas in the field and gives an account of our current knowledge of animal cytokinesis. It contains descriptions of division in different kinds of cells, as well as the proposed explanations of the mechanisms underlying the visible events. Experiments devised to test cell division theories are described and explained. The forces necessary to deform animal cells to the degree shown in cytokinesis now appear to originate from the interaction of linear polymers and motor molecules that have roles in force production and in the motion and the shape change that occur in other phases of the biology of the cell. The localization of the force-producing division mechanism to a restricted linear part of the subsurface is caused by the mitotic apparatus, the same cytoskeletal structure that ensures orderly mitosis.

All those with a particular interest in the cell cycle will find this book invaluable. Its clear readable style, however, will ensure that those working in other areas of cellular and developmental biology will also find this an interesting account of an event of central importance in the life of the cell.

DEVELOPMENTAL AND CELL BIOLOGY SERIES

EDITORS

P. W. BARLOW J. B. L. BARD P. B. GREEN D. L. KIRK

CYTOKINESIS IN ANIMAL CELLS

Developmental and cell biology series

SERIES EDITORS
Dr. P. W. Barlow, *Long Ashton Research Station, University of Bristol*
Dr. J. B. L. Bard, *Department of Anatomy, Edinburgh University*
Dr. P. B. Green, *Department of Biology, Stanford University*
Dr. D. L. Kirk, *Department of Biology, Washington University*

The aim of the series is to present relatively short critical accounts of areas of developmental and cell biology where sufficient information has accumulated to allow a considered distillation of the subject. The fine structure of cells, embryology, morphology, physiology, genetics, biochemistry, and biophysics are subjects within the scope of the series. The books are intended to interest and instruct advanced undergraduates and graduate students and to make an important contribution to teaching cell and developmental biology. At the same time, they should be of value to biologists who, while not working directly in the area of a particular volume's subject matter, wish to keep abreast of developments relevant to their particular interests.

OTHER BOOKS IN THE SERIES
R. Maksymowych *Analysis of leaf development*
L. Roberts *Cytodifferentiation in plants: xylogenesis as a model system*
P. Sengel *Morphogenesis of skin*
A. McLaren *Mammalian chimaeras*
E. Roosen-Runge *The process of spermatogenesis in animals*
F. D'Amato *Nuclear cytology in relation to development*
P. Nieuwkoop & L. Sutasurya *Primordial germ cells in the chordates*
J. Vasiliev & I. Gelfand *Neoplastic and normal cells in development*
R. Chaleff *Genetics of higher plants*
P. Nieuwkoop & L. Sutasurya *Primordial germ cells in the invertebrates*
K. Sauer *The biology of* Physarum
N. Le Douarin *The neural crest*
M. H. Kaufmann *Early mammalian development: parthenogenetic studies*
V. Y. Brodsky & I. V. Uryvaeva *Genome multiplication in growth and development*
P. Nieuwkoop, A. G. Johnen & B. Albers *The epigenetic nature of early chordate development*
V. Raghavan *Embryogenesis in angiosperms: a developmental and experimental study*
C. J. Epstein *The consequences of chromosome imbalance: principles, mechanisms and models*
L. Saxén *Organogenesis of the kidney*
V. Raghavan *Developmental biology of the fern gametophytes*
R. Maksymowych *Analysis of growth and development in* Xanthium
B. John *Meiosis*
J. Bard *Morphogenesis: the cellular and molecular processes of developmental anatomy*
R. Wall *This side up: spatial determination in the early development of animals*
T. Sachs *Pattern formation in plant tissues*
J. M. W. Slack *From egg to embryo: regional specification in early development*
A. I. Farbman *Cell biology of olfaction*
L. G. Harrison *Kinetic theory of living pattern*
N. Satoh *Developmental biology of ascidians*
R. Holliday *Understanding ageing*
P. A. Tsonis *Limb regeneration*

CYTOKINESIS IN ANIMAL CELLS

R. RAPPAPORT

The Mount Desert Island Biological Laboratory, Maine
and
Union College, New York

CAMBRIDGE UNIVERSITY PRESS
Cambridge, New York, Melbourne, Madrid, Cape Town, Singapore, São Paulo

Cambridge University Press
The Edinburgh Building, Cambridge CB2 2RU, UK

Published in the United States of America by Cambridge University Press, New York

www.cambridge.org
Information on this title: www.cambridge.org/9780521401739

First published 1996
This digitally printed first paperback version 2005

A catalogue record for this publication is available from the British Library

Library of Congress Cataloguing in Publication data
Rapppaport, R.
Cytokinesis in animal cells / R. Rappaport
p. cm. – (Developmental and cell biology series; 30)
ISBN 0-521-40173-9 (hc)
1. Cytokinesis. I. Title. II. Series.
QH605.R27 1996
591.87′62–dc20 95-37443
CIP

ISBN-13 978-0-521-40173-9 hardback
ISBN-10 0-521-40173-9 hardback

ISBN-13 978-0-521-01936-1 paperback
ISBN-10 0-521-01936-2 paperback

This book is dedicated to Barbara Nolan Rappaport,
who made it all possible, but declined co-authorship

Table of Contents

Preface

The study of the mechanisms of animal cell division is a mature field. Division has long been considered a central event in the life of cells and it has, as a consequence, stimulated much curiosity and thought, many observations, and numerous ingenious experiments and measurements. The literature of the field is rich and reliable, but it does not appear to have developed in a markedly systematic or orderly fashion; therefore, persons seeking a broad acquaintance with the available information can easily miss important work or ideas. The purpose of the book is to lead the reader into the literature. It was not possible to do full justice to the work of all the investigators who have made significant contributions.

Although the time period during which cytokinesis has been actively investigated is by modern standards lengthy, the number of persons actively contributing new information and ideas at any one time has been small, and the rate of progress has been slow. Whether they proved to be right or wrong, the work and thoughts of the older workers are important, and the gradual emergence of the personalities of these workers from the yellowing pages of their writings has been an enjoyable aspect of this protracted writing project. My emphasis on the importance of published literature may arise in some part from my entrance into the field without formal training or guidance. I was enthralled by the experiments of Yatsu, which demonstrated that cells are far more durable than I had imagined and that simple experiments can reveal fundamentally important facts about cytokinesis. Since then,

I have tried to follow a similar pattern of investigation involving experiments, many of which, I now realize, would have been possible with the equipment available by the early 1920s.

It is still possible to find out new important facts about cytokinesis by physical experimentation on living animal cells, and I hope that some of the readers of this book will be moved to adopt this method of analysis.

Acknowledgments

The original research described in this book and the writing of the book itself were supported by grants from the National Science Foundation, to whom the author extends sincere thanks. Extensive library support was provided by Ruth Anne Evans, David Gerhan, and other members of the Union College Library staff.

1

Normal Cell Division

> All we can do is to rely on visual phenomena and interpret them as best we may. Unfortunately, visual phenomena are seldom identical in any two types of cell. An observer is likely to attribute greater importance to a particular phenomenon if this is more strikingly obvious than another in the particular material he is observing: in another type of cell the clarity may be reversed and a different prospective is obtained. For this reason, it is not hard to reach a position more adapted to dialectic skill than scientific enquiry.
>
> J. Gray, *A Textbook of Experimental Cytology* (p. 211)

Early interest in cell division was fueled by recognition of its importance in both the development and maintenance of animal form and by the relatively simple and easy measures that enabled 19th-century biologists to witness its dynamic events. The realization that those events were clearly organized and controlled by each individual cell added to the attraction. In their thinking about cell division, biologists have tried to formulate a basic or fundamental form of the process that can, with minor adjustments, be used to explain the ways in which it occurs in all its variations. For reasons of convenience and history, the basic form of cytokinesis starts with a spherical cell containing a central mitotic apparatus. Division entails the formation of a uniform circumferential indentation on the surface that deepens in a plane that intersects with the midpoint of the mitotic apparatus and is oriented perpendicular to its axis. Division ends in the formation of two daughter cells of equal volume. This "typical" process closely resembles the early cleavage divisions of many invertebrate embryos that are convenient for observation, as well as for experimentation, and it is not surprising that embryologists were responsible for most of the early cell division studies.

In 1948, Jean Clark Dan wrote (p. 191): "The cleavage of marine eggs was one of the first loves of the early embryologists, and their successors in slender but persistent series continue to give earnest thought to the problem of how a cell divides. One of the most spectacular of biological phenomena, cell division is the more difficult to analyze because of the elementary nature of the structures involved; its apparent simplicity captures the imagination and baffles it; the tremendous range of variations in this process challenges and, at the same time, discourages the investigator."

Many of the qualities that made early cleavage divisions of marine invertebrate eggs the choice investigational subject of early workers are still considered important today. The eggs and sperm are plentiful in season and are relatively easy to obtain. Most eggs have a regular, symmetrical shape, and the consistency and relative synchrony of events that follow simultaneous fertilization of a large number of them has been an important convenience. Except for a cyclical tendency to round up and relax, the shape changes of cleaving embryonic cells are directly connected to the division process. Transparency and durability are common properties. The normal medium required for cell survival is plentiful and simple. And in a period when concern with artifact was pervasive, the fact that eggs are normally free and independent of other cells and are bathed by their normal medium obviated worries about damage and artificiality that plagued early observers of isolated tissue cells.

Observation of cell division and speculation about the mechanisms that underlie it seemed to go hand in hand. Few of the early observers and describers devoted themselves exclusively to the study of animal cell division, but the contributions of the many who recorded their thoughts on the subject make an appreciable body of literature. Many of the early insights, observations, and perceptions appear only in footnotes and discussion sections of papers that are primarily devoted to other subjects. This circumstance makes a complete survey nearly impossible, and this book can best be considered as a way to enter the literature. But thorough literature searches are essential for those who would avoid the embarrassment of being unknowingly the last in a long line of persons to propose an old hypothesis or, perhaps worse, championing an already disproven hypothesis. The idea that the mitotic apparatus is as important for the division of the cytoplasmic mass (cytokinesis) as it is for the sorting and transportation of chromosomes (mitosis) was considered general dogma many years ago (Gurwitsch, 1904). Yet, depending upon what one considers satisfactory explanation, many important aspects of the correlation between time and place of the two events are unresolved.

The relation between observation, hypothesis, testing of hypothesis, and reformulation has not followed a textbook pattern in this field. Reduction of confusion and convenience of exposition may be achieved by ignoring the chronology of discovery. Cytokinesis begins after most of the events of mitosis are finished. Both mitosis and cytokinesis are consequences of a complex of cyclical metabolic events that are beyond the scope of this book. This first chapter is concerned primarily with descriptions of the divisions of three cell types that have shaped both theory and the course of experimentation. The amount of detail in the descriptions will approximate that which served as the basis for most of the theorization. The following chapter contains accounts of the speculation and theory that arose from observation of the normal process. Subsequent chapters concern the experiments, measurements, and the refined and extended observations that have been used to evaluate assumption, conjecture, and theory.

The three cell types that were used repeatedly in early cell division studies are represented by cleaving echinoderm eggs, cleaving ctenophore and amphibian eggs, and dividing vertebrate tissue cells. Many echinoderm and ctenophore eggs are sufficiently large and transparent to permit detailed study of living cells with relatively simple microscopes and illuminators. Vertebrate tissue cells, on the other hand, were at first usually prepared by classical cytological methods. Fixed tissues containing actively dividing cells (like salamander testis) or relatively isolated cells (like embryonic blood cells) were favored. An accurate conception of the cytoplasmic activity during and immediately before cytokinesis in these cells came only with development of tissue culture methods.

Sea Urchin Egg Cleavage

Among the echinoderms there are many species of sea urchins and starfish whose cleaving eggs are nearly ideal subjects for observing cell division. Sea urchins, however, have become the standard experimental material because the techniques for obtaining ripe gametes are simple and the maturation divisions of the egg have been completed before spawning. In the early days, gametes were obtained by opening animals, removing the gonads and mincing them. At present, intact animals are spawned by measures that induce contraction of the saccular gonad walls. Judicious use of potassium chloride, acetylcholine, or alternating current permits repeated spawning of the same animal and its eventual return to the sea at the end of the breeding season.

Fertilization is normally external and, in the laboratory, suspensions of eggs and sperm are mixed in proportions that favor the penetration of each egg by a single sperm. Each egg is surrounded by a thin, closely fitting vitelline membrane and, external to that, a thicker layer of transparent, colorless jelly. Entrance of the sperm triggers the rupture of a population of cortical granules or vacuoles positioned close to the egg surface. The rupture places the vacuole contents outside the confines of the egg. Part of the vacuole contents combine with the vitelline membrane, and the reinforced structure (the fertilization membrane) is lifted by the inflow of water into the space between it and the egg surface. Another part of the vacuole contents remains associated with the egg surface and is the initial source of a closely fitting, transparent, extracellular hyaline layer. The thickness and toughness of the hyaline layer vary among the species. Relatively simple methods for removing these various layers and structures have been devised so that direct access to the egg surface is available to the experimenter. The transparency of sea urchin egg cytoplasm is variable, because yolk material and pigment contribute to opacity. For some micromanipulation experiments, sand dollar eggs are preferred because pigmented material is deposited in the removable jelly, leaving the egg cytoplasm relatively transparent.

The time between the introduction of sperm and first cleavage is consumed by fertilization as well as mitosis. The interval in the sand dollar (*Echinarachnius parma*, a flattened echinoid common in the cold waters of the western North Atlantic) is about 90 minutes. The internal events of fertilization that culminate in the fusion of sperm and egg nuclei take up the first 15 minutes. The events of fertilization have generated a large body of experimentation and literature and will not be elaborated in this book. In brief, the most striking visible structure that develops and facilitates the union of the nuclei strongly resembles in appearance and constitution one of the principal parts of the mitotic apparatus that waxes and wanes in each subsequent division cycle. This structure, the sperm aster, organizes around an extranuclear body, the centrosome, which entered the egg as part of the sperm. The radiate appearance of the sperm aster is based upon alignment of cytoplasmic granules and membranes along microtubules that extend toward the egg surface from the centrosome. As the sperm aster grows, the sperm nucleus is displaced toward the egg center. Eventually the sperm aster appears to fill the egg, and the egg and sperm nuclei touch and fuse. The centrosome splits and the daughter centrosomes take up positions on diametrically opposite sides of the nucleus. The radiate structure then fades and subsequent events appear to be open to several interpretations. According to Schroeder (1987), the sperm aster disassembles entirely, leaving the centrosomes, and no rebuilding of the aster begins until prophase. P. Harris, Osborn, and Weber (1980) believe that interphase asters centered on the centrosomes develop as the sperm monaster fades. The interphase asters grow and, before the nuclear membrane ruptures at the beginning of prophase, the rays are redistributed into a discoidal configuration that excludes cell particulates to the extent that, when viewed on edge, it appears to be a light streak. Schroeder does not discuss the streak. From this point on, the accounts are harmonious. Prophase begins shortly, and the nuclear membrane ruptures. During this period, the organization and expansion of the mitotic apparatus is the dominant visible event. The chromosomal portion of the mitotic apparatus originates in the nucleus; the remainder of the structure is organized by the centrosomes from other cytoplasmic components. The fibrillar matrix of the mitotic apparatus is constituted much like the sperm aster.

Although the spindle and asters of sea urchin eggs are large and clear, the chromosomes are not (Figure 1.1). This leads to some uncertainty in precise identification of the stages of mitosis in living cells. While the chromosomes progress through the familiar mitotic events,

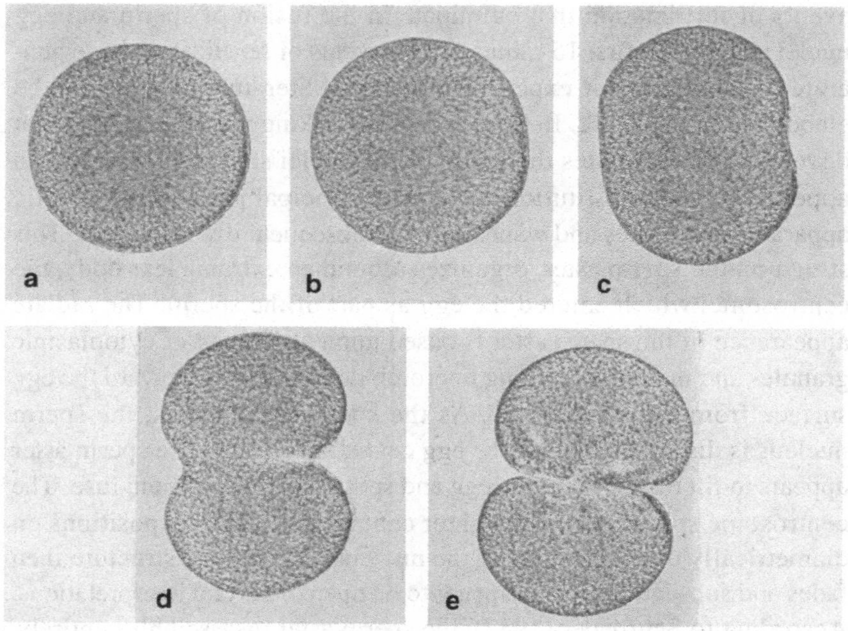

Figure 1.1 First cleavage of an unflattened sand dollar (*Echinarachnius parma*) egg.

important nonchromosomal events take place. The aster rays grow during metaphase and anaphase. Despite the size and clarity of the living mitotic apparatus in eggs with transparent cytoplasm, its dimensions and form at the maximum stage of development have been a subject of controversy. The astral centers are distinct, so the mitotic axis and spindle length are evident. However, the outer limit of the asters is unclear during stages when the mitotic apparatus appears to fill the cell. Many of the classical cytologists (reviewed, K. Dan, 1943b; J. Dan, 1948) and investigators who work with transparent, living cells (K. Dan, 1988) were convinced that astral rays extended across the equatorial plane (the plane of the chromosomes at metaphase) from both centrosomes. But an ultrastructural study of the microtubular arrays that are thought to form the physical basis of the radiate structure led to the conclusion that the asters are mutually exclusive hemispheres (Asnes and Schroeder, 1979). The point has theoretical significance and will be discussed later. The chromosomes move toward the poles and the spindle elongates during anaphase. In late anaphase, the centrosome region flattens into a plate oriented perpendicular to the mitotic axis and the shape of the spindle changes from fusiform to nearly cylindrical. In telophase, the centrioles within the flattened centrosome separate and duplicate. Centriolar duplication and distribution in the absence of a nucleus are normal (Sluder, 1990). As the spindle poles approach the underside of the polar surfaces, some of the intervening cytoplasm is displaced. It flows toward the equator where, according to some descriptions, the currents from the poles converge and are redirected toward the cell axis. Flow movements in sea urchin eggs are not striking. Changes in the appearance of rays during the mitotic cycle have frequently been described. Fry (1929) characterized the early prophase rays as "vague"; those of late prophase and early metaphase were "clear but very delicate." From late metaphase to late anaphase, they were "coarse," and in early telophase they were again clear but delicate. Fry used classical cytological techniques, but more recent techniques have yielded similar observations (P. Harris, Osborn, and Weber, 1980). Cell deformation begins as the chromosomes approach the spindle poles in telophase. The chromosomes assume vesicular form and the chromosomal vesicles, or karyomeres, fuse to form daughter nuclei. In some circumstances there is a brief period when the polar curvature decreases slightly, resulting in a degree of axial flattening or "oblation" (Usui and Yoneda, 1982; Schroeder, 1985a). Whether this event is observed or not, the first extensive deformation is a decrease in curvature of the equatorial surface that rapidly transforms that part of the cell into a short cylinder. Cell volume remains constant during division, so that the decrease in curvature at the equator displaces local cytoplasm and causes distension at the poles. The cylindrical contour is rapidly transformed into a shallow indentation as the cleavage progresses. The significance of the early cylindrical phase has been differently interpreted over the years. Some have considered it to be the beginning of cleavage, whereas others assumed it to be a separate, precleavage phenomenon caused by a different physical mechanism. The distinction can become important in interpretation of the phrase "beginning of cleavage" in the older literature.

The indenting cleavage furrow completely divides the egg in about 10 minutes, depending upon the species and the temperature. At first, the daughter cells are nearly spherical, but those that are surrounded by the normal extracellular investing materials soon flatten against each other and remain hemispherical until the next division. Daughter cells cyclically change from compressed shapes to near-spheres at the beginning of each successive cleavage (Figure 1.2). The phenomenon is very common in animal cells and has been a subject of great interest. Before second cleavage, the mitotic apparatus of each blastomere orients parallel to the plane of the first cleavage in the longest dimension of the cell (Figure 1.2). The two blastomeres are completely separate, but the events of mitosis and cytokinesis are normally synchronous. The plane

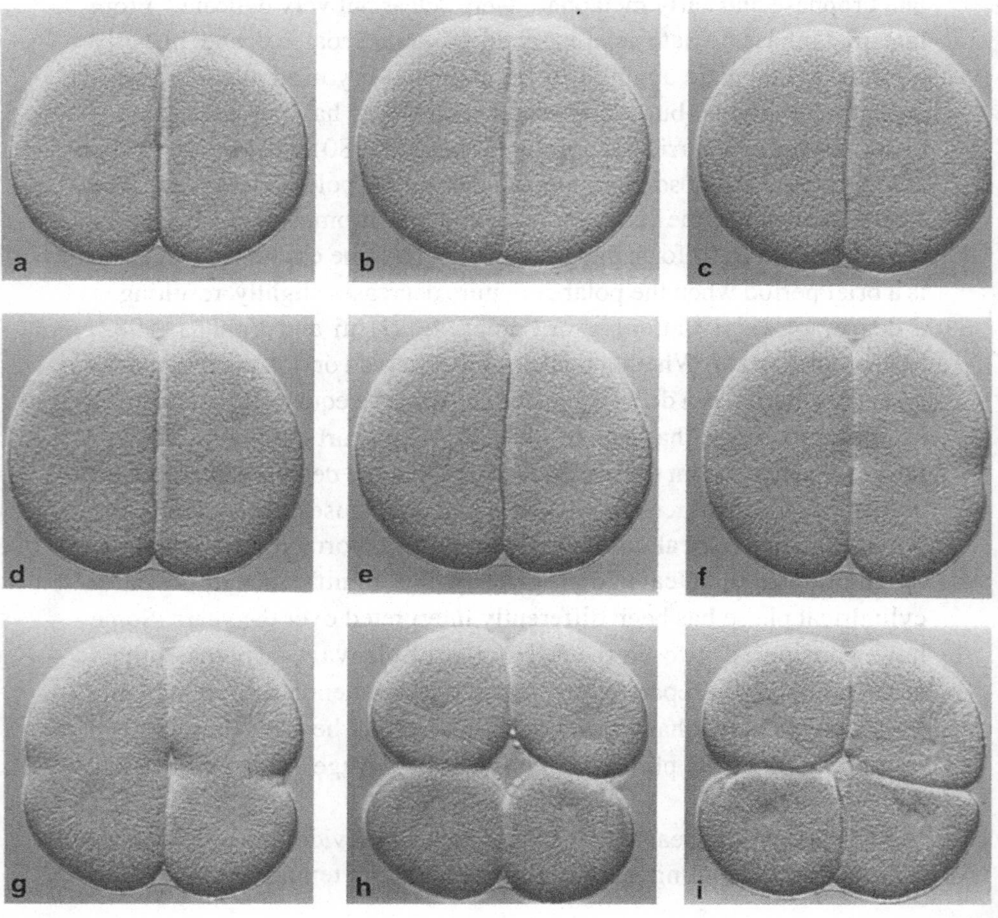

of the second cleavage furrows is perpendicular to the plane of the first, and the egg is divided into four equal blastomeres. The interval between the first and second cleavages in *Echinarachnius parma* is 48 minutes, and the division process itself requires 8 minutes. The longer, 90-minute period between insemination and first cleavage comprises the time necessary for fertilization and gamete pronuclear fusion, as well as mitosis. Blastomeres at second cleavage are often used in experiments where events of fertilization must be avoided and an accurate time control is required.

The division process in sea urchin eggs is usually represented as radially symmetrical. The mitotic apparatus is large, the asters are equal in size, and the midpoint of the spindle axis appears to coincide with the geometrical center of the whole egg. In this circumstance, the furrow appears simultaneously in all parts of the equatorial surface, and it progresses toward the mitotic axis at the same rate around the entire circumference. The inevitable consequence is that the last part of the cell to cleave is close to the mitotic axis. Such symmetry can be a great convenience for experiment and measurement, but it is not necessary for division, even among echinoderms. In many species of sand dollars, for instance (K. Dan and J. C. Dan, 1947), the mitotic apparatus is somewhat off center, so that the center of the spindle is not the same distance from all parts of the equatorial surface. The cleavage furrow

Figure 1.2 Second cleavage of a slightly flattened sand dollar egg .
(a) 21 min before beginning of second cleavage; nuclear membrane intact, small asters at diametrically opposite sides of the nucleus.
(b) 8 min before beginning of second cleavage, 7 min after nuclear membrane breakdown; small mitotic apparatus oriented parallel to first cleavage plane.
(c) 6 min before beginning of second cleavage; asters enlarged.
(d) 4 min before beginning of second cleavage.
(e) 3 min before beginning of second cleavage; surface flattened near subpolar region of upper aster in right blastomere.
(f) Beginning of second cleavage.
(g) 1 min after beginning of second cleavage; chromosomal vesicles present on the furrow sides of the telophase asters.
(h) 5 min after beginning of second cleavage; chromosomal vesicles more distinct and blastomeres are rounded.
(i) 15 min after beginning of the second cleavage; nuclei have reformed and persistent aster rays are present; blastomeres have flattened against each other.

develops in the same relation to the asters and spindle as in spherically symmetrical eggs, but the early indentation appears first in the region of the equator closest to the mitotic apparatus. The egg looks heart shaped when cleavage begins, but the furrow indentation gradually extends around the circumference until its initially unsymmetrical form is lost.

Unilateral Cleavage, Ctenophores, and Amphibia

The basic components of ctenophore and amphibian eggs are so differently arranged and proportioned from those of sea urchin eggs that they seem like a natural experiment. They clearly show that the near spherical symmetry and large mitotic apparatus of sea urchin eggs are not necessary for division. Adult ctenophores are pelagic, and in any coastal area their number is affected by local currents, weather, and other unknown factors. As a consequence, their availability may be sporadic, a serious disadvantage for an experimental animal. The transparent adults are functional hermaphrodites although barriers to self-fertilization appear to exist in some species (Carré and Sardet, 1984). Their spawning activity is affected by the diurnal light cycle, and early investigators placed them in jars and collected newly laid eggs each morning. They are apparently more sensitive to pollution than sea urchins, because Yatsu (1912a) remarked that the water of Naples Bay was toxic, so he kept them in water obtained near Capri. The eggs are exceptionally transparent. Yatsu (1912 a) described an egg organization consisting of a thin, semifluid outer layer; a thicker, clear, firm ectoplasm (which glows upon electrical stimulation in some species); and a large central endoplasmic region containing sizable, transparent vesicles of yolk (Figure1.3). The very small mitotic apparatus is embedded in the ectoplasm so that it is far from the egg center and close to the surface. Ziegler (1898b) described changes in the ectoplasm before cleavage. These changes included formation of a thickened ridge projecting slightly into the endoplasm where the furrow would later develop, and a pair of noselike surface bulges that contain the telophase asters. The furrow develops between the bulges, and the thickened ectoplasm is carried down between them as the "cleavage head" (Figure 1.4). Yatsu (1912b) and others described intense "spinning activity" in the furrow

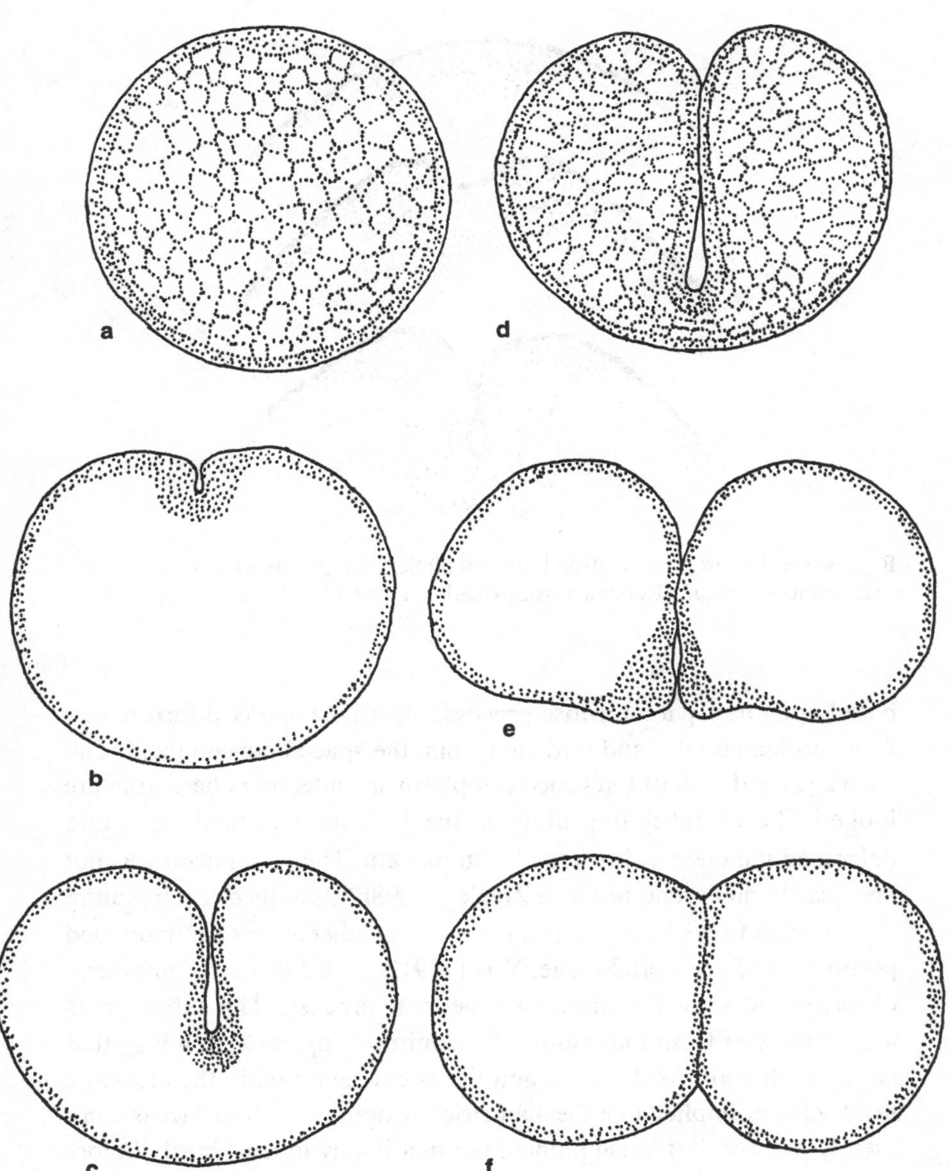

Figure 1.3 First cleavage of *Beroë ovata.* The drawings emphasize the difference between the endoplasm and ectoplasm. (Redrawn from Needham, 1950)

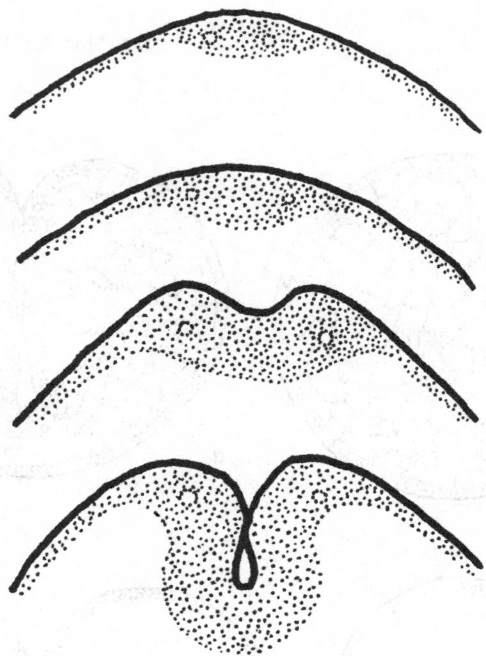

Figure 1.4 Diagram of relation between nuclei, ectoplasm, cleavage furrow, and surface contour. (Redrawn from Ziegler, 1898b)

based upon transparent, filose processes from the opposed furrow surfaces projected into, and retracted from, the space between them. The cleavage head with its thickened ectoplasm and internal radiate structure looked like an intriguing division machine as it pushed aside and deformed the clear yolky vesicles in its path. The endoplasm was not necessarily quiescent, because Ziegler (1898b) described reciprocating cytoplasmic flows from one blastomere to another across the undivided portion of the equatorial plane. Yatsu (1912a) used the term "unilateral cleavage" to describe the unsymmetrical process. The furrow cuts across the egg from the region of the mitotic apparatus in a way that suggests that all the dynamic activity is concentrated in the cleavage head. Ziegler implied that the diametrically opposite cell surface is completely passive, but Yatsu pointed out that it may indent shortly before division is completed. The later cleavages resemble the first in that the mitotic apparatus is relatively small and eccentric, the furrow is unilateral, and the separation between ectoplasm and endoplasm is distinct. Eventually the ectoplasm appears to be concentrated in a restricted part of the blastomere population, and furrows become symmetrical.

Ctenophore eggs attracted attention not only because of their internal structures, but also because of their size and durability. *Beroë ovata* eggs are about 1 mm in diameter, sufficiently large to suggest that simple freehand operations might be possible. When such operations were attempted, the physical consistency of the eggs and their favorable healing abilities ensured a satisfactory survival rate (Driesch and Morgan, 1895). Similar operations on dividing *Beroë* and other eggs followed (Yatsu, 1908, 1912b), but this experimental approach was not widely used until many years later. Unilateral cleavage is not restricted to ctenophore eggs. Many coelenterate eggs cleave unilaterally. Some are also relatively transparent and contain a small, highly eccentric mitotic apparatus.

In amphibians and some primitive fish eggs the relation between the mitotic apparatus and the unilateral furrow is similar to that in the ctenophore egg in the early cleavages; however, the cytoplasm, and in most cases the surface, is opaque, so that internal events, including mitosis, are obscured. The normal spherical shape of amphibian eggs is imposed by their surrounding membranes. When the membranes are removed, the eggs flatten against the substratum into a thick disk. Their large size (newt eggs are about 2 mm in diameter) permits observation of surface behavior with low magnification and reflected light and, for the practiced, the relatively slow furrow progress permits recording and mid-cleavage experiments by freehand operations. Methods for artificially inducing ovulation and maintaining healthy laboratory colonies of both frogs and salamanders can spare investigators the inconveniences of seasonal availability.

Contributing to surface opacity in many species is a concentration of melanin pigment granules embedded in the cortical and subcortical regions. The animal hemisphere surface usually appears darker than the vegetal hemisphere surface and, depending upon the species, the regional color difference may be slight or may be nearly as great as that between black and white. Immediately after insemination, before the surrounding jelly swells by hydration and while the vitelline membrane closely invests the egg surface, eggs are oriented randomly with respect to the animal–vegetal axis. But after eggs are free to rotate in the perivitelline space, their less dense animal hemispheres are uniformly positioned upward. The mitotic apparatus lies beneath the animal pole in an eccentric position. At the time of fertilization, the egg mitotic apparatus is in metaphase of the second meiotic division, and it then proceeds to second polar body formation before the haploid gamete pronuclei join. Depending upon the species and the temperature, cleavage occurs from an hour and a half to several hours after insemination. Cleavage begins with the formation of a restricted, elongate depression in the somewhat flattened animal hemisphere surface. The furrow is perpendicular to the mitotic axis and, as it deepens, minute folds or wrinkles appear in the adjacent surface; they terminate in, and are oriented normal to, the cleavage plane. The furrow progresses by simultaneously deepening and lengthening, so that it encompasses a progressively larger part of the surface. The wrinkles are smoothed, and in some forms the marginal regions between the pigmented and unpigmented hemispheres at the ends of the furrow appear to shift toward the animal pole as they are folded into the furrow. The appearance of the animal hemisphere surface seems to depend upon the species and upon the conditions imposed upon the egg. In *Amblystoma punctatum* (Harrison, 1969) and *Triturus alpestris* (Selman and Waddington, 1955) surrounded by intact membrane, the pigmentation in the animal surface adjacent to the furrow pales as the furrow deepens. In *Cynops pyrrogaster*, the paling is not visible while the egg membranes and jelly are intact, but it appears after demembranated eggs flatten on the substratum (Sawai, 1987). In *Rana pipiens* and *Rana sylvatica,* the dark brown pigmentation of the animal surface does not change as the eggs cleave within the vitelline membrane (personal observation).

As the furrow deepens, its ends progress toward the vegetal pole more rapidly than does its central region, so that in the intermediate stages, the base of the furrow describes an arc. Eventually the ends meet near the center of the vegetal hemisphere surface. During the 30–45 minutes required for the first cleavage, the two nuclei formed in the first mitosis enter the second cycle, and the second cleavage begins before the first is completed.

Observation of the internal events that precede and accompany cleavage was complicated by egg opacity in life or brittleness following fixation and embedding. The mitotic apparatus is similar to that of other animal cleavage cells, although its size relative to the cytoplasmic volume is not as large as that of echinoderm eggs. Associated with the mitotic apparatus in the future division plane is a thin plate of cytoplasm that is distinguished by a reduced concentration of stainable yolk particles. This structure, which was termed the diastem or diastema, has been observed primarily in the eggs of amphibia and in those of sturgeon and other primitive fish that employ a similar cleavage pattern. The diastema was first described many years ago (reviewed, Zotin, 1964), and similar structures were described in the future cleavage planes of other vertebrate and invertebrate eggs. Because some of the structures that were considered to be diastema proved to be artifacts, skepticism concerning their reality appears to have discouraged investigation of their function for some time. The diastema and the paling of the equatorial surface appear to be distinguishing features of amphibian cleavage, and both have been the subject of considerable speculation.

Cell division in egg cleavages is in some ways different from the type of cell division that occurs in the tissues both of later embryonic and of adult stages. The size of the mitotic apparatus as a whole, as well as the size of the asters in particular, is usually relatively greater in the cleavage stages. There is no growth during the cleavage period, so that successive divisions decrease the volume, and increase the proportion, of nuclear material in each cell. The factors that determine the length of the cell cycle in the different stages also appear to differ. However, the transition between the two division types that occurs during development in every multicellular individual is almost imperceptible, and the basic structures and activities that are directly related to cytokinesis appear to be similar.

Tissue Cell Division

Although Flemming (1895) described division of living epithelial cells at a very early date, most of the older studies of dividing tissue cells were based upon fixed, sectioned, and stained material. The older studies appear to have been satisfactory for understanding the basic events of mitosis, but they did not convey a clear picture of cytoplasmic activity and they provided only an approximate idea of timing. Tissue cell division activity is less regular than cleavage division. The time between divisions is more variable, as are the durations of the mitotic phases. Tissue cell shape in all phases of division is more irregular. Shape variations may in part be associated with the strong tendency of tissue cells to adhere to, and to form junctions with, nearby cells and substrata.

The invention of tissue culture methods (Harrison, 1907), the progressive improvement of these methods, and the use of time-lapse cinematography have greatly increased the accessibility of the process. Also important have been optical and computer methods that enhance the contrast between structures and regions that are characterized by small differences in index of refraction. The visible chromosomal phases of mitosis have been very convenient reference points for descriptions of all the activities within the cell during the division period. Correlations between the time when the chromosomes are in a particular configuration and the time when the cytoplasm engages in a particular activity should not, however, be assumed to be the consequence of a direct causal relation between the two events.

The earliest unmistakable signs of impending division appear in, and immediately around, the nucleus. The nucleoli fade and the nuclear matrix appears more granular. The chromosomes become more ribbon-like and condense into double-stranded form. The nucleus often swells. The centrosome divides and the halves, containing centriole pairs and surrounded by a radiate structure similar to that of sea urchin eggs, move to diametrically opposite points on the nuclear surface. The rays between the centrosomes elongate as spindle formation begins. The nuclear events of early prophase are accompanied by cytoplasmic events involving both attachments at the surface and the cell contour. Cells that may have been spread thinly on the substratum into a nearly two-dimensional form with an irregular marginal contour begin to round up. The peripheral cytoplasmic sheets and processes retract and the cytoplasm is displaced to the vicinity of the nucleus (Figure 1.5). The methods and cells used in early tissue culture studies created the impression that all tissue cells must round up and, in effect, convert themselves

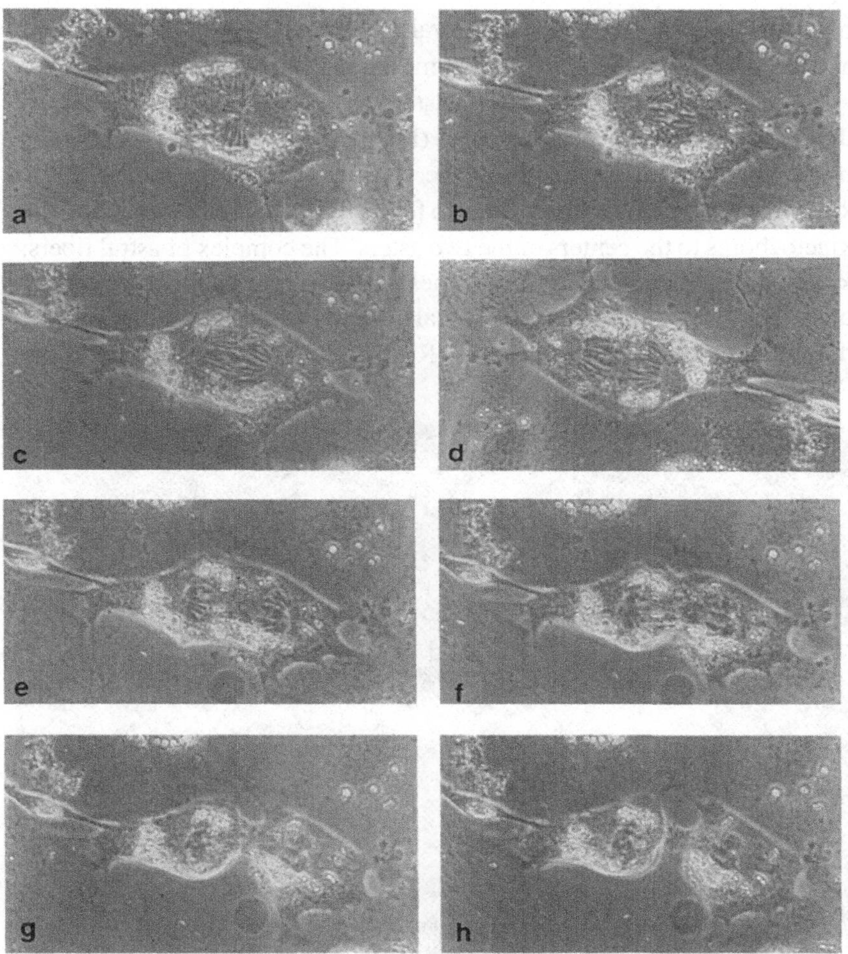

Figure 1.5 Dividing newt kidney epithelial cells. (Rappaport and Rappaport, 1968, and with permission of Wiley-Liss)

into small sea urchin eggs before the critical phases of division begin. This interpretation was a great convenience for theorists, but it is not true. With the introduction of different culture media, chamber surfaces, and tissue cell types (especially epithelia), the cell's capacity to divide despite starting from a great variety of shapes has been demonstrated clearly.

The breakdown of the nuclear envelope that signals the prometaphase begins near the centrosomes. The astral rays extend from the centrosomes, and the spindle organizes as the chromosomal centromeres associate with astral microtubules. As the mitotic asters grow, their rays interdigitate in the region between the centrosomes (Figure 1.6). After nuclear envelope breakdown, the microtubules in the forming spindle interact with the kinetochores on each chromosome to form kinetochore fibers that link the kinetochores to the centers of the two asters. The complex of astral fibers, kinetochore fibers, and chromosomes makes up the spindle, which not only bears the chromosomes, but also binds the asters into a single, durable unit, the mitotic apparatus (Rieder, 1990). The coherence of the

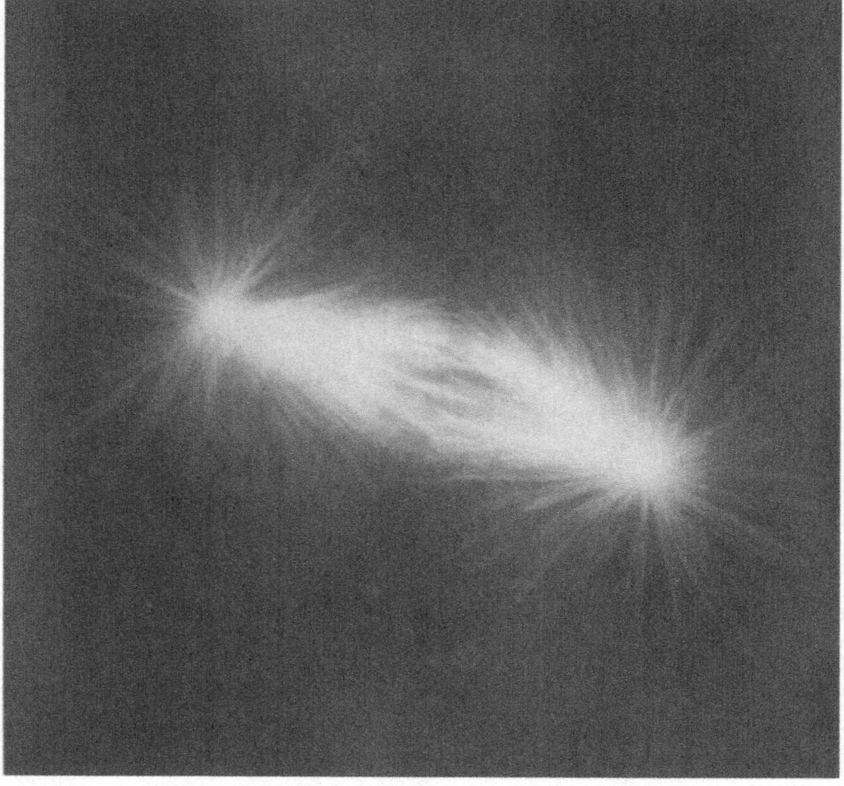

Figure 1.6 Antitubulin indirect immunofluorescent micrograph of a prometaphase newt lung cell (Photograph kindly supplied by Dr. Conly Rieder, and reproduced from the *Journal of Cell Biology*, 1986, 103:590, by copyright permission of The Rockefeller University Press)

spindle excludes cytoplasmic granules. Spindle–centromere interaction orients the chromosome so that each of its duplicate parts, or chromatids, faces one of the spindle poles. When the chromosomes are arranged so that each is equidistant from the spindle poles, the cell is considered to be in metaphase.

Anaphase begins when the chromosomes begin moving toward the spindle poles (Figure 1.5b). Their movement away from the spindle midpoint or equator has two components. "Anaphase A" consists of the movement of the chromosomes along the spindle fibers toward the poles. "Anaphase B" is the simultaneous increase in the length of the whole spindle. As the two chromosomal groups separate, the portion of the spindle that lies between them loses its capacity to exclude cytoplasmic granules, which then flow toward the cell axis (Figure 1.5d–e). In late anaphase, the chromosomes appear to swell and contact each other, and the nuclear envelope begins to re-form. At this time, cytokinesis also begins. The processes that attach flattened cells to the substratum tend to retract in the equatorial region, and the contour of the associated margin appears less flattened (Figure 1.5e–f). The equatorial cell diameter progressively shrinks. Early observers described an intense bubbling and blebbing at the polar surfaces during anaphase in some types of tissue cells. In some cases, this activity may have been caused by culture conditions.

The completion of nuclear envelope re-formation marks the beginning of telophase. At this point, chromosomal decondensation begins and the appearance of the nucleus reverts to the interphase form. Cytokinesis continues. The furrowing process reduces the equatorial diameter to a fine thread that may persist for some time after completion of other division-related events. The shape changes that comprise furrowing in tissue cells follow a predictable pattern at the equator, but they are variable in the polar and subpolar regions. Sometimes, the degree of polar flattening and attachment are scarcely changed; however, in other cells, the prospective daughter cells round up as the processes detach from the substratum and retract. During interphase, the tendency to round up is reduced, and the cell becomes more closely applied to its immediate physical surroundings.

Many of the basic phenomena of animal cell division were clearly described in the 19th century, but a pervasive preoccupation with problems of artifact created a kind of ferment. There was general acknowledgment that artifacts could be introduced by microscopical technique, by cytological stains and reagents, by maintaining cells in unphysiological media, by simple mechanical agitation, and by many other factors. Some structures that are now accepted as real, such as centrioles, spindle fibers, details of astral structure, and cyclical changes in centrosomes, were considered artifacts. Other structures that are now considered artifacts, or misinterpretations, such as vesicles in the future plane of cleavage, were considered real. Adding to the confusion were well-documented instances in which mitosis is not immediately followed by cytokinesis, as in insect development and the early cleavages of some coelenterate eggs. There were other instances where one could deduce that cytokinesis was not immediately preceded by mitosis, as in the development of some crustacea (Pereyaslawzewa, 1888). E. E. Just (1939), after describing the different relations that appeared to exist between nuclear division and cytoplasmic division, rejected the idea that the two events were directly linked.

In an atmosphere where some observations may be dismissed as cytological artifacts and where simple experimental results may also be dismissed as the artifactual consequence of an operation, it is difficult to accumulate a large body of generally accepted data. The amount of speculation possible is inversely proportional to the amount of conclusive data available. This relationship sustained a period of intense theorization.

2

Theories of Cell Division

Usually, it is easier to invent a new theory of cell division than to test an old one.

L. V. Heilbrunn, *The Colloid Chemistry of Protoplasm* (p. 256)

Accurate descriptions of the events of mitosis and cytokinesis were available at a time when fascination with the mechanical forces that shape cells and organisms was widespread. Although the genetic basis of development was acknowledged, the physical basis for the arrangements and the form that typify cells and developing organisms was considered an equally important area of investigation among those who were at the forefront of cell and developmental biology in the late years of the 19th century. The concept of Entwicklungsmechanik is generally considered to have originated among German scientists, and much of the early, speculative literature concerning the mechanism of cell division was written for German periodicals by investigators interested in animal development as well as cell biology.

The obvious importance of cell division, as well as the relative ease of observation and the apparent simplicity of the event, captivated the inquiring and imaginative mind and provided a seemingly irresistible temptation into speculation. At one time or another most of the leading cell and developmental biologists of the day either contributed a theory or commented in print on those contributed by others. The possibilities for theorization were enormous. Simple observation of the event in either living or fixed cells did not reveal whether events were active or passive. The chemical and structural basis of the mitotic apparatus and other visible organelles was unknown, and the significance of the form and proportion changes that both the mitotic apparatus and the whole cell undergo, as well as the obligatory increase in cell surface area, could be construed in many ways.

The correlation between the position and orientation of the mitotic apparatus and the furrow led to the assumption of a causal relation. The assumption was supported by the frequently made observation that in eggs flattened to such an extent that the mitotic apparatus was abnormally oriented, the furrows were also abnormally oriented with respect to the egg axis, but normal with respect to the mitotic apparatus.

Proponents of novel theories realized that any hypothetical division mechanism had to be consistent with an acceptable concept of the basic nature of submicroscopic cell organization, but the nature of protoplasmic organization and the way in which it formed the visible organelles were also open to speculation.

By 1895, many theories of mitosis postulated that spindle fibers moved chromosomes by active shortening, like muscle fibers. On the other hand, the changes in internal morphology that took place at that time in mitosis prompted Drüner (1894) to propose that, at the same time that contracting fibers pulled the chromosomes toward the poles, the similar-appearing fibers in the central region were actively pushing to produce the visible elongation of the spindle and the separation of the centrosomes. There were, however, others who favored the idea that the linear radiate elements of the mitotic apparatus were caused by dynamic activities and were based upon the interplay of forces and motion rather than organized stable structures. The different conceptions were often based upon the author's interpretation or convictions about the physical basis of cytoplasmic structure. The idea that the fibrillar achromatic part of the mitotic apparatus was based upon flow phenomena arose repeatedly. Both O. Hertwig and Roux postulated that the radiations were produced by exchanges between the nucleus and cytoplasm that involved some kind of attractive interaction (reviewed, Wilson, 1928). Platner (1886) favored the idea that the rays were hollow tubes through which clear cytoplasm flowed. Chromosome movements and shifting of cell parts during division he considered to be passive responses to cytoplasmic streaming patterns. Bütschli had proposed an alveolar theory of protoplasmic structure in which the basic arrangement was that of a foam (described in Wilson, 1928). In the cytoplasm, he postulated, the foam was formed by two immiscible fluids of different viscosity that formed a continuous and a discontinuous phase. The general configuration of the foam changed readily as different physical forces were brought to bear upon it. The theory gained wide acceptance because it was consistent with the microscopic appearance of both living and fixed cytoplasm, and because Bütschli was able to make lifelike models with emulsions of thickened olive oil and soapy solutions of mineral salts. The achromatic mitotic apparatus, Bütschli said, reflected the orientation of the alveoli in the foam caused by diffusion streams. During the growth of the asters, he proposed, the fluid uptake that occurred at the centers resulted in tensions in the surrounding cytoplasm that caused radiating figures.

From 1875 to the early 1900s, the period during which the majority of theories were proposed, there were few opportunities or attempts at evaluation. The intense speculative activity of the time did not result from rapid testing, rejection, and reformulation, but rather, it now seems, it was based upon the many different interpretations and rearrangements that were possible with the available information, and the desire of each author to formulate an original contribution. Because each of the theoretical mechanisms was designed to replicate the normal event, and because there was little useful information about the way in which cells divide under abnormal circumstances, one hypothesis appeared to be as rational as another. The suggestion that another new hypothesis might be little improvement over an untested old hypothesis is not to be found in the writings of the time; neither is the possibility that the intense expenditure of ingenuity and energy that characterized the speculative era may have in the final analysis been only minimally helpful in the understanding of the actual event.

Chronologically arranged discussions of progressive scientific inquiry are often the clearest method of exposition, but, in this case, because the order of publication does not usually reflect a rational process, it seems preferable to group the theories according to the general mechanism proposed.

Theories That Postulate a Physically Active Mitotic Apparatus

A simple and intellectually satisfying way to account for the apparently close linkage between mitosis and cytokinesis is to assume that both events are caused by the same physical mechanism. The idea that the achromatic parts of the mitotic apparatus could transport chromosomes to the end of the spindle and then deform the whole cell to the extent necessary to complete division is very appealing. Several attractive theories that involved active participation of different parts of the mitotic apparatus and different physical mechanisms were developed.

Traction fiber theories

The idea that the radiate structure of the asters and the fibrillar appearance of the spindle reflected arrangements of linear tension-generating or tension-transmitting elements had occurred to several early students of cell division, and, in 1887, Van Beneden proposed that all movements that accompany cell division are caused by the contractility of protoplasmic fibrillae that are generally responsible for cell deformation and locomotion. Wilson (1928) characterized these ideas as a "beautiful hypothesis" although he pointed out that the arrangement of rays that Van Beneden postulated does not exist in many cells.

Heidenhain (1897) presented a clear, carefully reasoned version of this theory that was typical of his time. It appears to follow logically from his conception of the basis of the physical structure of cells and from the particular cells that he chose to study. He maintained that the scaffolding, or cytoskeleton, that supported and controlled animal cell shape consisted of linear arrays of contractile molecules ("wie die Gleider einer Kette" – like the links of a chain) that permeated the cytoplasm at all times. The resulting cell form was not static, but dynamic and subject to change as elements contracted and relaxed, and the clearest evidence of its adaptable nature was the intracellular cytoplasmic flows that he had observed in isolated living cells such as amoebae or amoeboid cells. The radiate structure that he observed in stained cells in association with the centrosomes and mitotic apparatus were, he proposed, composed of contractile material. The cells upon which he based his speculations were embryonic, dividing, duck red blood cells. Interphase cells were described as resembling biconvex lenses that became spherical as mitosis approached. The nucleus was relatively large, so that when the daughter centrosomes took up diametrically opposite positions across the nucleus, they were very close to the polar surfaces. Heidenhain's cytological methods allowed accurate observation of changes in chromosomal configuration, the achromatic parts of the mitotic apparatus, and the external cell contour.

The assumptions of Heidenhain's model were:

a. The rays that originate at the centrosomes are contractile, and rays of equal length exert equal force.
b. Longer rays exert greater force.
c. The rays are firmly attached to the surface.

The movement of the daughter chromosomes was attributed to the contractile rays, and the subsequent attachments of the rays from both poles of the mitotic apparatus to the surface could be construed as a continuation or modest modification of that relationship. Cytokinesis resulted when the longer rays that were attached to the equatorial surface exerted greater traction and pulled the equator inward toward the mitotic axis. The subsequent elongation of the cell and the mitotic apparatus was considered by Heidenhain to be passive events resulting from the displacement of cytoplasm out of the equator that inevitably accompanies the local reduction in diameter. Although the physical basis of the shape change might not have been as simple as he supposed, the hypothesis was good in that it was consistent with his observations. It was based upon structures and activities that were reasonable. In other words, he proposed that division was carried out by structures and activities that were similar, or identical, to those that functioned throughout the cycle. The special nature of division was based primarily on the configuration of the astral rays and their degree of development, but not upon the acquisition of special properties.

A further intriguing aspect of Heidenhain's presentation of his traction theory was his detailed description of a two-dimensional mechanical model made of spring steel bands, rubber bands, hooks, and rings that replicated to his satisfaction some of the early events of cytokinesis (Figure 2.1). The impact of Heidenhain's work upon the investiga-

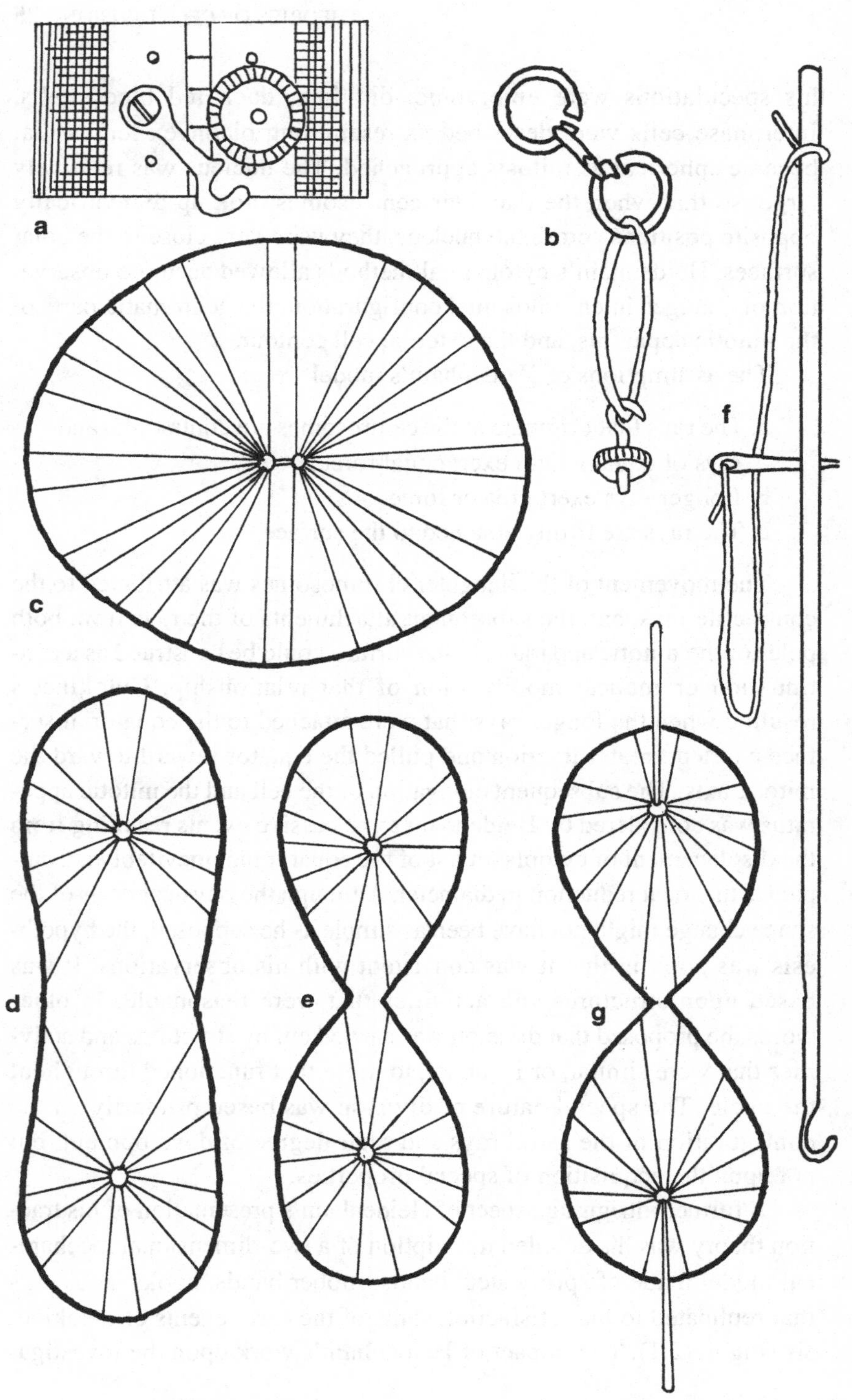
a
b
c
d
e
f
g

tors of his time is now difficult to assess. Model building was then a common way of trying to understand and explain complex physical phenomena. One might guess that it was impressive to noncytologists. Certainly his contemporaries pointed out that alternative mechanisms might work as well, but few presented schemes in such elaborate detail.

A modified, more complicated version of Heidenhain's fiber traction theory was presented by Rhumbler (1903). At the heart of his theory, too, was a mechanical model that may have been as important in shaping his ideas as the cell itself. Critics of Heidenhain's model pointed out that the hinges he put in the equatorial region of the steel bands that represented the surface of his two-dimensional model had no biological equivalents. The hinges made the physical properties of the model surface unacceptably different from the cell surface. Rhumbler avoided the difficulty by substituting a circle of rubber tubing for the steel band that represented the cell periphery, but otherwise he used a similar arrangement of radially arranged rubber bands and spring rings. The model simulated division activity in much the same way as did Heidenhain's, and its flexibility permitted a certain amount of experimentation that convinced

Figure 2.1 The Heidenhain (1897) model of cytokinesis. (Drawings are not all to the same scale.) The model consisted of two 22-mm-wide spring steel bands joined at their ends by hinges (a). Holes were drilled through the steel bands at regular intervals to accommodate small steel hooks, which were held in place by nuts. Rubber bands were placed on the hooks and threaded through rings (b). The two rings, which represented the aster centers, were held together, and when the hinge catches (a) were closed, the hinges could not be flexed. When the pair of rings was positioned midway between the hinges and when the rubber bands were installed so that they joined the hooks to the rings, the perimeter (surface) of the model became slightly ellipsoidal (c). After the short connection between the rings was removed, the rings shifted apart as the model snapped into an elongate shape with a slight constriction in the plane of the hinges, which were prevented from flexing (d). When the hinge catches were released, the hinges flexed and moved toward each other (e). If at this point the model were hung on a perpendicular surface, the distance between the hinges decreased. Heidenhain believed that in this circumstance gravity acted like a force generated by the anaphase B movement of the asters, which affected the surface contour through the radii and the tension they generate. Further constriction was obtained by drilling holes through the midpoint of both pieces of spring steel and inserting a hook on a shaft (f), engaging each of the rings, and pulling them closer to the surface (g). Heidenhain felt that the behavior of the model proved that cell deformation resulted from force generated by the anaphase movement of the spindle poles and by tension in the radial elements. (Redrawn from Heidenhain, 1897)

Rhumbler that the general concept of change in cell shape by traction of linear elements was correct. Rhumbler's theory differed from Heidenhain's in several respects that he considered important. Heidenhain's assumption that the radiate structure of the mitotic apparatus was based upon permanent chains of contractile protein that could shorten like muscle fibers was rejected in favor of a more complicated hypothetical structure based on Bütschli's theory of alveolar cytoplasmic ground substance. Accordingly, the rays were based upon centripedal channels of cytoplasm that flowed into the centrosomal area at the time of division. This interpretation was consistent with the observation that the clear central portion of the asters grew while the rays were present. Rhumbler postulated that this activity produced a tensile stress parallel to the direction of flow and a compressive stress perpendicular to it. The resulting forces constituted the most important agents of shape change.

Neither Heidenhain's nor Rhumbler's model simulated the changes in surface area (or model circumference) that accompanies cytokinesis, but Rhumbler expanded on the subject and incorporated it in his discussion. He pointed out that in the usual rounding up that occurred in division of tissue cells, a certain amount of excess surface became available and was temporarily stored. This, plus new surface that grew as needed during division, sufficed to cover both daughter cells. The additional surface, he thought, was not formed by simple stretching of original surface. The role of the rays, he posited, was to move the newly available surface and prepare the way as the cleavage furrow cut through the cell.

Ziegler (1895) pointed out that traction fiber theories were based on studies of stained, fixed cells. This line of speculation reached its apogee with Heidenhain and Rhumbler. The idea that traction fibers could explain the entire process of cytokinesis fell into disfavor, but the idea that their activity could account for some aspects of cytokinesis lingered in several theories that invoked multiple causal mechanisms.

Expanding spindle theories

The elongation of the spindle during mitosis was accepted as fact in the latter half of the 19th century, and some proponents of traction fiber theories attributed it to tensile stress exerted by fibers that bridged the space between the centrosomes and the polar surfaces. The visible events of cytokinesis did not, however, exclude the possibility that the elongation, rather than being a passive response to tensile stress, might reflect the activity of expansive forces located within the spindle itself. With the acceptance of this alternative interpretation of visible events, it requires only a short logical leap to propose that active spindle elongation may be the primary motive force for cytokinesis. Platner (1886) proposed a simple mechanism in which active spindle expansion elongates the cell, and then the cytoplasm divides by the action of surface forces. He likened cell division to the bisection that follows when a drop of mercury is stretched between two needles.

Although this apparently simple explanation of the entire process did not seem to attract much support, it reappeared as a contributing factor in several theories that attributed division to the operation of several different physical activities that operate simultaneously or sequentially.

Astral cleavage theories

In contrast with traction fiber theories, astral cleavage theories were based upon observation of experimentally manipulated cells. Teichmann (1903) produced sea urchin egg cells with more than two asters by inducing polyspermy and by selectively suppressing cytokinesis with cold, ether, and shaking. Because mitosis was not inhibited, each cycle doubled the number of asters present in a single cell so that when suppression was discontinued, he could see the way in which furrowing occurred when the number and arrangement of asters were abnormal. He found that furrows could form between asters that were not attached to poles of the same spindle, but furrows could not always form between every aster pair. He also stated that the depth of the furrow depended upon the distance between the asters. Wilson (1901a) had previously noted that, in similar circumstances, the depth was related to the size of the asters. Teichmann proposed that cytoplasm periodically develops a tendency to aggregate. Early in the mitotic cycle, he maintained, it aggregates around the nucleus, but later it gathers around the centrosomes. As the asters grow, they take up the material in the equatorial region and, as the chromosomes approach the astral centers, the cytoplasmic uptake removes so much material from the equator that the diameter of the equatorial region is progressively reduced. The furrow develops as a consequence of the separation between the asters. Morgan (1899) had briefly suggested a similar mechanism (Wilson maintained that his ideas were based upon inadequate observations), and Danchakoff (1916) simply assumed that events such as Teichmann proposed were responsible for the progressive changes she described in fixed and stained cleaving sea urchin eggs. She does not appear to have considered any other possible mechanism.

Chambers (1919) attempted to estimate the relative firmness or viscosity of different regions of dividing cells by probing them with needles. He concluded that asters were formed by accumulation of material into two semisolid masses at the expense of more fluid cytoplasm. The growth of asters in this fashion, he maintained, produced cell elongation along the mitotic axis. He elaborated on these ideas and described a hypothetical mechanism closely resembling that of Teichmann (Chambers, 1921). The astral rays, he proposed, were centripetally directed streams of fluid cytoplasm that caused microscopically visible transport of small granules and fat droplets. The liquid cytoplasm jelled in the asters and the asters grew at the expense of the fluid cytoplasm with the eventual consequences proposed by Teichmann. At that time, people were trying to explain protoplasmic behavior in terms of colloid chemistry, and Chambers emphasized the importance of cyclical sol–gel transformation in aster formation.

The basic ideas of the astral cleavage mechanism were seductive. Burrows (1927) felt that they satisfactorily accounted for the visible events of cytokinesis in cultured vertebrate tissue cells, although Gray (1931) thought otherwise. Gray incorporated the ideas of astral cleavage in a major cell division theory. It could account for (or imply the usefulness of) several activities that were observed in normal cleavage, such as increase in aster diameter, the centripetal movement of cytoplasmic particles along astral rays, the increase in size of the central clear centrosomal area in the asters, and the normal relation between the asters and the furrow. It implied that all the major active components in cytokinesis were visible microscopically. More impressively, it could be used to explain the furrowing activity in polyastral cells, which would be very difficult to do with traction fiber theories, and it was consistent with the results of Chambers's microdissection studies.

Gray was reluctant to speculate extensively on the division mechanisms of tissue cells. He termed the process "disjunctive" division. He was impressed by the anaphase bubbling at the polar surfaces and by the active drawing apart of the daughter cells by pseudopodial activity. The asters, which were the central division machinery in his hypothetical astral cleavage hypothesis, were not visible with the optical methods used to observe early cell cultures, and he doubted their existence although they were described and figured in earlier studies of sectioned and stained tissues. Similar ideas were suggested by Ziegler (1895), Gallardo (1902), Gurwitsch (1904), and Bonnevie (1906). If Wilson could consider Van Beneden's traction fiber theory beautiful, the astral cleavage hypothesis might be considered exquisite, although it did not explain everything, and it, too, was eventually disproven.

Theories That Postulate a Physically Active Surface

Deformation of the surface is characteristic of animal cell division, but the question whether the deformation is active or passive was long-lived and controversial. The traction fiber theories and the astral cleavage theory invoked simple, familiar physical mechanisms. They could be modeled, and their primary physical mechanisms were analogous, if not homologous, to familiar biological activities like muscle contraction or blood clotting. Traction fiber theories also required no additional explanation of the correlation between the position of the furrow and the position of the mitotic apparatus. For, although physically active surface theories have been relatively simple to explain, understand, and, in some instances, model, the process that establishes the necessary regional differences in cell surface behavior has introduced an additional set of mysteries. Yatsu (1909) maintained that surface theories that failed to account for the correlation between mitosis and cytokinesis were simply restatements of fact.

Equatorial contraction theories

Anyone who observes the division of a large, relatively transparent cell finds the idea that furrowing might be caused by a tightening band of equatorial surface material easy to accept. The minimal mechanism merely requires a means whereby surface contractility becomes regionally unequal so that the more contractile area is in the equator. The necessary difference could arise either by enhancing equatorial surface contractility or by reducing polar surface contractility. The end result would be the same, but the means by which the difference arises would have to be different.

EQUATORIAL TENSION INCREASE. In 1876, Bütschli published one of the earliest attempts to incorporate information concerning the visible events of cell division, protoplasmic organization, and cell mechanics into a comprehensive theory. In brief, he proposed that cells divide because the tension at the equatorial surface increases and that the increase results from the action of the mitotic apparatus. Fundamental to his hypothesis were certain convictions concerning the basis of visible astral structure. In observations of the activity of the contractile vacuole of the amoeba, he concluded that the associated radial canals were the consequence of fluid currents that converged upon the central vacuole. He assumed that such channels would look the same whatever the direction of flow, and he proposed that the visible rays of asters resulted from channels carrying clear fluid away from the centers. He further assumed that the effectiveness of the material transported in the rays would decrease with distance from the center, but he proposed that the ray pattern would distribute more material to the equator than to the poles because the effects of the asters were additive in that region. If the material increased the surface tension (perhaps by decreasing water content), then the cell would become a sphere in which the tension at the surface decreases from equator to pole and the form changes typical of cleavage would follow. Ziegler (1898b) applied Bütschli's theory to cleavage of ctenophore eggs with some modifications. Ziegler saw that the ectoplasmic thickening in the equatorial plane (see Chapter 1) was carried downward by the tip of the furrow. He reasoned that the ectoplasm exerted pressure on the egg contents in proportion to its thickness. The initial thickening in the equatorial plane could account for the early deformation at the beginning of cleavage, and the persistent thickening that surrounded the tip of the furrow continued to pull the surface through the yolk. According to this mechanism, the continuing contraction of the furrow base resulted from the persisting additive effect of the material that originated in the centrosomes. Ziegler's hypothesis attributed all of the force required for division to the contractile activity of the special band of ectoplasm in the furrow rather than to the simple physical surface tension changes proposed by Bütschli. There appears to be no suggestion by Ziegler that equatorial constriction could continue independently after it had been triggered, because he implied that continued centrosomal influence was required for the continuation of division. Ziegler (1903), who championed the virtues of studying living, dividing cells, also described an ectoplasm in echinoderm eggs similar in appearance and behavior to that in ctenophores. He was, however, confusing the echinoderm extracellular hyaline layer with the intracellular ectoplasm of ctenophores and cnidaria. He would not be the last to do so.

Just (1939) considered the most superficial region of the cell to be the site and cause of cell division activity. He also used the term "ectoplasm" to designate the region we now call the hyaline layer, and, although his thinking appears to have been shaped by studies on cleaving echinoderm eggs, his conception of the nature of the surface led him to the conclusion that ectoplasm is definitely a part of the living cell and that it is present in all animal cells. His interpretation of its structure also allowed him great latitude in the formulation of its hypothetical activities. His division theory appears to be in great part a redescription of the normal events of sea urchin egg cleavage. Up to anaphase, the thickness of the ectoplasm is uniform over the surface. In late anaphase, before deformation of the spherical cell begins, he asserted that the ectoplasm acquires a capacity for independent movement, and it flows from the poles toward the equator in a wavelike, amoeboid process. He considered ectoplasmic movement to be the cause of a superficial cytoplasmic streaming that develops at about that time (late anaphase). The ectoplasm that accumulated at the equator then was assumed to push the local surface inward toward the cell axis in the region where the cytoplasmic flows converged. The combination of ectoplasmic intrusion at the equator and a reduced resistance to deformation at the poles caused by local thinning of the ectoplasm would result in elongation. Just considered elongation unimportant, because it does not occur in the division of all cell types. He was uncertain whether, in the later phases of division, the surface that develops in the furrow region originated by separation of cytoplasmic cell plates (the diastema) or by extension of the ectoplasm inward. Just had little to say about events that set the process in motion. He dismissed nuclear division as a factor because, in his judgment, no constant relationship between the precise timing of mitosis and cytokinesis had been established.

The idea that the forces essential for cytokinesis were localized in a superficial gel layer was taken up again by W. H. Lewis in 1939, at a time when the colloid properties of cytoplasm were generally assumed and the possibility that the transition from sol to gel states could result in force production were generally accepted. Lewis argued that contraction of a superficial gel layer could explain cell shape changes, amoeboid locomotion, and cytokinesis (Figure 2.2). He stated that since

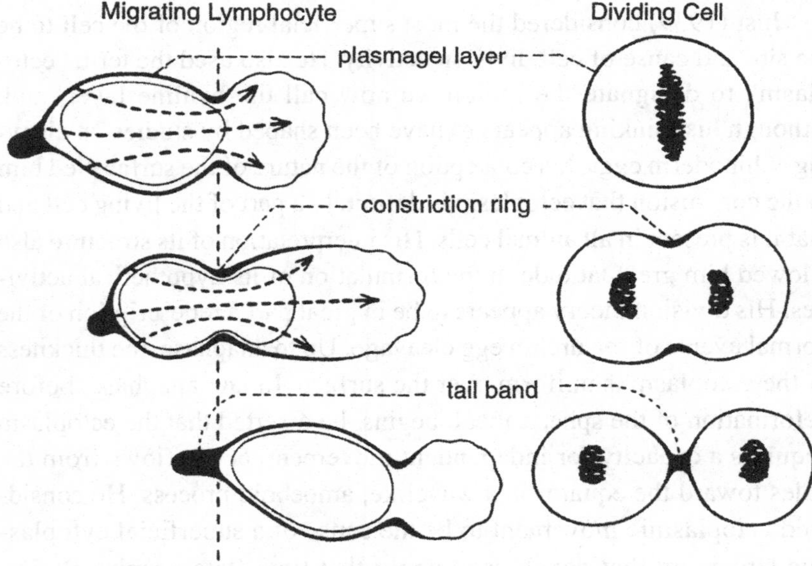

Figure 2.2 Diagrammatic comparison of the hypothetical role of superficial plasmagel layer contraction in lymphocyte migration and cytokinesis. The tension generated by the layer was assumed to vary with its thickness or density, which may differ locally. In the lymphocyte, the contraction of the gelated tail causes a pseudopodial projection in a weakened area of surface. When the ring of contractile material develops around the cell equator during cytokinesis, cytoplasm is displaced toward the poles and the cell does not move. The constriction continues until the cell is severed. (From W. H. Lewis, 1939)

most colloids exert contractile tension when they gel, he considered it probable that gelated cytoplasm acts likewise. In the 1930s, there was considerable interest in cytoplasmic viscosity, its regional differences, and cyclical changes. Viscosity was thought to be a consequence of the same molecular behavior that caused gelation, and a confusing tendency to equate viscosity, gelation, and contractility is sometimes encountered in literature of the period. Lewis maintained that the surface forces due to physical surface tension were negligible and that cell deformations were caused by changes in the superficial, elastic, contractile plasmagel layer. The initial reshaping of cells into elongate cylindrical form was attributed to the contraction of a broad, thickened, equatorial band of plasmagel layer. The functional furrow derived its contractile tension from a continuous gelation and thickening of a narrow region of the equatorial band. The special qualities of the equator were proposed to result from thickening of the plasmagel layer induced by underlying structures. Lewis based his hypothesis on observations and time-lapse studies of moving and dividing vertebrate cells in tissue culture.

Scott (1946) concurred with the equatorial contracting ring hypothesis and focused his attention on the factors that determine the placement of the ring. Studies of the effects of cold and hypotonicity on the shape of cleaving sea urchin eggs led him to propose that any deforming force that established an isthmus in a cell would result in the development of a contracting ring, provided that the egg was in a cleavage phase. He considered an actively elongating spindle, enlarging gelated asters, and endoplasmic flow as potential causes of deformation. In a subsequent modification of this theory, he proposed that the asters release an activator to contraction that affects the entire surface. Active contraction would proceed most easily at the aster periphery wherever (1) competing stresses and endoplasmic viscosity were favorable and (2) the initial furrow deformation would facilitate further furrowing. The early contraction that reshaped the equator into a cylinder, he suggested, could be caused by randomly oriented contractile units, but, as the furrow deepened, most of the active units would be circumferentially oriented.

The question of the nature of the contractile mechanism continued to occupy the thoughts of those who were won over by the equatorial tension theories, and Marsland and Landau (1954) elaborated further upon the idea that the necessary forces resulted from contraction of a superficial gel in the equator. There was a question concerning how the equatorial cortex could contract to the extent necessary to complete division, and Marsland and Landau proposed that there was a turnover of contractile material in the furrow that involved a continuous series of sol–gel reactions. The greatest contractile activity they attributed to gel mobilized along the furrow walls. As the contractile capacity of the gel diminished, the gel shifted to the region subadjacent to the furrow trough where it was "demobilized" as it presumably returned to a cytoplasmic pool. The contractile region progressed by turning over its constituents, as spent gel was replaced by freshly formed material. At the same time that gelation occurred in the furrow walls, solation took place under the polar surfaces as they were stretched. The hypothesis was durable and as new data appeared, they were incorporated with little change in the hypothesis. When the molecular nature of furrow composition became better known or strongly suspected, the idea that the cortical gel was an actomyosin-like protein fit easily. When polar relaxation was proposed as an important component of the division process and when the existence of "relaxing factors" was postulated, both phenomena could be incorporated as part of the polar solation phase. Although they attributed the distribution of relaxing factors to mechanical activity centered in and around the mitotic apparatus, Marsland and Landau did not speculate extensively about the factors involved in creating the necessary regions of very different activity in and near the surface.

POLAR TENSION DECREASE. In polar tension decrease theories, the hypothetically necessary regional difference in tension at the surface is achieved by a decrease in tension at the poles rather than by an increase at the equator. The first theories to incorporate this idea assumed that simple physical surface tension differences could divide cells. Although the forces that can be expected from such a mechanism are now considered insufficient, some of the ideas incorporated in surface tension theories are also applicable in other theories in which the tensile forces originate in other forms of molecular interaction. As an alternative to Bütschli's hypothesis, R. S. Lillie (1903) proposed that the necessary regional tension differences could arise if the anaphase chromosomes released a surface-tension–reducing substance. As the chromosomes approached the poles, the regional difference would increase until the equatorial surface behaved like a constricting band. McClendon in 1912 devised a convincing model in which a drop of oil, representing the cell, divided when the surface tension of diametrically opposite regions was rapidly reduced by applying sodium hydroxide. At the same time that the equator constricted, McClendon saw surface movement from the poles that converged at the equator and then returned poleward through the axis of the elongated drop. Spek (1918) restudied the currents in cleaving nematode eggs and emphasized the similarity of the patterns in the oil drop models and the living eggs. Because of this great similarity, it was tempting to assume that the boundary forces that operated in the eggs originated in the same way as those that divided the oil drop. But it became evident that boundary forces that originate in different ways may have similar effects. Greenspan (1977) described this group of hypotheses in concise analytical form and illustrates the operation of classical oil drop models (Greenspan, 1978).

In speculation concerning cytokinesis of amoebae, Chalkley (1935) proposed that the process was driven by regional differences in elastic gel strength rather than by differences in physical surface tension. The amoeba rounds up before it divides (Figure 2.3), and he suggested that beginning at anaphase the platelike arrangements of chromosomes release a substance that causes nearby surface to relax. As the chromosomes near the poles, their relaxing effect would cause the formation of polar pseudopodia that attach to the substratum. As a consequence of their normal locomotor activity, the polar pseudopodia would exert traction sufficient to tear the cell in half in the equatorial region, and the daughter amoebae would become permanently independent. According to this hypothesis, the force that divides the cell derives primarily from pseudopodial traction. Cytokinesis thus becomes a special case of amoeboid locomotion, and Chalkley preferred the term "fission" rather than "division" when applied to amoebae.

By 1960, when Wolpert proposed his detailed version of an astral relaxation theory, data from measurements and experimentation, as well as simple observation, were available. The existence of a cyclical increase in tension that appeared to involve the entire surface had been recognized in several cell types. Wolpert proposed that as the asters grow they retain their initially spherical form, so that when they are suf-

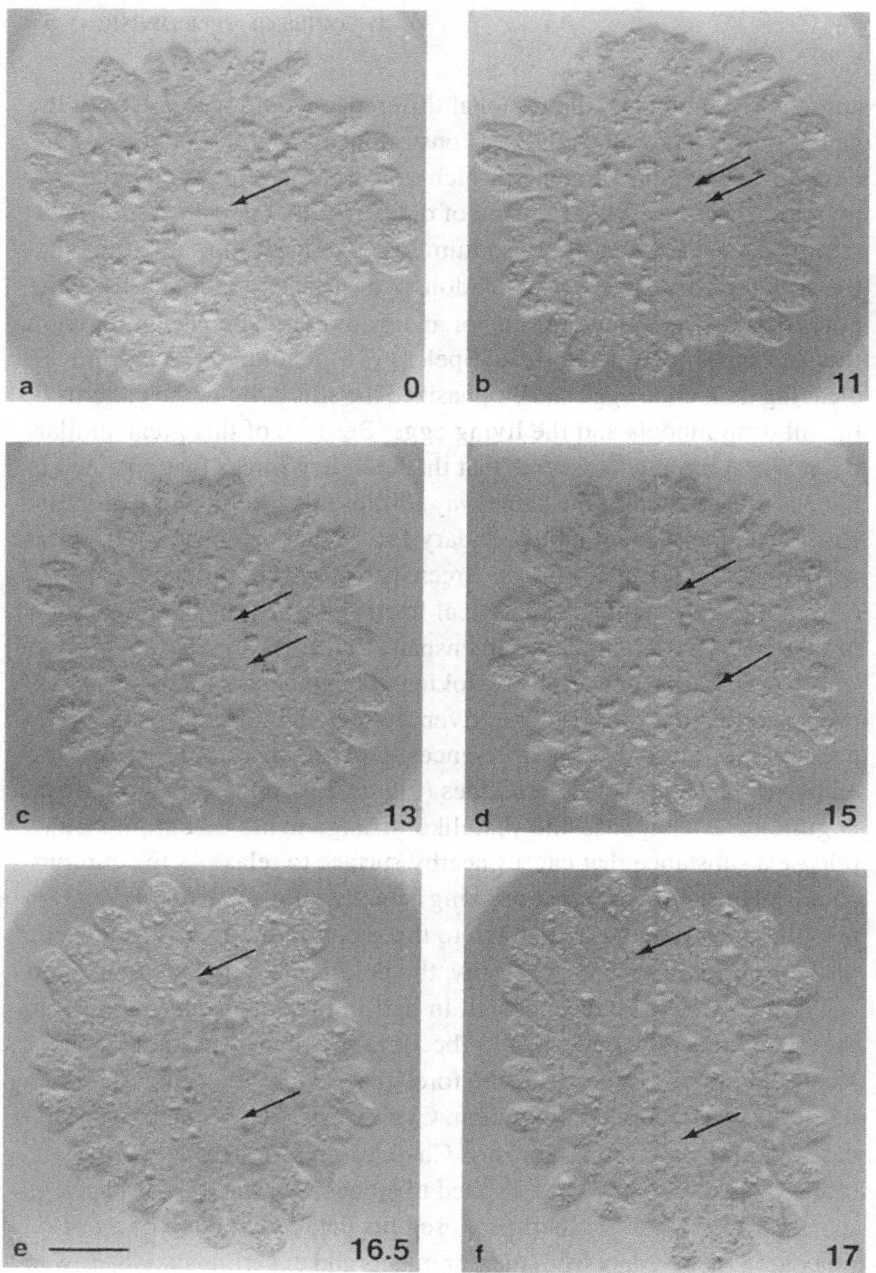

Figure 2.3 Cytokinesis in *Amoeba proteus*: mitotic movement of chromosomal plates. (a) Prophase nucleus (arrow) with adjacent contractile vacuole. (b–d) Progressive separation of chromosomal plates. (e–f) Telophase elongation. Time (min) indicated in the lower right corner of each frame. Bar = 50 μm. (From Rappaport and Rappaport, 1986, and with the permission of Wiley-Liss)

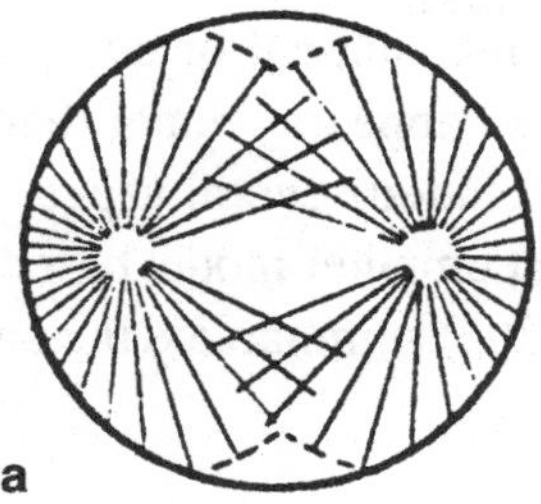

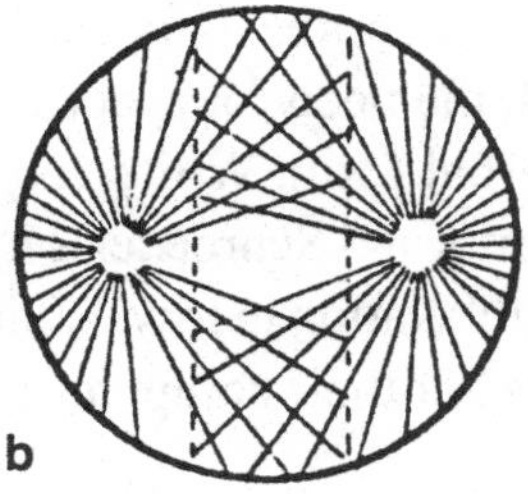

Figure 2.4 Diagrammatic representation of cleavage stimulus patterns. Stimulation is proportional to spatial frequency of membrane–aster contacts.
(a) Polar stimulation (b) Equatorial stimulation
(From Rappaport, 1971b, and with the permission of Wiley-Liss)

ficiently large to contact the polar surfaces, they do not reach the equatorial surface (Figure 2.4). This configuration ensures that a substance moving out of the aster at equal rates in all directions will reach the pole first and, if the asters cease to grow at that time, the substance might never affect the equatorial surface. If the substance causes the polar surface to relax, then regional tension differences at the surface will be established. In addition, if the asters do not have any effect on the equatorial surface, the increase in tension that had previously characterized the surface as a whole can continue, and the differences between the polar and equatorial regions will be further enhanced. The necessary new surface, Wolpert assumed, was provided by stretching preexisting membrane. After Wolpert's theory had declined in acceptance because it incorrectly predicted the outcome of several experiments, its main points reappeared in hypothetical mechanisms proposed by Schroeder and by White and Borisy.

Schroeder (1981a) postulated, as had Wolpert, that the generalized increase in tension at the surface that is not directly associated with mitosis constitutes an important motive force. He proposed that, as in Wolpert's hypothesis, the regional differences arise by a relaxing effect of the asters on the poles. The tension inequality subsequently causes equatorial constriction. It also causes a repositioning of a hypothetical cortical contractile lattice away from the vicinity of the aster and toward the region of the furrow. An event of this kind would be expected to further enhance regional tension differences, and its existence had been suggested previously by W. H. Lewis (1942) and Greenspan (1977).

When a hypothesis is cast in the form of a mechanical or computer model, the need for specificity, rigor, and detail appears greater than in the verbal form. White and Borisy (1983) included several aspects of Wolpert's and Schroeder's hypotheses in their computer model based on astral relaxation theory. White and Borisy (p. 302) summarized the basic features of their computer model as follows:

"1. Cleavage is a consequence of differential cortical tension.
2. Cortical tension is generated by an initially even distribution of randomly oriented linear tension producing elements that are free to move in the plane of the cortex.
3. These elements are brought into a state of uniform activation.
4. The presence of the asters modifies this activation by reducing the tension of elements in their proximity."

A rigorous description of the conditions that distribute the relaxing effect of the asters is necessary, and, in the original form of the theory, it was assumed that (1) the stimulus from the asters induces a relaxation of cortical tension proportional to the magnitude of the stimulus; (2) the stimulatory effect of each aster is spherically symmetrical and obeys an inverse power law; and (3) the level of stimulation at each point on the surface is the scalar sum of the stimulation from each aster. It was also necessary to assume that the stimulatory effect of the asters ceased by the time that cell deformation began. Cessation is required because the changes in shape that normally occur at the equator bring the equatorial surface so close to the asters that their assumed relaxing effect would reduce equatorial tension and thus stop division before it was completed. In its original form, the computer model simulates division of spherical cells. It is less successful in simulating division of cells that are not spherical. White (1985) later modified the astral relaxation theory by eliminating the assumption that the astral effect is spherically symmetrical. Although this change brought the assumptions closer to reality, its effect on the operation of the model was not discussed.

EQUATORIAL TENSION DECREASE OR POLAR TENSION INCREASE. The idea that an increase in equatorial surface tension or a decrease in polar surface tension can divide cells with easily deformable surfaces now seems simple and logical, but a rarely referred-to footnote in the history of this general subject illustrates how easily one can be led astray. The essence of cytokinesis in animal cells is the relocation of cell material out of the equatorial plane. Any mechanism that appears to move material in the opposite direction would seem to be working against the division mechanism. The earliest oil drop models of the cell surface that were developed at the end of the 19th century revealed that streaming movements were produced when drops of soap solution were applied to restricted parts of the oil drop surface. The streaming was directed away from the region where the soap lowered the surface tension. Robertson (1909) felt that the demonstrated direction of fluid movement argued against the possibility that cells divide by increased equatorial tension, because the model predicted that the flow away from the region of lower tension at the poles would cause accumulation of material in the equator and local increase in surface curvature. He predicted that the cell, rather than cleaving, would become more discoidal as the curvature at the equator increased and as that at the poles decreased. He proposed that in the cell, both telophase nuclei release surface-tension–reducing substances that attain their highest concentration at the equator because their effects are additive. Robertson devised an oil drop model in support of his hypothesis. He reported that a floating drop of oil divided when a thread impregnated with substances that decrease oil surface tension was gently brought in contact with a diameter of the drop. The division plane of the model coincided with the contact area of the thread. The accompanying streaming was directed away from the thread toward the poles. McClendon (1912) was unable to repeat Robertson's oil drop experiments and devised a model of his own (described earlier), in which an oil drop divided when diametrically opposite surfaces were exposed to surface-tension–lowering sodium hydroxide.

Notwithstanding McClendon's response, R. S. Lillie (1916) adopted Robertson's concept of a division mechanism based upon higher tension at the polar surfaces, but he attributed the difference to polar increase rather than equatorial decrease. In brief, he proposed that electrical fields produced by the asters depolarized the polar surfaces so that their tension was increased. Because local increases in surface tension can cause increased surface curvature, he suggested that the curvature at the poles would increase while that at the equator would appear to decrease. In this way, the cylindrical form of cells immediately before cleavage would arise. Subsequent deformations would follow from continuation of the same process.

Local surface growth and expansion theories

The nature of the relation between the inevitable increase in cell surface that takes place at cytokinesis and the forces that deform the cell illustrates the cause–effect dilemma that has faced the framers of cell division theory. The mechanisms thus far described have assumed (when the matter was considered at all) that new surface was formed by the stretching of old surface by stresses that hypothetically originate in a contractile ring, or in the traction fibers, or in the elongating spindle, and so forth. It is not difficult, however, to imagine that the roles may be reversed so that new surface formation becomes the active, dynamic phase of the process and the physical separation of the daughter cells is a relatively passive event requiring little investment of energy or other special activity. Theories incorporating the idea of active surface formation have been championed because they appear to be consistent (or not inconsistent) with observation of different kinds of living and fixed cells and because some investigators maintained that they had observed the process in living cells.

NO FORCE IMPLIED: DIASTEMA AND VESICLES. The idea that nascent surface may be preformed in the equatorial plane before division begins has been durable. Different forms of the hypothetical presurface condition have been suggested and described, but, in general, the transition of surface from nascent to functional states involves fusion of the preformed surface with the underside of the preexisting surface in the equatorial plane, followed by splitting of the nascent material beginning at the periphery. In this way, the reduction in equatorial diameter is interpreted as the result of progressive, organized exposure of the apposed new surfaces of the forming daughter cells. The apparent simplicity of the theory has been attractive. It could be considered a "morphological" theory, because cytokinesis is explained without the need for postulating associated force-producing mechanisms that must be created, activated, operated, turned off, and then made invisible. Perhaps equally appealing is its resemblance to the way in which cytokinesis is carried out in higher plant cells. Cytokinesis is assumed to be an ancient, as well as a fundamental, process, and the idea that plant and animal cells use a similar active surface formation has intellectual attractiveness.

The term "diastem" was first applied to a shelf of different-looking cytoplasm that appeared in the equatorial plane of sectioned cells. In 1897, G. F. Andrews – a highly respected cytologist whose descriptions of some of the events that take place on the surfaces of living cells appear astonishingly accurate today, considering the relatively simple equipment that was available to her – published a detailed account of diastema formation in an echinoderm egg. She pointed out that asters do not look like the drawings that Bütschli had previously published, and she emphasized the dynamic activity of astral rays that extended, shortened, ramified, and anastimosed. In the equatorial plane, she described a fusion of the tips of the rays that originated in both asters. In the place where fusion occurred, she described a deposition of a continuous layer of material that began as a crooked network and then became a distinct platelike thickening in the future cleavage plane. The plate was described as double in origin (because the rays from both asters contributed to its formation) but optically single. Then there occurred a rearrangement of alveolar cytoplasmic material that caused the cytoplasmic structure on either side of the diastema to resemble that underlying the existing surface. Although the rays appeared to be fused to the plate, she stated that they did not contribute to the wavelike shrinking apart of the doubled layers of the plate. The foregoing is, of course, not a hypothesis, but a description. It appears to contain all the structural elements and activities that would be necessary to divide a cell by a diastema mechanism and, in fact, it seems more complete than most of the theoretical mechanisms of this kind that were subsequently presented. Her observations, however, have not been confirmed, and it is unclear what they were based upon.

The existence of diastema in a variety of cell types appears to have been widely accepted in the early years of the 20th century. Conklin (1902) described a diastema in cleaving *Crepidula* cells. Gurwitsch (1904) noted that some cells appeared to be able to divide either by constrictions or by plates. He suggested that diastema form when the cytoplasmic streams that originate at the poles converge over the entire equatorial plane, but when the streams affect only the equatorial surface, the cell divides by constriction. Yatsu (1909) was also convinced of the existence of the diastema, which he considered to be in part composed of nuclear fluid. He suggested that the diastema plays a part in reducing the density of cytoplasm in the division plane, and he felt that the problem of its formation must be included in any explanation of cell division. Motomura (1940) doubted that the surface of echinoderm eggs is pliable enough to be drawn down into the cleavage furrow, and he became convinced that the furrow surface is newly formed in the division process. Over the years he published a series of papers describing the activity of a population of cytoplasmic granules and vacuoles with special staining characteristics. According to his description, the granules and vacuoles of the sea urchin egg began to gather in the region of the equatorial plane when the cell elongated along the mitotic axis. After furrowing began, the granules and vacuoles appeared to organize in the equatorial plane and associate with the surface of the advancing furrow. After cleavage was completed, he found the stainable substance between the newly divided cells and on their opposed cell surfaces. He proposed that the diastema were formed by an assemblage of granules in the equatorial plane. The granules were transformed into vacuoles by swelling. The vacuoles then fused to form the new cell surface between the daughter blastomeres. Motomura did not speculate about the mechanism that aligns the granules and vacuoles in the equatorial plane.

Diastema activity appeared especially likely during the cleavage of the large, yolky, pigmented eggs of some amphibia and primitive fish. For not only has the existence of prefurrow platelike structures in the equatorial plane been reported in this cell type for many years, but the relatively unpigmented surface in the deepening furrow of some cells has also supported the idea that the furrow surface is newly formed and not simply stretched precleavage surface. Selman and Waddington (1955) proposed that, in the newt egg, the early diastema, which they observe first at late anaphase, expanded until it comprised about one-sixth of the equatorial plane, at which time it contracted. Because the diastema was attached to the underside of the upper animal surface, the contraction was responsible for the characteristic superficial early linear indentation typical of amphibian cleavage. By the time the diastema seemed to fill two-thirds of the equatorial plane, it appeared to be composed of two apposed parts. The original, pigmented surface then parted in the division plane, and the laminar walls of the diastema separated and opened outward to expose the new, unpigmented surface. All the new surface required for cleavage was presumed to form in the diastema.

The idea that animal cells generally use division mechanisms involving prefurrow surface assembly in the equatorial region later received support from electron microscope studies. Buck and Tisdale (1962) reported the platelike arrangements of vesicles in the equatorial plane of cultured vertebrate tissue cells. They speculated that fusion of these vesicles was responsible for new plasma membrane formation in the apposed surfaces of the daughter cells. Similar arrays of vesicles, which were similarly interpreted, were described in mussel eggs (Humphreys, 1964), lymphocytes (Murray, Murray, and Pizzo, 1965), and fish eggs (Thomas, 1968). Tahmisian, Devine, and Wright (1967) described equatorial, ringlike deposits of polyribosomes that later polymerized into helices. They proposed that the polyribosomes were engaged in new surface formation and that the lessening of free ribosomal surface that accompanied tightening of the helical assembly could turn off new surface synthesis at the end of cleavage.

After G. F. Andrews's time, most of the direct evidence for the existence of the diastema and diastema-like arrays of vacuoles and vesicles was obtained by studies of fixed, sectioned material. With changes in cytological methods, the evidence dwindled, as did the acceptance of this mechanism as an explanation of cytokinesis in animal cells in general.

FORCE IMPLIED: ACTIVE EXPANSION. The idea that an actively expanding surface can exert enough compression stress to deform cells appears first in early theories as a contributing factor and then later as the principal driving force. Schechtman (1937) proposed that cleavage of salamander eggs is a two-step process. In the first step, linear contraction in the furrow site deforms the surface and pulls local cortical material into a ridgelike accumulation that bulges toward the interior. In the second phase, the accumulated cortical material expands and pushes into the cell interior. The region of surface indentation in the primary phase becomes a site of intense surface growth, and the furrow surface lightens as it incorporates unpigmented subsurface cytoplasm. The necessary additional firm gelated material that supports the expanding furrow surface is assumed to be provided by subcortical cytoplasmic streaming directed toward the furrow. In this way, wall growth occurs at the expense of central sol. At the same time, Chambers (1938) proposed that a similar mechanism divides echinoderm eggs, and he emphasized the role of visible currents in relocating potential gel material to the furrow region. Marsland (1938, 1942) also adopted and slightly extended Schechtman's hypothesis.

Although cleavage by surface growth localized in the equator is an apparently workable mechanism, it implies a pattern of new surface formation that is not found in all cells. Swann and Mitchison (1958) proposed that an alternative mechanism, in which active surface formation occurs outside of the equatorial region, operates in cells. Their hypothesis requires that the dimensions and physical properties of the surface and cortex enable these regions to generate and transmit the compressive stress necessary to push the equatorial surface toward the cell axis. To satisfy this requirement, Swann and Mitchison postulated that the interphase cell possesses a relatively rigid cortex with a molecular constitution that results in expansion of the surface when cortical organization is disrupted or disoriented. The expansion provides the force necessary for division. Disorientation would, in turn, be caused by a substance that originates in the chromosomes. The geometrical relation between anaphase and telophase chromosomes and the spherical cell surface ensures that a diffusible substance that originates in the chromosomes affects the polar surface regions but not the equatorial surface so that, during division, the actively expanding poles push the passive, nonexpanding surface toward the cell axis.

Hybrid Theories

The mechanisms so far described have in common the idea that cytokinesis can be accomplished by a single, relatively continuous activity, like contraction or expansion or relaxation or growth. In a way, they are minimal mechanisms that attempt to explain visible events in the simplest possible way. The custom of accepting as most likely the simplest hypothesis that seems consistent with all the data imposes a welcome orderliness on theoretical deliberations, but it is well to keep in mind that cells may not be bound by this dictum. It has from time to time appeared possible that cytokinesis is caused by several different coordinated activities in synchrony or in series. In fact, the coincidence of events in the normal process makes the idea attractive. The possibility of multiple causes raises the specter of the immense number of complex theories that can be devised by permutation and recombination of the simple theories. Most theories invoking complex combinations of different kinds of physical activity were presented before the laws of parsimony became ingrained. Hybrid theories by their nature are nearly impossible to categorize. They are interesting examples of ingenuity and, in some cases, they present for the first time mechanisms that were later expanded to comprehensive theories.

Drüner (1894), in commenting upon an earlier version of Heidenhain's theory, agreed with the idea that the visible fibrils in dividing animal cells were the most important agents of shape change, but he presented an alternative interpretation of their role. He divided the time that they were visible into two periods: a period of expansion (or progressive growth and development) and a period of diminution (or regressive development). The changeover from expansion to diminution, he proposed, occurs at the time the chromosomes separate and move out of the plane of the metaphase plate. This interpretation is in rough agreement with the generalization that the radiate structure of the mitotic apparatus appears largest at anaphase and then gradually shrinks. Drüner assumed that the physical nature of the rays enables them to exert either traction or pressure, depending upon the circumstances. During the period when the rays lengthen, he proposed, they exert pressure upon objects and regions they contact. By reference to the dual roles of the rays, he explained both the congression of the chromosomes at the metaphase plate, as well as spindle elongation. The central spindle therefore forms a support structure that maintains polar surface contour and elongates the cell. The expansion of polar fibers pushes the mitotic apparatus into the center of the cell. During the period of diminution, the same fibers that had changed the arrangement of cell parts by expanding produce additional rearrangement of cell parts by contracting. As they shorten, the fibers that had pushed chromosomes to the metaphase plate move them toward the poles. Fibers that extend from the centrosomal region toward the subpolar surfaces perpendicular to the mitotic axis elongate the cell as they pull the surface toward the mitotic axis so that cytoplasm is passively displaced toward the poles. Drüner had little to say about cleavage, but his ideas of the way in which the astral fibers, during his second stage, could deform the surface to which they are attached are not greatly different from the ideas of Heidenhain.

A study of sectioned, stained, salamander testis cells prompted Meves (1897) to propose that active elongation of rays that join the asters to the subequatorial surfaces pushes the spindle poles apart and stretches the polar surfaces. Because the cell volume remains constant, the equator must be pulled in when the poles are pushed out. This event leads to the early indentation of the furrow. After furrowing starts, the furrow walls actively grow and push the tip of the furrow through the deeper cytoplasm until the cell is bisected. Meves proposed that in cleaving cells, where rays extend toward the polar surfaces, active ray elongation facilitates the pushing out of the polar surfaces and increases movement of cytoplasm away from the equator.

Utilizing a different interpretation of the physical nature of the astral rays, Bonnevie (1906) proposed that the rays carry cytoplasmic fluid to the aster centers. The fluid then moves into and distends the spindle. Spindle elongation causes anaphase chromosome movement and pushes the polar surfaces of the cell apart. At this point, the direction of flow is reversed. Fluid moves out of the equatorial region and into the aster, causing it to expand. From then on, her version of cytokinesis is like the astral cleavage mechanism of Teichmann (see section on Astral Cleavage Theories).

Yatsu (1912a) explained ctenophore egg cleavage by combining equatorial constriction and astral cleavage mechanisms. Additive effect of the centrosomes, he proposed, causes a surface tension increase along the cleavage plane. The tension increase causes a linear gathering of cortical ectoplasm that becomes the "cleavage head." Then, the fluid endoplasm gathers around the centrosomes and the movement of cytoplasm out of the equatorial region draws the cleavage head through the cell. In this hypothesis, the cleavage head is not an active agent.

K. Dan (1943b) proposed that the deformation and elongation of the early phase of cytokinesis and the later stages of constriction and completion are caused by two different, smoothly coordinated mechanisms. He assumed that before division the aster rays are attached to the spindle poles at one end and to the cell cortex at the other end. This assumption was combined with observations of spindle elongation and the extension of astral rays across the equatorial plane. Taken together, the assumption and observations suggested that the driving force for the early elongation phase is active elongation of the spindle. Assuming no change in the length or attachment of the rays directed toward the equator, the equatorial surface would be pulled toward the spindle; cytoplasm would be shifted toward the asters and the cell would elongate. In an early version of the theory, K. Dan (1943b) proposed that an astral cleavage mechanism like those of Teichmann (1903) and Gray (1931) was responsible for constriction and completion. The fluid cytoplasm of the equatorial zone was carried toward the aster centers between the rays. As the volume of the equatorial region shrank, a kind of "suction" caused the local surface to collapse toward the cell axis. A later revision (K. Dan, 1988) incorporated the same spindle elongation mechanism to explain the early elongation phase, in which the equatorial surface is pulled toward the mitotic axis, followed by the formation and operation of a constriction ring mechanism at the furrow base. In this version, the mitotic apparatus serves both as a mechanical agent and as the source of a cleavage stimulus. The stimulatory activity results in the assembly of actin filaments that compose the contractile ring. The mechanical activity elongates the cell and contributes to the forces necessary to cause the initial constriction.

Most hypotheses referring to division mechanisms located in the surface have been parsimonious, in that they propose that only part of the surface is altered and active, while the unaltered parts are relatively unchanged and passive. Mascher (1989) proposed that active equatorial constriction and active astral relaxation may take place simultaneously in different regions of the same cell. She pointed out advantages of a pair of coordinated supplementary activities.

Study of living and fixed cleaving nematode eggs led Tadano (1962) to propose that two different mechanisms in succession are needed for complete division. In the early phase, a standard constriction ring mechanism divides the cell. The blastomeres adhere closely and vesicles crowd against the inner surfaces of the apposed membranes. The membranes disappear and the vesicles fuse to form a new membrane. Finally, the blastomeres are separated by a space containing the former vesicle contents, and the apposed blastomere membranes are formed by a mosaic of vesicle membranes.

In 1948, Rashevsky proposed that the forces that set cell division in motion originate from interaction between the flow of cytoplasmic substances and the resistance of the cytoplasmic medium to the flow. The flow, he proposed, results from metabolic activity that creates regional differences of substances and consequent flow of these substances from regions of high concentration toward regions of low concentration. The forces thus created were assumed to act within cells that behave like liquid drops, and the forces elicit deformations due primarily to surface tension forces. Although Rashevsky (1952, p. 293) described his ideas as "the only quantitative theory of cell division which has been developed somewhat systematically," the theory had little impact on the direction of research, and it was rarely cited by biologists who appear to have been discomfited by its neglect of available observations and experimental evidence, as well as by its heavy emphasis on mathematical formulation.

Hypothetical Stimulus Mechanisms

Cell division theories that do not assume a mechanical role for the mitotic apparatus must provide an explanation for the temporal and geometrical correlation between mitosis and cytokinesis. Diastema and vesicle fusion theories also require an explanation of the localization and formation of these prefurrow structures but, as Yatsu (1909) remarked, it is not usually provided. Theories that attribute cell division to a force that originates outside the mitotic apparatus assume that the mitotic apparatus participates in cytokinesis by changing the physical properties of another part of the cell and that the change precipitates division. The affected part of the cell is assumed to be the surface. Although all parts of the surface can respond (that is, furrows can form anywhere on the surface), it is necessary to include a provision in the system for distributing the effect of the mitotic apparatus in a definite pattern. Among the hypothetical pattern-forming factors that have been suggested are the different distances from different parts of the mitotic apparatus to the nearest surface, assumed differences in the ability of different parts of the mitotic apparatus to produce the necessary effect, and assumed regional interference by parts of the mitotic apparatus with the action of a uniformly distributed change, and cytoplasmic current patterns. The event that causes the change is assumed to be invisible. The hypothetical division-precipitating agent has been called a stimulus. The term conveys the idea that the stimulus can change local conditions in a way that elicits a response; it implies nothing about the nature of the change. The need for explaining how a force production is activated at the right place and time may complicate theorization, but it has the virtue of offering more opportunities for testing, because a force production mechanism can operate only if the conditions necessary for its creation exist.

Global stimulation combined with mitotic apparatus interference

In global stimulation hypotheses, an agent that promotes contraction of the entire surface permeates the cell, but the region that actually contracts is regionalized by the asters. Painter (1918) proposed that a factor that originates in the nucleus can cause swelling of the ectoplasm (hyaline layer in this case) and changes in surface tension and flow of cortical cytoplasm. The effects of the factor are restricted to prescribed areas by the denseness of the asters. Scott (1946) suggested that the asters release a general contractile activator that affects the whole surface. The subsequent position of the furrow would be determined by competing surface stresses and regional viscosity differences associated with the asters.

Regionally restricted stimulation mechanisms

Most hypothetical stimulus mechanisms assume that differences in surface behavior result from unequal initial distribution of the stimulating agent. The minimal division mechanism requires only that the equatorial surface contractility or tension differs from that of the poles. Because the effect of the stimulus is to precipitate division, the hypothetical pattern of its distribution must be consistent with its hypothetical immediate effect. It would not be logical, for instance, to propose that a division mechanism that postulates polar surface relaxation can be activated directly by an agent that causes surface contraction.

POLAR STIMULATION. Most polar stimulation mechanisms assume that the active agent causes surface relaxation or tension decrease. The pattern of agent distribution would cause an inequality of tension, with the tension being lowest at the poles and highest at the equator. R. S. Lillie (1903) proposed that the chromosomes release a surface-tension–reducing substance and, similarly, Chalkley (1935) suggested that the fission of an amoeba resulted from an increase in polar cytoplasmic fluidity caused by a substance released from the plates of chromosomes as they moved toward the poles. Alternatively, Roberts (1961) suggested that a polar relaxing factor might be produced by an interaction between the centrioles and the centromeres. Wolpert (1960), White and Borisy (1983), and Schroeder (1981a) also proposed that asters have the property of causing surfaces to relax. They assumed that the relaxing effect decreases with distance, so that in a spherical cell with a large mitotic apparatus (like a sea urchin egg) the polar surfaces relax, because they are closer to the aster centers. Because of the normal cell geometry, the aster effect is absent or reduced at the equator (Figure 2.4). Older polar relaxation theories assumed that the regional tension differences are sufficient to drive cytokinesis. Schroeder, as well as White and Borisy, proposed that the aster-induced effect is amplified by the shifting of a subsurface contractile web or the accumulation of mobile force producing units.

Lillie (1916) attributed the apparent radial structure of asters to diffusion fields whose electrical properties align movable cytoplasmic granules in a way analogous to the similar-looking alignment of iron filings in a magnetic field. The electrical field of the aster "compensated for" or depolarized the polar surfaces. The depolarization was thought to cause a local increase in surface tension that pulled cytoplasm away from unaffected areas and thus decreased the diameter at the equator.

The polar expansion mechanism of Swann and Mitchison (1958) also invoked a polar stimulation pattern. They proposed that chromosomes release a substance that causes disorganization of the interphase cortex. The disorganization of the cortex was assumed to result in its active expansion. The movement of the chromosomes toward the pole was considered to concentrate their effect in that area.

EQUATORIAL STIMULATION. Equatorial stimulus mechanisms require a means for concentrating the effect of the mitotic apparatus in the region where the furrow develops. They assume that the mitotic apparatus directly causes the tension increase at the affected surface. Bütschli (1876) and Ziegler (1898b, 1903) pointed out the potential for an additive astral effect in the equatorial plane. Bütschli suggested that equatorial tension increase could be caused by converging currents of material that might cause local tension increase by dehydration. Ziegler's observations of cleaving ctenophore eggs led him to suggest that the immediate effect of additive astral influence was the formation of a thickened equatorial ectoplasmic ridge that exerted increased pressure on the underlying cytoplasm. Their ideas originated from study of cells with large asters and spindles both located at some distance from the surface, but that circumstance is not essential for equatorial stimulation. The alternative possibility – that in cells with small asters and a large spindle located near the surface, the spindle may be the most active component – has also been suggested (Rappaport and Rappaport, 1974; Kawamura, 1977). The two ideas are not mutually exclusive. The idea that asters can act additively is based on a relationship between the mitotic apparatus and the surface that is less obvious than that proposed to explain polar stimulation. The hypothetical additive aster effect also lacks a generally agreed-upon morphological basis and it raises another set of questions about just what is accumulated in the equatorial plane and, if material is moved, how it is transported. Customarily, it has been proposed that the additive effect changes the equatorial milieu by adding something to it. But it is equally possible that the milieu may be changed by removing something from it.

Mascher (1989) urged consideration of the possibility that the mitotic apparatus affects all parts of the surface in a way that has qualitatively different effects in different parts of the surface. According to this scheme, the basic ideas of polar and equatorial stimulus mechanisms are both functional, so that a relaxing effect is concentrated at the poles, and a contractile effect is concentrated at the equator. The two different effects would originate from a single pair of asters.

Most polar stimulation and equatorial mechanisms rely only upon the visible events of formation and growth of the mitotic apparatus. Cytoplasmic currents that develop in the later stages of mitosis have also been accorded an important role. Cornman and Cornman (1951) proposed that the nucleus releases a furrow-forming substance when the membrane dissolves and the substance is brought to the surface by the asters and then moved toward the equator. Henley and Costello (1965) associated furrow-forming activity with stainable cytoplasmic granules that accumulate near the centrosomes. The cytoplasmic currents move the granules toward the polar surfaces. They are redirected by the outer limits of the cell into a "fountain streaming" pattern and move toward the equator. Because the flow patterns from the cell halves are symmetrical, the streams meet and the granules are distributed in the equatorial plane. The furrow forms where the streams converge, and the cell divides by constriction.

Durham (1974) proposed that the effect of the mitotic apparatus upon the surface resulted from its mechanical activity. Incorporating the idea that the expanding mitotic apparatus pushes upon the poles, he reasoned that the ensuing deformation would cause greater stretching in the equatorial surface than elsewhere. He predicted that the ion permeability barrier would be reduced in the maximally stretched region and that a resulting local calcium ion influx would enhance actomyosin contraction. In this fashion the contractile ring would be established.

The mitotic apparatus is the only large, well-organized subsurface structure visible in the cells usually studied at the time of division. Diastema have, however, been described in sections of dividing primitive fish and salamander eggs. Sawai and Yomota (1990) and Zotin (1964) proposed that these structures are the immediate source of the stimulus for equatorial constriction. The mitotic apparatus, they suggest, creates the lucent, yolk-sparse plate of cytoplasm in the plane formerly occupied by the metaphase chromosomes, and the plate, in turn, either transports furrow-inducing cytoplasm to the surface or represents a region where this special cytoplasm is localized. Andreassen et al. (1991) proposed that in mammalian cells, a material of chromosomal origin that accumulates in the equatorial zone may interact with the surface to establish a local contractile region.

It now seems apparent that a detailed knowledge of the nature and scheduling of the normal events of cell division did not greatly limit the number of potential logical explanations. The difficulty in understanding lay not in an inability to explain the process, but rather in the overwhelming number of reasonable possibilities. Deeper understanding required the consideration of experimental facts obtained by observing the course of events when division occurs under imposed, unusual circumstances. Despite occasional zealous assertions of proprietary rights, the originator of a theory and even its previous existence were sometimes forgotten. There followed a cyclical reemergence of the same idea as it occurred to a succession of authors. The impetus for the idea's reappearance sometimes appeared to be its perceived freshness and "originality" rather than the development of new supporting data.

At times the gap between experimentalists and theorists appeared unbridgeable, and the two groups took little account of each other's work. It was as though the former regarded the latter as bright enough but a little lazy (because they did not keep up with experimental literature), and inclined to use free play of the imagination unfettered by data because they were constitutionally incapable of designing good experiments. The latter, however, appeared to relegate the former to the ranks of the merely clever who were incapable of appreciating the beauty of a hypothesis apart from its relation to reality.

It will subsequently be brought out that the very early theory of Bütschli now appears to have been closer to the mark than many that followed it. This argues against the idea that successive theories approached ever closer to some kind of perceived reality. Lacking was the experimental testing that could have prompted and directed reformulation. Ziegler, Morgan, and a few others emphasized the need for experimentation and made suggestions for possible experimental designs. But, for some years, experimentation on dividing cells was sporadic and the results appeared to have little theoretical impact.

3

The Site of the Division Mechanism

The theories described in the previous chapter are certainly numerous and varied and, upon a little reflection, we must expect that most of them are wrong. The principal activity that divides the cell was placed in different regions by the different groups of hypothetical mechanisms; therefore, some progress in elimination of alternative hypotheses is possible by identification of the cell regions or structures that are essential for division. With few exceptions, identification of essential or nonessential components was not possible by simple observation of the normal process, and the need for experimentation was recognized. In the most successful of the early experiments, the importance of a structure or region was assessed by removing or disrupting it. If the cell divided, the affected region was, by definition, not essential. The results of such an experiment are clear and certain, and the experimental design constituted a method of proceeding by disproof. Interpretation of division failure as a result of experimentation is neither simple nor clear. Failure may be due to removal of an essential component, or it may be due to some other deficiency or damage that the experimenter unknowingly inflicted. In such cases, the design of control experiments is important. All the major regions of the cell – the mitotic apparatus, the surface, and the region between the mitotic apparatus and the surface – were, in different theories, considered to play an essential role, and evaluation of the degree of dependency of division upon the integrity of a particular region provided a way to test hypotheses.

The Mitotic Apparatus

The correlation in time and place of the events of mitosis and cytokinesis supported the idea that they are causally related, and the possibility that the mitotic apparatus plays a direct, mechanical role has the attractiveness of simplicity, parsimony, and ease of understanding. But mechanisms like those of Heidenhain (1897) and Rhumbler (1897, 1903) met with skepticism. T. H. Morgan (1899) pointed out that in amphibian eggs the second cleavage begins before the first is completed. He remarked (p. 521), "The results show that a division once started may complete itself independently of the rays around the astrosphere [aster] since these, meanwhile, have undergone changes leading to the second division before the first division has completed itself." These remarks appear to have been lost in the wilds of a baroque discussion section in his paper on the actions of salt solutions on invertebrate eggs. The comments did not deter Heidenhain or Rhumbler or Platner or many others from subsequent refinements or new inventions of theories that invoked physical activity of the astral rays as an essential component of cytokinesis. Not satisfied with simply pointing out the implications of this pattern of cleavage, Morgan continued, "I think it would be possible to test experimentally these mechanical hypotheses in the following way. It would be possible to cut off during the first cleavage of *Beroë,* or of some of the Hydrozoa, those parts of the first two blastomeres containing the nucleus and astral center. If then the cleavage continued, as I think it might do, in the non-nucleated part the result would demonstrate the uselessness of the astrospheres as mechanical centers of division of the protoplasm." Morgan's thinking was unusual for the time. He not only designed a decisive experiment, but he also suggested that it be done on the best experimental material, considering that the operation had to be done freehand. He had previously worked with Driesch at Naples on "germinal localization" experiments on *Beroë* eggs. One of the purposes of such investigations was to determine when the developmental fate of different parts and embryonic structures becomes irreversibly fixed. A method for obtaining the information was to remove pieces of the egg or early embryo at different times and then look for anatomical deficiencies in the later larvae. Some investigators removed parts of the egg by first flattening it on a microscope slide in a thin film of water and then slicing off a piece with a keratomy knife or a needle with a flattened tip that had been ground to a sharp edge. The slide and egg were subsequently submerged in a shallow dish of water, where they separated, so that the slide could be removed and further development of the egg could be observed. Morgan knew that *Beroë* eggs could survive this treatment, but he did not do the experiment. A young Japanese investigator, N. Yatsu, who was working in E. B. Wilson's laboratory, did similar germinal localization experiments on the eggs of the nemertine worm *Cerebratulus lacteus* and also cut cleaving eggs in ways that separated part of the furrow from the spindle, the chromosomes, and a major part of the asters. He found that the furrows were usually completed (Yatsu, 1908) (Figure 3.1), and he concluded that furrow progress requires neither the nucleus nor the aster, "provided it be in a state of division activity" (p. 269), and he formed the idea that the eggs were divided by a diastema mechanism. He then traveled to Naples, where he did similar experiments on *Beroë ovata,* and he may have been able to obtain information about both germinal localization and cell division mechanisms from a single operated egg. His first cleavage experiments followed Morgan's design, and his results were as Morgan expected (Yatsu, 1912a). Despite amputation of the portion containing all of the mitotic apparatus, the non-nucleated portion continued to divide, provided it contained the cleavage head (Figure 3.2). Yatsu considered his results entirely inconsistent with the Heidenhain–Rhumbler mechanism. He was also skeptical of Ziegler's suggestion that a partial contractile ring was operating, because in some cases the furrow tip of operated eggs curled back upon itself. He found no indication that the ectoplasm alone was an active cleavage agent. He proposed that the cell centers or asters caused the ectoplasm to form the cleavage head, which was a passive structure. The principal division activity, he felt, resulted from the rounding up of the cytoplasm around the centers. In other words, he favored an astral cleavage mechanism. Yatsu did not mention Morgan's suggestions, and Wilson (1928) did not mention Yatsu's experiments in an extensive discussion of the mechanisms of animal cell division. Chambers (1919) repeated Yatsu's *Cerebratulus* experiments on sea urchin (*Arbacia*) eggs with similar results. He interpreted them as support for an astral cleavage mechanism. He does not appear to have taken into account the results of Yatsu's *Beroë* experiments, in which the furrow progressed even though it was completely separated from the asters.

Chambers and Kopac (1937) introduced large oil drops into sea urchin eggs by pressing an expanding oil drop against the egg surface. The surface tension relation between the oil drop and the egg surface caused the two to coalesce rapidly, and the oil drop was included in the egg cytoplasm. When an oil drop was located in the equatorial plane of

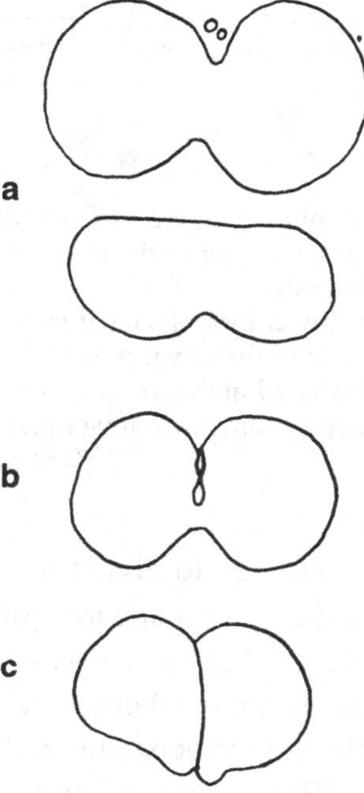

Figure 3.1 *Cerebratulus* egg cut horizontally at anaphase; enucleated fragment below, nucleated fragment above. (a) Enucleated fragment dividing (7 min later). (b) Enucleated fragment 2 h 50 min later. (c) The enucleated fragment has completed division. (Redrawn from Yatsu, 1908)

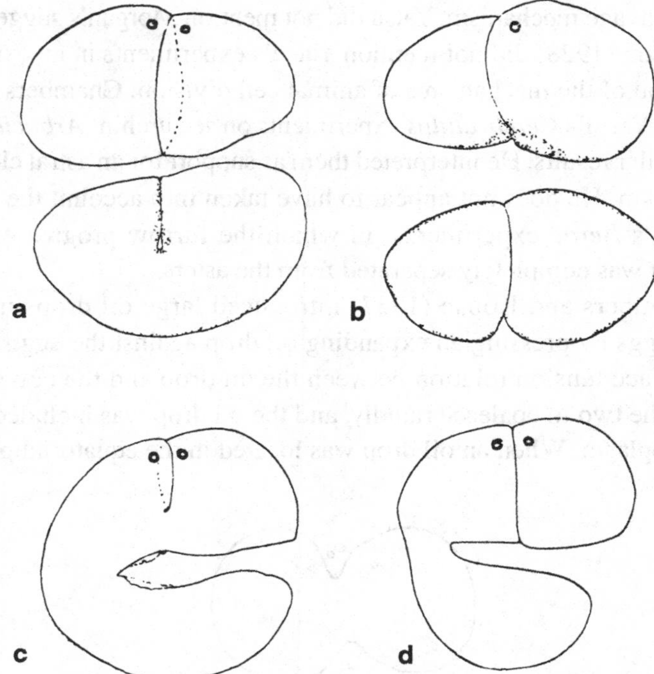

Figure 3.2 Separation of the mitotic apparatus from the cleavage furrow in *Beroë ovata* by cuts made perpendicular to the cleavage plane. Circles indicate positions of the telophase asters.

(a) Cleavage head in the lower piece isolated from the mitotic apparatus.
(b) Completion of cleavage in the lower piece 1 h 15 min later.
(c) Cleavage head not included in the lower piece.
(d) Same egg 23 min later, no furrow in lower piece.

(Redrawn from Yatsu, 1912a)

a cleaving egg, the egg continued to divide until the drop was constricted to form, in many cases, two spheres joined by a thin stalk. Because the oil drop displaced – and probably disrupted – the mitotic apparatus, the egg must have continued cleavage in the absence of the normal relation between the mitotic apparatus and the surface.

Beams and Evans (1940) took advantage of the relatively recent advent of the use of colchicine to disrupt the mitotic apparatus. At the concentrations they used (2×10^{-4} M), colchicine prevented formation of the mitotic apparatus when administered early and, when administered later in the mitotic cycle, also caused the disappearance of already formed astral rays. Although the mitotic apparatus was always affected, cleavage took place when the mitotic apparatus was disrupted at late anaphase or later. Earlier disruption blocked cytokinesis. The experiments of Beams and Evans clearly demonstrated that the entire progressive deformation of one sphere into two spheres could occur in the presence of a visibly disrupted mitotic apparatus. They concluded that the asters are important, but that their main role in the cleavage process occurred before the late stages in the mitotic cycle.

Swann and Mitchison (1953) observed that lower concentrations of colchicine caused disappearance of the asters and spindle when cells were observed by polarized light microscopy. They found that treatment in early anaphase permitted some cell elongation and early furrows that receded. They also confirmed Beams and Evans's finding that treatment in late anaphase did not block cytokinesis.

Direct physical removal of the mitotic apparatus by sucking it out with a pipet overcame any doubt that the mitotic apparatus is completely dispensable for the normal physical events of cytokinesis. Hiramoto (1956) inserted a pipet and aspirated the mitotic apparatus from eggs at different times in the mitotic cycle. Removal at any point up to anaphase blocked cleavage, but the same operation done at anaphase or later did not. Hiramoto later (1965) displaced or disrupted the mitotic apparatus by injection of oil or sea water at different times in the cell cycle. Again his results showed that from anaphase onward, cleavage proceeded without regard to the presence or state of the mitotic apparatus.

It is now clear that animal cells that have been subjected to this kind of experiment can successfully divide after the mitotic apparatus is removed *while cells are spherical.* The physical presence of the mitotic apparatus is thus not required for any division-related shape change. There is a clear thread of reasoning, observation, and experiment in support of this statement, beginning with Morgan in 1899, and culminating with Hiramoto in 1956. Since that time, few have tried to attribute a physical role to the mitotic apparatus. In this field it is not unusual that the final disproof of a hypothesis follows a series of observations or experimental results that point in the same direction. It may be that those who would like to believe otherwise require maximally dramatic disproof.

Acceptance of mitotic apparatus dispensability eliminates the traction fiber theories of Heidenhain and Rhumbler; the expanding spindle theory of Platner; the astral cleavage theories of Teichmann, Chambers, and Gray; and, because they attribute some essential component of cytokinesis to the physical activity of the mitotic apparatus, the hybrid theories of Drüner, Meves, Bonnevie, Yatsu, and K. Dan.

The Region between the Spindle and the Surface

The findings described in the previous section are usually assumed to indicate that the physical mechanism is located in the surface, but they do not convincingly eliminate the possibility of an important structure or activity between the spindle and the surface. The idea that organized subsurface activity, such as new surface formation, is possible persisted after Hiramoto's 1956 experiments. Most speculation concerning important subsurface contribution implies or requires that subsurface organization be stable during cytokinesis. Formation of furrows by vesicle fusion (Buck and Tisdale, 1962), for instance, requires in the equatorial plane a thin sheet of discontinuous vesicles arranged in close regularity that permits their fusion to form a pair of opposed membranes. Evidence cited in support of mechanisms of this type originated largely in morphological studies of fixed material. Other than G. F. Andrews (1897), no one claimed to see special organization of the equatorial subsurface in living cells. Two questions arise concerning this kind of mechanism: (1) Does the cell behave as though its division is dependent upon subsurface stability? and (2) Are the subsurface structures that were reported real?

Subsurface structures with sufficient strength, organization, and coherence to divide a cell might be expected to interfere with, or be disrupted by, movement or displacement of the cytoplasm in the equatorial plane. There have been numerous reports of reciprocating cytoplasmic flow across the equatorial plane during division. Ziegler (1898b) saw such flows in normal cleaving ctenophore eggs, as did Yatsu (1912a). Conklin (1908) observed them in cleaving *Linerges* eggs, and Chambers (1938) in cultured human epithelial cells. Consistent with these observations of dividing normal cells are the results of experiments involving disturbance and displacement of the equatorial cytoplasm during division. Chambers (1937, 1938) tore the polar surface of a dividing sea urchin egg and reported that furrowing continued, despite the flow across the constricting neck joining the incipient blastomeres. Chambers and Kopac (1937) observed that a large oil drop located in the cleavage plane was pinched in two by the furrow. During part of the process, the oil drop occupied the entire equatorial plane within the cortex. Hiramoto (1965) confirmed Chambers and Kopac. He further observed cleavage following disruption of the mitotic apparatus and equatorial cytoplasm after injection of a relatively large amount of sea water. Mechanically induced flows across the equatorial plane by alternate flattening of the poles of dividing sand dollar eggs also fail to interrupt furrowing, as does using a needle to stir equatorial cytoplasm in sand dollar and several other cleaving invertebrate eggs (Rappaport, 1966). Cultured newt kidney epithelial cells continue to divide when the mitotic apparatus and subsurface are driven back and forth through the equatorial plane by rapid alternate compression and release of the polar regions (Rappaport and Rappaport, 1968). Hiramoto (1968) also demonstrated that there was no barrier to free diffusion of an injected dye solution across the cleavage plane of a dividing sea urchin egg.

The experimental evidence suggests that in the early phases of the mitotic cycle the integrity of subsurface structures and organization is essential for normal division. But during the period beginning shortly before the furrow appears and ending at its completion, the removal, displacement, or chronic disruption of subsurface cytoplasm does not interfere with the process. These results would not be expected if division were caused by concerted action of organized subsurface elements.

Judgement of the reality of subsurface structures involves information based upon morphological, rather than experimental functional, studies. Early descriptions of subsurface structures were based upon fixed material. Diastema were described with the light microscope; aligned sheets of vesicles, with the electron microscope. The describers emphasized the possibilities for a role for the vesicles in the physical process of cell division, with little discussion of existing contradictory information. Schroeder (1970) pointed out that all the electron microscope studies that showed equatorial vesicles used fixation procedures previously shown to cause artifactual vesiculation of cell membranes. He suggested that the vesicular arrays may have originated by osmium-induced vesiculation of already formed furrow surface. Whatever their cause may have been, vesicles were not subsequently described in ultrastructural studies of dividing animal cells. Diastema in sturgeon and amphibian eggs appear to be a different matter. Zotin (1964) found that their fragmentation was accompanied by multiple furrows. Sawai and Yomota (1990) reaffirmed the presence of the diastema in frog and salamander eggs; by disruption experiments, they showed that the diastema is not necessary during cleavage but is required earlier. They propose that, like the mitotic apparatus, its necessary function is completed before cleavage, when it induces furrowing in the overlying cortex (see Chapter 7).

The Surface

The elimination of the mitotic apparatus and the region between the mitotic apparatus and the surface as possible sites for essential mechanical activity during cytokinesis focuses attention on the remaining cell surface and cortex. The fluid properties of the plasma membrane further suggest that its role in cell deformation may be limited, so that the site of dynamic activity is presently considered to be the underlying cortex; the analysis of cytokinesis has, in effect, become a series of investigations directed toward a clearer understanding of the structure and function of the cortex, the factors that affect its behavior, and the ways in which its properties become regionally differentiated. The idea that cells are bounded by a peripheral layer of material with special physical properties is old and is based upon the ways in which that region responds to mechanical disturbance, rather than based upon recognizable organization. Early studies of living protozoa suggested that the region immediately beneath the surface is firmer (or more viscous, in the terminology of the day) than the deeper cytoplasm. Chambers (1917a) proposed that the existence of a transparent cortex is common among animal cells. He based his opinion on simple dissection experiments on various cells, including invertebrate eggs, in which he found that after the fluid cytoplasm was forced out through a surface tear, a firm transparent rind was left behind. W. H. Lewis (1939), in summarizing years of observations of moving and dividing cultured tissue cells, deduced the presence of a "superficial plasmagel layer" that controls form, motion, and division by exerting tension in organized patterns. Motomura (1935) found that pigment granules or vacuoles located close to the cell surface were not shifted by centrifugation, and he concluded that they are embedded in the cortex. After Chambers and Kopac (1937) injected an oil drop in the path of an active furrow, they found that the drop was deformed when the cell surface was 4–5 μm away. They concluded that the gap was produced by firm, transparent cortex and that they had in effect measured its thickness at the furrow tip. Hiramoto (1957) substituted a rounded needle tip for an oil drop and, after correcting for the index of refraction of the cytoplasm, he concluded that the cortex was uniformly 3 μm thick in an unfertilized sea urchin egg. After fertilization, the cortex increased to 4 μm, and as furrowing began, it increased at the equator and decreased at the poles. Higashi (1972) induced localized elliptical cytoplasmic flow patterns with a vibrating needle inserted into a sea urchin egg. He studied the orbits of the moving cytoplasmic particles (the relation between velocity profile and distance from the surface). He estimated the thickness to be 5.6 μm shortly after fertilization, decreasing to 3.6 μm at 60–65 minutes. At metaphase the thickness was 3.2 μm in the polar region and 2.4 μm in the prospective furrow region. The thicknesses of these two regions did not change during cleavage. Higashi's method should have minimized disturbance to the normal organization. It appears that the existence of a cortical region is not doubted, but its dimensions vary according to the method used, as well as varying regionally and temporally. Chambers (1938) demonstrated regional differences in cortical stiffness during cleavage. He tore the surface of one pole and found that the dimensions of the opposite pole shrank as the cytoplasm flowed outward. The diameter in the furrow did not decrease at the same time, and he concluded that the internal rigidity of the furrow cortex maintained its dimensions while the rest of the surface collapsed as the internal pressure fell.

Despite early recognition of the importance of the cortex, there is much that is still unknown about the structural basis of cortical properties. The classical cytological techniques that were available when the cortex first attracted interest lacked adequate resolution, and for many years its physical properties, even its very existence, were most convincingly demonstrated by the results of microdissection experiments like those described earlier. Further complicating any discussion of the subject are technical problems that still raise questions about the adequacy of preservation of cortical structures in some cell types.

The superior optical properties of cultured tissue cells and their ability to create internal cytoplasmic flow during locomotion and division permit observations that led W. H. Lewis (1939) to the conclusion that a cortical layer of cytoplasm characterized by elasticity, firmness, and contractility was responsible for tissue cell form, movement, and division. In a study involving light and electron microscopy, Buckley and Porter (1967) subsequently demonstrated a dense, feltlike mesh of 75 Å filaments immediately beneath the cell membrane of cultured rat embryo cells. They pointed out that a meshlike configuration could account for the properties and behavior of the superficial cytoplasm described by Lewis and other investigators. This region of the cytoskeleton has been extensively investigated (surveyed by Poste and Nicolson, 1981). The microfilaments that are its principal constituent are actin polymers (Stossel, Hartwig, and Yin, 1981), and the complex as a whole rapidly responds to imposed tension by realignment (Kolega, 1986).

The preservation of cortical structure in cleavage cells appears to have been less satisfactory than in tissue cells. Mercer and Wolpert (1962) found no structural evidence of a cortical layer in normal and centrifuged fertilized and unfertilized sea urchin eggs, and they attributed the concept of a cortex to misinterpretation of the results of microdissection experiments. Although methods have improved, Schroeder (1981b) remarked that there are no accepted diagnostic ultrastructural criteria for the cortex. P. Harris (1968) first described 40–50 Å microfilaments that form the core of the microvilli that project from the egg surface. She discussed their similarity to those that were found in the gel portions of amoeba and slime molds and suggested that their presence might have implications for cytokinesis. Burgess and Schroeder (1977) confirmed the presence of core microvillar microfilaments and demonstrated their actin nature. Spudich and Spudich (1979) estimated from studies of cortical fragments that the cortex contains 15–30% of the total egg actin. Electron microscopy of prefertilization fragments revealed few actin filaments, but they were more numerous in fragments of fertilized eggs, an observation that was confirmed by Mabuchi, Hosoya, and Sakai (1980). In a review of the role of the egg cortex in sea urchin and starfish development, Schroeder (1986) distinguished between the microvillar and submicrovillar cortex. Both regions of fertilized eggs appear to be well supplied with actin microfilaments, but, for reasons that will be discussed later, he considered the submicrovillar region to be more important in cytokinesis. Most studies of submicrovillar cortical organization have revealed a pattern and quantity of filamentous material that seem insufficient to account for the physical properties and dimensions that are implied from the results of other techniques of investigation. The adequacy of preservation and methods of detection have been questioned (Schroeder, 1986). However, in an exceptional investigation of two Japanese sea urchin species, Usui and Yoneda (1982) were able to demonstrate, in sections of whole eggs, an extensive, thin network of 70–90 Å filaments under the entire egg surface from mid anaphase until shortly before division. By early telophase, the filaments were found only in the anticipated furrow region. Comparable observations have not been reported in other sea urchin species. Bits of cortex isolated from living or fixed cells usually reveal more filamentous structure, as well as satisfactory preservation of microvilli, but the potential for disruption in the submicrovillar region is great. In brief, attempts to relate the considerable amount of biophysical and behavioral information about the sea urchin and starfish cortex to its ultrastructural organization in a detailed fashion await further methodological improvement.

The importance of method in investigations of cortical ultrastructure is illustrated by Forer and Behnke's (1972) investigation of meiotically dividing crane fly spermatocytes. Conventional methods had revealed no filaments and only a few microtubules and some amorphous material under the cell periphery. Following glycerination, which disrupts the plasma membrane to the extent that it allows the passage of large molecules, Forer and Behnke placed the cells in a solution of heavy meromyosin, which adheres specifically to filamentous actin in a characteristic pattern. After fixation and sectioning, extensive arrays of filaments were visible beneath the plasma membrane. The technique revealed that the arrangement of the filaments changes as the cell approaches the time of division. In early prometaphase there are few filaments at the periphery. The number increases between prometaphase and anaphase, and during that period the filaments are oriented parallel to each other and to the surface in such a way they extend in pole-to-pole arcs, like the stripes on a melon, in a layer 0.5–1 μm thick immediately beneath the plasma membrane. When the furrow appears, the filaments under the equatorial region are arranged circumferentially. Forer and Behnke suggest that their success in demonstrating cortical actin filaments may have resulted from their stabilization with heavy meromyosin, so that the filaments are less damaged by the reagents used in fixation and embedding. Reorganization of microfilaments during cytokinesis in glycerinated mouse blastomeres has also been reported (Opas and Soltynska, 1978). At metaphase, a randomly oriented layer 0.1–0.25 μm thick is present under the cell surface in all regions. In elongated prefurrow blastomeres, the polar filaments remain randomly oriented, but those located elsewhere are oriented parallel to the cell's long axis. After furrow formation, the filaments in the furrow are oriented circumferentially, whereas the orientation of filaments under the rest of the surface is unchanged.

Investigations of the fertilized, uncleaved amphibian egg cortex generally agree that there is a 0.1–0.2 μm thick, electron-dense, felt-like layer immediately beneath the plasma membrane, but further details of its structure remain elusive. Selman and Perry (1970) reported that within the cortex, fibrillar material could sometimes be distinguished, and Perry, John, and Thomas (1971) found no filaments that could be coated with heavy meromyosin outside of the furrow. Bluemink (1972) noted that the dense layer of *Xenopus* eggs excludes droplets and pigment granules, and he found that a thickened meshwork of short filaments rapidly formed at the margins of a healing wound. Schroeder (1974) elicited massive and rapid contractions of the animal surface of uncleaved *Rana pipiens* eggs by immersing them in a calcium ionophore. In the ensuing rapid surface shrinkage, the surface ruptured, and the cytoplasm was forcefully expelled. Within the isolated surface, the folded plasma membrane was underlain by a band of 35–70 Å microfilaments that were randomly oriented and had chemical properties similar to actin. No such filaments were visible in control cortical material. Schroeder concluded that the filaments could form and perform mechanical work in less than 30 seconds.

Investigations of cytokinesis have focused upon the processes by which cortical material is converted into division machinery, the physical nature of the mechanism, and the way in which it is established in a restricted part of the surface. Although some of the general characteristics of division mechanism organization are now known, present ultrastructural information does not provide a clear picture of the cortex from which the mechanism is assembled or the nature of the assembly process.

4

The Nature of the Division Mechanism

The observations and experiments described in Chapter 3 demonstrated that no deep subsurface structures are required for division and infer that the force that drives the mechanism is located in the surface or cortex. The findings lead immediately to two further questions: First, how does the division mechanism work? and, second, how is division activity correlated in time and position with the events of mitosis? This chapter is concerned with the first question.

The typical deformations of dividing animal cells require an unequal distribution of forces, and parsimonious hypotheses require only a single force exerted in a pattern of active and passive areas. In spherical cells, either active expansion or active contraction appear to be the logical alternatives. Cell constitution and behavior seem to admit the possibility that different cells could use different division mechanisms, but at this time only one mechanism has been identified in all the animal cells that have been subjected to experimentation. For purposes of speculation, it has been customary to assume that all cells are reshaped into spheres before the events of division begin, but this generalization is not true, especially in tissue cells. Different force-producing mechanisms may require that the site of major physical activity be located in different regions. A contractile mechanism, for example, could be located only at the equator, whereas an expansion mechanism could be effective in any part of the surface.

Expanding Surface Mechanisms

Active surface expansion mechanisms postulate that the process that furnishes the increased surface required by division is also the source of the force that deforms and divides the cell. The stress in the cortex is assumed to be compressive (that is, a bit of the surface would expand if it were freed of the resistance of its surroundings), and division is accomplished as the passive endoplasm is pushed aside in the equatorial region. The mechanism seems both possible and parsimonious. The cell surface clearly increases during division, but the question is whether the increase is an active or passive process.

Early expanding membrane theories placed the actively expanding part of the surface in the furrow region (Meves, 1897; Chambers, 1937, 1938; Marsland 1938, 1942). Schechtman's (1937) hypothesis was based upon observations of newt egg cleavage, in which the paling of the furrow walls strongly suggests local stretching or expansion, but no such clues exist in other cells. J. M. Mitchison (1952) later proposed that the surface is at first in a state of slight tension and that, beginning at the poles, the tension is released. The regions of released tension spread toward the equator and develop into regions of active expansion. As the surface in the furrow region expands, it pushes the base toward the mitotic axis and divides the cell. These ideas were formulated before the pattern of surface increase in dividing cells was well known. K. Dan and his co-workers (K. Dan, Yanigata, and Sugiyama, 1937; K. Dan and Ono, 1954; Hiramoto, 1958), by meticulously recording the movement of particles embedded in the sea urchin egg surface, found that surface in the furrow region at first shrank and then stretched so that there was little net gain. The expanding membrane hypothesis was then changed to restrict the active expansion to the region outside the furrow; the furrow surface was assumed to be passively pushed inward (Swann and Mitchison, 1958). In its revised form, the expanding membrane hypothesis was in reasonable agreement with the documented pattern of surface increase during division, as well as with the normal cyclical increased resistance to deformation that J. M. Mitchison and Swann (1955) attributed to increased surface stiffness.

A division mechanism based upon active surface expansion does not permit tension in the surface or positive pressure in the endoplasm (J. M. Mitchison, 1953). The hypothesis was tested by determining whether the physical requirements of the mechanism are met and whether the circumstances required to put it into operation exist. The question of tension at the cell surface presented difficulties. The predivision rounding up of both tissue cells and embryonic blastomeres had been observed for many years and had been attributed to increased tension at the surface. Chambers (1938) tore the pole of a dividing sea urchin egg and observed that, as the interior cytoplasm flowed outward, the diameter of the intact blastomere decreased as its contents flowed across the constricted furrow region. He concluded that the pressure inside the egg was initially greater than that of the surrounding medium and that the intact blastomere surface shrank elastically. In an analysis of motion pictures of the event, Sichel and Burton (1936) deduced the presence of positive membrane tension. J. M. Mitchison (1953) considered Chambers's experiments inconclusive. He reported that the event occurred only in cytolizing *Arbacia* eggs immersed in a KCl solution that he considered unphysiological. Mitchison also described a group of microdissection experiments that were designed to further explore division-related activity. In one experiment he pushed a needle through both poles so that it was oriented parallel to the mitotic axis. He stated that as the base of the furrow deepened, it passed through the needle without any tendency to gape, and cleavage was completed. The experiments were done at low magnification. These results, Mitchison felt, could not be reconciled with a division mechanism that consisted of a contracting equatorial ring, but they might be expected with an expanding membrane mechanism. When the experiments were repeated (Rappaport, 1966), the magnification used by Mitchison was found to be too low for satisfactory observation of events at the furrow base. At higher magnification it was apparent that the furrow never passed through the needle. Rather, the furrow stopped where it touched the needle, and it was completed by further intrusion of the unblocked parts. When two needles were inserted in opposite directions so that there was a space between them, they stopped the furrow at diametrically opposite points on the equator. In the eggs of four invertebrate genera, cleavage was never completed, and the incipient blastomeres remained connected by a flattened tube of surface and endoplasm with dimensions determined by the needle diameters and distance between them (Figure 4.1). When the base of the furrow was purposely cut by inserting a needle and then moving it outward through the equatorial surface, division ceased and the furrow usually regressed (Rappaport, 1966).

Smooth, spherical cells with a very limited repertoire of mechanisms for autonomous shape change have many conveniences for both experimenters and theorists, but simple inspection reveals little of the physical state of their surfaces. Tissue cells, especially those in culture, are, by their changeable shapes, more informative. The tendency for retraction at the margins of flattened cells was described in the early days of cell culture (Harrison, 1908; Chambers and Fell, 1931; W. H. Lewis, 1939). The retraction of detached parts of the cell margin can be demonstrated in all stages of division of flattened cultured cells (Rappaport and Rappaport, 1968). Tendency of a cell region to retract implies that it is in a state of tension, or at least is not actively expanding.

More quantitative information concerning the physical state of the cell surface is clearly desirable. The number of parameters that can be measured in healthy intact cells is, however, clearly limited, and the interpretation of the data may be complicated and difficult. Cole (1932; Cole and Michaelis, 1932) compressed *Arbacia* eggs using a flat gold fiber against a parallel glass plane. Because the bending moment of the gold fiber was known, Cole and Michaelis could determine the force needed for different degrees of egg flattening. Because the unflattened egg was a sphere, the deformation increased the surface. The nature of the resistance to deformation, they reasoned, revealed something of the physical nature of the surface. Had the surface behaved as a fluid interface, the force required for deformation would not have changed as the degree of deformation was increased. But they found that the surface force increased as the surface area increased, and they concluded that the surface behaved as a stretched elastic membrane. By extrapolation, they calculated that the internal

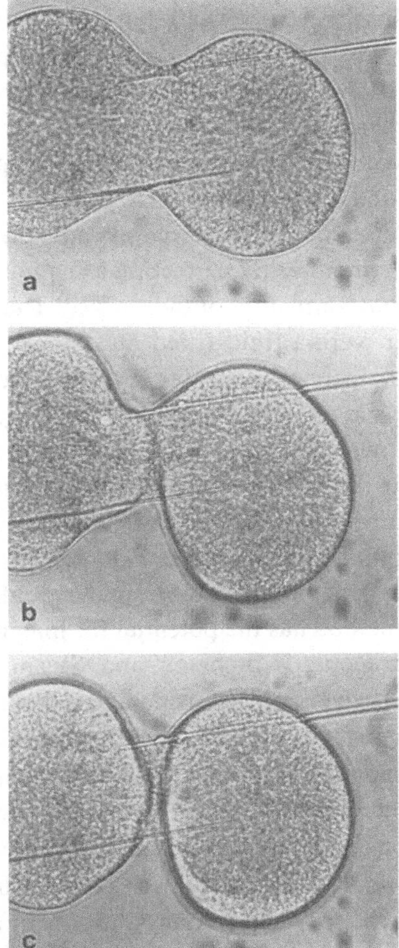

Figure 4.1 Cleavage of an *Echinarachnius* egg with two needles inserted through the division plane. (From Rappaport, 1966, and with the permission of Wiley-Liss)

(a) Shortly after the insertion of the needles.

(b) Furrow has deepened and pushed needles together. Surface contour in the furrow walls is altered. Although the blastomeres later round up, the furrow does not deepen.

(c) End of cleavage. The needles have been moved apart slightly to show the flattened tube of granular cytoplasm that still connects the blastomeres.

pressure of the spherical egg was 40 dyne/cm^2. J. M. Mitchison and Swann (1954a, 1954b) felt that the extrapolation procedure used by Cole and Michaelis was not justified, and they pointed out that the resistance to deformation that was revealed by the compression method might have been due to the physical properties of internal structures, such as the mitotic apparatus.

Mitchison and Swann (1954a) circumvented the problem of interference by internal structures by perfecting a suction method for studying the resistance to deformation of the surface. The basic components of their elastimeter were a fluid-filled glass pipet with an orifice of about 50 μm and an open reservoir, the level of which could be altered to control the hydrostatic pressure in the system (Figure 4.2). When the reservoir level is lowered to create a negative pressure, touching the pipet tip against the cell surface (with the aid of a micromanipulator) causes a small part of the surface to bulge upward inside the nozzle. Both the pressure in the system and the extent of the bulging deformation are measurable. In addition to obviating the problem posed by internal structures, the method has the potential for making regional as well as overall measurements. Mitchison and Swann (1954a, p. 449) remarked that "the conversion of this data to information on the properties of the cell surface presents considerable difficulties." They accepted Cole's contention that the surface is an elastic system, but they pointed out that the convex theoretical curve that represents the relation between tension and internal pressure in surfaces that are thin elastic membranes did not resemble the linear experimental curve from the measurements (Mitchison and Swann, 1954b). In a series of model experiments on scaled rubber balls of different wall thicknesses, Mitchison and Swann found that over a considerable range of wall thickness the pressure–deformation curves were linear. They concluded that the membrane is sufficiently thick to resist deformation by its own rigidity and proposed that the 1.6 μm thick cell surface and cortex behave more like the wall of a tennis ball than that of an inflatable elastic balloon, in that their strength and rigidity allow the dimensions to be independent of internal pressure. Mitchison and Swann attributed the cell's resistance to flattening to the "stiffness" of the surface. The stiffness was determined by the slope of the pressure–deformation curve and was directly proportional to the elastic modulus.

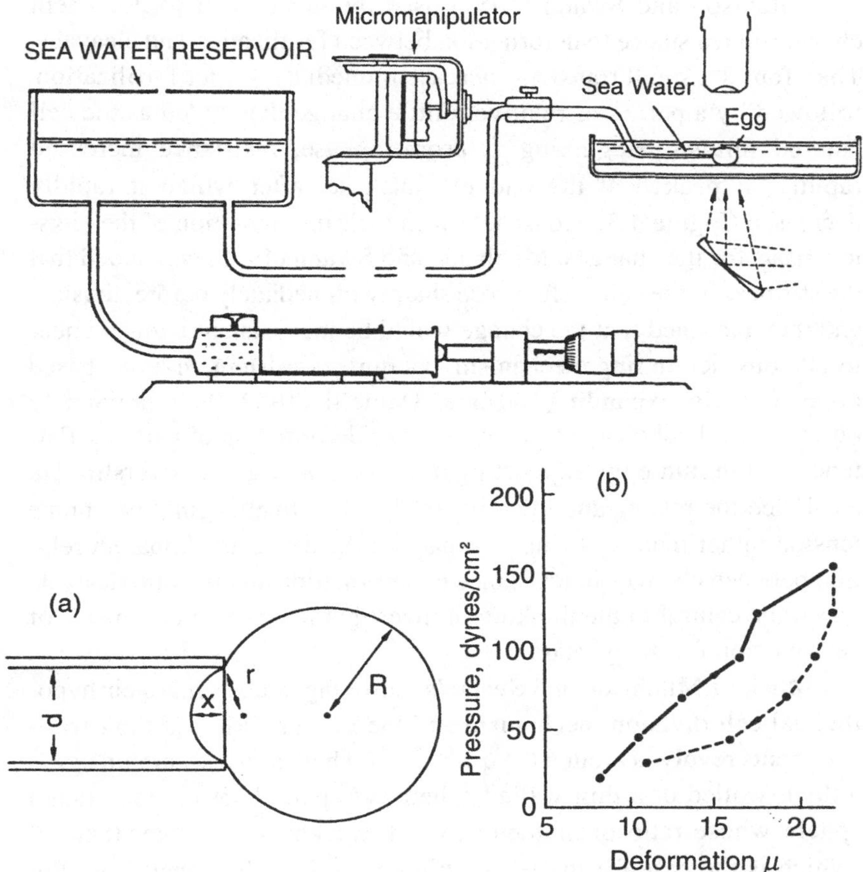

Figure 4.2

UPPER PANEL: General arrangements of the cell elastimeter (not to scale). (From J. M. Mitchison and Swann, 1954a, and with the permission of the *Journal of Experimental Biology*)

LOWER PANEL: Use of the elastimeter.

(a) Diagram of the deformation of the egg by suction. (From Hiramoto, 1970, and with the permission of Pergamon Press)
d = internal diameter of the micropipet
R = radius of curvature of the egg surface outside the pipet
r = radius of curvature of the egg surface inside the pipet
x = deformation of the cell surface

(b) Pressure–deformation curve in the suction method obtained by increasing negative hydrostatic pressure (solid line) and by decreasing pressure (broken line). (From J. M. Mitchison and Swann, 1954a, and with the permission of the *Journal of Experimental Biology*)

Mitchison and Swann (1955) used the elastimeter to document changes in resistance to deformation between fertilization and cleavage. They found a small transient increase immediately after fertilization, followed by a period of relatively little change that ended as the cell entered mitosis. Beginning at prometaphase, resistance increased rapidly; it peaked at the end of anaphase, after which it rapidly decreased (Figure 4.3). Consistent with their interpretation of the physical basis for the changes, Mitchison and Swann (1955) concluded that the stiffness of the cell surface rose sharply immediately before division, and they reasoned that the change would be inconsistent with an equatorial constriction ring mechanism, but quite consistent with one based upon actively expanding surface. Danielli (1952) had previously observed cyclical changes in resistance to deformation of partially flattened sea urchin eggs supporting a piece of thin glass coverslip. He attributed the raising and lowering of the glass to changing membrane tension rather than to stiffness. Whatever the cause, the apparent relation between changes in resistance to deformation and the mitotic cycle was to be central in the thinking of investigators of the mechanisms of cell division for many years.

Much of Mitchison and Swann's reasoning concerning their hypothetical cell division mechanism and the interpretation of their measurements revolved around the question of whether the sea urchin egg is a thick-walled or a thin-walled sphere. Wolpert (1960) argued that a sphere whose ratio of diameter to wall thickness is greater than 60 (which is the case for the sea urchin egg) should be considered thin walled. He also pointed out that Mitchison and Swann's data (1954a, Figure 11) showed that tests on model rubber balls with a diameter/wall ratio of about 60 reveal that the stiffness increases linearly with wall thickness. If the balls were to be considered thick walled, he pointed out, stiffness should increase as the cube of thickness, because resistance to flexure is proportional to the cube of thickness. Since resistance to extension of the model surface was directly proportional to wall thickness, Wolpert concluded that the effect of membrane flexure in a thin-walled object, like a sea urchin egg, could be neglected.

Hiramoto (1963) proposed that the linear relation in the pressure–deformation curve reported by Mitchison and Swann (1954b) might have been caused by slippage of the egg or the rubber ball surface

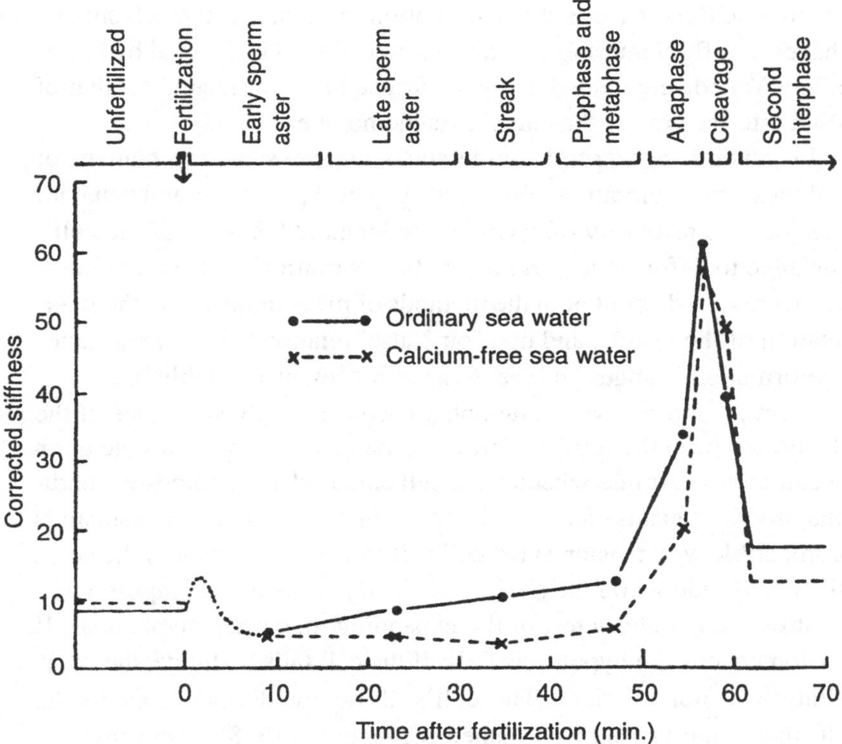

Figure 4.3 Stiffness changes during first interphase and division: eggs of *Psammechinus miliaris.* Corrected stiffness is in dyne/cm^3/μm deformation for 100 μm diameter egg and 50 μm diameter pipet. (From J. M. Mitchison and Swann, 1955, and with the permission of the *Journal of Experimental Biology*)

around the edge of the pipet nozzle, and he concluded that resistance to deformation of the endoplasm and bending of the cell surface were negligible in both the compression and the elastimeter methods. Yoneda (1964) reexamined Cole's experiments and pointed out (as had Mitchison and Swann, 1954b) that the necessary accurate measurement of the areas of contact between the cell and the parallel, compressing flat surfaces is very difficult. Yoneda avoided the problem by calculating the tension by an equation that instead employed the total cell surface area. The tension calculated by this method remained constant, irrespective of the changes in surface area (that is, the degree of cell flattening). If the cortex and surface behave as an elastic membrane, the tension would increase as the deformation increases. If the cell surface behaves as a fluid interface, its tension would not be changed by deformation. Yoneda concluded that cell surface behavior resembles that of a fluid interface rather than an elastic membrane.

In general, attempts to understand the physical mechanism of cytokinesis by exploration of the cell's physical properties and behavior were focused primarily on trying to understand the basis of the cell's resistance to deformation and its cyclical variation. But serious questions were raised about both the methods of measurement and the interpretation of the results, and no clear causal relation between resistance to deformation changes and the division process was established.

A simple alternative to attempts to deduce the physical state of the cell surface from the results of measurements is to impose a state upon the cell and determine whether the cell can, under the imposed conditions, divide. Because an actively expanding membrane mechanism is incompatible with tension at the cell surface, determination of the cell's ability to divide while the surface is clearly in a state of imposed tensile stress constitutes a test of the expanding membrane hypothesis. If division occurs, the hypothesis fails. If the cell fails to divide, the interpretation is not so clear. The cell's ability to divide when in the deformed state is well established. Chambers (1938) found that sea urchin division was not impaired when the eggs were stretched with pairs of needles parallel or perpendicular to the mitotic axis. J. M. Mitchison (1953) also demonstrated that compression of the poles perpendicular to the axis is not inhibitory. These manipulations would be expected to increase the tension in an elastic membrane. The cell's response to such deformation is not precisely like that of a rubber balloon or ball. When the cause of deformation is removed, the cell does not rebound quickly into its original form. Its viscoelastic properties result in a relatively slow recovery. The ostensible yielding of the shape-maintaining components led to questions concerning the distribution and magnitude of forces in static deformation and suggested that a study of the cell's ability to divide at the same time that it supported a significant weight could produce clearer results (Rappaport, 1960).

The experimental design took advantage of the increased adhesivity of the echinoderm egg surface that follows brief treatment with acid sea water (Allen, 1954). Sand dollar eggs were treated with acid sea

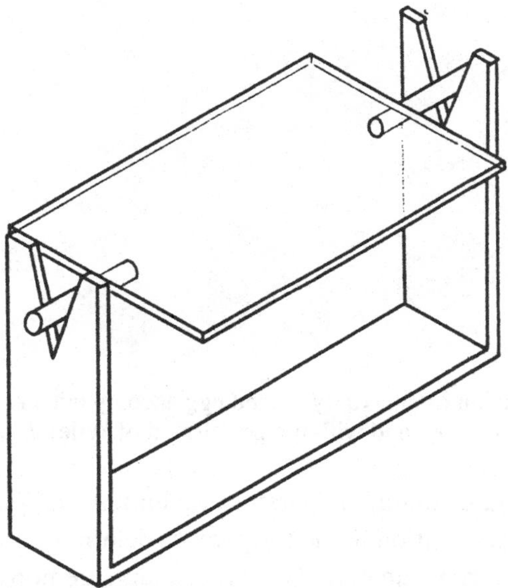

Figure 4.4 Tilting table used in subjecting cells to tensile stress. This unit is immersed in a small aquarium and viewed through a horizontal microscope. (From Rappaport, 1960, and with the permission of Wiley-Liss)

water and then pipetted to the surface of a tiltable glass table that was immersed in a small aquarium (Figure 4.4). The table top and attached eggs were sprinkled with small glass beads, and by chance some of the eggs adhered to both the table top and a glass bead. The fluid surrounding the eggs was replaced with calcium-free sea water, and 30 minutes before cleavage was expected, the table top was inverted so that eggs were on the underside and any attached glass beads were suspended freely in the medium. The weight of the beads rapidly stretched the cells into cylinders with a length about nine times greater than the diameter (Figure 4.5). The extraneous material had been removed from the surface, and the cell contents are generally considered to be fluid and incapable of supporting weight. The weight of the bead was therefore borne by the several-micrometers-thick combined surface and cortex. The mitotic apparatus expanded and oriented parallel to the cylinder axis, and at the normal time, the furrow developed in the normal relation to the asters. Division was completed, and one blastomere was carried to the bottom of the chamber by the falling bead. The uniform relation

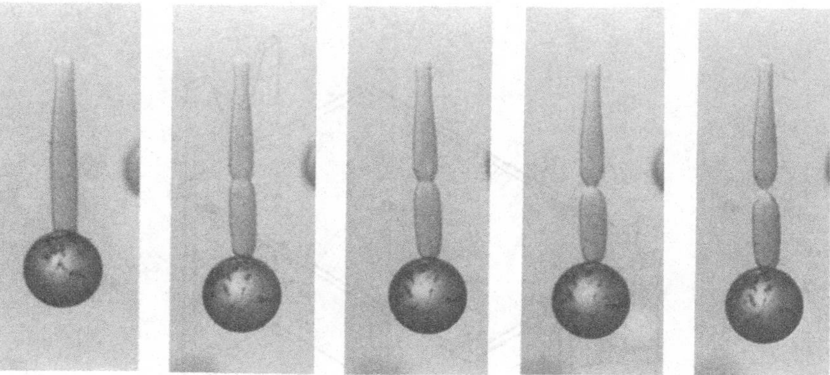

Figure 4.5 Division of an axially loaded egg accompanied by elongation. (From Rappaport, 1960, and with the permission of Wiley-Liss)

between cell shape, mitotic apparatus orientation, and plane of division indicates that the position of the furrow was determined after the weight was suspended from the cell. The experiments demonstrated that the division mechanism can be established and then function to completion under constant tensile stress. This observation cannot be reconciled with the expanding membrane hypothesis.

Equatorial Constriction Mechanism

Equatorial constriction is a logical alternative to an expanding membrane hypothesis. In fact, it constitutes an antihypothesis in that the driving force is located in the equatorial region and the surface is in a state of tensile stress. It is now evident that division occurs when the tension at the equatorial surface exceeds that of the rest of the cell surface. There are different ways in which this simple relationship could be created, and these different ways may require different processes to set them in operation. The similarities among these mechanisms can make discrimination among them difficult. Before cleavage, the tension at the surface is uniform and isotropic. The regional tension differences that accompany division arise from mitotic apparatus activity. The basis of the difference can be explored by studying surface structure and activity and by analyzing the way that the mitotic apparatus positions the furrow. The discussion that follows emphasizes the first approach. The possibility that the necessary regional tension difference could arise either by polar surface decrease or by equatorial increase has been part of the literature of the field for many years. These alternatives assume vastly different responses of the surface to the mitotic apparatus.

Polar tension decrease

Early speculation concerning the tension at the cell surface assumed that it approximated the forces at an oil–water interface. But as information accumulated, it became apparent that the actual values are far smaller. Permeability studies of fat-soluble substances supported the concept of a lipoid plasma membrane, but further studies of non–lipoid-soluble substances encouraged the idea that cholesterol and protein were also membrane constituents that would add to the complexity of the interface. Microdissection studies revealed that the plasma membrane is intimately associated with the underlying cortical layer, which has the properties of a protein gel (Chambers, 1917a). E. N. Harvey and Shapiro (1934) determined that the interfacial tension between a naturally occurring oil drop and the surrounding cytoplasm in a mackerel egg averaged 0.6 dyne/cm with no marked indication of elasticity. In vitro measurement of interfacial tension of mackerel oil against water gave values of about 10 dyne/cm. These results suggested that some surfactant substance was present in the living system and Danielli and Harvey (1935) identified it as protein. The growing realization of cell surface complexity and the low measurable forces diminished support for a simple surface tension cell division mechanism. E. N. Harvey (1954) pointed out that the surface forces in a naked cell were the sum of the tension of the elastic, gelled cortex and the surface tension of an overlying, thin semiliquid coat. The actual surface tension of this combination might be proportionally small, and he suggested that the term "tension at the surface" be applied to the surface forces. He further remarked that in a small object with a high surface/volume ratio, surface energy becomes an important factor; cell shape change and cell division become more difficult, because they work against surface energy. He suggested that high surface energy might be as disadvantageous to the cell as large volume.

Cell division mechanisms that operate by polar tension decrease could utilize a source of tension other than the tension at fluid interfaces.

In Wolpert's (1960) aster relaxation theory, the tension generated by the elastic properties of the surface was postulated to be the main driving force. He assumed that the cyclical increase in resistance to deformation described by J. M. Mitchison and Swann is due to uniform increase in elastic tension. When the asters interact with the surfaces outside of the equatorial region, the surfaces relax. The relaxing effect of the asters fails to reach the equatorial region because of the geometrical relation between the mitotic apparatus and the surface (Figure 2.4). Because the equatorial surface is not affected, its tension continues to increase and the equatorial diameter actively decreases. All the other documented changes in dimensions, area, and physical properties are assumed to be passive. The equatorial surface becomes in effect a contractile ring. The principal reservations about this theory concern the assumed relaxing effect of the asters and the ability of the cell to complete division using this mechanism. White and Borisy (1983; White, 1985) devised a computer model that incorporated the assumptions of the Wolpert mechanism and found that it failed to simulate completed division for several reasons. Although the postulated enhanced aster effect at the poles seems reasonable while the cell is spherical, the diameter decrease that takes place during division brings the equatorial surface closer to the mitotic apparatus so that the aster-induced relaxation would affect the equator in the same way that it was presumed to affect the poles. Division activity would then stop before completion when the actively contracting region relaxed. This difficulty confronts all theories in which the region of the furrow is designated by the effect of the asters on surface regions outside the equator. White and Borisy (1983) tried to circumvent it by assuming that the asters ceased to affect the surface immediately after they alter the poles. This assumption is testable, and it was found to be invalid (see Chapter 7). White (1985) suggested another difficulty concerning division completion according to Wolpert's mechanism. Wolpert's theory assumes that tension at the surface is isotropic, and White pointed out that in its final stages, the furrow is a saddle figure with a positive and negative curvature (Figure 4.6). When the two radii have similar values, the forces of tension at the surface cannot generate inwardly directed pressure, so that the deformation taking place in the furrow would stop before division is completed. It is also true that isolated bits of equatorial surface can divide despite absence of regions of the surface that, according to the hypothesis, must be present if furrowing

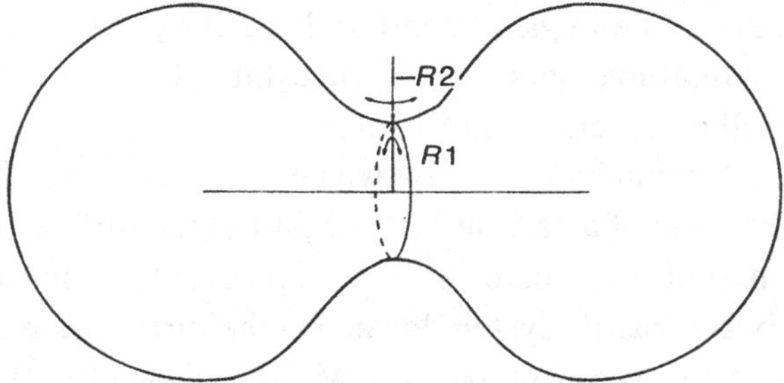

Figure 4.6 A high level of isotropic surface tension in the equatorial regions cannot by itself produce cleavage. A cleavage furrow is a saddle surface; the two principal radii of curvature of the surface, R1 and R2, will therefore have opposite signs. The inward-directed pressure due to surface tension forces (T) is T/R1 + T/R2. In the configuration drawn, R1 is the same size as R2 but is of opposite sign. Therefore there can be no inward-directed force from surface tension in this situation. (From White, 1985)

is to succeed (Rappaport, 1969a). Dividing sand dollar eggs subjected to constant tensile stress fail to show a consistent pattern of stretching or yielding in the surface near the asters (Rappaport, 1960). But the greatest weakness of this and other forms of the astral relaxation hypothesis concerns the basic postulate that the asters designate the position of the furrow by causing the nearest surface to relax. This essential component of the overall mechanism has failed many experimental tests (see Chapter 7).

Equatorial tension increase

Theories of equatorial tension increase assume that equatorial tension increases following mitotic apparatus–activated enhancement and that division is caused by a ring of contractile material concentrated at the base of the furrow. There is a clear divergence of ideas about the way in which the pattern of tension at the surface changes from uniform to local enhancement. The older hypotheses of Bütschli, Ziegler, W. H. Lewis, and Scott postulate that tension at the equator is directly enhanced and that the events that take place outside the equator are the passive results of active constriction in the furrow. These may be called the astral contraction hypotheses. Although astral contraction and astral relaxation mechanisms may be viewed as opposites, their differences lie not in the way the division mechanism functions, but in the means used to create the zone of enhanced contractile activity. Experimental testing of hypothetical mechanisms for the establishment of the division mechanism will be described in Chapter 7.

Equatorial tension increase – by whatever means – can divide a cell by active shortening of a narrow, beltlike part of the surface that encompasses all or part of the equatorial plane. The events of normal division imply that any contractile system located at the furrow base can shorten until the system literally disappears. J. M. Mitchison (1952) considered the idea of such an event so incongruous as to be an argument against an equatorial constriction mechanism. Certainly the degree of shortening must exceed that of any known stable contractile system.

Division mechanisms that operate by active shortening bring to mind a tightening belt or elastic band analogy and suggest that division would cease if the furrow were cut. Yatsu (1908) and Chambers (1919) cut at a right angle or diagonally across the furrow of *Cerebratulus* and *Asterias* eggs and found that furrowing activity was not disrupted. Both concluded that division resulted from subsurface activity. Yatsu favored a diastema, and Chambers an astral cleavage mechanism. The nature of the events that ensue when a cell is cut between a hard, flat surface and the shaft of a needle or the edge of a knife was not often discussed in the reports of early microdissection studies. When such cuts are made, the surface in contact with the tool is pushed closer to the surface in contact with the flat surface, and the intervening cytoplasm is displaced. The two inner surfaces touch and apparently fuse as the tool passes through the cortex and surface membrane region and contacts the flat surface (McNeil, 1991). The cell contents are not exposed to the medium, and the requisite new surface appears to form almost instantaneously in response to the force that deforms the cell. The persistence of the furrow in the experiments of Yatsu and of Chambers could have resulted from the successful fusion of opposite cell surfaces. The operation may not have actually cut the furrow base, but rather fused it into loops of smaller diameter. This possibility can be avoided by cutting the furrow in a different way. When a needle is inserted through the pole and across the division plane of a dividing cell, it is not in a position to push opposed surfaces together. As it moves through the endoplasm toward the periphery, it cuts through the cortex and surface, and then contacts the flat surface. When dividing sand dollar eggs were cut in that way, the furrows rapidly regressed around the entire equatorial circumference (Rappaport, 1966). Similar cuts made elsewhere on the surface did not affect cleavage. These results supported the idea that a structure located at the base of the active furrow is necessary for the constricted deformation that characterizes a dividing cell.

Force measurements

In the mid 1960s, several ultrastructural studies that appeared to demonstrate diastema or vesicle sheets in the equatorial plane (Buck and Tisdale, 1962) implied that division took place by the simple separation of newly formed opposing membranes. This mechanism would require the application of little or no force in the equatorial region, and it might not be impeded by needles or other mechanical blocks placed in its path. Despite a body of experimental information inconsistent with this view, the idea of cytokinesis by vesicle fusion in the equatorial plane appeared to be favored by many electron microscopists of the time.

Investigation of cleavage furrow behavior revealed that a pair of needles inserted through the division plane stopped furrow progress after the furrow base contacted the needles (Rappaport, 1966). When very slender needles were used, they were bent after the furrow contacted them, and the distance between the two needles decreased before cleavage was finally blocked (Figure 4.1). This observation indicated that the base of the furrow exerts a force and that by using a needle with a known bending moment, it would be possible to measure that force directly. Calibrated needles had previously been used to measure very small forces in other investigations. Standard needles were calibrated by measuring their deflection after loading with small pieces of thin paper of known weight. The standard needle was then used to calibrate more sensitive needles. The first measurements were made by inserting two paraxially oriented needles in opposite directions through the poles and across the equatorial plane (Figure 4.7) (Rappaport, 1967). One of the needles was thin, flexible, and calibrated; the other needle was thicker, too stiff to be bent by the furrow, and uncalibrated. The deepening furrow contacted the needles and deflected the calibrated needle. The regions of the furrow that did not contact the needles continued to deepen until the equatorial surface resembled a short flattened tube with parallel sides (Figure 4.1). The cal-

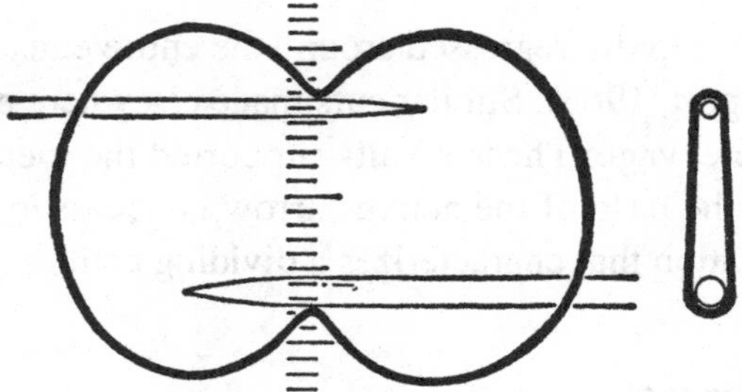

Figure 4.7 Arrangement of cell and needles for determination of tension exerted by the sea urchin cleavage furrow. Left: Upper calibrated needle is deflected downward during cleavage. Lower holding needle does not move. Right: Schematic section through, and parallel to, the cleavage plane during isometric contraction. Diameter of the needles is exaggerated. (Rappaport, reprinted with permission from *Science,* 156:1241–1243, copyright 1967, American Association for the Advancement of Science

ibrated needle bent until its resistance to deflection equalled the force exerted by the furrow, when further progress was blocked. The amount of deflection was measured with an ocular micrometer. The forces measured for *Astriclypeus manni* (a Japanese sand dollar) and *Pseudocentrotus depressus* (a Japanese sea urchin) were, respectively, 3.04×10^{-3} dyne and 2.00×10^{-3} dyne. This method does not measure the actual force that the furrow must exert in order to divide the cell, but rather the maximum force that the furrow can exert. The same method later showed that the sand dollar (*E. parma*) first cleavage furrow exerts 1.58×10^{-3} dyne at first cleavage, and the force exerted by the second cleavage furrow is not significantly different (Rappaport, 1977). The force exerted by the furrow is independent of its length. The results of measurements made on the same furrow at two different lengths in which the shorter was about two-thirds of the longer were not significantly different. The cleavage process in sand dollar eggs requires about 8 minutes, and at completion, no vestiges of the division mechanism remain. To determine whether this relatively brief life-span is predetermined, duration of the period of maximum contraction exerted by one of the two blastomeres at second division was measured, and the other blastomere was used as a time control. Furrows exerted maximum tension for as long as 9 minutes after the furrow of the companion blastomere (control) was completed. When the needles used to make the measurement prevented completion, the furrow exerted maximum force for a period that was twice the entire normal time span of the division mechanism. The normal longevity of the division mechanism must be affected by its function, in addition to other limiting factors.

Hiramoto (1979) modified the method of measurement by adding a reference needle, so that the force could be measured while the length of the active furrow remained constant (Figure 4.8). He found that the force exerted by *Clypeaster japonicus* furrows averaged 4.2×10^{-3} dyne. Yoneda and Dan (1972) estimated the force of constriction at the furrow by calculations involving the width of the stalk connecting the incipient blastomeres after the beginning of cleavage, the initial diameter of the cell, and the tension at the polar surface. They arrived at a maximum

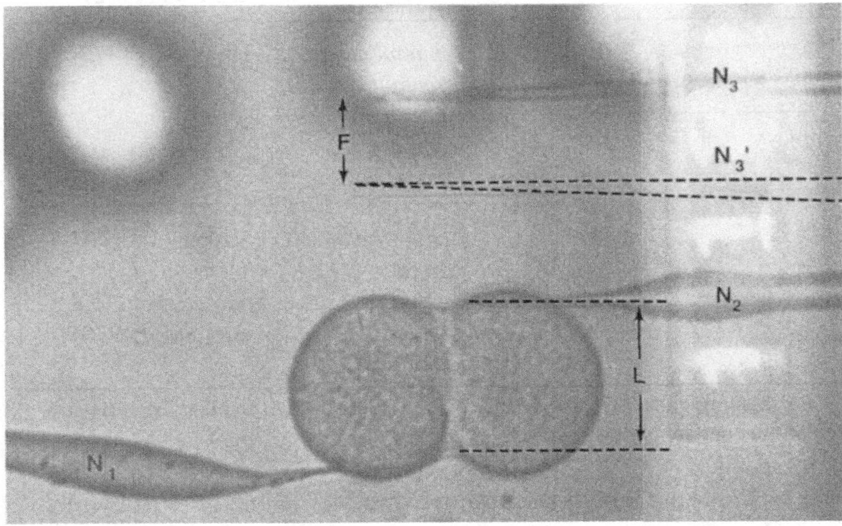

Figure 4.8 Measurement of the force exerted by the cleavage furrow in a sea urchin egg. Two microneedles (N_1 and N_2) maneuvered by micromanipulators were thrust through the opposite polar surfaces of a dividing sea urchin egg until they passed through the equatorial plane. The third microneedle (N_3), which was held by the micromanipulator holding N_2, was used as a reference. When the force exerted by the cleavage furrow acted on the tip of N_2, the distance (L) between N_1 and N_2 tended to decrease owing to the bending of N_2. The position of the micromanipulator holding N_2 and N_3 was changed so that the distance L remained unchanged and, in consequence, the distance between N_2 and N_3 increased. The force exerted by the cleavage furrow could then be determined, since the change (F) in distance between N_2 and N_3 was proportional to the bending of N_2. (From Hiramoto, 1979, and with the permission of the University of Tokyo Press)

Table 4.1 Constricting force and tension generated by the contractile ring in echinoderm eggs (mean values)
Modified from Hiramoto, 1979

FORCE (10^{-3} dyn)	TENSION (10^5 dyn/cm^2)	SPECIES	REFERENCES
3.04		*Astriclypeus manni*	Rappaport (1967)
2.00	2.5	*Pseudocentrotus depressus*	
6	4.5		Yoneda and Dan (1972)
9 4	3.0 1.3	*Temnopleurus toreumaticus*	Hiramoto (1975)
2	0.7	*Clypeaster japonicus*	
1.58	0.781*	*Echinarachnius parma*	Rappaport (1977)
4.2	1.3	*Clypeaster japonicus*	Hiramoto (1970)

* Value assuming that the constricting force is 2.5 x 10^{-3} dyn and that the cross-section of the contractile ring is 3.2 μm^2.

of 6 x 10^{-3} dyne, which, assuming a contractile band 6.5 μm wide, is equivalent to a tension of 9 dyne/cm. Because this value was so much greater than the tension developed at the polar surface (3 dyne/cm), they suggested that furrowing is not simply a passive result of polar relaxation but may involve, in addition, an autonomous generation of constricting force at the equator.

The observation that injected oil drops are deformed and may be cut into two by the advancing furrow (Chambers and Kopac, 1937; Hiramoto, 1965) prompted Hiramoto (1975) to devise another method for determining the force exerted by the furrow. The principle was to determine the amount of force that was applied to a drop of ferrofluid from the amount of deformation of the drop that took place during cleavage and from the interfacial tension between the ferrofluid and the protoplasm. The virtue of the ferrofluid is that the interfacial tension between it and the protoplasm can be determined from the degree of deformation of an injected drop in a controllable magnetic field. After the surface tension of the drop in the protoplasm is known, the constricting force can be calculated from the radii of curvature and surface areas of the two parts of the drop lying on either side of the constricted region and from the circumference of the cross section of the drop. By this method, the maximum forces developed by the furrows of *Clypeaster japonicus* and *Temnopleurus toreumaticus* were, respectively, 2.0×10^{-3} dyne and 3.9×10^{-3} dyne.

The forces generated by the cleavage furrow in several echinoderm eggs when measured or estimated by several different techniques are similar (Table 4.1), although both species differences and the different methods may contribute some variation. The measurements by themselves do not clearly identify the way in which the force originates. It is conceivable that needles could be bent and oil drops deformed as a result of active surface expansion as well as localized contraction. The identification of the constriction mechanism requires analysis of both the physical properties of the surface and the behavior of the furrow under experimentally imposed conditions.

At the time of the earliest direct force measurements (Rappaport, 1966), polar relaxation, equatorial constriction, and vesicle fusion were actively championed or considered likely possibilities by one author or another. The likelihood that the force results from active constriction suggested the desirability of comparison with better-known contractile mechanisms, which in turn required information about the cross-sectional area of the supposedly active part of the surface. In 1966, the best estimates of the dimensions were about 0.1 μm thick (Mercer and Wolpert, 1958; Weinstein and Hebert, 1964) by 5 μm wide (Wolpert, 1960). Subsequent ultrastructural studies modified the dimensions to 8 μm wide by 0.2 μm thick (Schroeder, 1972). The tension generated by the "average" echinoderm egg cleavage furrow is 2.5×10^5 dyne/cm^2 (Table 4.1). The tension generated by an actomyosin thread in isometric contraction is 2.45×10^5 dyne/cm^2 (Portzehl, 1951). These values are about 1/20 of the force generated by vertebrate striated muscle (Huxley, 1972).

5

Positioning the Division Mechanism

Cell division theories that assume that the mitotic apparatus is part of the physical division mechanism have an appealing simplicity, because both the origin of the motive force and the reason why the furrow is positioned as it is are explained at the same time. When the possibility that the mitotic apparatus plays a physically active role was disproven, a new set of problems concerning the basis of the correlation between mitosis and cytokinesis required solution. It was necessary to explain how the division mechanism normally develops in the surface at the right time and place and in the right orientation. The realization that there is an apparently separate establishment process adds further complication, and it now appears that the formation of the mechanism contains more unknowns than its function. But the complications have provided more opportunities for experimentation. Investigations have yielded information about the division mechanism as well as the events that put it in operation. The basic assumptions associated with the process are simple: (1) Before the division mechanism functions, the characteristics of the surface are uniform; (2) the effect of the mitotic apparatus upon the surface is not uniform, and it produces regionally different characteristics; (3) the surface characteristics of the region that are most affected by the mitotic apparatus are the ones changed the most; and (4) the mitotic apparatus–dependent change precipitates division. Investigations have attempted to answer the following general questions: (1) Which parts of the mitotic apparatus are required for this activity? (2) How is division mechanism establishment temporally related to mitosis? (3) Which parts of the surface are most affected by the mitotic apparatus? (4) What is the immediate effect of the mitotic apparatus upon the surface? (5) What are the necessary geometrical relations between the mitotic apparatus and the surface? In all of these questions there is an added assumption that the effect of the mitotic apparatus is carried across some part of the distance between the mitotic axis and the surface, but the chemical basis of the effect has been primarily a subject of speculation.

Analysis of the relation between the mitotic apparatus and the surface at the time of division mechanism establishment has been impeded by several factors: The constituent phenomena have few measurable properties; there is no decisive chemical information; there is little or no useful ultrastructural information; there are no distinctive changes before, during, or immediately after the critical period of their interaction and, as in light microscopy, ultrastructural changes are rarely, on the basis of their appearance alone, identifiable as active or passive.

How Is the Position of the Mitotic Apparatus Determined?

The position and orientation of the furrow are determined by the mitotic apparatus. In cleavage cells, the orientation of the mitotic apparatus is fixed considerably before its interaction with the surface by processes that are separate from cytokinesis. The final orientation of the mitotic apparatus can be affected by events that occur before or after its formation.

Orientation during formation

The orientation of the mitotic apparatus is determined by the positions of the centrosomes, which organize the spindle and the astral cytoskeleton.

The importance attached to the centrosome and the degree of consensus concerning its structure has varied over the years (reviewed, Wilson, 1928; Mazia, 1984). The sea urchin sperm carries two centrosomes into the egg. One serves as the basal body of the sperm flagellum and the other is associated with the sperm mitochondrion. After pronuclear fusion, the basal body centrosome is firmly attached to the zygote nuclear envelope; the mitochondrial centrosome is more loosely associated with the envelope. The centrosomes are separated by granular, osmophilic centrosomal material that appears to serve as microtubule organizing centers (Paweletz, Mazia, and Finze, 1987a, 1987b). The basal body centrosome remains fixed while the mitochondrial centrosome is moved by microtubule-mediated spreading of centrosomal material until it is located diametrically opposite the basal body centrosome. At first, the basal body centrosome contains a normal centriole, whereas the centriole in the mitochondrial centrosome is in precursor form. By the time of mitosis, the centrosomes are indistinguishable and the spindle forms between them. At anaphase the centrosomes begin to flatten, and by telophase they are reshaped into thin plates within which the centrioles are separating (Paweletz, Mazia, and Finze, 1984). During the subsequent interphase, much of the osmophilic material is dispersed in the cytoplasm. Fluorescent antibodies to centrosomal components permit observation of the components' recruitment to the nuclear envelope's outer surface before the next cleavage (Leslie, 1990). As the egg prepares for the second cleavage, the centrosomal components concentrate in a thin layer on the nuclear envelope in the region of the centrioles. The spreading cap of centrosomal material bisects, then it accumulates at diametrically opposite poles, where it concentrates into two spheres. In this and subsequent divisions, both centrosomal regions resulting from the bisection move as they assume antipodal positions. In studies of induced multipolar mitosis caused by fertilization of procaine-treated sea urchin eggs with X-irradiated sperm, Czihak et al. (1991) found that the positioning of centrosomes is consistent with the predictions of a computer model based upon Mazia's hypothesis that the two thin, skullcap-shaped areas of centrosomal material expand equally and simultaneously. Because they are assumed to be mutually exclusive, the centrosomal pair must take up diametrically opposite positions on the nuclear envelope.

The nature of centrosomal reproduction and the rearrangement of centrosomal material that accompanies each mitotic cycle ensures that the mitotic axis that forms between the centrosomes is shifted about 90° in successive mitotic cycles. This phenomenon has been important in the determination of embryonic cleavage patterns.

Reorientation of the formed mitotic apparatus

The behavior of the assembled mitotic apparatus during the period of its enlargement depends upon its physical consistency and shape, and the shape of the space in which it is confined.

Early investigators observed that in eggs with supernumerary sperm, the centers of the multiple sperm asters became more distant as their diameters increased, suggesting a kind of mutual repulsion. Teichmann (1903) proposed that the phenomenon could result from condensation of cytoplasm around the aster centers. If the growing asters were unable to fuse and their consistency were firmer than that of the surrounding cytoplasm, the movement of the centers would follow. Chambers (1917b) confirmed Teichmann's interpretation in microdissection experiments on sand dollar asters. Chambers attributed the more gelated quality of the asters to rigid projections of cytoplasm that extend toward the aster center among the hyaline rays, which he regarded as liquid (reviewed, Chambers and Chambers, 1961). This interpretation of the physical structure of asters and their surroundings allowed the possibility that the asters could exert force against surrounding structures, which might result in cell deformation or division or movement of the aster within the cell. As an example, Chambers pointed out that the pattern of sperm nucleus movement toward the egg center could be explained as a consequence of sperm aster expansion within the spherical egg (Chambers and Chambers, 1961).

The elongate form of a mitotic apparatus with well-developed asters, combined with its relative rigidity and its capacity for expansion before division, means that its growth within a confined space can cause its realignment, and this activity provides a basis for the generalization that the mitotic apparatus is usually oriented parallel to the cell's longest axis at the time of division (Czihak, 1973). In cells with relatively small asters, the mitotic apparatus is more nearly spherical, so that its long axis may not correspond so precisely to the long axis of the surrounding cytoplasm.

After microtubules were shown to occupy the central hyaline core of astral rays, they were favored as the basis for aster consistency because their appearance suggests an inherent stiffness. Estimates of their modulus of elasticity indicate a value similar to those of Plexiglas and other rigid plastics (Gittes et al., 1993). This degree of rigidity, when considered together with the large number of microtubules in the asters and the force that could be generated by polymerization in each microtubule, appears to provide a simple mechanism for relocation of the aster and mitotic apparatus. Bjerknes (1986) wrote that, in this circumstance, the growing aster would be displaced by the first tubules to contact the surface, and its equilibrium position would be a complicated function of its original position, the force generated by the microtubules, the viscosity of the cytoplasm, and the cell's shape. This relationship is probably what early observers had in mind when they speculated about aster relocation. But Bjerknes (1986) also called attention to the fact that elongate, thin structures of microtubular dimensions buckle relatively easily. Thus microtubules are unable to exert the full force of polymerization because their buckling force decreases with length squared. The equilibrium position of an aster is the position where the net force acting on the aster is zero. The proposed relationship implies that the equilibrium position of the aster depends only on cell shape and is not affected by the initial conditions. Bjerknes described possible experimental tests of the theory. The effects of crosslinker structures on the buckling of linear elements within the mitotic apparatus are unknown.

Analysis of the factors that position the mitotic apparatus may also be complicated by its changeability. Early microscopists observed the elongation of the spindle and the astral rays, the changes in the structure and staining properties of the rays, the dissolution of the interzone of the spindle in anaphase and telophase, and, finally, the near-complete fading of the astral rays after completion of division. Less evident by conventional microscopy, and more recently described, is the phenomenon of dynamic instability (reviewed, Cassimeris, 1993). The majority of spindle and interphase microtubules were found to exist in phases of elongation or rapid shortening by adding or losing subunits. Within a microtubule population, most are in the slower elongation phase and relatively few are in the rapid shortening phase. The transition between phases is abrupt and its mode of regulation in vivo is not known. These findings imply that at the same time that the overall dimensions of the spindle are increasing by microtubule polymerization, some of its radial linear elements are shortening by depolymerization. Microtubule depolymerization can promote particle movement in vitro (Coue, Lombillo, and McIntosh, 1991), so that two-way traffic within the mitotic apparatus may be based upon dynamic instability as well as motor molecules.

Which Parts of the Mitotic Apparatus Are Necessary?

Because the major visible parts of the mitotic apparatus have been known for a very long time, the attempt to determine their necessity for cytokinesis has a long history. The reasoning behind the deductions and experimentation is simple: A part is unnecessary if cells can divide without it. Although it is easy to determine whether or not a cell divides, it may not be so easy to determine whether the successful division is in all respects normal; therefore, factors that are contributing, but not essential, may not be revealed.

Chromosomes

Chromosomes appeared to have great potential for a role in cytokinesis. At one point in time, they are arranged in the future cleavage plane, and later they are close to the centers of the prospective daughter cells. An active role could be therefore invoked by those favoring the essential interaction at the equatorial surface as well as by those postulating interaction at the polar surfaces.

Boveri (1897) fertilized enucleated egg fragments of *E. microtuberculatus* with *S. lividus* sperm. He noted that in some cases the nuclear material was segregated in one blastomere at the first cleavage. In subsequent division cycles, the nucleated blastomere cleaved normally, and the enucleated blastomere continued the cyclic reproduction of asters but failed to divide. He concluded that the regular interval between bisection of the centrosomes, separation of the daughter centrosomes, and the formation and disappearance of the radiate structure is independent of the nucleus, but the nucleus is essential for division of echinoderm eggs.

Ziegler (1898a) observed a normally fertilized egg in which all the nuclear material went to one pole of the mitotic apparatus and was subsequently confined to one blastomere. The unnucleated part cleaved more slowly and less regularly, but the furrows were permanent. He concluded that there is no direct causal relation between chromosomes and cytokinesis. The "division energy" of chromosome-free cells appeared to be weaker, but chromatin did not appear essential.

The chemical treatments that can start development in unfertilized eggs (parthenogenetic development) can in some circumstances produce a number of extra asters in the cytoplasm. These cytasters appear identical to the asters that normally develop at the spindle poles and their mode of origin, as well as that of the centrioles they contain, has been investigated off and on over a long period. Their importance for this discussion lies in the fact that, after the breakdown of the nuclear membrane that takes place during the meiotic divisions, their formation is independent of nuclear material. E. B. Wilson (1901a) observed that, following parthenogenetic treatment and the formation of cytasters in addition to those that formed in association with egg centrosomes, furrows developed between both kinds of asters, but in the great majority of cases division was completed only between asters that were connected with chromosomes (Figure 5.1). After remarking on the probability that complete division may sometimes occur about astral centers unconnected with chromatin, he concluded that the activity of cytasters differed only in degree, not in kind, from that of the nuclear asters.

Soon after Wilson's work, mechanical devices were applied to the problem. McClendon (1907, 1908) employed a screw-activated "mechanical hand" to control a pipet under a stereomicroscope. An unfertilized starfish egg was maneuvered so that its visible incipient polar body lay next to the pipet nozzle. The capillary flow of water pulled the egg against the nozzle. By sucking gently on the end of a rubber tube attached to the pipet, McClendon pulled the nucleated part of the oocyte into the pipet (Figure 5.2). The attenuated region that connected the nucleated to the unnucleated region was broken by introducing a second pipet with a larger orifice, close to the immobilized egg,

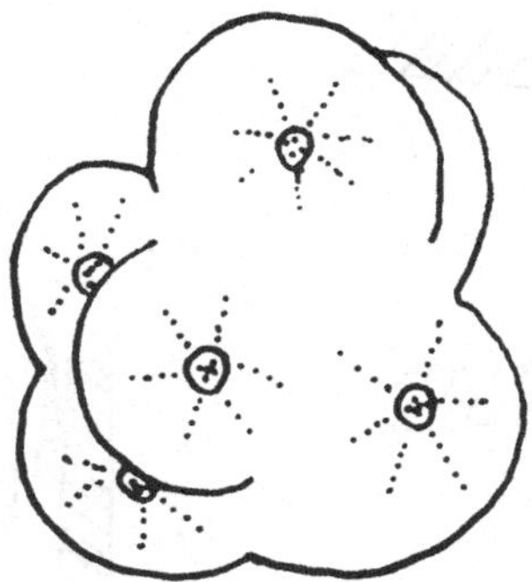

Figure 5.1 Multiple partial furrows formed in association with nuclear asters and cytasters that resulted from prolonged treatment of a sea urchin egg with a parthenogenetic agent. (After Wilson, 1901a)

and allowing the capillary current to sweep the large, enucleated part of the egg up into the lumen. Following parthenogenetic treatment, the eggs cleaved irregularly. After fixation, sectioning, and staining, they were inspected for chromatin. McClendon reported that not all the operated eggs were completely devoid of chromatin, but division took place whether chromatin was present or not. McClendon's experiments required ingenuity, perseverance, dexterity, and judicious timing, but they do not appear to have convinced many skeptics.

Fry (1925) considered McClendon's method unreliable and reinvestigated the subject in a slightly different way. With the side of a hand-held glass needle, he bisected unfertilized sand dollar eggs and, in many cases, the nucleated and enucleated parts remained together in the same jelly hull. Following parthenogenetic activation with butyric acid, 11% of the enucleated pieces cleaved irregularly. The completion of division depended upon the size, number, and position of the cytasters. Like Wilson, Fry concluded that asters associated with chromatin are more effective and that differences between cytasters and nucleated asters were ones of degree rather than kind. Ethel Brown Harvey (1936) found that the cytasters that form in enucleated red halves of centrifuged *Arbacia* eggs elicit furrows. Direct aspiration of the sea urchin egg nucleus (Lorch, 1952) did not stop aster multiplication, and temporary furrows developed between aster pairs. Lorch did not describe the size or position of the asters. Seven to 8 hours after enucleation, operated cells

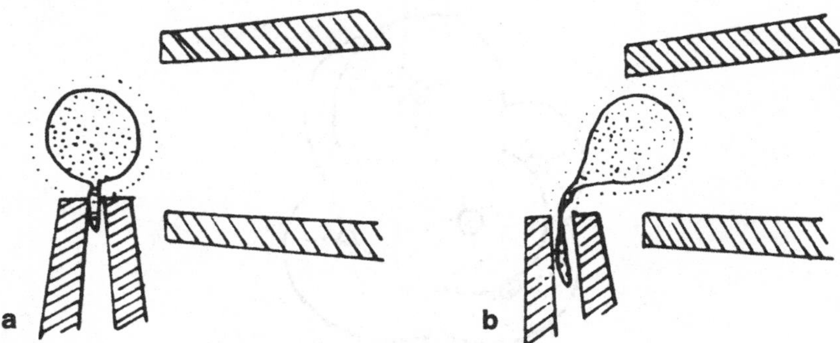

Figure 5.2 Early method for enucleation of a starfish oocyte. (a) By capillary flow, the oocyte was held against the nozzle of the vertical pipet so that the region containing the mitotic apparatus was at the orifice. By mouth, the nucleated region was sucked into the pipet. (b) By capillary flow into the larger horizontal pipet, the nucleated region was attenuated and eventually torn away from the bulk of the cell. The enucleated region was subsequently parthenogenetically activated, and furrows formed in association with the cytasters. (From McClendon, 1908)

rapidly broke up into many small spheroidal bodies. Lorch termed the phenomenon "pathological cleavage" and pointed out that it occurred in the absence of asters. However, her description and illustration strongly suggest that she was observing a form of cytolysis that occurs in old, overheated, or physically damaged echinoderm eggs.

The requirement that cytoplasm be mixed with "nuclear sap" before aster formation can occur is not immediately related to the events of cytokinesis, but a postulated role for chromosomes in the alteration of surface properties is. The early reports of division activity without chromosomes did not always deter cell division theorists who could, with some justification, point out that furrowing that began in the absence of chromosomes was not consistently completed. The inconsistency was correlated with differences in distance between the asters, differences in aster size, and differences in distance from the asters to the surface. The differences were uncontrollable, and their possible significance was not analytically discussed. Theories that proposed a specific, essential role for chromosomes usually postulated that at metaphase, anaphase, or telophase, the chromosomes as a group released a substance that affected the behavior of the nearest surface. For this reason, chromosomes and their normal pattern of redistribution were accorded an essential role in cytokinesis. There was a need to control the geometrical relations among the potentially active components of the system, and the first goal was to determine the furrow-producing ability of a normal aster pair positioned at the normal separation but without an intervening spindle and chromosomes. Simple aspiration presented the problem of possible incomplete removal, and so the relation was obtained by simple rearrangement of the organelles (Rappaport, 1961). When a glass ball or the rounded tip of a glass rod is pushed through the center of a fertilized sand dollar egg, the sphere is converted to a torus or doughnut. The mitotic apparatus that subsequently develops in the curved cylindrical cytoplasm appears normal, as does the furrow that divides the cell at the midpoint between the aster centers (Figure 5.3). Although division is complete, the cell is binucleate because there is no division activity in the region diametrically across the perforation from the furrow. As the two mitotic apparatuses form and grow preceding the second cleavage, they orient parallel to the axis of the cylindrical surface and elongate so that the asters farthest from the first cleavage plane approach each other. In the second cleavage, furrows develop in the normal relation to each mitotic apparatus and, if the asters are close enough, a third furrow appears between the asters that were brought closer as the mitotic apparatus elongated. The third furrow is located in an area that never contained a spindle or chromosomes, yet it is complete and often synchronous with the other furrows in the same cell. Anyone who repeats this experiment must be impressed by the importance of the distance between the asters. When the distance approximates that of an intact mitotic apparatus, furrowing is normal, but the greater the distance, the more incomplete the furrowing activity.

Hiramoto (1971a) subsequently aspirated different parts of the mitotic apparatus before the position of the furrow was established and confirmed that the presence of nuclear material or the spindle is not required for formation of the cleavage furrow or for complete cytoplasmic division. About 23% of eggs that were enucleated while the nuclear membrane was intact developed asters and cleaved. Total or partial spindle removal in the period between middle metaphase and early telophase resulted in cleavage of about two-thirds of the operated eggs. The experiment demonstrated that the events that occur, and the struc-

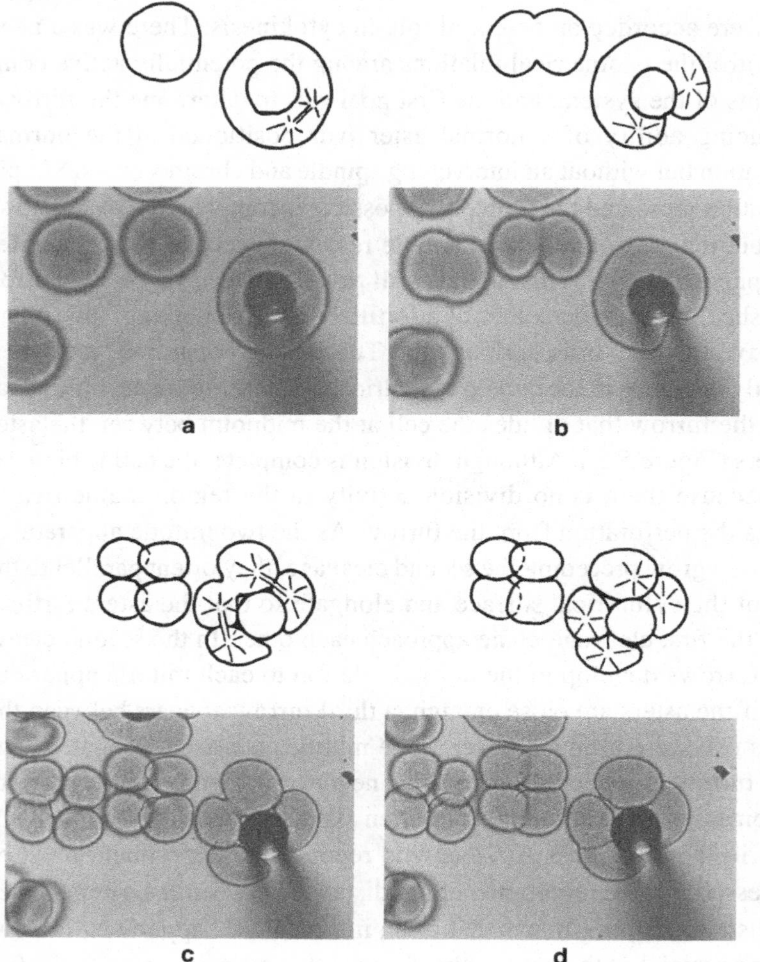

Figure 5.3 Cleavage of a torus-shaped sand dollar egg. Condition of the mitotic apparatus is shown in the line drawings. The position of the spindle is indicated by a double line. Note synchrony with controls. Timing begins at fertilization. Initial temperature 19.5 °C.

(a) Beginning of first cleavage. 69 min.

(b) First cleavage completed. The binucleate cell shape resembles a horseshoe. 79 min.

(c) Second cleavage. Two cells have divided from the free ends of the horseshoe, and the binucleate cell is furrowing between the polar ends of the asters. 142 min.

(d) Second cleavage completed. Each blastomere contains one nucleus. 144 min.

(From Rappaport, 1961, and with the permission of Wiley-Liss)

tures that normally appear, between the normal aster centers of echinoderm eggs are not necessary for the eggs' division, but they do not shed any light on the problem of reduced "division energy" in asters that develop out of contact with nuclear material, because both types of asters develop in the presence of nuclear material.

Sluder, Miller, and Rieder (1986) aspirated the nucleus of sea urchin eggs in early prophase, leaving one centrosome in place, and thus avoided the geometrical confusion of multiple cytasters. Observation of these cells through several cycles revealed that the asters cycled normally, but they were abnormally far apart when they first appeared. The furrows that subsequently formed were incomplete and usually receded in a manner reminiscent of that observed in the early investigations.

Because the removal of nuclear material and the stimulation of cytaster formation also resulted in uncontrollable changes in the distances between the interacting parts, an aura of uncertainty around the role of nuclear material in cleavage furrow establishment was preserved. The topic was studied again under circumstances that permitted greater control. Easily deformable tubes of capillary dimensions were cast in transparent silicone rubber (Rappaport, 1986a). The lumen was flattened so that the division-related organelles of a sand dollar egg were forced into a single optical plane after the egg was inserted in the tube. The thin elastic roof of the capillary permitted cell puncture or deformation with glass tools; the other regions of the capillary wall, which were not deformed, controlled the cell dimensions. With this arrangement it was possible to observe the effect of each aster on the adjacent surface and avoid the variability that arises when the distance between the asters cannot be satisfactorily controlled. When a glass ball was pushed downward on the capillary roof, it made a large hole in the cell. When the hole was positioned in the middle of the cell, the subsequent relation between the nucleus and the asters was determined by chance. When the asters straddled the hole, their relation to the adjacent surfaces was identical, but only one of them developed in the presence of nuclear material. The surface contractile activity was greater in the presence of nuclear material (Figure 5.4). Repeated extrusion of sand dollar eggs from a pipet while the nuclear membrane is intact usually results in separation of the asters. The nucleus often remains closely associated with only one aster, so that asters with and without nuclear association reside in the same cell. Insertion of such cells in flattened

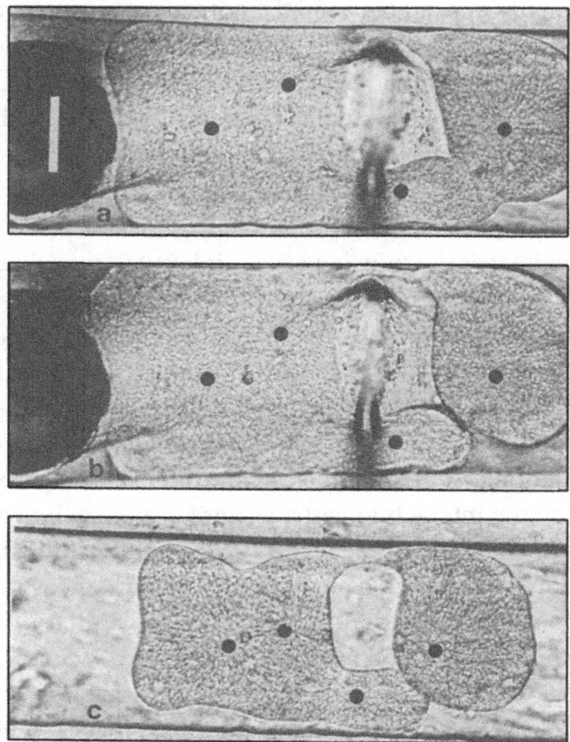

Figure 5.4 Sand dollar egg with four asters during the second cleavage cycle confined in a capillary with a flattened lumen. The aster pairs are separated by a perforation. The nucleus is associated with the right aster pair. (a) Furrow near nucleus-associated pair. (b) Nucleus-associated furrow completed. (c) Furrow associated with the non-nucleated aster pair is shallow and ephemeral. Dots indicate aster centers. Bar = 50 μm. (From Rappaport, 1991, and with the permission of Wiley-Liss)

capillaries improves observation, and it is clear that, after the first cleavage, permanent furrows are never formed in the enucleate blastomere and are always formed in the nucleated blastomere (Figure 5.5), as Ziegler had previously reported. Normal aster activity in cytokinesis requires a contribution from within, or from the immediate vicinity of, the nucleus. The identity of the contribution is unknown, and the nucleus contains many things beside DNA. The ability of starfish eggs to continue cleavage in the presence of aphidicolin in concentrations that block

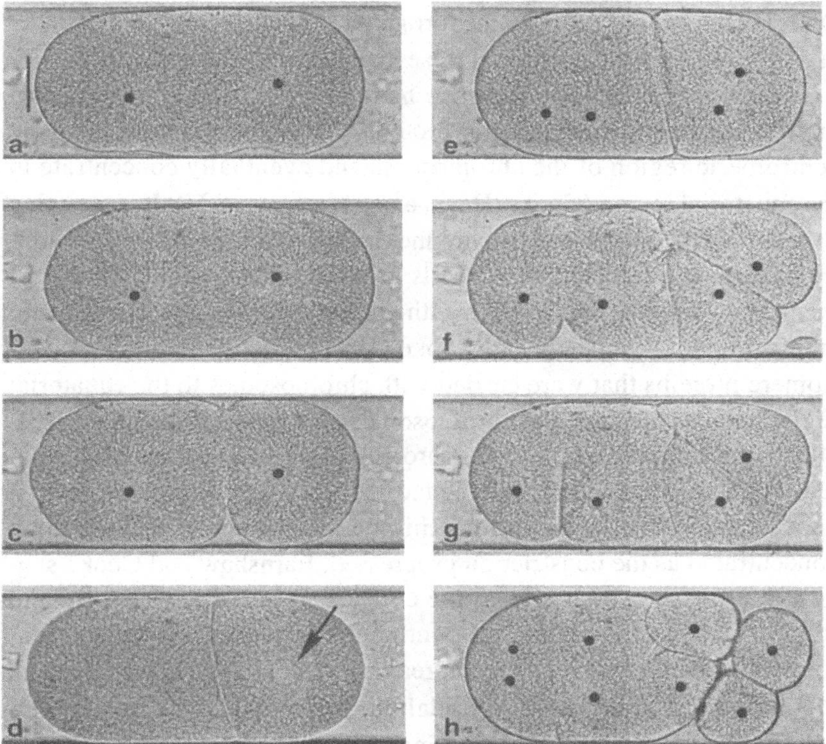

Figure 5.5 Division activity in a flattened, cylindrical sand dollar egg with widely separated asters. The nucleus is associated with the right aster.

(a) Astral constrictions in the planes of both aster centers on the lower margin.

(b) The constriction near the right aster deepened and appeared to slide a short distance into the zone between the aster centers.

(c) Constriction deeper.

(d) Cell divided. Nucleus visible in right cell (arrow).

(e) Asters for second cleavage formed.

(f) Furrowing in nucleated blastomere completed. Unilateral furrow in left cell.

(g) Maximum extent of furrowing in left blastomere. Furrow later regressed.

(h) Third cleavage cycle. Right: Nucleated cells completed division. Left cell contains four asters, but there is no furrowing activity. Dots indicate position of aster centers.

Bar = 50 μm.(From Rappaport, 1991, and with the permission of Wiley-Liss)

DNA synthesis and chromosome formation (Nagano et al., 1981; Saiki et al., 1991) strongly suggests that the essential component is not DNA.

Immunofluorescence methods have revealed in mammalian cultured tissue cells the presence of proteins that at first associate with the centromeric region of the chromosome and eventually concentrate in the equatorial region (Cooke, Heck, and Earnshaw, 1987). It is tempting to associate the proteins with cytokinesis, despite the demonstrated ability of cleavage cells and tissue cells to divide after removal, displacement, and chemical interference with normal chromosome arrangement and formation. Earnshaw and Cooke (1991) described two inner centromere proteins that were carried with chromosomes to the equatorial region, where they left the chromosomes and appeared in streaks parallel to the spindle axis. As sister chromatids separated, some of the protein became detectable at the equatorial cortex, although most of it appeared to remain closer to the mitotic axis, where it became more concentrated as the constriction progressed. Earnshaw and Cooke suggest that the proteins move to the cortex and interact with actin and tropomyosin in the equatorial contractile network. A different centromere protein, described by Andreassen et al. (1991), accumulates in the equatorial plane in a discoidal configuration that extends well beyond the spindle before furrowing. Myosin is also demonstrable in association with the disc, and Andreassen et al. speculate that the disc concentrates and aligns the myosin in the region where the furrow will later form. The localized interaction between the disc-associated myosin and cortical actin produces a narrow contractile ring. Were these centromere proteins playing essential roles in cytokinesis, the process should be more vulnerable to the kinds of chromosomal manipulation previously described than it apparently is.

Spindle

The methods used to remove chromosomes or to prevent their formation also result in spindle elimination or in reduction of the spindle to an indeterminate state (Nagano et al., 1981; Saiki et al., 1991). Because the cells divide despite simultaneous ablation of both spindle and chromosomes, it is logical to conclude that neither plays an essential role in furrow establishment. In most early embryonic cells, only the asters remain to be considered, and in cleaving echinoderm eggs – with their spherical form, large asters, and centrally located spindles far removed from the equatorial surface – the idea that asters alone are adequate is easy to accept. In somatic cells, however, asters are typically smaller, and the spindle is relatively larger and closer to the equatorial surface. It is also true that the microtubular component of the spindle superficially resembles that of the aster, and because astral configurations of various origins have the ability to induce furrows or furrowlike activity, there is cause to wonder whether the apparent absence of a spindle role in cleavage cells is a consequence of its constitution or its geometrical circumstances. The matter was studied by changing the relation between the spindle and the equatorial surface in ways that were calculated to increase interaction with the spindle and decrease interaction with the poles. Sand dollar and sea urchin eggs were flattened and notched or perforated in the equatorial plane adjacent to the spindle, so that surface was positioned in apparent contact with the midregion of the spindle (Rappaport and Rappaport, 1974). In similar experiments, the equatorial surface of dividing newt kidney epithelial cells was pushed toward the mitotic axis to a degree that the chromosomes were displaced (Rappaport and Rappaport, 1974). In both sets of experiments, furrows formed on the surfaces in contact with the spindle (Figure 5.6). Similar deformations of dividing grasshopper spermatocytes caused by exposure to X-rays (Ris, 1949) and micromanipulation (Kawamura, 1977) also showed that furrows form in surfaces where contact with the spindle is apparent and where interaction with the reduced asters is difficult to imagine (Figure 5.19).

Asters

The results of experimental investigations that diminished the possibility that the spindle or chromosomes play an essential role in positioning the division mechanism focused attention on the importance of the asters. It is also true, however, that the cells that are for several reasons best suited for investigations on this subject have large asters, and the possibility of attributing greater importance to a particular phenomenon if it is more strikingly obvious in the material studied (described by Gray [1931] at the beginning of Chapter 1) cannot be ignored. The data indicate that

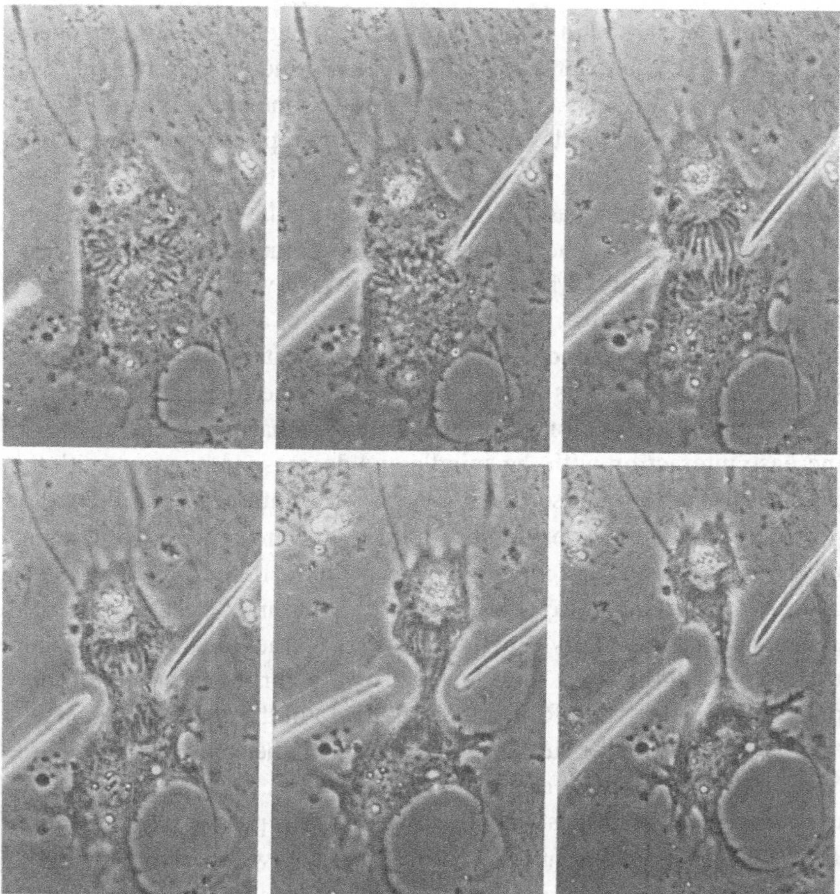

Figure 5.6 Furrowing in a flattened, cultured adult newt kidney cell. The equatorial margins were moved toward the spindle until peripheral chromosomes were displaced (middle picture, top row) and held in that position until furrowing began. (From Rappaport and Rappaport, 1968, and with the permission of Wiley-Liss)

the achromatic, linear elements of the mitotic apparatus are essential. In the region between the aster centers, the relation between the equatorial cortex and the nearest radial mitotic apparatus structures appears similar whether or not the asters are prominent (see Figure 1.6 and Figure 7.6). Until evidence to the contrary is presented, it seems logical to assume that the same mechanism operates in both circumstances.

Presence of paired asters at the ends of the oblong mitotic apparatus is the norm, and it appears inevitable in view of the centrosomal cycle and the mode of formation of the mitotic apparatus. A single aster is unable to establish a furrow in a spherical cell (Hiramoto, 1971a), but it can do so when the cell is reshaped into a cylinder (Rappaport and Rappaport, 1985a). Its failure in the spherical cell is associated with its tendency to center itself, with the probable consequence that its effect on the cortex is uniform, and the necessary restricted zone of higher contractility fails to develop. A single aster cannot be equidistant from all parts of a cylindrical surface. The important requirement that must be met is that the effect of the mitotic apparatus must be concentrated in a restricted circumferential region of the cortex. The nature of the geometrical configuration that is utilized to meet that requirement has little effect on the nature of the cortical response. The relative difference in size and distance among the interacting cell structures ensures that, in somatic cells, furrows are caused by the spindle. In cleavage cells, furrows are caused by the asters. In development, when cleavage divisions give way to somatic cell divisions, the transition would require no change in the fundamental mechanism.

The diastema

The spindle, chromosomes, and asters are the principal components of the mitotic apparatus in cells customarily observed and manipulated in cell division experiments, but a fourth structure, the diastema, has been described for many years in large, yolky, holoblastically cleaving eggs. That the diastema is the site of localized, *de novo* surface formation in the deeper region of the furrow now appears unlikely, but the results of investigations suggest that, where it is present, its role is essential. The subject is confused by older reports of diastema that now appear to have been based upon artifact or upon misinterpretation of light microscope images. The presence of the diastema in sturgeons and amphibians is substantiated, and the following description is based upon investigations of cleavage in their eggs. In section, the diastema first appears at anaphase as a thin plate in the plane and position formerly occupied by the metaphase chromosomal plate (Zotin, 1964; Sawai and Yomota, 1990). It is distinguished by a paucity of yolk granules, so that it appears as a light-colored streak. Sawai and Yomota (1990) found that 30 minutes before cleavage in *Cynops* (a salamander), the mitotic apparatus degenerates and the diastema becomes recognizable. The diastema expands in the direction of the animal pole and enters the region immediately below the cortex in about 15 minutes. In *Xenopus,* the diastema appears in late anaphase and reaches the vicinity of the animal cortex about 5 minutes before cleavage. Zotin (1964) found that the appearance of multiple furrows following heavy water treatment was accompanied by fragmentation and dispersal of the diastema. Sawai and Yomota (1990) found that colchicine injection could cause mitotic apparatus disruption, failure of diastema formation, and blocked cleavage. They propose that the asters form the diastema at their periphery while the mitotic apparatus degenerates at the centers. Diastema expansion by a secondary process would bring the aster-originated or accumulated material closer to the surface, where furrow establishment could occur despite the initially great distance between the mitotic apparatus axis and the surface.

How Is Division Mechanism Formation Temporally Related to Mitosis?

The assumption that mitosis and cytokinesis are causally related encouraged speculation about the way that the visible events of mitosis are fitted to the physical events of cytokinesis. Enthusiastic espousal of hypothetical division mechanisms was sometimes associated with a lack of precision in the description of mitotic events that were important to the hypothesis. Chambers (1924) and Gray (1931) supported the idea of astral cleavage with the statement that echinoderm cleavage begins when the asters reach maximum size. Just (1939) pointed out that the asters are largest before cleavage and begin to shrink when furrowing begins.

The regularity of the mitotic events and the beginning of cleavage in a single species often led to the assumption that the events observed in one cell type were typical of all species and, secondarily, that the existence of the relation was based upon some fundamental aspect of the division mechanism. A review of the differences among these relationships in different cell types led Just (1939) to conclude that nuclear and cytoplasmic division are entirely separate phenomena, and he implied that searching for a connection was misguided and would ultimately prove fruitless. Not surprisingly, he made no attempt to relate his theory of cytokinesis based upon accumulation of hyaline substance at the equatorial surface to mitosis or any other event or activity that might explain why the furrow develops when and where it does. The idea that the two events are entirely independent was not generally adopted, and the variability cited by Just implied that either different cells divided by somewhat different mechanisms or the linkages between the two events are flexible.

When is the position of the furrow established?

Operationally, the division mechanism is considered established or formed when surface division activity can proceed independently of the mitotic apparatus. Experiments usually involve ablation of the mitotic apparatus, followed by observation of the cell's capacity for division. Beams and Evans (1940) placed *Arbacia* eggs in 2×10^{-4} M colchicine at 2-minute intervals beginning 10 minutes after fertilization, and the treatment was continued past the normal division time. Treatments that began during the first 22 minutes after fertilization completely blocked division. At least 50% of the eggs treated between 26 and 30 minutes divided. Beams and Evans determined by cytological methods that at 22 minutes, eggs were in prophase; at 26 to 30 minutes, they were in prophase, metaphase, or anaphase. Colchicine caused the normal radiate aster structure to disappear at all stages. These results suggested that by the time of the 26- to 30-minute treatment, the asters in about half of the eggs had completed their chief function in cytokinesis, and the authors concluded that asters are not required after anaphase. The essential role of the mitotic apparatus appeared to be largely confined to the middle stages of mitosis. The results indicated that the mitotic apparatus plays no physical role in cytokinesis and that the essential aspects of any other role it may play are completed before cytokinesis begins.

Subsequent experiments in which the mitotic apparatus was aspirated from sea urchin eggs close to the anticipated time of cleavage also showed that cleavage can occur when removal takes place at anaphase or later (Hiramoto, 1956) (Figure 5.7). Lorch (1952) previously showed that aspiration of the nucleus and centrosomes while the nuclear membrane is intact blocks cleavage. Y. Hamaguchi (1975) injected colchicine in one of the blastomeres resulting from the first cleavage and used the other blastomere as a time control. This method increased precision by rapidly destroying the mitotic apparatus. He used the midpoint of control cleavage as a timing reference point. The earliest time when injected colchicine failed to block cleavage was 10 minutes before the midpoint in *Clypeaster japonicus* and 8 minutes in *Temnopleuris toreumaticus*. In both species these times correspond to the beginning of anaphase. In Hamaguchi's experiments, cells were deformed when needles were inserted, and the interaction between the mitotic apparatus and the surface was affected by the distance that separates them. In a further attempt to improve precision, cell shape was controlled by confining the two blastomeres that result from first cleavage in a short length of capillary tube (Rappaport, 1981). The imposed cylindrical cell

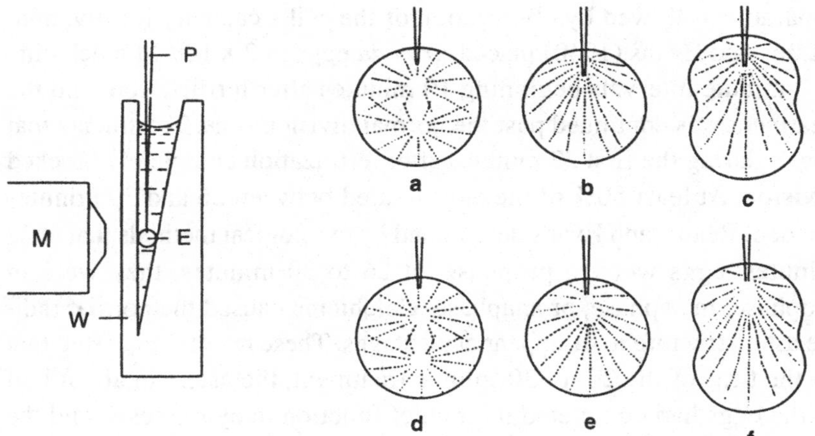

Figure 5.7 Hiramoto's method for mitotic apparatus removal.
Left: Manipulation chamber. E = egg; M = horizontal microscope objective; P = pipet; W = coverslip chamber wall.
Right: Relation between mitotic figure and cleavage plane.
(a–c) Successive stages when the spindle is removed.
(d–f) Successive stages when the spindle is displaced by removal of part of the egg cytoplasm. In both series, the cleavage planes were independent of the position of the asters at the time of the operation.
(From Hiramoto, 1956, and with the permission of Academic Press)

shape positioned the mitotic apparatus of both cells in the capillary axis and ensured that the distances from the mitotic axis to the surfaces were the same. The beginning of cleavage, when the equatorial surface separated from the capillary wall, was clear. The mitotic apparatus was sucked out of one of the cells by inserting a pipet through the polar surface. When the mitotic apparatus was removed 4 minutes or less before the furrow appeared in the control, the operated cell divided. When the operation was done 5 minutes or more before control cleavage, there was no division. The results indicate that the necessary, or minimal, amount of interaction is completed 4 minutes before the furrow appears. Comparable data concerning the time when the furrow position is determined were obtained in large, yolky, opaque cleaving eggs of frogs, salamanders, and sturgeon by removing the mitotic apparatus or screening it from the surface (Kubota, 1966), or by compressing the eggs between glass sheets parallel to the animal–vegetal axis at different times in the mitotic cycle (Zotin, 1964; Selman, 1982; Sawai and Yomota, 1990). Compression oriented the mitotic apparatus parallel to the glass sheets, so that the furrow established when the mitotic apparatus was in that position was oriented perpendicular to the plane of the glass sheets. When the plane of the furrow was significantly different from the perpendicular, the furrow was not established while the mitotic apparatus was oriented parallel to the glass sheets, but rather before it was reoriented by compression. By compressing eggs at different times in the mitotic cycle, it was possible to learn at what time the orientation of the furrow was independent of the changed position of the mitotic apparatus and was, therefore, determined. In salamander, sturgeon, and frog eggs, the position of the furrow is determined in the majority of cases in the anaphase–telophase period, as is the case in echinoderm eggs (Zotin, 1964; Kubota, 1966; Selman, 1982; Sawai and Yomota, 1990). In yolky eggs, the real time between furrow determination and the beginning of cleavage is appreciably greater than in echinoderm eggs. Selman (1982) compared the determination times among the different species in terms of Detlaff units (Detlaff and Detlaff, 1961). A Detlaff unit corresponds to the time between the beginning of first and second cleavages. The unit facilitates comparisons among eggs of different species that develop at different rates and temperatures, and excludes the events of fertilization. Detlaff found that the interval between determination and cleavage in echinoderm eggs is about 0.1 Detlaff units, whereas in amphibian and sturgeon eggs, it is about 0.4 Detlaff units (Table 5.1). He suggested that the difference may reflect the activity of the diastema in furrow determination among the species in which the interval is greater. In subsequent experiments, Sawai and Yomota (1990) disrupted the mitotic apparatus and diastema with colchicine, or aspirated them at various times in the second mitotic cycle in frog and salamander eggs, and found that the position of the furrow was not determined in the cortex at the time when the cleavage plane was fixed by the mitotic apparatus. They found that the formation of the furrow in the cortex is determined just before the onset of cleavage, and they propose that the event is dependent upon the formation and action of the diastema. In this type of division, the principal role of the mitotic apparatus in cytokinesis would be the formation of the diastema, which in turn establishes the division mechanism in the cortex (Zotin, 1964).

The results of this group of experiments have been expressed in mitotic time as well as in chronological time. Although mitotic time has the virtue of relating the process to other division-related events, it has the disadvantage of seeming to suggest a direct, functional relation between the position of the chromosomes at anaphase and the mechanism that establishes the furrow. It is necessary to keep in mind that the normal presence of both chromosomes and spindle between the asters in the cells that were used in these experiments is superfluous.

What is the duration of the stimulus period?

The length of the period during which furrows can be established depends upon how long the mitotic apparatus can affect the surface, how long the surface can respond, and how long the necessary geometrical relation between the surface and the mitotic apparatus is maintained. The cryptic qualities of the establishment process limit its experimental accessibility. It is not accompanied or accomplished by morphologically distinguishable events. One can determine only that the process has occurred successfully when furrow activity subsequently develops. The question whether the interaction period is brief or lengthy has a bearing upon certain theories of cell division.

As yet, there is no way of knowing when the interaction period normally begins. At that point in the mitotic cycle, the asters are expanding and the distance between their centers is increasing; there are no clearly defined morphological events that could serve as a signpost at any level of organization.

It is possible to get an idea of the least time necessary to transform nonfurrow surface into active furrow by rapidly bringing nonfurrow surface into the vicinity of a mitotic apparatus whose establishment activity is operative. The experimental design used to measure the least time incorporated the observation that eggs with a very eccentric mitotic apparatus cleave only from the side closest to the mitotic apparatus (Bergh, 1879; Ziegler, 1898b). The same type of furrow can be induced in eggs with normally central nuclei and symmetrical furrows by pushing the mitotic apparatus out of the center close to one part of the cell margin (Rappaport and Conrad, 1963). The effect of displacing the mitotic apparatus can be reversed by pushing the distant part of the surface close to the equatorial part of the mitotic apparatus (Figure 5.8). If the originally distant surface is released immediately after it is pushed in, the margin resumes its original contour and the surface shows no lasting effect. If the surface is held close to the mitotic apparatus longer, it develops a furrow after it resumes its original contour. The minimum time necessary to establish furrowing activity in this way is 1 minute (Rappaport and Ebstein, 1965). After 1 minute, the surface returns to its original contour and then the furrow develops 2.5 minutes later. Furrowing thus begins 3.5 minutes after the surface is brought close to the mitotic apparatus. The period cannot be shortened by prolonging the stimulus period. Deformations of the egg surface at some distance from the mitotic apparatus produce no local furrowing activity (Rappaport, 1988). These results suggest that effective stimulation does not immediately give rise to an active mechanism and that intervening, time-consuming events are required. The brevity of the interaction period was also revealed when the mitotic apparatus was moved when furrowing began. When cells are reshaped into cylinders, the mitotic apparatus can easily be moved after the early appearance of the furrow. The shape ensures that both the orientation of the mitotic apparatus and its distance from the surface do not change when the mitotic apparatus is displaced. In this circumstance, the time between displacement and appearance of

Table 5.1 Estimate of determination times before furrow formation

SPECIES	°C	DETLAFF (in min)	ESTIMATE	STAGE	METHOD	REFERENCE
Triturus alpestris (newt)	18–22	70–117	0.46 D	Anaphase	Horizontal compression	Selman, 1982
Ambystoma mexicanum (axolotl)	8–19	10	0.2–0.4 D	Late anaphase–early telophase	Horizontal compression at 1st cleavage	Zotin, 1964
			0.2–0.6 D (0.4±0.1 D)			Zotin & Pagnaeva, 1963
		100	0.25 D		at 2nd cleavage	
Acipenser güldenstadti (sturgeon)	(16.2)	(60)	0.2–0.4 D	Late anaphase–early telophase	Horizontal compression	Zotin, 1964 Zotin & Pagnaeva, 1963
Rana nigromaculata (frog)	19	45	(0.42–0.51 D) 19–23 min	Metaphase–mid anaphase	Removed mitotic apparatus	Kubota, 1966
			(0.38–0.45 D) 17–20 min	Early anaphase–late anaphase	Displaced mitotic apparatus by paraffin	Kubota, 1966
Echinarachnius parma (sand dollar)	18	45	(~0.09 D) 4.0 min	Late metaphase–early anaphase	Removed mitotic apparatus by micromanipulation	Rappaport, 1981
Clypeaster japonicus (heart urchin)	28	(33)	(0.09 D) 3 min (0.06 D) 2.5 min	Late metaphase–anaphase	Removed mitotic apparatus	Hiramoto, 1956

Clypeaster japonicus (heart urchin)	28	(33)	(0.09 D) 3 min (0.06 D) 2.5 min	Late metaphase–anaphase	Removed mitotic apparatus	Hiramoto, 1956
Parechinus microtuberculatus (sea urchin)	18–19	(43)	(0.1 D) 2–6 min		Displaced mitotic apparatus by centrifugation	E. B. Harvey, 1935
Clypeaster japonicus (heart urchin)	20	45	(0.16 D) 7 min	Early anaphase	Injected colchicine	Y. Hamaguchi, 1975
Temnopleurus toreumaticus (sea urchin)	20	38	(0.13 D) 5 min	Early anaphase	Injected colchicine	Y. Hamaguchi, 1975
Psammechinus miliaris (sea urchin)	18	31	(0.16 D) 5 min	Mid anaphase	External colchicine to disrupt mitotic apparatus	Swann & Mitchison, 1953
Arbacia punctulata (sea urchin)	22–24.4	(_26)	(~0.46 D) 12 min (~0.3 D) 8 min	Metaphase Anaphase	External colchicine to disrupt mitotic apparatus	Beams & Evans, 1940
Chortophaga viridifasciata (grasshopper)				Late anaphase	Displaced mitotic apparatus by micromanipulation	Carlson, 1952

Figures in parentheses are estimates and not direct quotations from the references cited. (After Selman, 1982)

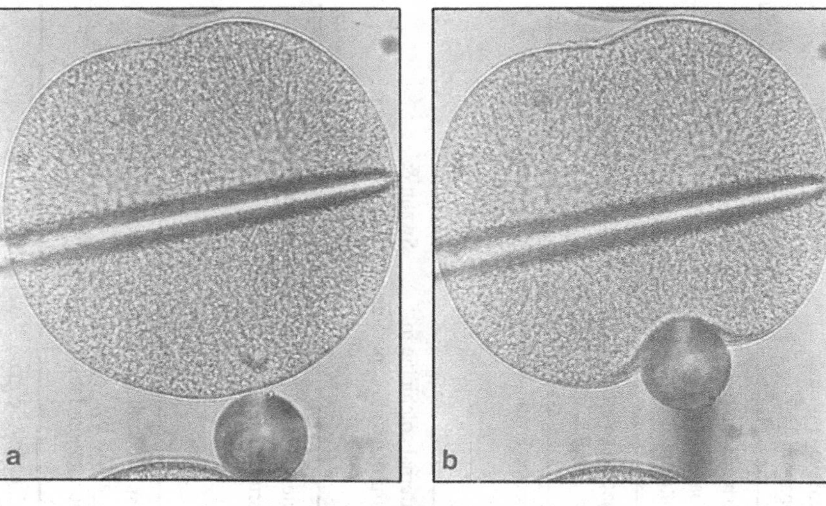

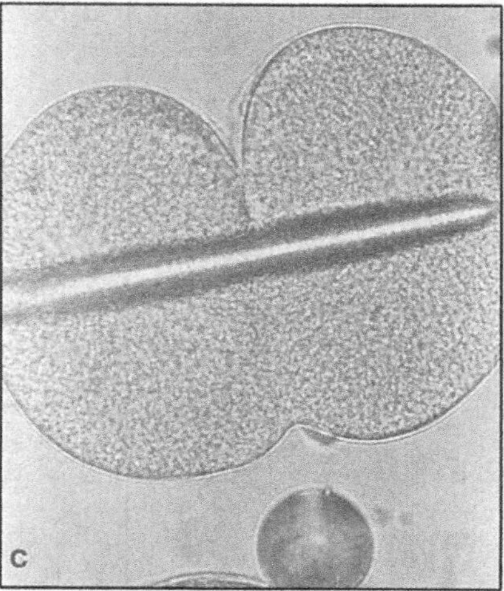

Figure 5.8 Establishment of the furrow in experiments involving displacement of both the mitotic apparatus and the surface. (a) Formation of the unilateral furrow on the margin closest to the mitotic apparatus. (b) Diametrically opposite, more distant margin is pushed closer to the mitotic axis. (c) Active furrowing following successful stimulation. (From Rappaport and Ebstein, 1965, and with the permission of Wiley-Liss)

a new furrow that corresponds to the new geometrical circumstances is about 2 minutes (Rappaport, 1975, 1985).

These experiments established the minimum necessary time, but they did not reveal the length of the period during which effective interaction is possible. The apparent precision and regularity of division events have suggested the operation of a series of subprocesses with well-defined beginnings and endings. Better understanding has resulted from further experiments in which the normal geometrical relations have been changed. The effect of the mitotic apparatus appears to move outward toward the surface. Following this reasoning, it should be possible to create precocious furrows by reducing the distance between the mitotic apparatus and the surface, because the stimulus would travel a shorter distance. The idea was tested by operating on one of the two blastomeres that result from the first cleavage and using the other as a time control. Parallel glass needles were used to cut away cytoplasm so that the mitotic apparatus was isolated in a narrow cylinder. Division of the operated cell was completed as much as 7 minutes before the controls (Rappaport, 1975; Rappaport and Rappaport, 1993); the average was about 5 minutes. The components of the system are therefore capable of interaction at least 5 minutes earlier than normal.

When cytokinesis is completed, the asters are separately confined in nearly spherical daughter cells. Although their radiate structure persists, these single asters seem incapable of further furrow establishment activity. The absence of division activity is comparable to that of a cell from which one aster has been removed (Hiramoto, 1971a). In order to determine how long astral activity persists, the aster pair must be kept together and positioned near the cortex. By moving the mitotic apparatus each time a furrow appears, this requirement can be satisfied simply. In several investigations involving cylindrical sand dollar eggs, a single mitotic apparatus proved capable of producing up to 13 furrows, and the maximum time between the first furrow and the last was 24.5 minutes (Rappaport, 1975, 1985).

Further experiments on the second and third cleavages of sand dollar eggs revealed that whereas normal control cleavage required an average of 8 minutes from beginning to completion, the mitotic apparatus could induce new furrows for an average of 18 minutes. The chronic manipulation did not significantly affect the events of the cell

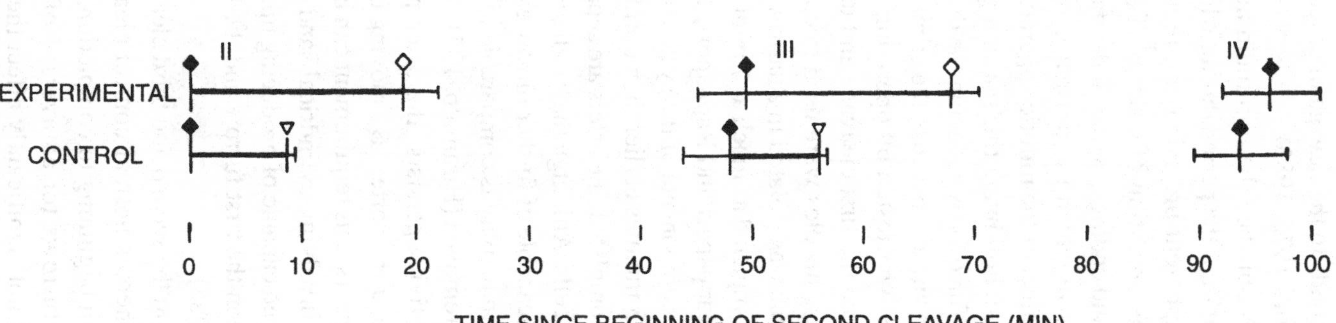

Figure 5.9 Comparison of durations of cleavage activity in control cells and experimental cells in which the mitotic apparatus was repeatedly shifted during the second, third, and fourth cleavages. The thick bar indicates the period of furrowing activity. The thin extending bars indicate standard deviation. (From Rappaport and Rappaport, 1993, and with the permission of Academic Press)

cycle, because the time between the beginning of second and third, and third and fourth, cleavages was the same in operated and control cells (Figure 5.9) (Rappaport and Rappaport, 1993).

The level of interaction between the mitotic apparatus and the surface necessary to create a localized region of furrowing activity can be achieved in about 1 minute, but the period when they are capable of interaction is far longer. The successful interactions demonstrated 5 minutes before, and 24 minutes after, the normal event suggest that the duration of the competent period is about 30 minutes. In terms of the cell cycle, 30 minutes is lengthy. The time from fertilization to first cleavage in sand dollar eggs is 90 minutes, and the time between the first and second, second and third, and third and fourth cleavages is 48 minutes. How, then, is the beginning and end of the normal interaction period determined? Because the beginning of the period can be advanced by simply bringing the surface closer to the mitotic axis, it is logical and parsimonious to suggest that normal beginning is also determined by reduction of the distance between the mitotic apparatus and the surface. In unoperated cells, the distance between the mitotic axis and the surface does not change before division, but the distance between the periphery of the mitotic apparatus and the surface decreases because of the expansion of the asters that normally precedes division. The end of the normal interaction period is apparently not determined by the loss of competence of either of the interactants, but rather it may result from the dismantling of a necessary geometrical configuration of mitotic apparatus and surface. Each aster is confined near the center of a nearly spherical cell when cleavage is completed, and the aster's effect would tend to be uniformly distributed under the surface. Absence of furrowing activity at that time would result not from failure of interaction, but from failure to attain a necessary degree of nonuniform pattern of the effects of the interaction. The cause of the eventual loss of competence is unknown.

All polar or astral relaxation theories require that the stimulatory activity of the mitotic apparatus end with the deformation at the beginning of cytokinesis, otherwise the equatorial surface would relax as it neared the mitotic apparatus, and completion would be impossible (White and Borisy, 1983). The experimental production of multiple furrows that follows repeated relocation of a single mitotic apparatus

(Rappaport, 1975, 1985) demonstrates a persistence of stimulatory activity and responsiveness that is inconsistent with those theories. These findings, in fact, indicate that the stimulatory events that create the furrow persist throughout division; the asters are active, the surface is responsive (Aimar, 1988), and, during the first half of the process at least, the geometrical arrangement appears to be satisfactory. If the contractility of the equatorial cortex were enhanced by prolonged exposure to, and diminished distance from, the mitotic apparatus, then some reinforcement of the initial contractility would be expected. However, the previously described ablation experiments (Hiramoto, 1956; Y. Hamaguchi, 1975; Rappaport, 1981) demonstrate convincingly that any stimulatory effect that the mitotic apparatus exerts later than about 4 minutes before the cell begins to change shape is not essential to the process. But both Hiramoto (1975) and Y. Hamaguchi (1975) noted that after ablation, furrows tended to progress slowly and sometimes remained incomplete, and Hiramoto (1958) found that furrow surface behavior was not normal in the absence of the mitotic apparatus.

What Geometrical Relations between the Surface and the Mitotic Apparatus Are Necessary?

The presence of mitotic apparatus components in the cell does not guarantee furrows. In early experiments involving cytasters (Wilson, 1901a) and reduced asters, Wilson (1901b) showed that the size and the distance between the components also affected furrow formation. In early commentary, reduced or absent furrow activity was correlated with reduced aster size, the distance between the asters (Wilson, 1901b), and the distance between the mitotic apparatus as a whole and the surface (Bergh, 1879; Ziegler, 1898b). These observations passed with little discussion or further analysis. The questions of the geometrical relations among the interactants and the identification of the most affected surface regions are closely related, and the results of many experiments have had implications for both. The previously discussed findings that cleavage cells can dispense with spindles, whereas somatic cells appear to rely upon them, suggest that although the mitotic apparatus of both cell types contains the same parts capable of qualitatively similar activities, the geometrical relations normally required for normal cytokinesis may be different. The invisibility of the cleavage furrow establishment process imposed a need for analysis by experimentation. Analysis often involved deductions based upon surface behavior after normal configurations had been changed.

Cleavage cells

The process of division mechanism establishment in cleavage cells appears to be complicated because the centers of the asters (the only necessary structures) are farthest from the place on the surface where the division mechanism develops. From the beginning, Bütschli (1876) assumed that the mechanism develops where the effect of the asters is maximal and that a higher level of aster effect at the equator results from their additive effect in that region. The popularity of the idea has been cyclical over the many years since it was proposed, not because of contradictory evidence, but because newer alternative ideas appeared more interesting.

HOW DOES DISTANCE AFFECT THE MITOTIC APPARATUS–SURFACE INTERACTION? Unilateral cleavage appears to be the result of a natural experiment that demonstrates the failure of the mitotic apparatus effect to reach distant surfaces, and Ziegler (1898b) and Yatsu (1908, 1912a) so interpreted it. But another interpretation is possible. Conklin (1908) proposed that unilateral cleavage was due to differences in the resistance of the central endoplasmic region to the progress of the furrow. The interchangeability of unilateral and symmetrical furrow appearances was demonstrated by experiments involving repositioning of the mitotic apparatus (Rappaport and Conrad, 1963). The normally symmetrical furrow of sand dollar egg develops as a unilateral structure when the mitotic apparatus is held in an eccentric position at the time when the division mechanism is established. Such furrows not only look like those of *Beroë* (Yatsu, 1912a), but they also respond in the same way to operations. If the egg is cut parallel to the mitotic axis so that the "cleavage head" remains in the same region with the mitotic apparatus, the enucleated part does not divide, but if the cleavage head is isolated in the enucleated part, division continues in the absence of the mitotic apparatus. On the other hand, the normally unilaterally cleaving blastomeres of a coelenterate, *Hydractinia echinata*, cleave symmetrically when the mitotic apparatus is repositioned to the center by amputation of peripheral cytoplasm (Figure 5.10). These results also show that the effectiveness of the mitotic apparatus decreases with distance and that, within a cell, distance can prevent interaction with competent surface.

This relationship seems to provide at least a partial explanation of the way that some chemicals block cell division. Ether, ethyl urethane, and colchicine and its derivatives reduce the size of the mitotic appara-

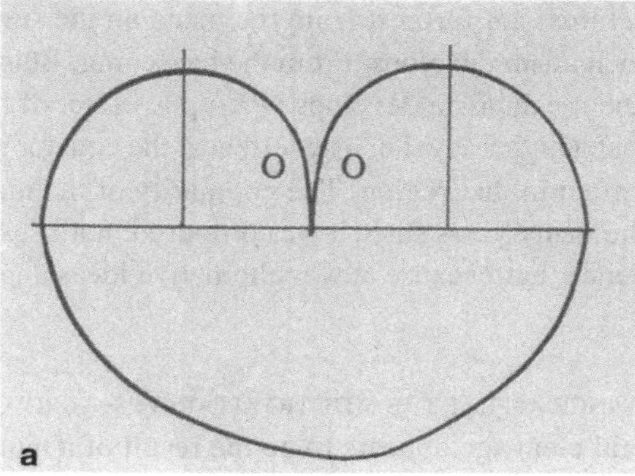

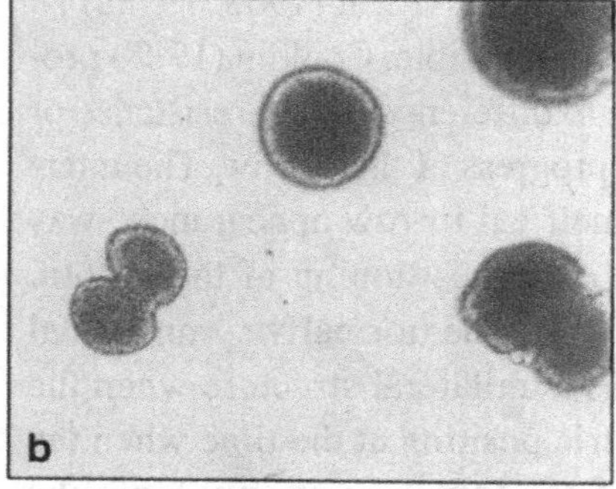

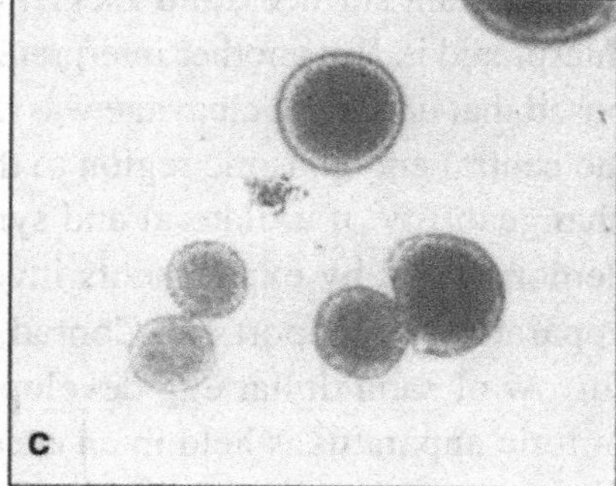

Figure 5.10 *Hydractinia echinata.* (a) Diagram of first cleavage showing positions of cuts that isolate each nucleus in about one-fourth the normal volume of cytoplasm. (b, c) Symmetrical furrows established by second cleavage nuclei following removal of about three-fourths of the normal amount of cytoplasm. (From Rappaport and Conrad, 1963, and with the permission of Wiley-Liss)

tus at the same time that they block cleavage. Wilson (1901b) and Kobayashi (1962) found that the reduced asters resulting from treatment with, respectively, ether and demecolcine continue to divide in the absence of cytokinesis. As the asters multiply, some shift closer to the surface. When they near the surface, furrows appear. The phenomenon could be studied in more detail by changing the distance between the mitotic apparatus and the surface in individual cells. Concentrations of ether or ethyl urethane that completely block cleavage in spherical sand dollar eggs are ineffective when the distance from the mitotic apparatus to the equatorial surface is decreased by forcing the egg part way through a small hole in the fertilization membrane (Rappaport, 1971a) or by constricting it with an 80 μm diameter glass ring (Rappaport and Rappaport, 1984) (Figure 5.11). The chemically reduced mitotic apparatus can also establish a unilateral furrow when it is forced against the margin of a flattened egg with the side of a needle (Rappaport, 1971a) (Figure 5.12). In the concentrations used, ether and ethyl urethane affected only, or primarily, division mechanism establishment, and they did so by decreasing the range of mitotic apparatus effectiveness. The decrease was caused by a simple reduction in mitotic apparatus size. When the dimensional relations were readjusted to compensate for the reduction, both the mitotic apparatus and the surface could carry out their normal activities. Mitotic onset, on the other hand, can be affected by drugs that do not necessarily cause net microtubule depolymerization. Vinblastine and nocodazole in low concentrations can block mitosis in tissue cells in the presence of a full complement of spindle microtubules (Jordan, Thrower, and Wilson, 1992), suggesting that these substances block mitosis, primarily by inhibiting the dynamics of spindle microtubules rather than by simple depolymerization.

The distance between the mitotic apparatus and the surface during the period of their interaction affects the rate of progress of the furrow that subsequently develops (Rappaport, 1982). The distance between the mitotic axis and the equatorial surface in normal sand dollar eggs is about 70 μm. The distance can be manipulated by flattening eggs with a small piece of glass early in the mitotic cycle, when the nucleus is normally eccentric. In this way, the eccentric mitotic apparatus must act upon diametrically opposite cell margins that are different distances from the mitotic axis. There is a positive correlation between the dif-

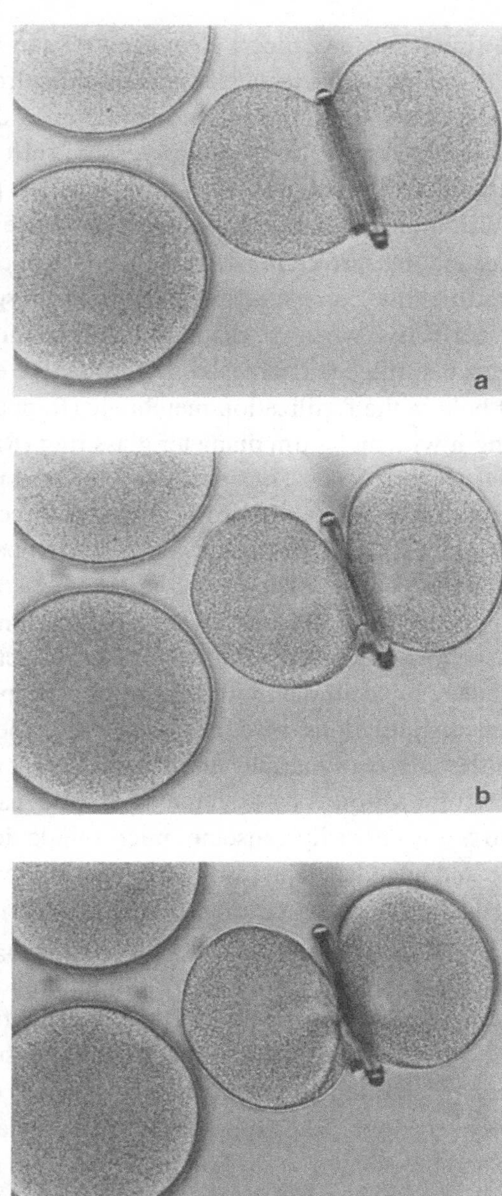

Figure 5.11 Cleavage of a urethane-treated sand dollar egg in which the reduced mitotic apparatus straddles, and is roughly centered in, the constriction plane. None of the unconstricted urethane-treated control eggs cleave. (a) Before cleavage. (b) Partially cleaved. (c) Cleavage completed. (From Rappaport and Rappaport, 1984, and with the permission of Wiley-Liss)

ferences in initial distances and the difference in the rates of furrow progression. When the distance during furrow establishment was between 100 and 110 μm, the rate averaged 5.2 μm/minute; when the distance was between 65 and 75 μm, the rate averaged 10.8 μm/minute. The resistance to deformation of the opposite margins was the same, and the difference in rates may reflect differences in exerted force. Greater force in furrows formed closer to the mitotic apparatus suggests the presence of more contractile material, which could, in turn, result from a larger affected area or greater activity per unit area of the affected region.

The aster centers in spherical cells are closer to the polar surfaces than to the equatorial surfaces, and this difference has provided a basis for speculation about the way that the mitotic apparatus plays its role in cytokine-

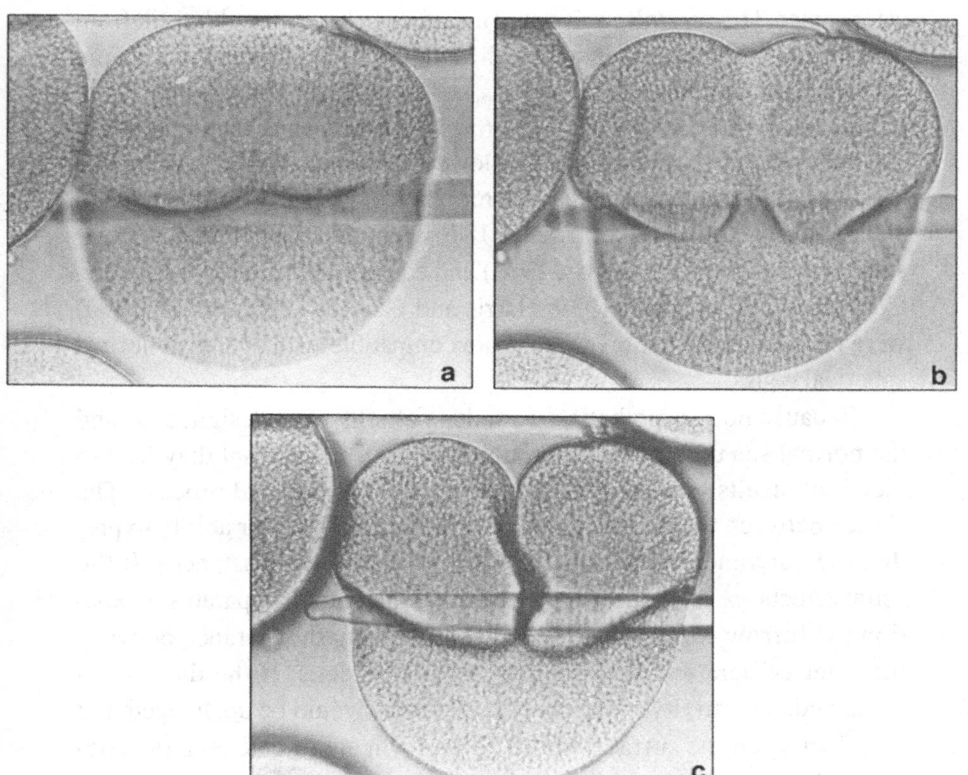

Figure 5.12 Cleavage of a urethane-treated sand dollar egg. The furrow forms only after the reduced mitotic apparatus is pushed close to the cell margin. (From Rappaport, 1971a, and with the permission of Wiley-Liss)

sis. The partition of the cell surface into actively contractile and passive, non- or less-contractile regions results from the inequality of mitotic apparatus effect, and the basis of that inequality has been persistently debated. Two alternative schemes that require different hypothetical stimulus patterns have been championed. According to one, the intensity of mitotic apparatus effect upon the surface varies because the distances vary, and the intensity is assumed to be inversely proportional to some function of the distance between the mitotic apparatus and the surface. This system places maximal aster effect at the poles, and it is incorporated in the family of astral relaxation theories. It implies that the effect of each aster is symmetrical and, because the asters are assumed to be mutually exclusive, their effects are not additive. The intensity of stimulatory activity at any point on the surface of the mitotic apparatus is determined by its distance from the aster center. The nonuniformity of cell surface behavior would result from the normal, measurable distance differences. The alternative scheme assumes that the asters, or their effects, can be additive. The intensity of the effects may decrease with distance from the aster centers, but the possibility that they may be additive is allowed. These assumptions can place the maximum aster effect at the equator. They have been incorporated into equatorial constriction theories since 1876. Computer models of both polar stimulation (White and Borisy, 1983) and equatorial stimulation (Devore, Conrad, and Rappaport, 1989; Harris and Gewalt, 1989; Harris, 1990) were devised, and their predictions are compatible with events in normal spherical cells.

Because both hypothetical stimulus patterns were designed around the normal sea urchin egg, it should not be surprising that they lead to identical results when they are used to model the normal process. The choice between them, therefore, must be based upon their ability to predict the outcome of division activity in unusual circumstances. If the astral effects are not additive, the ability of the mitotic apparatus to position the furrow would be affected by equalizing the distance between the aster centers and the different surface regions. If the distance is equalized, the activity of the mitotic apparatus would be unchanged, but its effect upon the surface would be uniform and the necessary functional differentiation would not occur. The distance between the aster centers and different points on the aster centers can be changed by constricting the cell in the equatorial plane. Equatorial constriction simultaneously distends the polar surfaces, because the cell volume is unchanged, as is the distance between the aster centers. Consequently, the normal difference in distances can be relatively easily erased or reversed during the interaction period. The first artificial constriction experiments were done by mechanically removing the fertilization membrane of sand dollar eggs in a way that favored their extrusion part way through a hole in the membrane (Rappaport, 1964). Although the desired geometrical relation was the result of chance, it was clear that varying degrees of equatorial constriction did not block cleavage. By inserting eggs in glass loops about 80 μm in diameter, similar constrictions could be obtained under controlled conditions that allowed measurement of the distances between the aster centers and the polar and equatorial surfaces (Figure 5.13) (Rappaport and Rappaport, 1984). When the constricted cell is compared with the normal spherical cell, it is apparent that constriction reverses the distance relations so that the equatorial surface, rather than the polar surfaces, would have been maximally affected (Figure 5.14). Constriction did not interfere with division, and furrowing in constricted cells frequently began earlier than in spherical controls.The experiments show that furrows form in the equatorial surface whether it is closer to the aster centers than to the polar surfaces, or farther from the aster centers than from the polar surfaces. We may assume that the normal cleavage of artificially constricted eggs results from a normal arrangement of more-contractile and less-contractile surface regions. The normal arrangement is not a simple consequence of distance differences. Constrictions by themselves result in cleavage activity only when they are close to asters (Rappaport, 1988).

IS THE NORMAL OPPORTUNITY FOR INTERACTION BETWEEN THE SURFACE AND ALL PARTS OF THE MITOTIC APPARATUS ESSENTIAL? The reasoning thus far presumes that the existence of more- and less-actively-contractile regions is the direct or indirect result of regional differences in aster effect. The constriction experiments that showed that the distance from the aster to the polar surface is unimportant also imply that any interaction that may occur there is also unimportant. The opportunity for interaction can also be hindered or prevented in a more precise and controllable way by interposing physical barriers between the interactants.

K. Dan (1943a) perforated sea urchin eggs with a 15 μm diameter

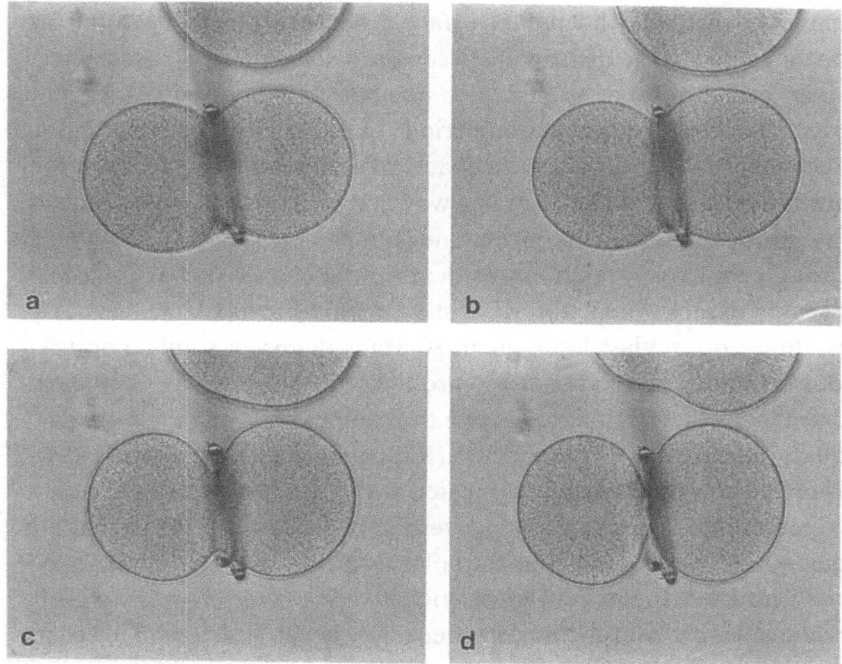

Figure 5.13 Cleavage of an otherwise normal, artificially constricted sand dollar egg. (a) 12 min before cleavage. The mitotic apparatus straddles the plane of an artificial constriction unequally, as there is more of it to the left of the constriction than to the right. (b) 6 min before cleavage. The enlarging mitotic apparatus has shifted so that the constriction plane bisects the interval between the asters. (c) Cleavage has begun. (d) Cleavage nearly completed. Note progress of the furrow in the control cell. (From Rappaport and Rappaport, 1984, and with the permission of Wiley-Liss)

flat pipet tip. The operation brought the opposed cell surfaces together and resulted in their local fusion where the tip of the pipet pressed against the surface of the chamber. The perforation formed a cylindrical barrier that changed the surface configuration without hampering division activity (Figure 5.15). Dan found that aster rays that normally extend to the cell periphery were always absent in a fan-shaped region distal to the hole "like the shadow of a screen in front of a light source" (1943a, p. 299). He also found that when the perforation was in the equatorial plane, the furrow formed on the proximal surface of the perforation, but not on its distal surface or on the cell margin distal to it

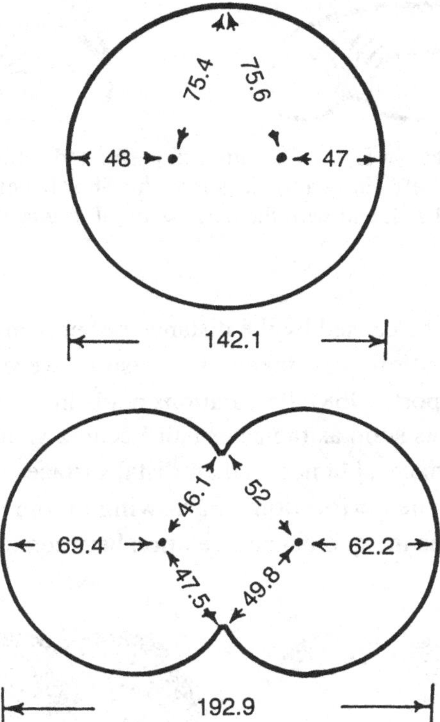

Figure 5.14 Upper: Diagram (to scale) of the positions of the astral centers relative to the poles and the equator in normal spherical sand dollar eggs 5 min before beginning of cleavage. The distances (μm) are the means of the measurements of 22 eggs. The distances from the astral centers to the equator were calculated. Lower: Diagram (to scale) of the positions of the astral centers relative to the poles and the equator in otherwise normal, artificially constricted sand dollar eggs 5 min before the beginning of cleavage. The distances (μm) are the means of measurements made directly on 16 eggs. (From Rappaport and Rappaport, 1984, and with the permission of Wiley-Liss)

(Figure 5.16). Dan interpreted these and other perforation experiments as the result of suction, aster fiber traction, and other consequences of the operation of his hypothetical spindle elongation mechanism of cytokinesis.

Dan's method is excellent for putting barriers between the mitotic apparatus and the surface. The experiment requires that the surface lie within the normal distance to the mitotic apparatus. Otherwise, failure

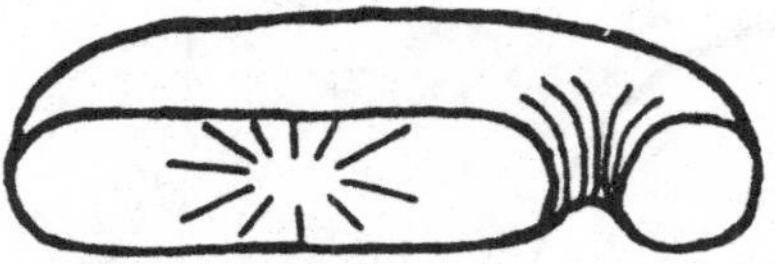

Figure 5.15 Changes in form of a flattened, perforated echinoderm egg, schematic section. Left: Early amphiaster. Right: Shortly before cleavage. (From Rappaport, 1968, and with the permission of *Embryologia*)

of furrowing may be caused by the distance rather than the obstruction. When this precaution was taken, the results were much as Dan described (Rappaport, 1968). Perforations made in the equatorial plane of flattened cells as soon as the aster pair became visible furrowed on their proximal surfaces, but not on their distal surfaces or on the normal margins distal to the perforations. Furrowing on the unoperated side was normal, as was the geometrical relation between the asters and the

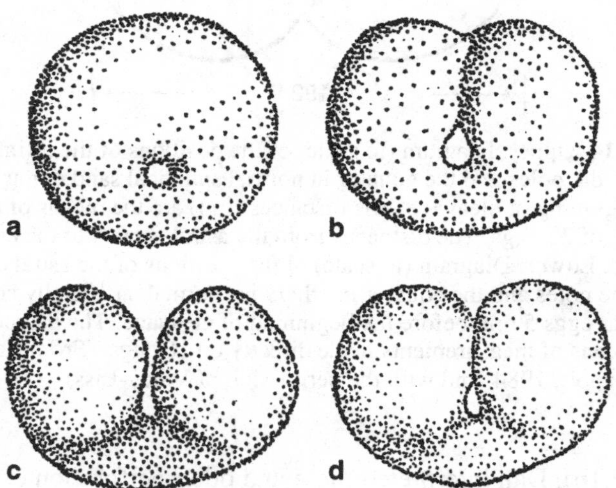

Figure 5.16 Cleavage-associated form changes of a perforation in the equatorial plane of sea urchin egg. (a) 10 min after perforation. The hole is elongating toward the mitotic axis. (b) 25 min. The furrow is visible in the margin of the unoperated side. (c) 29 min. Cleavage is nearly complete. The furrow has joined the perforation, and the distance between the perforation and the nearest cell margin is unchanged. (d) 34 min. The apposed blastomere surfaces are flattened against each other and cleavage activity has ended. (Redrawn from K. Dan, 1943a)

polar surfaces. Very small perforations slowed but did not completely block furrows in the distal surface. When the mitotic apparatus was straddled by the perforations, the furrows that formed on their proximal surfaces joined in the mitotic axis, but no furrows formed in the distal margins.

In further experiments, the size, shape, and position of the blocks were varied under better control by using slit-shaped punctures, glass rods, and oil drops (Rappaport and Rappaport, 1983). The experiments showed that large equatorial blocks were more inhibitory than small blocks and that blocks of the same size located outside the equatorial region did not affect furrowing. If the blocks in the polar region had had no effect, this experimental design would have proved nothing. The interference that occurred only when the block was in one region strongly implies that division requires interaction between the parts that were separated. The experiments demonstrated the effectiveness of blocking, as well as the locus of the essential interaction.

The idea that there is a surface region where interaction with the mitotic apparatus is essential for cytokinesis implies that there are other regions that may be disrupted or removed without affecting cytokinesis. Consistent with this interpretation are the findings that isolated furrows and furrow fragments of sand dollar eggs can complete division (Rappaport, 1969a) and that locally administered calcium chelator that prevented or caused abnormal cleavage when applied to the equatorial plane of sea urchin eggs resulted in normal cleavage when applied at the pole (Timourian, Clothier, and Watchmaker, 1972). Jura (1975a), on the other hand, found that the ultraviolet lesions on the polar surface of actively cleaving snail eggs caused furrow regression and early death, but similar treatment of the equatorial surface had no effect.

IS THE NORMAL DISTANCE BETWEEN THE ASTERS IMPORTANT? By the early 1900s, techniques for producing eggs with multiple asters, multiple mitotic apparatuses, and multiple cytasters made possible a number of observations of the relations between asters, spindles, chromatin, and furrowing. From those investigations there arose an impression that the degree of development of the furrow was increased by larger asters, closer proximity of the asters to the surface, and the presence of nuclear material. The same polyastral cells also made possible the investigation of the effect of distance between the aster centers and the degree of development of the furrow, but no studies were reported. Compressoriums with circulating sea water that were then available made it possible to flatten cells to the extent that all the asters were in the same plane. There were discussions of size and distance, but there was a notable lack of simple linear measurement in all investigations. The absence of measurement in the reports of division-related investigations during this period is puzzling. Conjecture about the effects of distance go back to the beginnings of thoughtful analysis of the process. Asters and spindles were clearly seen and drawn. In 1879, R. S. Bergh faithfully recorded the dimensions of the spindles and the blastomeres in a simple description of early development of a coelenterate. But at a later time, when structures that might have had some role in the division process were described as larger or smaller, or closer or farther apart, there is a remarkable lack of numbers. Perhaps the frequent use of the camera lucida led to an impression that measurements were unnecessary. Teichmann (1903) made a nonquantitative study and concluded that furrowing activity was reduced when the asters were close.

The mitotic apparatus appears stable but its dimensions change. Anaphase B, the increase in spindle length, continues in both spherical (Rappaport and Rappaport, 1984) and cylindrical (Rappaport, 1981) cells while furrow establishment is in progress. When the asters are experimentally detached from the spindle before they attain maximum size, the centers shift farther apart than normal (Boveri, cited in Wilson, 1928). This observation suggests that the spindle holds the asters together, rather than pushes them apart. The first two cleavages of torus-shaped eggs provide an opportunity to study the effect on furrow activity of the distance between the asters (Rappaport, 1961). There are two possible places for furrows in the first cleavage of torus-shaped cells – at the spindle and in the diametrically opposite region – but the furrow develops only near the spindle. Because the spindle and chromosomes are not required, it appears likely that the difference in the distance between the asters in the two regions is a factor. At the second cleavage, formation of a furrow between the polar surfaces of the asters depends upon the distance between the asters' centers. A large glass sphere produces a large hole in the torus and, in such cases, the distance between the asters that are not joined by a spindle is great. No furrow forms at the second cleavage, and the phenomenon is postponed until a later cleavage (Rappaport, 1961). In torus-shaped cells, the shortest distance from the aster centers to the surface is fairly uniform, so that the distance between aster centers seemed to be the most likely critical relation.

Flattened, dispermic sea urchin eggs permitted more detailed study (Rappaport, 1969b). Flattening puts all the asters in one plane, and the four asters that develop when two sperm enter the egg increase the variety of interastral distances. The ability of the aster pair to establish furrows depends upon both the distance between their centers (interastral distance) and the distance from the spindle or mitotic axis to the surface (spindle-to-surface distance). The presence or absence of a spindle between the asters has no effect. In flattened, dispermic *Hemicentrotus pulcherimmus*, the distance between the centers of the asters that are attached to the spindle is 32.5 μm. Asters 32.5 μm apart produced furrows when the spindle-to-surface distance was between 33 and 48 μm. When the interastral distance was increased to 35 μm without changing the spindle-to-surface distance, furrows were rare (Rappaport, 1969b). Abnormally distant asters appear to have a reduced ability to induce furrows. Abnormally short spindle-to-surface distances can be obtained by punching a hole in the center of the flattened cell. In this circumstance, the rarely effective aster pairs separated by 35 μm or more produce furrows without exception when the spindle-to-surface distance is 20 μm or less (Figure 5.17). These results may be explained by assuming that the astral effect is radially symmetrical and distance limited, and that furrow establishment requires the additive effect of the asters upon the equatorial surface. As the distance between the aster centers is increased, the diameter of the zone of their additive effect decreases and can fall short of the surface (Figure 5.18). Decreasing the spindle-to-surface distance by moving either the mitotic apparatus or the surface would put the surface in the zone where the effects are additive (A. K. Harris and Gewalt, 1989). The effect of increasing the distance between the asters can also be reversed by pushing them together. Sluder and Begg (1983) cut the spindle in flattened sea urchin eggs, and the asters and half-spindles moved apart to the extent that no furrow formed. When the same operation was fol-

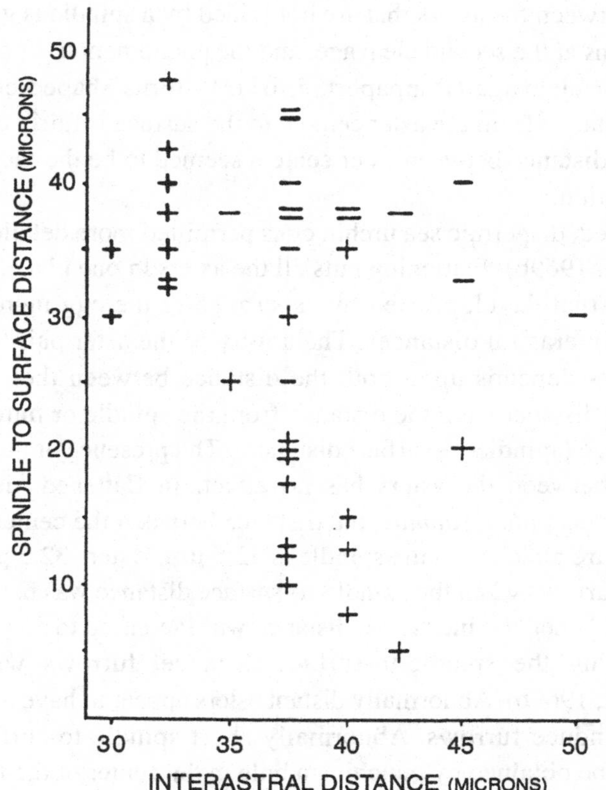

Figure 5.17 Summary of data from perforated and unperforated cells. When the spindle-to-surface distance is 35 μm or more and the interastral distance is 35 μm or more, furrowing almost invariably fails. If the spindle-to-surface distance is reduced to 20 μm or less, furrowing occurs in conjunction with a 35-μm interastral distance. + indicates a furrow formed adjacent to the asters, - indicates no furrow formation in any location. (From Rappaport, 1969b, and with the permission of Wiley-Liss)

lowed by manipulation that pushed the asters back together, the halves reassociated to form a normal spindle, and furrows followed. In cylindrical sand dollar eggs, no furrows form when the asters are more than 68 μm apart, but asters that are as far apart as 137 μm establish furrows after they are pushed together until they are 50 μm apart (Rappaport and Rappaport, 1985a).

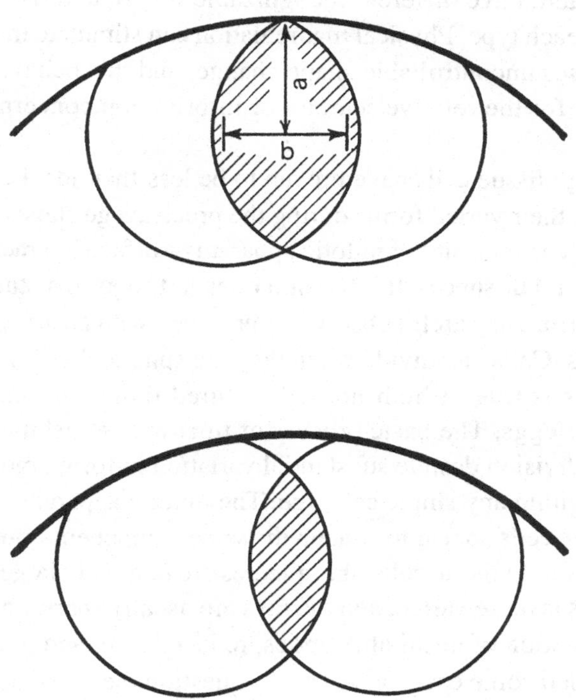

Figure 5.18 Aster–equatorial surface relations that would obtain if furrow establishment were a consequence of joint action of the asters. In the upper figure, the area affected by both asters reaches the surface. The lower figure shows the effect of moving the asters apart. Astral diameter and distance from the astral center to the surface are unchanged. (a) Spindle-to-surface distance measured parallel to the flattened surface. (b) Interastral distance. (From Rappaport, 1969b, and with the permission of Wiley-Liss)

Tissue cells

Experimental analysis of the essential geometrical relations between the mitotic apparatus and the surface has usually involved forcing the cell components into abnormal configurations during the interaction period. The imposed arrangement must be stable, and the inability of most cleavage cells to move or autonomously change shape makes them ideal experimental materials for this analysis. Different tissue cell types, on the other hand, have different recognizable forms, and there is variation within each type. Physical manipulation can stimulate motile activity that causes uncontrollable shape change, and this behavior may be responsible for the relative scarcity of information concerning tissue cells.

Although tissue cells have proven to be less-than-ideal experimental subjects, their varied forms during the precleavage stages have been instructive. The intensity of mitotic apparatus–surface interaction is distance related, but successful division does not require a standard precleavage form. The careful observer is provided with an array of natural experiments. Cells can divide when they are spherical or flat or spindle shaped; this is true of both normal cultured tissue and manipulated invertebrate eggs. The basic process of furrow establishment ensures successful division despite substantial variation in form among the cell types and within any single cell type. The mitotic apparatuses of tissue and cleavage cells appear to contain the same components, but their proportions differ. In tissue cells, the spindles are relatively larger, the form of the asters is more varied, and the rays are usually sparser and shorter. The combination of unsuitable optics, reduced size, and poor viewing conditions led some early observers to question the existence of asters in living cultured cells.

Most of the relevant experiments on other than cleavage cells were done on grasshopper spermatocytes and neuroblasts, which differ from many tissue cells in their near-spherical predivision form; they are similar to typical tissue cells in that the spindles are relatively larger than in cleavage cells, and their asters are more variable. Ris (1949) found that the spindles of X-rayed grasshopper spermatocytes tended to form lateral bulges that pushed the cell surface outward, thereby forming long processes oriented perpendicular to the original mitotic axis. Subsequently, a constricted area that resembled a furrow formed where the surface was near the spindle bulge, and the distal part of the process was cut off (Figure 5.19). Carlson (1952) used micromanipulators to move and reorient the grasshopper neuroblast mitotic apparatus and found that displacement of the mitotic apparatus was followed by comparable displacement of the furrow. Asters were not visible in his preparations, but they could in rare instances be seen in fixed stained cells. Preparations of isolated beetle testicular tissue in which cells containing two to four spindles were regularly found allowed Roberts and Johnson (1956) to study the relation between spindles, spindle poles, and furrows in unusual configurations. They found that furrows formed between pairs of spindle poles whether or not they were joined by spindles. Furrows that formed between spindle poles that were closer together than normal tended to regress. The authors made no mention of asters. Kawamura (1957) made a careful study of the mitotic apparatus in the spermatocytes of several grasshopper species. He found that the distance between the equatorial surface and the spindles or equatorially directed astral rays was short. In some species, no astral rays were directed toward the polar surface, but in all species the aster rays were directed toward the equator in anaphase and telophase. Kawamura (1960) later physically manipulated the spindle position and orientation in grasshopper spermatocytes and concluded that the midregion of the spindle can establish furrows in any part of the cell surface. He also reproduced the same geometrical configuration that Ris (1949) had observed after X-radiation and got the same results.

These observations and those, previously discussed, made on dividing newt epithelial cells (Rappaport and Rappaport, 1974) strongly suggest that the spindle region of tissue cells can directly establish the division mechanism in the nearby equatorial surface. The strong probability that tissue cells and cleavage cells normally depend upon different parts of the mitotic apparatus for furrow establishment does not mean that those parts not normally used have lost their competence. The spindle of sea urchin and sand dollar eggs can establish furrows (Rappaport and Rappaport, 1974), as can the spindle poles of multi-

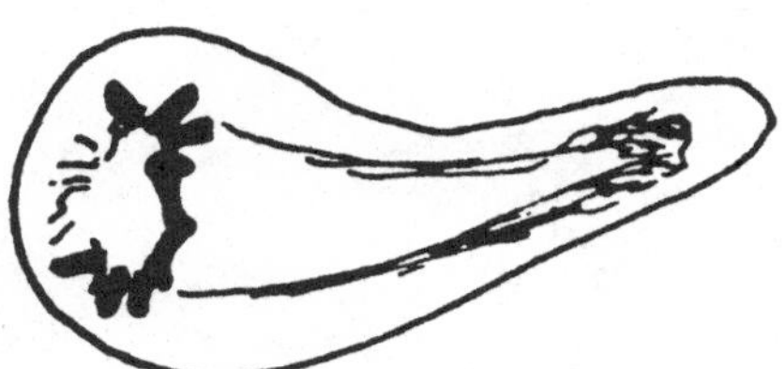

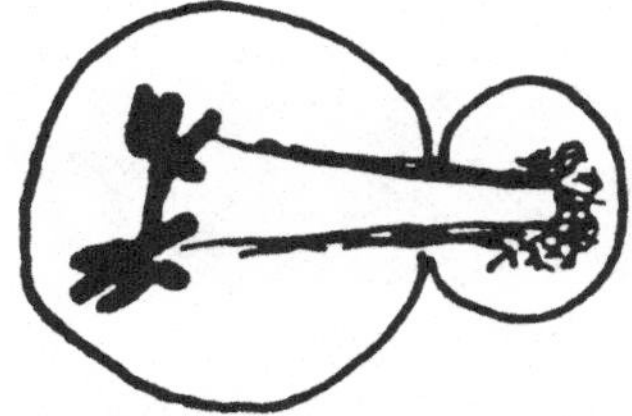

Figure 5.19 Abnormal lateral stretching of the spindle in *Chortophaga viridifasciata.* Primary spermatocytes after X-ray–induced sticking of the chromosomes followed by furrowing activity (right) in the adjacent surface (From Ris, 1949, and with the permission of the *Biological Bulletin*)

spindle cells of beetle testicular tissue (Roberts and Johnson, 1956) and multinucleate PtK_1 tissue cells (Ghosh and Paweletz, 1984).

Experimentation has confirmed and extended the historically early expectation that the mitotic apparatus is instrumental in the positioning of the division mechanism in the surface region. In cells that do not develop diastema, essential aspects of the positioning process are microtubule dependent but not chromosome dependent, and a factor associated with the nucleus is required for complete, permanent furrows. The event that establishes the division can be relatively brief, and it normally occurs (in the small number of cell types that have actually been tested) at metaphase–anaphase. In favorable circumstances, the necessary mitotic apparatus–surface interaction can take place during a greater proportion of the cell cycle than is normally used, so that the capacity to divide is less vulnerable to delay than it might be if it were limited to a stricter schedule of short-term events. The furrow forms at the midpoint between an aster pair whether or not a spindle lies between the asters. This relationship holds whether the equatorial surface is most distant from the aster centers or closest to them. These findings are important for subsequent discussions of the immediate effect of the mitotic apparatus on the surface. In cells that develop diastema, the presently limited amount of experimental information suggests that the role of the mitotic apparatus in division mechanism formation is to organize the diastema, which then directly affects the equatorial surface.

6

Formation of the Division Mechanism

Unlike multicellular contractile systems, the division mechanism may have no resting state. The contractile ring is exerting tension at the time it becomes ultrastructurally demonstrable, and it is gone when the division has been completed. Its very organization may be the consequence of local contractility. In this circumstance, formation and function may prove to be different phases of the same process and distinctions between them may simply make exposition more convenient.

The many changes in organization, structure, and behavior that immediately precede and accompany division have been carefully studied for clues about the mode of formation and function of the division mechanism, and the results of such studies form an important part of the body of information about cell division. Some of the phenomena that occur during the period of division activity were carefully described before their relation to the process was understood, apparently in the hope that, as the details were revealed, the connections would become clear. These expectations have not always been realized, and the significance of some of the most striking and best-studied events remains enigmatic.

Prefurrow Phenomena

Stiffness changes

The cyclical increase in resistance to deformation of the whole cell has previously been described and discussed. The earliest attempts to quantify the phenomenon were made on the eggs of sea urchin species in which the stiffness increase begins about 10 minutes before cleavage and then falls rapidly when cleavage begins (Figure 4.3). The apparent synchrony of stiffness change and cytokinesis strengthened convictions concerning their linkage, and led to several ingenious cell division theories and the occasional implication that other theories that did not in some way incorporate the phenomenon were thereby deficient. But, as more sea urchin species were studied, the variability of the relationship became apparent: In some species, the eggs develop two stiffness peaks; in other species, stiffness decreases before division begins; in yet others, stiffness increases before division. Among echinoderm eggs there is no demonstrable fixed relation between cyclical stiffness changes and cytokinesis (Hiramoto, 1990) (Figure 6.1). The basis of the fluctuation is also poorly understood. The early assumption that it was driven by the mitotic cycle was inconsistent with Wilson's observation (1904) that the isolated polar lobes of mollusks that contain no nuclear material continue their rounding up and relaxing cycle in synchrony with the cleaving embryonic cells from which they were detached. Swann and Mitchison (1953) found that the cycle persisted in the presence of colchicine concentrations that obliterated the asters, and Bell (1962) reported that enucleate ascidian egg fragments behaved like the isolated polar lobes described by Wilson. Yoneda, Ikeda, and Washitani (1978) measured the deformability of enucleated, parthenogenetically activated sea urchin fragments and found cyclical changes very similar to those of normal fertilized eggs. Yoneda and Schroeder (1984) determined the stiffness of sand dollar and sea urchin eggs in the presence of relatively high concentrations of colchicine. They found that 2×10^{-3} M colchicine prolonged the intermitotic phase of the sand dollar (*Dendraster excentricus*) but did not block the stiffness cycle. In response to 1×10^{-3} M colchicine, the sea urchin (*Strongylocentrotus purpuratus*) maintained a high stiffness level typical of metaphase through the next mitotic cycle without significant fluctuation. Yoneda

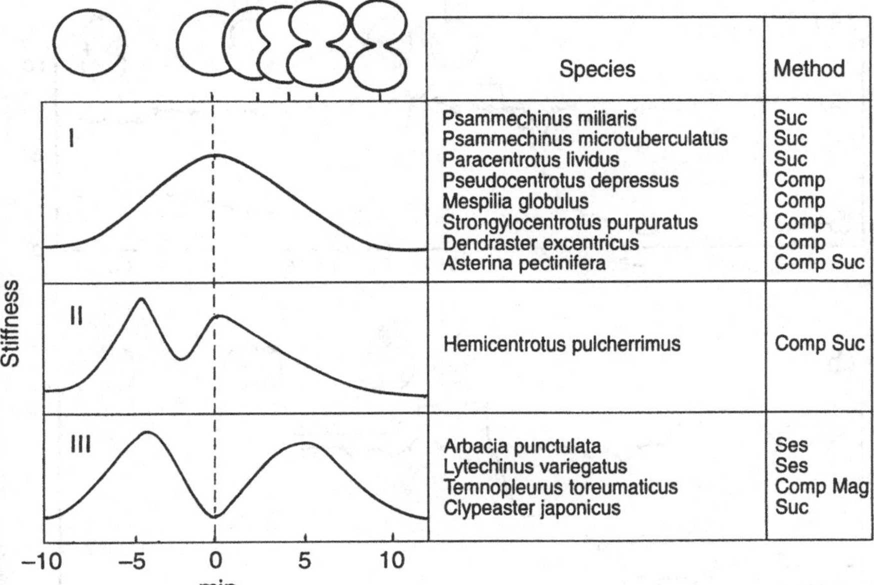

Figure 6.1 Stiffness changes before and during cleavage in echinoderm eggs. Abscissae: time after the onset of cleavage. Ordinates: Stiffness of the cell. Comp = compression method; Mag = magnetic particle method; Ses = sessile drop method; Suc = suction method. (From Hiramoto, 1990, and with the permission of the New York Academy of Sciences)

and Yamamoto (1985) manually bisected sea urchin eggs and found that, relative to the nucleated part, the cyclical changes in the stiffness of the enucleated part were prolonged by about 30%. At present, the role of the cytoplasmic stiffness cycle in cytokinesis is unknown. It is tempting to propose that the high points in the fluctuation are in some way associated with the period when the surface can respond to the mitotic apparatus, but the period of competence to respond spans the time when determinations indicate that the stiffness is nearly minimal as well as maximal (Figure 8.2).

The results thus far recounted were obtained by methods that demonstrated changes in stiffness or tension at the surface of the cell as a whole; they reveal little of regional differences. Using the elastimeter method, Ohtsubo and Hiramoto (1985) simultaneously determined the stiffness of the polar and equatorial regions of sea urchin and starfish eggs. In the sea

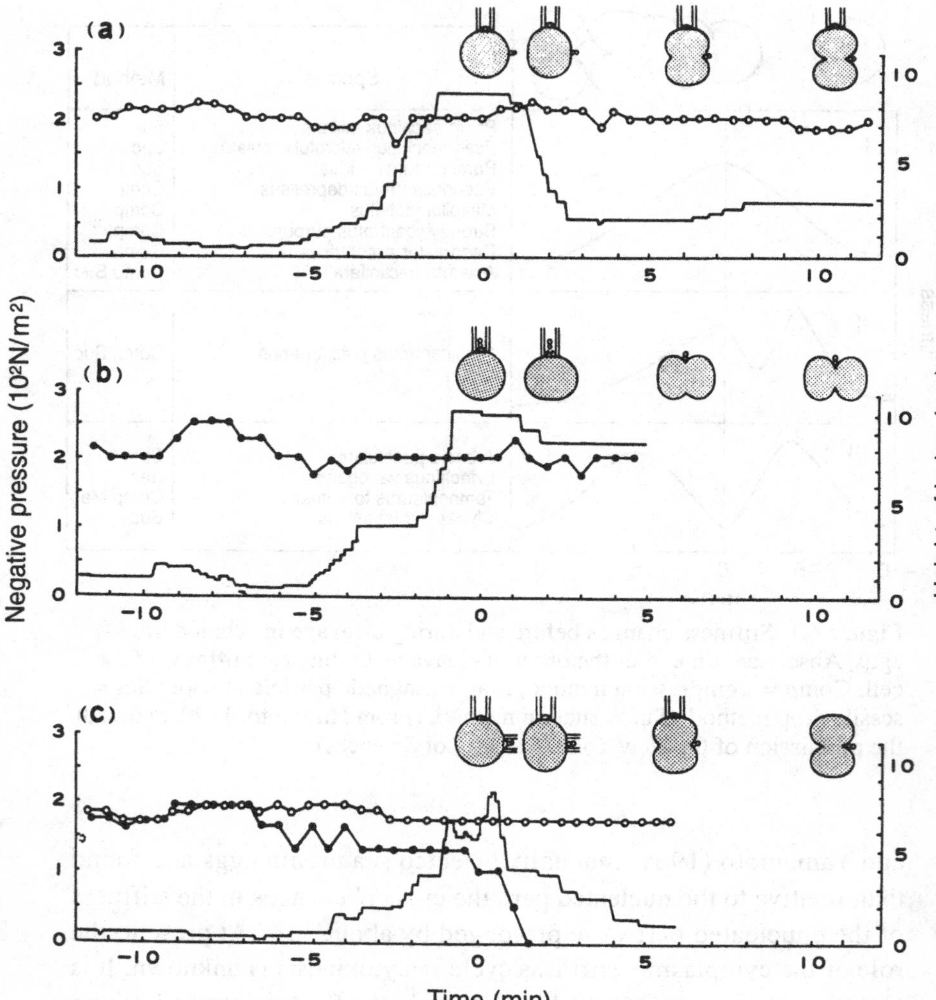

Figure 6.2 Stiffness changes of the cell surface in sea urchin eggs before and during cleavage. The stiffness is represented by the negative pressure (continuous lines) required to keep the height of the bulge (circles connected by line segments) at the cell surface in the micropipet constant. Insets indicate stages of cleavage. (a) The stiffness is measured at the polar surface. (b) The stiffness is measured at the furrow surface. (c) The stiffness is measured at both furrow and polar surfaces. Open circles represent the height of the bulge at the polar surface and solid circles represent the height of the bulge at the equatorial surface. The origin of the time scale represents the onset of cleavage. (From Ohtsubo and Hiramoto, 1985, and with the permission of *Development, Growth & Differentiation*)

urchin egg, the stiffness changes in the two regions occur simultaneously and to an equal degree until cleavage begins. At that point, the stiffness at the equator exceeds that in the poles. The contour of the surface bulge in the pipet suggested that the central part of the furrow was especially rigid. Similar determinations on starfish eggs revealed a slight but significant difference between polar and equatorial surface stiffness beginning about 5 minutes before cleavage (Figure 6.2). The greater stiffness in the equatorial region may reflect changes in structure that occur during formation of the contractile mechanism.

Changes in cortical gel strength

Another cyclical property of the sea urchin egg cortex is its ability to resist the displacement, by centrifugal force, of embedded visible pigment granules or vacuoles. After fertilization, the vacuoles in the endoplasm are readily stratified, but those in the cortex are resistant to great forces (Brown, 1934). When eggs are centrifuged at high hydrostatic pressures, the force necessary for stratification is much reduced. High pressure has been shown to block cell activities like amoeboid locomotion, which in the 1930s was attributed to reversible sol–gel transformations. High pressure also reversibly blocked cytokinesis (Marsland, 1938). These findings suggested a connection between sol–gel transformation, cytokinesis, and cortical capacity to resist pigment vacuole displacement. Zimmerman, Landau, and Marsland (1957) determined the centrifugation time required to reach a standard end point of centrifugal displacement at intervals between fertilization and second cleavage. They found that after several fluctuations, the required time rose to a maximum at about 30 minutes before division and fell rapidly when the process began. Subsequently, peaks appeared shortly after the completion of the first cleavage and shortly before the second cleavage (Figure 6.3). Marsland (1939) described a prefurrow regional difference in cortical gel strength that, under certain conditions of pressure, time, and centrifugal force, resulted in the displacement of the cortical granules in the polar regions while those in the equatorial cortex were not moved, so that they persisted as a distinct belt. This apparent localized region of higher gel strength led Marsland and his colleagues to devise several cell division theories based upon gel and contractile systems

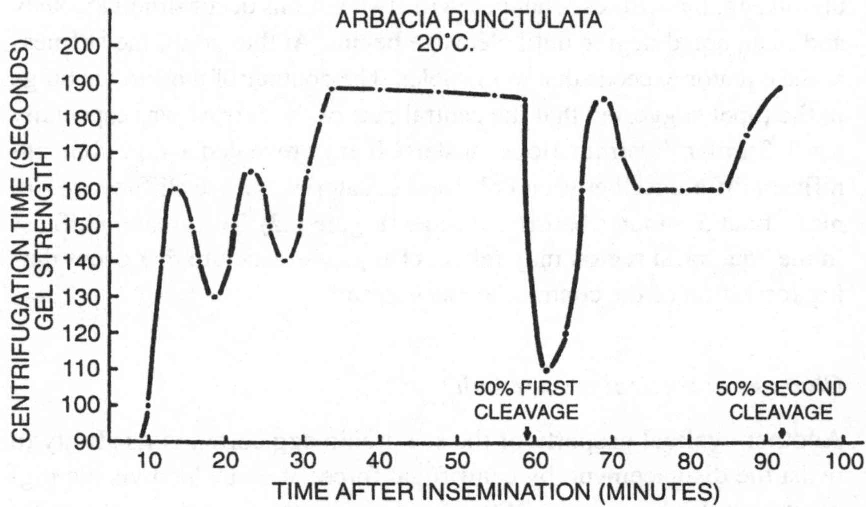

Figure 6.3 Cortical gel strength changes following insemination. Pressure-centrifuge measurements of the structural state of the plasmagel layer of the *Arbacia* egg made at high pressure (8,000 lb/in^2) and high force (41,000 x G). (From Zimmerman, Landau, and Marsland, 1957, and with the permission of Wiley-Liss)

(discussed in Chapter 2). Wolpert (1960) pointed out that the relation between the properties revealed by pressure-centrifugation and the mechanism of cell division are unclear. But the demonstration of regional differences (Marsland, 1939) suggests the operation of an early event in a sequence that culminates in the contractile ring. It may be evidence of an organizational change that precedes the shape change, but we are presently ignorant of its structural basis.

Resistance to bursting

The osmotic swelling that occurs when animal cells are immersed in a medium more dilute than the cytoplasm results in bursting when the difference is sufficiently large. The point of rupture in echinoderm eggs is localized. Because local rupture easily can be imagined as a consequence of local weakness in the cell periphery, study of rupture sites under controlled conditions has been used as a way of showing patterns of regional difference in surface strength. Just (1922) concluded from osmotic bursting experiments on sand dollar eggs that, immediately before division, ruptures invariably occurred in the polar surfaces. Just did not remove the hyaline layer because he considered it the material that divides the cell. He related bursting at the poles to a previously described prefurrow migration of hyaline substance from the poles to the equator and interpreted both phenomena as components of a cell division mechanism based upon ectoplasmic activity (see Chapter 2). The sea urchin egg's ability to divide successfully after removal of the hyaline layer discouraged support for his theory, but Just's results were cited years later in support of astral or polar relaxation theories (Wolpert, 1960; Schroeder, 1981a). Wolpert and Schroeder, however, assumed that Just's results were caused by differences in the cortex rather than the hyaline layer. Because good data that document regional differences in physical properties of prefurrow surface are scarce, Just's experiments were duplicated as faithfully as possible on the same species of sand dollar with the intent of comparing the bursting characteristics of eggs with and without the hyaline layer (Rappaport and Rappaport, unpublished results). About 10 minutes before the anticipated time of cleavage, the eggs were placed in a mixture of 40% sea water and 60% fresh water, and the responses of the cells were recorded by videotaping. Direct observation and the tapes revealed that the point of rupture is not localized in any particular cell region. When eggs with or without jelly and membranes were immersed in dilute sea water, the percentages of ruptures in the surface regions roughly approximated the percentage of the total surface that the region comprised. Just's results may have been caused by his method and timing of observations, which he did not clearly describe. The outflow of cytoplasm that follows the rupture causes an internal cytoplasmic rearrangement that usually results in the reorientation of the mitotic apparatus, so that its axis is parallel to the direction of outflow. Shortly after rupture, a pole of the mitotic apparatus is positioned close to the point of rupture. To someone who did not observe the instant of rupture, the outflow would appear to have originated at one of the poles. It now seems that there is no convincing evidence of regional differences in bursting strength of the prefurrow egg surface.

Mitotic apparatus–induced surface contractility

The cytoplasmically controlled cortical stiffness cycle makes it difficult to study the effect of the mitotic apparatus alone on the surface with any precision, but it is possible to compare the behavior of surfaces in the presence and absence of the mitotic apparatus of an echinoderm egg in a way that, although not quantitative, is instructive. Normally the mitotic apparatus virtually fills the spherical cell, so it can affect the entire surface. But when sand dollar eggs were artificially constricted so that the mitotic apparatus was confined to one side of the constriction, the direct effect of the mitotic apparatus was restricted to half of the surface (Rappaport and Rappaport, 1988). The necessary degree of constriction was attained by snaring eggs between opposed overlapping hooks (Figure 6.4). Differences in ten-

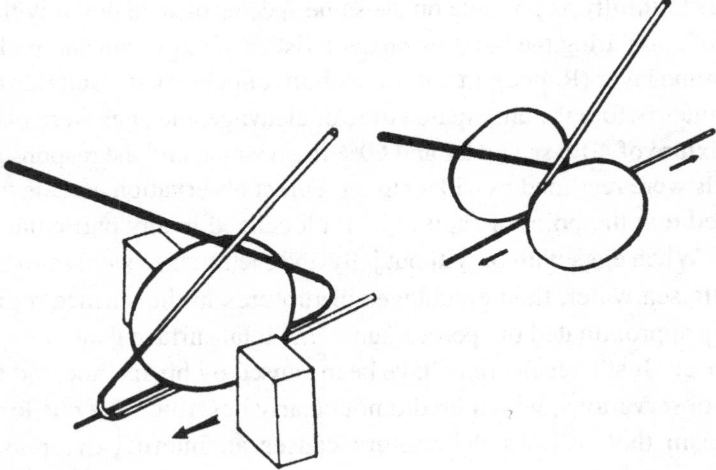

Figure 6.4 Method of egg constriction. Sand dollar egg was wedged between blocks so that it was suspended above the chamber floor. Hooks were positioned with micromanipulators on either side of the egg (left) and then moved toward the egg center to constrict it slightly. Then the chamber floor to which the rubber blocks adhered was moved aside with the mechanical stage (left, large arrow), leaving the egg suspended between the hooks. The hooks were then moved part way past each other (right, arrows) to reduce the diameter of the constricted zone. Not to scale; the size of the blocks is reduced for clarity. (From Rappaport and Rappaport, 1988, and with the permission of Wiley-Liss)

sion at the surface caused pressure differences, and differences in pressure between the two halves were revealed by the flow of cytoplasm across the connecting neck. The diameters of the two parts changed as the cytoplasm and mitotic apparatus moved from one part to the other. After the asters relocated, the direction of flow reversed. The reversal of flow and relocation of the mitotic apparatus usually occurred several times before division. The time from the beginning of flow in one direction to the beginning of flow in the opposite direction averaged 7.7 minutes. The flow was always directed out of the nucleated part, and when the mitotic apparatus was removed, there was no flow. The phenomenon did not occur in a concentration of cytochalasin B sufficient to block cleavage, although the mitotic apparatus appeared to develop normally. The previously described cortical stiffness cycle may be assumed to be operative in both cell parts because it is independent of the nucleus. The mitotic apparatus must have caused the tension at the surface in the two parts to become different, and it must have caused the tension in the nearest surface to increase, otherwise the flow would have been in the opposite direction. In many experiments the flow began before the mitotic apparatus was distinct and large, and this observation raises a question about the factors that start the phenomenon. The time between fertilization and the first cleavage is consumed by fertilization and the fusion of the male and female haploid nuclei, as well as by the mitotic cycle. By repeating the same kind of experiment on one of the blastomeres that resulted from the first cleavage, the premitotic events of fertilization were omitted. Reversing flow was observable at all times between completion of the operation and the beginning of the second cleavage (Rappaport, 1990). It occurred in the presence of the astral structure remaining from the first cleavage; it occurred when there was no radiate structure visible, and also in the presence of the astral structures associated with the second cleavage. When the times that flow began or reversed were related minute by minute to the period between the completion of the first cleavage and the time when the division mechanism for the second cleavage was established, there were no gaps. These findings suggest that the two basic components of the system that produces the furrow – the contraction promoter and responsive surface – are present throughout the cell cycle. The cyclical division activity may result from the fluctuating state of the mitotic structure that redistributes the contraction promoter. Unequal distribution of the contraction promoter could cause the regional tension differences that produce the furrow. These experiments demonstrate that the mitotic apparatus increases tension at the cell surface in a way that is superimposed on the cytoplasmically driven cycle. The sensitivity of both phenomena to cytochalasin B strongly suggests that they are both actin based, but are activated differently. A correlation between the site of initiation of surface contraction waves and the position of the nucleus has also been described in *Xenopus* eggs (Shinagawa et al., 1989).

Cortical birefringence

The cyclical changes in physical properties of the cell that have been associated with cleavage are generally attributed to events in the cortex that involve manipulation of macromolecular aggregates. The events are too small to be studied by light microscopy, and apparently too labile to withstand the methodology associated with electron microscopy without significant disruption. Some cell structures have anisotropic optical properties (that is, their properties vary according to the direction of measurement), and the resulting birefringence permits study of changes at the molecular level in some parts of the living cell. In early studies of birefringence in the living cell cortex (summarized in Shôji, Hamaguchi, and Hiramoto, 1981), radial birefringence was attributed to microvilli, and tangential birefringence was attributed to the hyaline layer. Shôji, Hamaguchi, and Hiramoto (1981) demonstrated cortical birefringence in the tangential plane of the sea urchin egg cortex. The birefringence increased over the entire surface before cleavage in parallel with the increase in resistance to deformation. In the early cleavage stages, birefringence decreased at the equator and increased at the poles; in later stages, it increased at the equator and decreased at the pole. No birefringence was found at the base of the furrow. Fukui and Inoué (1991) detected faint birefringence parallel to the equator in dividing *Dictyostelium* amoebae. Results indicated changes in cortical properties, but their relation to the formation of the division mechanism is not yet clear.

Prefurrow cortical response to chemical and physical agents

To the observer of the living cell and cells fixed by most methods, the prefurrow equatorial surface appears much like that of the rest of the cell. The apparent uniformity of appearance permitted the assumption that there is nothing special or important about the equatorial cortex. While there was still debate whether the furrow cortex was an active or a primarily passive participant in cleavage, the possibility that it might have properties that distinguished it from the rest of the cell cortex was explored by determining whether its visible response to various agents was distinctive. The results revealed regional differences in surface response that were, in some cases, demonstrable before cleavage began. Kojima's systematic studies (Kuno,1954; Kuno-Kojima, 1957) showed that the prefurrow equatorial surface region of echinoderm eggs is especially susceptible to heat and exposure to various chemicals. High temperatures caused blister formation in the equatorial region before and during cleavage. Sodium tungstate caused cells lacking fertilization membranes and hyaline layer to burst in the same region at the same times. Dinitrophenol, sodium azide, and sodium isomytal (sodium 5, 5-isoamyl ethyl barbiturate) caused circumferential ridges at the equator before cleavage, but not after cleavage began. Both 50% sea water and 0.5 M urethane caused thickening and a color change in the equatorial cortex. Dilute H_2SO_4 (0.02 N) caused the equatorial cortex to thicken and blacken before and during all stages of cleavage. The darkening effect appeared to be closely linked to division activity. Concentrations of urethane and colchicine that blocked division when administered early in the division cycle also blocked formation of the darkened areas, but the furrows that developed when the same concentrations were applied immediately before division were underlain by darkened material. In polyspermic eggs, the blackened areas appeared where the multiple furrows were expected, and in eggs that were pretreated with urethane or colchicine, which prevented spindle formation, no regional difference could be found. Surfactant compounds that block cleavage without blocking mitotic apparatus development also prevented formation of the blackened areas. These results indicate that the prefurrow surface differs from the surface of the rest of the cleaving echinoderm egg. The appearance of regional differences before division suggests that before the level of physical activity has increased to the degree that the cell is deformed, the equatorial surface is different from the rest of the surface, and that the differences in response to chemical and physical treatment reflect the changes that transform the equatorial surface to a dynamic structure. The changes are associated with the mitotic apparatus, and their duration is, in some cases, brief.

Prefurrow behavior changes in the normal equatorial surface

Pigment granules (or vacuoles) that are firmly fixed in the entire cortex of many echinoderm eggs immediately before division (Brown, 1934; Marsland, 1939) can reveal patterns of cortical movement. Several early accounts of invertebrate development describe the formation of a more densely pigmented stripe in the equatorial surface before cleavage (*Polychoerus caudatus* in Gardiner, 1895; *Arbacia pustulosa* [= *lixula*] in Fischel, 1906). The increased pigment density results from decreased distance between the granules, and, in the absence of independent granule motion at that time, the possibility of local shrinkage of the unpigmented cortical matrix has been assumed. Scott (1960b) carefully documented equatorial pigment behavior in *Arbacia punctulata* by analysis of photographs. In the equatorial region, he found two-dimensional shrinkage in both the latitudinal and longitudinal directions while the cell was still spherical. He estimated that the contractile band was initially about 15 μm wide and that it comprised about 10% of the entire cell surface. The shrinkage and compaction of granules continued after cleavage began. In a subsequent study of the same phenomenon in *Arbacia lixula* (Rappaport and Rappaport, 1976), the early events of stripe formation were emphasized using Scott's method. In this species, the prefurrow shrinkage occurs in a band 22 μm wide that comprises about 32% of the egg surface (Figure 6.5). The width of the band decreases to 15 μm, which implies a shrinkage of 34%. At the same time that the intervals between pigment granules in the stripe area decrease, the intervals in the adjacent subequatorial areas increase,

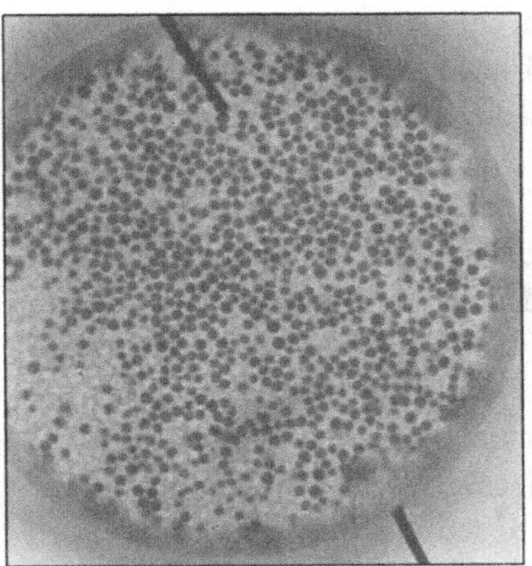
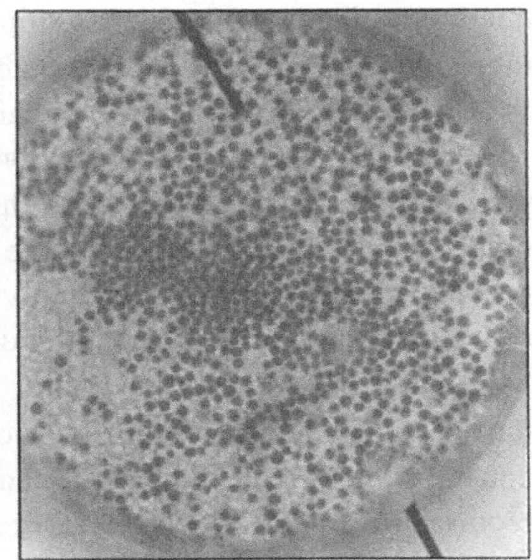

Figure 6.5 Development of the pigment stripe before furrowing in the eggs in *Arbacia lixula*. (From Rappaport and Rappaport, 1976, and with the permission of *Development, Growth & Differentiation*)

implying that equatorial shrinkage causes subequatorial stretching. The first movements of the pigment granules are slight. Despite many previous observations and experiments on *Arbacia punctulata* cleavage, no one appears to have observed the phenomenon until Scott made his photographic record. Schroeder (1981b) failed to observe the prefurrow stripe in *A. lixula* seen by previous workers and maintained that it develops after, not before, furrowing begins. He did not, however, describe his method for recording granule position during the prefurrow period.

Prefurrow distribution of contraction-related molecules

There was a fortunate near-coincidence between the general acceptance of the equatorial constriction mechanism of animal cell division and striking advances in the understanding of striated muscle chemistry, biophysics, and structure. At the time that alternative division mechanisms fell from favor, an array of techniques and concepts that had evolved in muscle studies promised equal effectiveness in the analysis of the contractile mechanism that divides animal cells. Current conceptions of the molecular basis of furrow function borrow heavily from muscle studies. Whether the relatively poorly organized and fragile division mechanism will permit the same detailed analysis that was necessary to create the present level of understanding of the more orderly and durable striated muscle is uncertain. But at a simple level, immunofluorescent and fluorescent staining methods have provided ways of using the light microscope to visualize the distribution of contraction-associated proteins in individual cells. Relevant to the formation of the division mechanism is the question whether its constituent contractile proteins congregate in the equatorial region before the cell changes shape. The brevity of the time required to convert nonfurrow cortex to active furrow may have created problems in the study of fixed, stained cells, but the injection of fluorescently labeled probes permits observation in living cells of both timing and distribution during division (Sanger, Mittal, and Sanger, 1990). The method shows that both actin and myosin concentrate in the equatorial region of cultured PtK_2 (kidney epithelium) cells before cytokinesis begins (Figure 6.6). Actin and myosin are associated throughout division. During interphase they are found in stress fibers and ruffled membranes. By metaphase, most stress fibers have disappeared and actin and myosin are distributed in the spindle and the cytoplasm. The concentration in the former appears greater; it is high at the beginning of anaphase and appears to spread to the equatorial cortex 4–6 minutes after anaphase begins. By the time cytokinesis begins, actin and myosin appear to be concentrated in the equatorial cortex and, in cells where the mitotic apparatus is eccentric, the vigor of furrowing is, in the same cell, correlated with the degree of their concentration.

Mabuchi (1990, 1994) studied the reorganization of cortical actin filaments during establishment of the contractile ring in echinoderm eggs by fluorescent staining of whole eggs and isolated cortices. Fixation preceded staining, so the real-time relations were lost in a population of whole cells but, when the method was combined with another that fluorescently stained whole chromosomes, Mabuchi could relate events at the equator and the deformation that marks the beginning of cytokinesis to the mitotic schedule. During anaphase–telophase transition, at the beginning of chromosomal vesicle or karyomere formation,

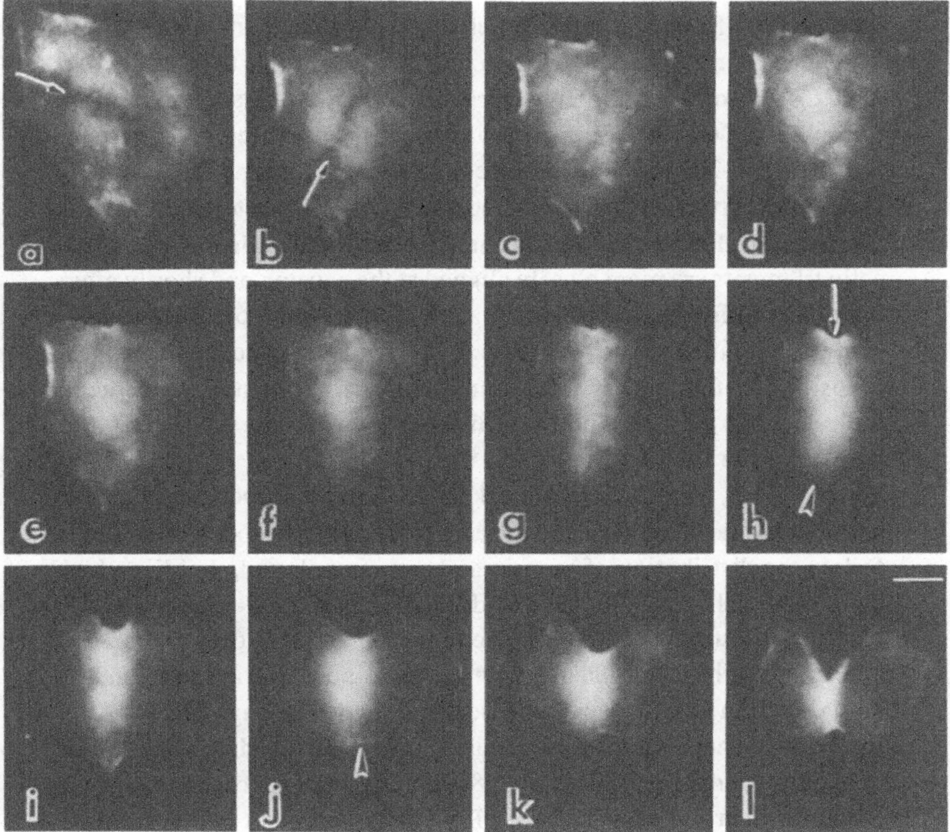

Figure 6.6 PtK_2 cell injected during interphase with fluorescent IAR-myosin light chains and recorded during cell division. Myosin is concentrated in the metaphase spindle (a, b), which rotated 90 degrees in the 34 min between time points a and b (arrows point to the metaphase plate alignment of chromosomes). During early anaphase (c, d), myosin remains concentrated in the spindle, and the first indication of a concentration in the cleavage furrow before contraction is apparent (e). As cleavage occurs, fluorescence appears bright (f–i), with three small strands present in one time point (arrowhead, h). In this cell, the furrow appeared first on, and contracted at a faster rate on, the side of the cell closer to the spindle, where myosin fluorescence was brighter. The tilt of the spindle with respect to the focal plane resulted in only one set of chromosomes being in focus in the micrographs. Minutes before [-] and after [+] onset of anaphase are (a) -37, (b) -4, (c) +2, (d) +3, (e) +5, (f) +6, (g) +9, (h) + 13, and (i) +16. Bar = 10 µm. (From Sanger, Mittal, and Sanger, 1990, and with the permission of the New York Academy of Sciences)

cortical actin filaments are distributed in 5 μm diameter clusters. When the chromosomal vesicles begin to fuse, the cell begins to elongate parallel to the mitotic axis and the actin in the clusters appears more fibrous. The density of fibrous actin is slightly increased in an equatorial band 15–20 μm wide. After completion of chromosomal vesicle fusion, the cortical actin filament bundles begin to disintegrate except in the equatorial zone. When the early furrow indentation appears, the actin filaments are more closely aligned at the equator, and shortly afterward, as the furrow deepens, the actin-rich band becomes more concentrated as it narrows (Figure 6.7). Immunofluorescent staining of myosin in combination with fluorescent phalloidin staining of actin indicated that the distributions of myosin and actin are coincident during furrow formation (Mabuchi, 1994).

Fujiwara, Porter, and Pollard (1978) used fluorochrome-labeled antibodies against myosin and the actin-crosslinking protein ∂-actinin to follow the distribution of myosin and actin in cultured, dividing chick embryonic cells. They found that during mitosis, the ∂-actinin was diffusely distributed in the entire cytoplasm, whereas myosin appeared to be present in higher concentration in the spindle. During cytokinesis, both ∂-actinin and myosin were localized in the cleavage furrow. After midbody formation, antimyosin staining in the furrow diminished while ∂- actinin staining persisted. In 1987, Sanger et al. followed changes in ∂-actinin distribution in cultured PtK_2 cells in greater detail. In interphase cells, ∂-actinin was located in the stress fibers; it became diffusely distributed in the cytoplasm when the fibers disassembled as the cells entered mitosis. By anaphase it was concentrated in the interzone of the spindle, and by the beginning of cytokinesis, it was concentrated in the broad area between the separating chromosome groups. Subsequent studies (Sanger, Mittal, et al., 1989; Sanger, Dome, et al., 1994) traced actin, myosin, filamin, and talin from stress fibers and their attachment plaques to the furrow region. In an ultrastructural study that stands alone among investigations of sea urchin furrow formation, Usui and Yoneda (1982) found a redistribution of microfilaments before division begins. Beginning at mid anaphase, the number of microfilaments at the equator increased as mitosis advanced, they reported, and during telophase the microfilamentous network was concentrated at the equator and absent at the poles.

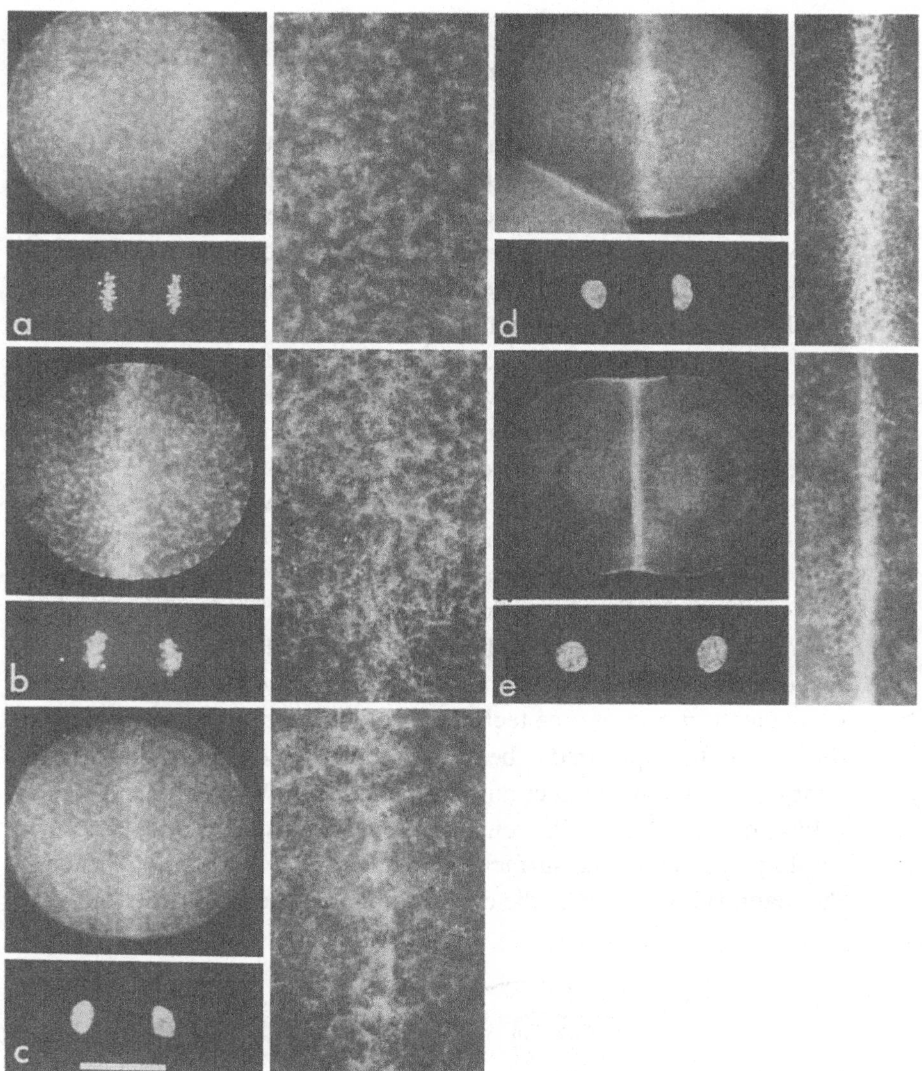

Figure 6.7 Fluorescent visualization of actin filaments in whole, fixed *H. pulcherimmus* eggs. The set of photographs of each division stage consists of a whole rhodamine-phalloidin stained egg, DAPI staining of the chromosomes in the same egg, and rhodamine-phalloidin staining of the equatorial cortical region at higher magnification. (a) End of anaphase. Chromosomes beginning vesiculation. (b) Early telophase. Chromosomes are vesiculating. (c) Completion of chromosomal vesicle fusion. A slight elongation of the egg in the pole-to-pole direction is apparent. (d) Onset of furrowing. (e) Advanced furrow stage. Bar for whole eggs and DAPI staining pictures, 50 μm; for higher magnification of equatorial region, 10 μm. (From Mabuchi, 1994, and with the permission of the Company of Biologists)

The Beginning of Cleavage

The special properties of the prefurrow surface are relatively subtle and short lived, but the differences between the base of the active furrow and the rest of the cortex are more marked and have been recognized for many years. The "cleavage head" of dividing ctenophore and coelenterate eggs is recognizable before division as a thickened ridge of equatorial ectoplasm. It persists in the furrow until the process is completed. There is no question of its existence before and during division, but there was some debate about whether it functioned as a restricted contractile zone (Ziegler, 1898b) or as an inactive, stiffened ridge that was pulled passively through the egg by forces that originated elsewhere (Yatsu, 1912a). Flemming (1891) published a drawing of a dividing larval salamander oral epithelium cell that showed a dark region in the furrow following haematoxylin staining (Figure 6.8) that Rhumbler proposed was the site of new surface formation. Kojima (Kuno-Kojima, 1957) and Kallenbach (1963) found dark material at the furrow base in, respectively, sea urchin eggs and sectioned rat tooth bud cells. Some early electron microscope techniques produced a subsurface electron-dense layer that appeared to be composed of tubules. It was most prominent in the furrow (Mercer and Wolpert, 1958). Weinstein and Hebert (1964) observed a similar dense layer and Weinstein (1965) described its shape as that of the surface of an unparted hyperboloid. Although the magnifying power of the electron microscope facilitated measure-

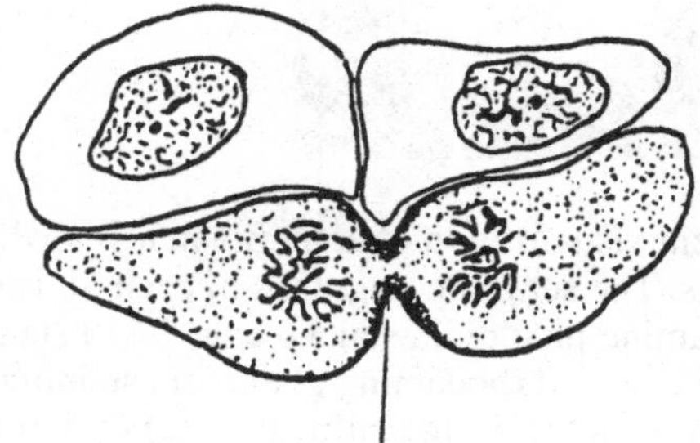

Figure 6.8 Dividing epithelial cell from the floor of the mouth of a salamander larva. Heavily stippled region indicates the distribution of material that stains strongly with haematoxylin. (From Flemming, 1891)

ments of the special region at the base of the furrow, it did not at first satisfactorily resolve the region's structure.

As methods and skills improved, so did the understanding of the distinctive region at the base of the furrow close to the plasma membrane. Arnold (1968) , Schroeder (1968), and Goodenough, Ito, and Revel (1968) independently demonstrated filamentous structure in a band 5–10 μm wide and 0.1–0.2 μm thick in, respectively, squid, coelenterate, and sea urchin egg furrows. Szollosi (1968) found similar 50–70 Å filaments in the furrow of a different coelenterate and an annelid, and Tilney and Marsland (1969) also reported them in cleaving sea urchin eggs. In squid and coelenterate eggs, furrowing normally begins in a small part of the surface near the mitotic apparatus and extends at both ends. This circumstance makes it possible to study a sequence of stages of furrow formation in the same cell. In the squid (Arnold, 1969), the electron-dense, filamentous layer extends beyond the tip of the incurved, apparently active part of the furrow. Subsequent ultrastructural studies revealed similar filamentous bands in the furrows of many different dividing animal cells (summarized in Schroeder 1970, 1972). The most characteristic feature of the bands is their microfilamentous composition. Vesicles, vacuoles, microtubules, and other structures that are often found in the cortical region are excluded. The filaments of an active contractile ring are predominantly circumferentially oriented, but a single filament spans only a small part of the entire circumference (Figure 6.9). The filaments are closely packed and, in cross section, number about five thousand (Schroeder, 1986). The dimensions of contractile ring microfilaments in different cells are similar, and (as will be discussed in detail later) they are the filamentous form of actin. The ring's thinness (0.2 μm) may account for its faint birefringence. Its width varies between about 5 μm and 15 μm, and it is not obviously proportional to cell or mitotic apparatus size, so that the 8–10 μm wide contractile ring of a 75 μm diameter sea urchin egg occupies about 13% of the egg surface in early cleavage, whereas the 10 μm wide structure in the 20 μm diameter cultured HeLa cell occupies 50% of the surface. Apparently the width of the contractile ring varies from point to point around the furrow circumference (Schroeder, 1972).

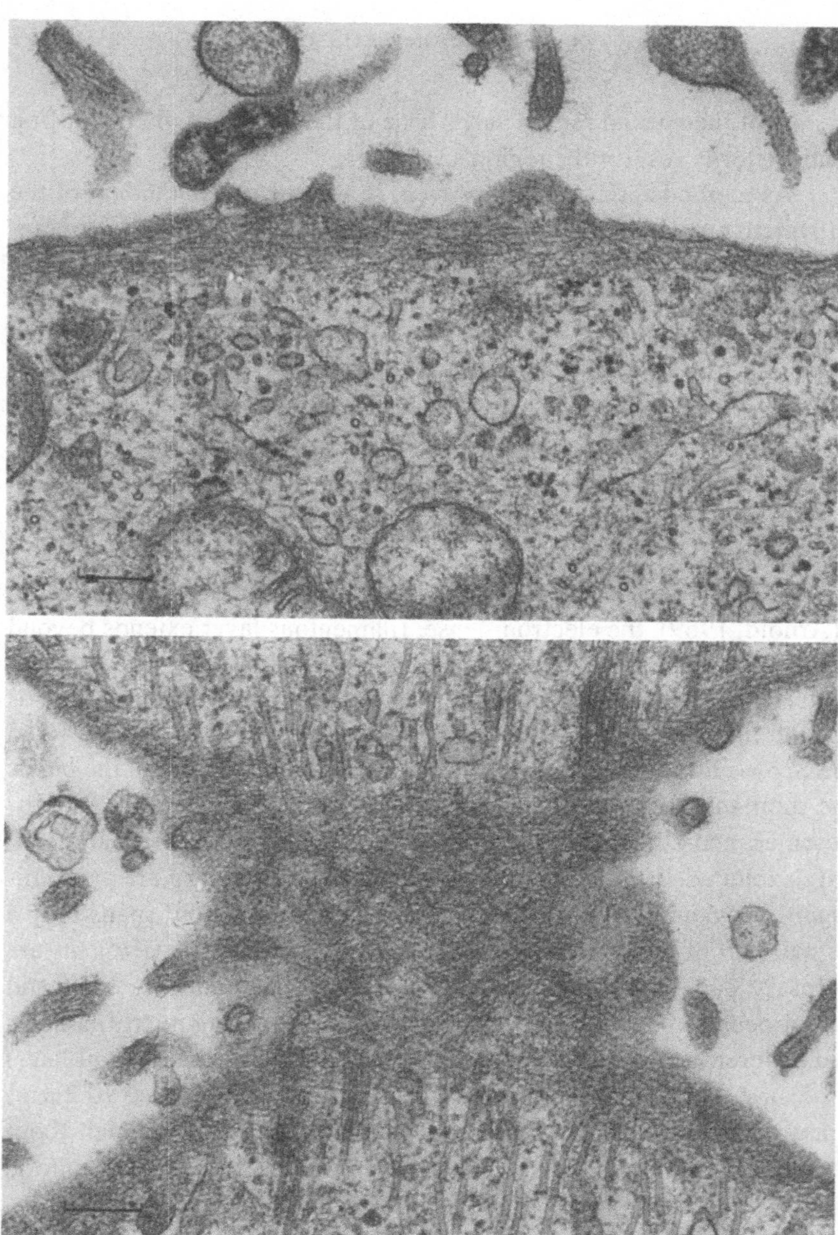

Figure 6.9 Upper: Part of a cross section of the furrow base of a half-divided HeLa cell, showing thin contractile ring filaments beneath the surface. Because equatorial sections are perpendicular to the axis of division, microtubules of the mitotic apparatus are sectioned transversely. Scale bar = 0.2 μm. Lower: This parameridional section passes through the base of a cleavage furrow perpendicular to the plane of the upper figure. It demonstrates that the

Sources of contractile ring material

The processes that concentrate and orient the contractile ring components to form the functional and ultrastructurally recognizable contractile ring are often counted among the remaining major mysteries of cytokinesis. The ultrastructure of the contractile ring is based primarily on its filamentous actin content, and there has been some question whether all the actin of the ring originates from within the equatorial cortex, or whether additional actin is recruited from other cell regions. The special appearance of the contractile ring might to some extent result from factors that do not require additional actin. Its apparent density may be enhanced by its exclusion of nonactin vacuoles and organelles. The nonrandom orientation of the microfilaments could make the ring more noticeable without any increase in the number of microfilaments. The actin in the contractile ring may be complexed with other polymers that make it more resistant to the deleterious effects of fixation and staining than the actin in the cortex outside of the furrow. Forer and Behnke (1972) found that treatment with heavy meromyosin revealed a layer of oriented actinlike filaments in the cortex of insect spermatocytes. The filaments, which were absent when heavy meromyosin was omitted, were oriented meridionally before division, but, in late anaphase, those in the equatorial region were reoriented to the typical circumferential direction perpendicular to the filaments outside the furrow region. The apparent concentration of actin in electron microscope studies could, then, have resulted from poor preservation of those filaments outside the equator that were involved in less contractile activity.

On the other hand, computer and mathematical models have suggested a need for an increase in contractile material at the equator. White and Borisy's (1983) model predicted that in a cortex containing a uniform distribution of tension elements, relaxation at the poles would not be sufficient to permit complete division at the equator. This prompted White and Borisy to explore the effectiveness of hypothetical mobile

---

predominant orientation of the contractile ring filaments is circumferential (horizontal to the page). Light and dense zones bordering the contractile ring presumably represent face-views of extracellular coating material and plasma membrane proper, respectively. Scale bar = 0.2 μm. (From Schroeder, 1970, and with the permission of Springer-Verlag)

tension units that migrate up the tension gradient that would result if the poles relaxed. With this addition, their model cell appeared to divide. White (1990) recognized that mobile tension units would also migrate in a gradient caused by enhanced contractility at the equator. He further suggested that the units could be reoriented at the equator by anisotropic tension as well as by the flattening in the early equator. Akkas and Engin (1981) pointed out that, after furrowing begins, circumferentially oriented contractile filaments would tend to slide from the edges of the equatorial zone toward its midline, and the bundling would increase the contractile ring force. A. K. Harris (1990) estimated from the behavior of his computer model that the contractile strength in the furrow must be at least 18 times that of the rest of the surface to produce complete division.

The idea that the concentration of actin, myosin, and other contraction-related proteins is greater in the contractile ring is now generally accepted, and there is a feeling that concentration is prima facie evidence for function, but information concerning the sources of the added material is presently incomplete. Usui and Yoneda's (1982) finding – that the actin meshwork diminishes before cleavage under the polar surface of echinoderm eggs but increases under the equatorial surface – has not yet been confirmed in other echinoderm eggs. In contrast with Usui and Yoneda's electron microscope studies, preparations of the isolated cortex reveal at the light microscope level that the formation of the contractile ring takes place without noticeable depletion of actin or myosin outside of the equatorial region (Yonemura and Kinoshita, 1986). The isolated cortex preparations used in these studies were made by a technique that takes advantage of the tightness of the adhesion between the surface of denuded sea urchin eggs and glass or plastic surfaces coated with polylysine or protamine sulfate. After the eggs stick to the surface, they are sheared with a jet of saline so that the cortex and surface remain attached and the remainder of the disrupted cell is dispersed in the medium (Allen, 1954; Vacquier, 1975). The cortex is fixed and stained and can be studied by several methods of microscopy. The preparation has the advantage of good optical properties: It is flat, thin, and transparent. But, in addition to all the problems of fixed material, there are lingering questions about what may be washed out when the cell is sheared. The structure and dimensions of the normal cortex *in situ* cannot yet be confidently described, and this treatment could be expected to add further complications. Nonetheless, when combined with F-actin–specific staining with nitrobenzoxadiazole phallicidin, the method has revealed a meshwork of actin bundles all over the isolated sea urchin egg cortex at anaphase (Yonemura and Kinoshita, 1986) that is visible with light microscopy (Figure 6.10). At telophase, after furrowing has begun, the bundles appear to be oriented at, and concentrated at, the equator (Figure 6.11). In what appears to be an intermediate stage, the actin bundles are concentrated, but not oriented, at the equator (Figure 6.12).

The microfilamentous cores of microvilli have also been suggested as a source of contractile ring material. Szollosi (1970) demonstrated the filaments in the microvilli on the surface of a cleaving coelenterate egg and further observed that microvilli were absent from the surface near the furrow. He proposed that as the microvilli adjacent to the furrow are stretched out of existence by the contractile mechanism, their microfilament core material becomes available for incorporation in the contractile ring. In this way, the microvilli could furnish both preformed surface and contractile material. Schroeder (1981b) has countered this opinion by pointing out that in the sea urchin egg, there is no evidence that any microvilli are consumed during cleavage and that the calculated amount of actin necessary to form the contractile ring would require more microvillar core filaments than are available in the immediate vicinity of the furrow. Scanning electron microscopy has also revealed that furrows form and function equally well on cell surfaces that contain few or no microvilli (Beams and Kessel, 1976; Schroeder, 1988).

In contrast with the observations made on fixed, cleaving invertebrate eggs, injected fluorescent probes for contractile proteins have revealed changing distributions of the proteins in living, dividing cultured tissue cells (Sanger, Mittal, and Sanger, 1990). In interphase, PtK_2 cells, actin, and myosin are localized in stress fibers. At prophase, the stress fibers dissociate, and the fluorescent probes are distributed diffusely throughout the cell. At metaphase, their concentration is higher in the mitotic spindle. At the beginning of anaphase, actin and myosin are concentrated in the spindle, and 4–6 minutes later they are also visible in the equatorial cortex; the first indication of their concentration develops before any indication of contraction (Figure 6.7). Active furrowing

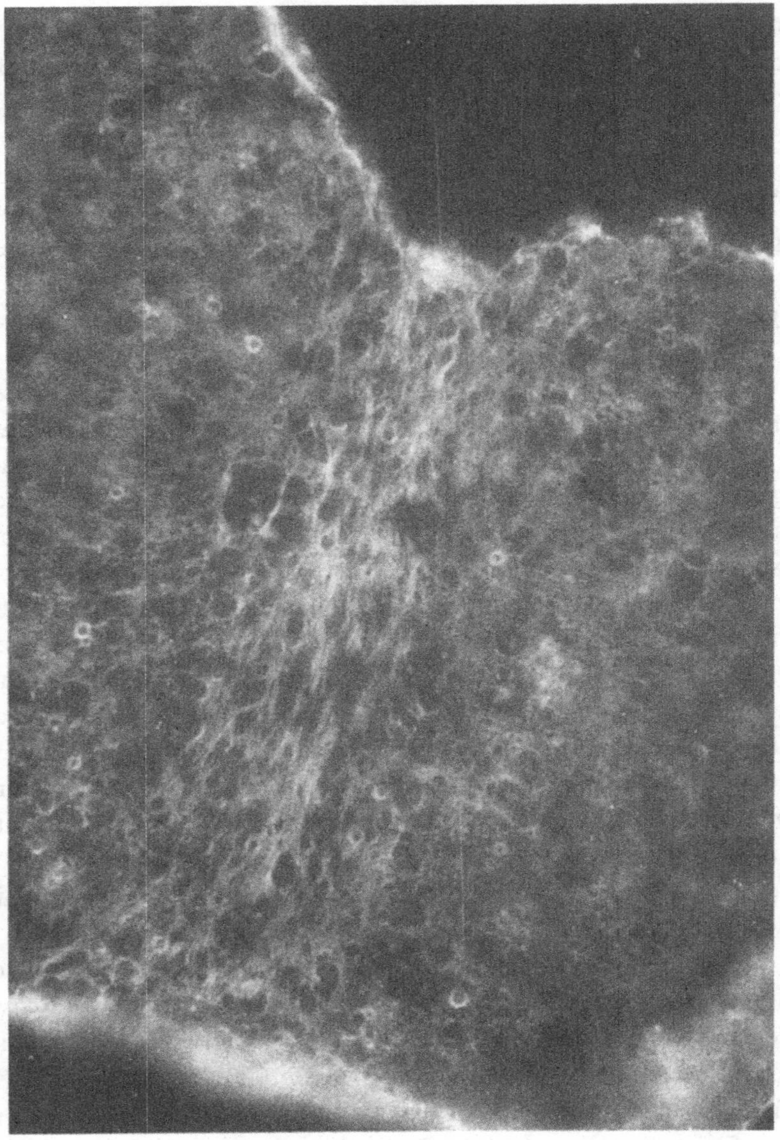

Figure 6.10 A fluorescence micrograph of an egg cortex isolated at anaphase (90 min after fertilization) and stained with actin-specific NBD-phallicidin. The fluorescent meshwork is distinct. X1610. (From Yonemura and Kinoshita, 1986, and with the permission of Academic Press)

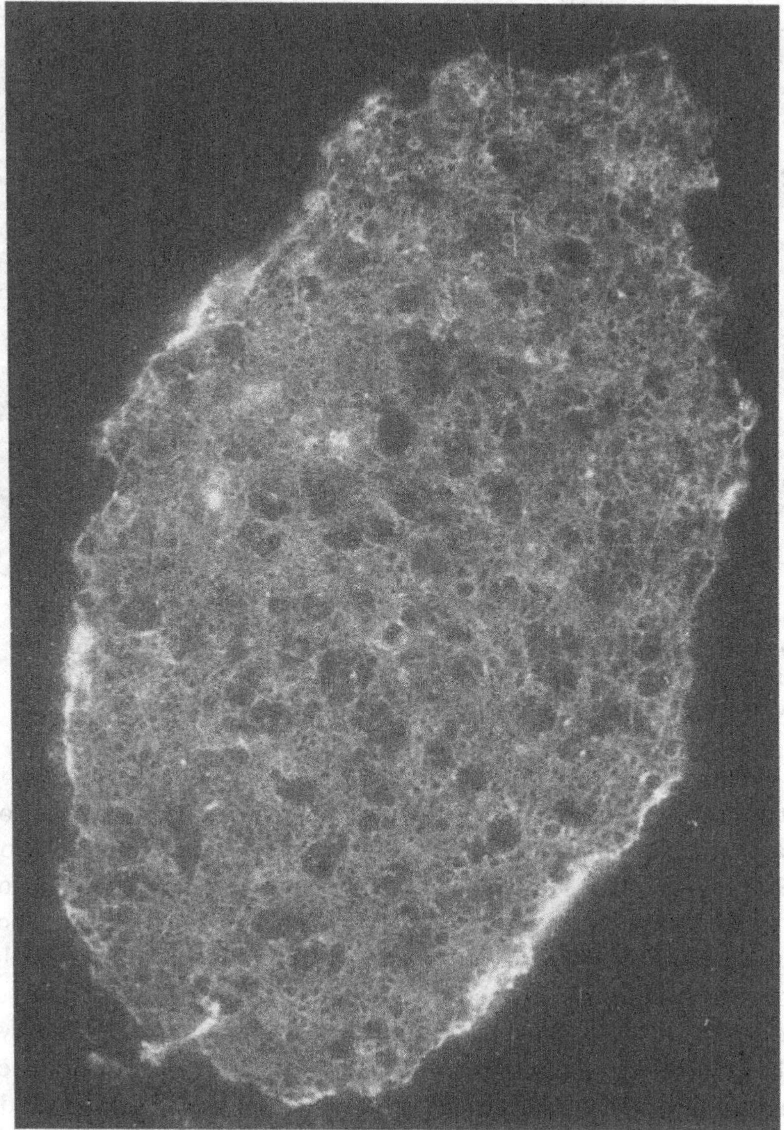

Figure 6.11 A fluorescence micrograph of sand dollar (*Clypeaster japonicus*) isolated egg cortex at telophase, stained with NBD-phallicidin. Fluorescent oriented fibers and fluorescent meshwork are distinct. X1610. Time after fertilization, 98 min. (From Yonemura and Kinoshita, 1986, and with the permission of Academic Press)

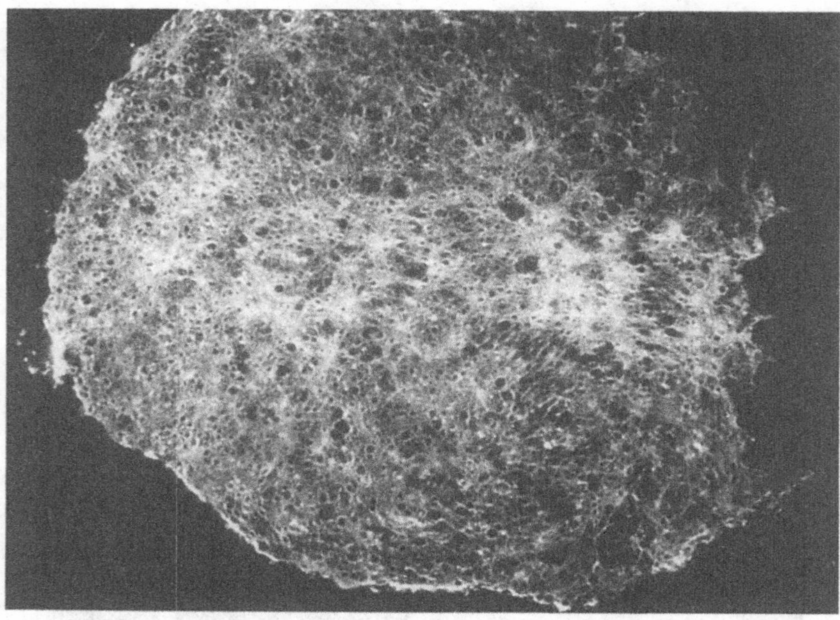

Figure 6.12 An isolated cortex stained with NBD-phallicidin showing a band of strongly fluorescent meshwork. X830. Time after fertilization, 96 min. (From Yonemura and Kinoshita, 1986, and with the permission of Academic Press)

is confined to regions where the actin and myosin probes accumulate. In PtK_2 cells, other fluorescent proteins that are not associated with contraction never concentrated in the furrow. Subsequent investigations (Sanger et al., 1994) of large PtK_2 cells emphasized similarities in constitution and organization in stress fibers and cleavage furrows. In both structures, actin and myosin can be in the form of striated fibers. Talin, which is found in the attachment plaques at the ends of stress fibers, is present as local densities sometimes arranged in rows in the furrows. The actin-binding protein filamin is incorporated in a striated pattern in organized stress fibers and has a similar organization in the fibers that form in the furrow. Stress fibers and cleavage furrows do not exist in the same cell. When the cell enters mitosis, the stress fibers disassemble and their constituent proteins become diffusely redistributed in the cytoplasm. When the contractile proteins assemble in the contractile ring, their presence in other cell regions often appears diminished. As division is completed, small rings of the proteins persist at the ends of the midbody, and stress fibers reappear. After the daughter cells separate, the normal stress fiber arrangement is reestablished. This pattern suggests that stress fibers in PtK_2 cells are important sources of the contractile proteins that cyclically develop in the cleavage furrow. Actin and myosin do not appear to have been transported from more distant regions of the cortex. They seem to have been recruited from the deeper regions to the equatorial cortex and concentrated there. PtK_2 cells are advantageous for this type of work, because they frequently remain flattened and attached to the substratum throughout all but the final stages of division, so they are relatively thin and transparent, and the mitotic events are easily observed. Comparable observations on living invertebrate eggs are hampered by their spherical form, the crowding of inclusions in the cytoplasm, and the small chromosomes.

Factors involved in assembly of the contractile ring

The striking changes that appear in the ultrastructure of the equatorial cortex may involve reorientation and redistribution of its constituents. It is unclear whether all the changes that have been described are directly related to contractile force production, and it is also unclear which of the changes are active and which are passive. Assumption that any substances that accumulate in the furrow are essential to the division process may not be warranted. The expectation that the concentration of contractile material that develops in the furrow results from transport from other regions has aroused much curiosity concerning both the source of the material and the mode of transport. Observations consistent with movement in both the cortex and the subcortical regions have been reported.

CORTICAL MOVEMENT. The location of the division mechanism immediately beneath the plasma membrane and the similarity between its composition and that of the nonfurrow periphery suggest a cortical origin for the contractile ring. The possibility that the ring may be assembled from nearby material in the equatorial region is appealingly simple but appears difficult to reconcile with an 18-fold increase in contractility that A. K. Harris (1990) considers necessary. The cortical flow hypothesis described in Chapter 2 is an attractive explanation. The phenomenon has been described in other aspects of cell biology (Bray and White, 1988), but for this discussion it is necessary to demonstrate that it is essential to cytokinesis. As Schroeder (1990, p. 83) remarked, "Just because it is theoretically possible and appealing, however, does not make it true." It will, in the first place, be necessary to determine which movements of the peripheral cytoplasm are caused by cortical flow phenomena. In studies of cortical flow in dividing animal cells, it will be essential to discriminate between movements due to the interactions among mobile force-producing units postulated by White and Borisy (1983) and movements associated with the passive shifting and stretching of the surface and cortex that occurs during the equatorial constriction and new surface formation discussed in detail in Chapter 8. It is also highly desirable to demonstrate that precleavage cortical flow is responsible for the precleavage accumulation of actin at the equator characteristic of both echinoderm eggs and tissue cells. Cao and Wang (1990b) reported that injected, labeled actin filaments moved from the deeper cytoplasm of cultured tissue cells toward the cortex during anaphase. The filaments became associated with the cortex during early cytokinesis, and they subsequently moved toward the deeper part of the furrow. Wang, Silverman, and Cao (1994) found that at the same time that the described actin filament movement occurs, fluorescent latex beads attached to the surface in the central region of cultured tissue cells shifted toward the equatorial plane in a fashion reminiscent of the movements of cortical pigment granules in precleavage *Arbacia* eggs (Scott, 1960b; Rappaport and Rappaport, 1976). Wang, Silverman, and Cao (1994) concede that the surface movement may simply reflect the network nature of the cortex and a gradient of cortical contraction that peaks at the equator, but they also suggest that it may reflect the movement of cortical actin and myosin toward that region. Such movement might be expected to result in depletion of actin and myosin in cortical regions distant from the equator, but on this point the evidence is inconclusive. Usui and Yoneda (1982) and Cao and Wang (1990a) have reported evidence of depletion in, respectively, ultrastructural studies of echinoderm eggs and fluorescence studies of tissue cells, but neither Yonemura and Kinoshita (1986) nor Schroeder and Otto (1988) detected depletion in fluorescence studies of isolated sea urchin egg cortices. Hird and White (1993) observed that, during the second cleavage of

Caenorhabditis elegans, flow of granules near the cortex occurs in the P_1 blastomere at the future posterior end of the embryo, but not in the AB blastomere at the future anterior end. The difference between the two blastomeres could be interpreted as an indication that either cortical flow is not essential for cytokinesis, or granule movement in this nematode egg is not dependent on cortical flow. Cortical flow is a generally accepted phenomenon in the life of cells (Bray and White, 1988), and techniques for direct observation of the actin cytoskeleton in thin, transparent living cells are available (Edds, 1993). But experimental demonstration of an essential role for cortical flow in division mechanism formation or function is an as yet unrealized goal.

SUBCORTICAL MOVEMENT. Contractile ring components that originate in the subcortical regions could move directly to the equatorial region or indirectly to cortex outside of the equatorial region and from there, by cortical flow, toward the equator as did the actin microfilaments injected by Cao and Wang (1990b). There are few details available about the process, but it is reasonable to suspect the existence of subcortical movement in circumstances where no cortical involvement is observable. Sanger et al. (1994) have proposed that in PtK_2 cells, components of the actin–myosin cytoskeleton are used alternately in the construction of stress fibers at interphase and in the formation of contractile ring during cytokinesis. Alternate use requires the shuttling of components between deeper cytoplasm and cortex. In support of this interpretation, Sanger et al. described the redistribution of injected myosin light chains from stress fibers to a diffuse uniform distribution in the cell, followed by recruitment to the equatorial surface where the intensity of the label increased both before and during division. Other than the narrowing of the band of labeled material in the equatorial cortex, they reported no cortical shifting comparable to that described by Cao and Wang (1990b). In a previous investigation (Sanger et al., 1989) they had reported that, after stress fiber disassembly, the diffusely distributed myosin appeared to be most concentrated in the mitotic spindle. Movement of myosin or other contractile proteins from the mitotic apparatus to the equatorial surface could also act as a stimulus mechanism for organization of the contractile ring (Rappaport, 1971b; Mabuchi, 1986).

Ultrastructural Changes at the Beginning of Cleavage

In an era of expectation that compositional, physical, and behavioral changes should be reflected in ultrastructural changes, it is surprising that diligent searching has not yet revealed undisputed ultrastructurally distinctive qualities in the equatorial surface before division begins. The problems of fixation and staining are still not trivial. Even among the eggs of a relatively homogeneous group like the sea urchins, there are pronounced differences in the clarity of the contractile ring components among the different species.

It is difficult to ignore the previously described differences in composition, behavior, and physical properties that distinguish the prefurrow equatorial surface and cortex from the rest of the cell. These differences permit speculation about the enhanced localized contraction that occurs before electron microscopy reveals specialized structure. Most cells that undergo deformation comparable to that of early elongation and very early furrowing do not develop localized contractile ring–like morphology, although alignment would increase the efficiency of the system. These circumstances are consistent with the idea that the ultrastructural qualities of the contractile ring are not absolutely essential for its function and that filament alignment may be as much an effect as a cause of contractile activity (Rappaport, 1975; Pollard et al., 1990). When contraction of the equatorial zone begins in a symmetrically dividing cell, the circumferential ring is in a sense contracting against itself and encountering resistance. When D'Haese and Komnick (1972a) clamped the ends of unstretched actomyosin threads and induced contraction with ATP, isometric contraction produced microfilament alignment comparable to that of a similar thread passively stretched to 140% of its original length. Microfilaments in cultured epidermal cells realign parallel to the direction of imposed mechanical tension within a few seconds (Kolega, 1986). In dividing cells, it is reasonable to assume that equatorial tension increases shortly before division. As the force increases, both deformation of the cell and alignment of filaments would be expected. Because deformation of the cell margin at the equator, as well as filament alignment, can result from applied tensile stress, their nearly simultaneous manifestations may result from similar force requirements. In this case, the characteristic structure of the contractile ring may indicate that the force and the dimensional changes of the contractile region have achieved a certain magnitude. Realignment would be facilitated by crosslinker turnover (Pollard et al., 1990). Inoué (1990) proposed that circumferential alignment could be caused by compression of the cortex resulting from stress extended by criss-crossing astral rays. The mechanism requires an intact, elongate mitotic apparatus firmly attached to the equatorial cortex. Mabuchi (1986) has described other alternative hypothetical mechanisms for actin filament alignment, including rearrangement of polymerized filaments by an undescribed effect of a stimulus from the mitotic apparatus; severing or depolymerization of randomly oriented filaments in the equatorial region followed by annealing or repolymerization into a parallel array of long filaments; or polymerization of previously unpolymerized actin directly in the circumferential orientation and increased concentration characteristic of the contractile ring.

Whether the contractile ring develops before or after furrowing begins depends upon how the contractile ring is defined. If it is defined anatomically as a microfilamentous region of predictable dimensions and structure in a particular relation to the plasma membrane, then it appears after contraction begins. But if it is defined by its function as a ring of contractile material, it must be considered already formed, or at least partially so, when it first expresses its special function. The period when the contractile ring is ultrastructurally demonstrable would then constitute a phase in the brief life of the division mechanism. This circumstance could explain Schroeder's (1972) estimation that the contractile ring forms in less than 20 seconds.

Contractile Ring Changes during Cytokinesis

The filaments that distinguish the contractile ring are packed roughly parallel in a relatively thin (0.1–0.2 μm), wide (5–20 μm) band located immediately beneath the plasma membrane at the base of the deepening furrow. They appear in some cells to be arranged in small bundles and sometimes appear to radiate from electron-dense plaques located on the membrane (Maupin and Pollard, 1986). The orientation of the filaments is predominantly circumferential, and they are attached to the surface or plaques by their barbed ends. Within the ring, they are oriented antiparallel in approximately equal numbers, and in tissue cells they may show a banded pattern (Sanger and Sanger, 1980) and may be localized in fibers. During division, the contractile ring's thickness does not change but, in the sea urchin egg, its width may increase somewhat until the egg is about half divided and then decrease to the original dimension for as long as the ring is present (Schroeder, 1972). The absence of ring structure in interphase cells implies that the ring has disappeared when its function has been completed, and Schroeder's measurements show that during at least half of its period of existence, its disappearance accompanies its function.

Although the visible events of cytokinesis in cleavage cells and tissue cells are strikingly similar, the ways in which final separation takes place may be different. In cleavage divisions, no remnant of the ring remains when the cells separate, and the connection is rapidly and completely broken (Pochapin, Sanger, and Sanger, 1983), but in tissue cells, concentrations of contractile protein at the former site of connection may persist for hours (Sanger et al., 1989).

Composition of the Contractile Ring

Actin

The first demonstrations (Perry, John, and Thomas, 1971; Forer and Behnke, 1972; Schroeder, 1973) that the characteristic microfilaments of the contractile ring are actin were made possible by the use of heavy meromyosin, which is a soluble fragment of muscle myosin. The heavy meromyosin complexes specifically with filamentous actin to form an ultrastructurally visible "arrowhead" pattern with a characteristic repeat structure that reveals both the nature and polarity of the filament. Heavy meromyosin does not penetrate intact cell membranes, so a preliminary step involving permeabilization with glycerol or alcohol is necessary. The method is presently unique in its ability to identify a specific contractile ring protein at the ultrastructural level. Actin is by no means confined to the furrow region, and about half of the actin in a cell is in the depolymerized state. Its distribution during division is presently under investigation. There is some controversy concerning the stability of the components of the contractile ring during its functional period. It is possible that contractile activity requires only the material that is present in the furrow at the beginning or, alternatively, that there may be a constant influx of contractile components accompanied by a return flow directed from the furrow tip toward the cell axis and thence toward the poles. The second pattern requires the disposal both of components that are brought to the equator during division, as well as of those that are already present. Cao and Wang (1990a, 1990b) injected fluorescently labeled actin into cultured tissue cells in early mitotic stages and followed its redistribution during contractile ring formation and division. They observed that labeled filaments became associated with the cell cortex and became highly concentrated in the furrow as they appeared to be depleted at the poles. Injection of fluorescently labeled actin during contractile ring formation resulted in only slight concentration in the furrow. From these results, Cao and Wang concluded that the ring is formed by preformed filaments that originate outside of the equator. A similar cortical flow of labeled actin filaments was observed during division, but they saw no poleward flow near the cell axis. This pattern of movement is consistent with a generalized cortical flow activity postulated by Bray and White (1988), but it is not completely consistent with the results of investigations in echinoderm eggs in which cortical protein distribution is determined differently. Yonemura and Kinoshita's (1986) isolated cortex preparations revealed no polar depletion of actin filaments after staining with NBD-phallicidin. Schroeder and Otto (1988) stained similar preparations with immunofluorescent stains for both actin and myosin and found no obvious depletion or gradation of cortical content outside of the furrow region, and they concluded that lateral recruitment of preexisting formed elements of the cortical cytoskeleton is unlikely.

The essential nature of actin participation in cytokinesis has been confirmed by experiments in which actin's disruption or complexing is accompanied by failure of cytokinesis. Cytochalasin B, a fungal antibiotic, in concentrations of the order of 1–5 μg/ml causes rapid disappearance of the contractile ring in many, but not all, cells, accompanied by relaxation and eventual disappearance of the already formed furrow. In susceptible cells, the rapidity of its effects, which include cessation of cleavage and disappearance of contractile ring structure, and the absence of changes in the mitotic apparatus, support the idea that it affects the division mechanism, rather than an event leading to formation of the mechanism. The cytochalasins affect filamentous actin by binding to the barbed end and preventing polymerization there (reviewed, Mabuchi, 1986). The barbed end has higher association and dissociation constants than the slower growing pointed end (Pollard and Mooseker, 1981) and, at steady state, net addition at the barbed end can be balanced by net loss at the pointed end. Reduced binding at the barbed end results in net depolymerization at the pointed end. Mabuchi (1986) questioned whether this circumstance fully explains the rapid disappearance of the contractile ring that accompanies cytochalasin treatment and suggested that, because filamentous actin attaches to the plasma membrane at the barbed end, detachment may also be an important consequence of treatment.

The effect of cytochalasins on dividing cultured mammalian cells is less clear-cut than the effect on cleaving invertebrate eggs. In similar concentrations of the inhibitor, the furrow forms in mouse fibroblasts and constriction progresses, but it is not completed. Eventually, the furrow relaxes, and the cell remains binucleate. The process may be repeated for several cycles, which produces multinucleate large cells (Krishan, 1971). The effect resembles that which occurs in cleaving invertebrate eggs subjected to chronic high temperatures, chemical treatment, or physical manipulation that interferes with either the normal establishment or the normal function of the division mechanism. The response of cultured mammalian cells to cytochalasins requires further investigation, but there are several simple possible explanations. If cytochalasins simply slowed the furrow process so that it could not be completed during the period when the surface is competent to engage in constriction, then the effect would be incomplete division similar to that produced in sand dollar eggs when the equatorial surface was stretched so that its length was greater than a normal furrow could traverse in its normal functional period (Rappaport and Rappaport, 1993). Cytochalasins also inhibit cell locomotion, and it may be that in cultured tissue cells, final separation or detachment results from shape changes or from locomotion of the sister cells. Even in cleaving invertebrate eggs, failure to sever persistent connections that are barely visible with the light microscope is followed by complete regression of the furrow.

Another selective inhibitor of normal filamentous actin activity is phalloidin, a peptide from the poisonous mushroom *Amanita phalloides*. Following injection of phalloidin into cleaving sand dollar eggs, division abruptly ceases and the partially divided cell holds its shape for an extended period (Y. Hamaguchi and Mabuchi, 1982). The mechanism of phalloidin inhibition is not well understood. Mabuchi (1986) suggested that it may prevent normal interaction between actin and other proteins, or that it may operate by blocking depolymerization during cleavage.

Actin in animal cells has been characterized by Satterwhite and Pollard (1992) as "intrinsically inert" and its dynamics have been attributed to a number of associated regulatory or modulating proteins. Mabuchi (1986, 1990) has described a number of such proteins that may conceivably play a role in cytokinesis. Actin-depolymerizing proteins may be responsible for the normally large globular monomeric actin (G-actin) pool that characterizes animal cells. Profilin, for instance, binds to G-actin and prevents polymerization, and it also induces filamentous actin depolymerization. Depactin also prevents polymerization, and it also has calcium-independent filament-severing activity.

Calcium-insensitive capping proteins bind to the ends of actin filaments independently of the concentration of physiological calcium. Beta-actinin caps the pointed end; other proteins cap the barbed end. Capping proteins could determine the direction of extension of polymerizing fragments. Actolinkin caps the barbed end and could attach actin to the surface. Immunofluorescent techniques reveal that actin is concentrated in the furrow, suggesting that it may anchor the contractile ring to the equatorial surface (Mabuchi, 1990). Leukosialin (also called sialophorin), an integral membrane protein, is concentrated in the cleavage furrow through direct or indirect interaction with actin filaments (Yonemura et al., 1993), as is radixin (Sato et al., 1991).

Severing proteins can sever and cap the barbed end of actin fragments in the presence of calcium. Fragmin and gelsolin are examples of this category of actin-modulating protein, as is villin, which severs actin in the presence of calcium and promotes actin bundle formation in the absence of calcium.

Crosslinking proteins consist of two polypeptide antiparallel chains that bear actin-binding domains near their ends. Alpha-actinin, spectrin, dystrophin, and filamin are included in this group. They differ in length according to the number and types of repeats in the chains.

Proteins that attach cytoskeletal elements to the plasma membrane are considered necessary for cell deformation. The best-studied linking structure is the attachment plaque that secures stress fibers to the membrane. Alpha-actinin (Pavalko and Burridge, 1991) and talin (Burridge et al., 1990) are major components of attachment plaques; stress fibers, in their constitution and organization, have much in common with the contractile ring.

This brief listing of some of the actin-regulating and actin-modulating proteins indicates the rich array of possible participants in the events that prepare filamentous actin for its normal role in cytokinesis and then modify and dispose of it as events progress. The number and variety and the potential for complex interaction suggest that discrimination between potentiality and actuality may prove challenging.

Myosin

There are several types of myosin among animal cells and many isoforms. Its location at the ultrastructural level cannot be determined as certainly as that of actin. Schroeder estimated that the amount present in the contractile ring of nonmuscle cells need be only about 1/50 of that in the myofilament of muscle cells. Its participation in division is corroborated by the presence of myosin-like structures in the contractile ring (Sanger and Sanger, 1980; Maupin and Pollard, 1986) and by its concentration in the cleavage furrow, which is demonstrable by immunofluorescent techniques. Myosin antibody also blocks division of starfish blastomeres, depending upon when in the cycle it is introduced (Mabuchi and Okuno, 1977). Injection during interphase blocks all subsequent cleavages, but it is less effective when carried out near the onset of cleavage. Injection after the beginning of cleavage does not affect the functioning furrow, but it blocks subsequent divisions. Mabuchi and Okuno (1977) point out that these results permit several interpretations of the role of myosin in division. Most likely is the possibility that the antibody binds to myosin molecules when they are in the cytoplasm so that they cannot subsequently interact with actin in the contractile ring. Failure of inhibition following injection during cleavage could result from inaccessibility of myosin after interaction with actin begins. On the other hand, the antibody could affect the furrow establishment process, rather than furrow function, by altering the architecture of the mitotic apparatus or by interfering with the signal or stimulus that is distributed by the mitotic apparatus.

Immunofluorescent methods show that shortly before and during cytokinesis, antimyosin staining of cultured tissue cells is most intense in the equatorial surface of fixed (Fujiwara and Pollard, 1976) and living cells (Sanger, Mittal, and Sanger, 1990). Changes in myosin distribution at first appeared to parallel closely the contemporaneous changes in actin distribution, but more recent investigations revealed a clear specificity in the handling of different myosin isoforms during cytokinesis. Cultured chick cardiac myocytes contain myosin-enriched sarcomeres, stress fibers, and, during cytokinesis, contractile rings (A. H. Conrad, Clark, and Conrad, 1991). Localization of cardiac myosin and cytoplasmic isoforms in dividing cardiomyocytes by double-label immunofluorescence showed that the two isoform classes are not mixed in the contractile ring. Stress fibers and myofibrils are disassembled in the central region of the cell, where their components may be pooled in the dispersed state. Only cytoplasmic myosin concentrates in the equatorial region before and during division. Cardiac myosin epitopes appeared to be diffusely distributed in the central regions in early division stages. As constriction proceeded, cardiac myosin was diminished and then excluded from the furrow region, but it was visible elsewhere in the cell (Figure 6.13). These results imply a well-developed specificity in the mechanism that redistributes motility-associated proteins.

Schroeder and Otto (1988) used immunofluorescent methods to study the distribution of myosin in precleaving and cleaving isolated sea urchin cortices and found that myosin distribution at the time of nuclear membrane breakdown was reticulate and nearly coincident with that of actin. Early stages of contractile ring formation appeared as slightly enriched bands of actin and myosin. As the contractile ring developed and furrowing began, myosin and actin staining in the equatorial cortex became more intense while that in the remainder of the cortex remained relatively unchanged and uniform. Absence of any evidence of contractile protein depletion in nonfurrow cortex appears to be inconsistent with the idea that the substance of the contractile ring is recruited

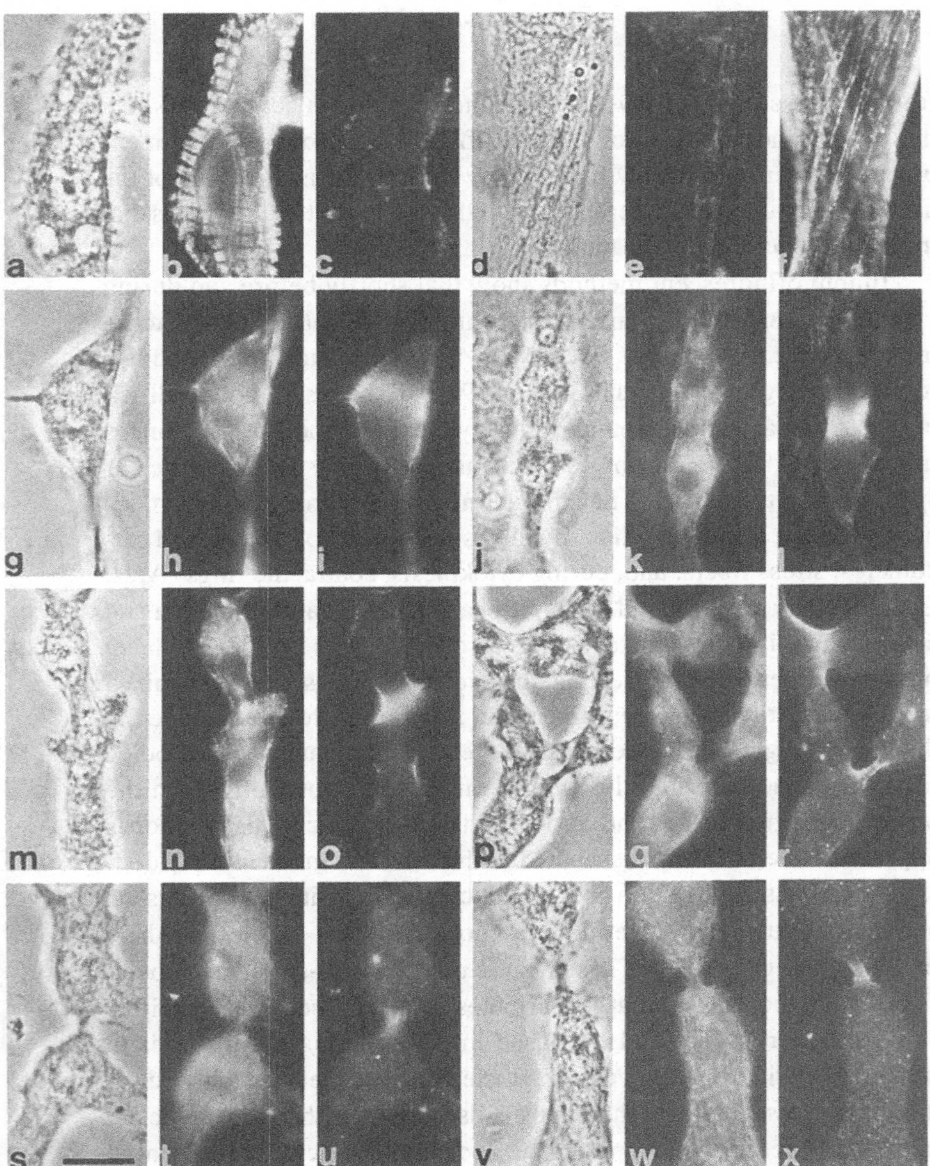

Figure 6.13 Simultaneous localization of cardiac myosin and cytoplasmic myosin isoforms in dividing cardiomyocytes visualized by immunofluorescence techniques. Antisarcomeric myosin (b, e, h, k, n, q, t, w) and anticytoplasmic myosin (c, f, i, l, o, r, u, x) are differently distributed. Individual cells are shown in sets: phase contrast – fluorescein epifluorescence – rhodamine epifluorescence: a/b/c, d/e/f, g/h/i, j/k/l, m/n/o, p/q/r, s/t/u, v/w/x. Bar = 10 μm. (From A. H. Conrad, Clark, and Conrad, 1991, and with the permission of Wiley-Liss)

by cortical flow of preformed elements that originate outside of the equatorial region (White and Borisy, 1983; Cao and Wang, 1990a).

Attachment proteins

The evidence implies that the general cell deformation that occurs during cytokinesis is caused by the contractile ring and that such a mechanism requires a means of applying the force generated in the ring to other parts of the cell. The necessary mechanical connections are demonstrable between the contractile ring and the confluent cortical cytoskeleton of the subequatorial region, and between the actin components of the ring and the overlying furrow plasma membrane. The substantial nature of the connections between the ring and the adjacent cytoskeleton are apparent in the isolated cortical preparations of Yonemura and Kinoshita (1986) and Schroeder and Otto (1988). The preparations convey the clear impression that the contractile ring is a thin, condensed part of the cortex that is distinguished by enriched contractile protein content and, after deformation begins, by distortion of the fibrous network. The suitability of this arrangement for the transmission of stresses that originate in the contractile ring is striking but not often noted. It is, after all, the cortex that is primarily responsible for the resistance to deformation of the cell.

More attention has been given to the connections between elements of the contractile ring and the immediately overlying plasma membrane. Attached actin filaments are oriented so that their barbed ends are toward the surface (Begg, Rodewald, and Rebhun, 1978), and electron-dense attachment sites positioned where actin microfilaments impinge upon the plasma membrane have been described in various cells (Maupin and Pollard, 1986; reviewed, Mabuchi, 1986). The attachment plaques that connect stress fibers to the plasma membrane appear similar to actin attachment sites and their chemistry has been investigated. Actin-binding proteins are important components of attachment plaques and two of the best studied are talin and alpha-actinin, which also bind to integral membrane proteins (Otey, Pavalko, and Burridge, 1990). Alpha-actinin is concentrated in the furrow (Fujiwara, Porter, and Pollard, 1978) and talin appears in localized densities in the presumptive and active furrow regions (Sanger et al., 1994), where it appears to bind to fibers along their lengths. The idea that the attachments between the contractile ring and the furrow plasma membrane transmit significant forces to cell regions distant from the furrow is not easy to reconcile with the fluid nature of the cell surface that was demonstrated by micromanipulation experiments (reviewed, Chambers and Chambers, 1961) and other means (Singer and Nicholson, 1972). Attachments between the membrane and the cortex might ensure that the distance between them does not change during deformation. Comparisons of the relations in different regions of the dividing cell appear necessary. Mabuchi (1986) found no regional differences in cleaving newt eggs. A ring of jellylike substance that constricts the mitochondrial bundle that surrounds the spindle occasionally appears in preparations of living grasshopper spermatocytes. It appears to be a contractile ring without an effective attachment to either the surface or the cortical cytoskeleton (Mota, 1959).

Organization and Force Production

Hypothetical cell division mechanisms have usually reflected currently prevalent models that were devised to explain other, similar cell activities. In an era when many cell activities were attributed to reversible sol–gel phenomena, Marsland (1938, 1939) first adopted Schechtman's (1937) idea that the furrow is pushed inward by local augmentation of cortical gel material, but after a particularly vivid remonstration by W. H. Lewis (1942), he (Marsland, 1950; Marsland and Landau, 1954) favored a reversible sol–gel mechanism that utilized the contractile force of gelation to divide the cell. He reconciled the limited contractility of cytoplasmic gels with the furrow's capacity to contract to the point of total disappearance by proposing a turnover of cortical gel in the furrow that involved a shifting of gel that possessed maximal contractile capacity down the furrow walls toward the base, followed by a localized demobilization at the base following contraction.

The discovery of actin-sized filaments in the furrow (Arnold, 1968; Schroeder, 1968), the demonstration that the shape of the dividing cell depended upon the integrity of the cortex and surface at the base of the furrow (Rappaport, 1966), and indications that the force generated by the division mechanism resembled that generated by an actomyosin thread (Rappaport, 1967) made an actomyosin system an attractive possibility. Earlier and contemporaneous investigations on the mechanism of striated muscle contraction (reviewed, Huxley, 1969) led to the formulation of a sliding filament hypothesis in which force production is generated by interaction between actin and myosin filaments. Although the existence of the mechanism is generally accepted, the fundamental nature of the mechanochemical activity on which it is based is not well understood (Vale, 1993). The dimensions of the filaments remain constant, and the contractile structure changes length as the parallel, overlapping units slide past each other. In this interaction, actin serves as a relatively inert track, whereas myosin acts as an enzyme that can use the energy stored in ATP to move along the actin filaments. The myosin II molecule is somewhat tadpole shaped, with two globular head domains and a thin, helically coiled tail domain. The sites for ATP hydrolysis and actin binding are in the head domain. When myosin II molecules aggregate into bipolar filaments, the tails are bundled into antiparallel arrays that form the axis of the filament, and the heads project from the sides where they mechanically interact with actin to produce motion. When the motion is combined with molecules that crosslink the actin filaments and with other molecules that link the force-producing system to the cortex or surface, the system becomes a likely mechanism for cell deformation. Janson and Taylor (1993) have proposed that actin-based gels composed of actin, myosin, and actin-binding proteins could be a component of several contractile events, including cytokinesis. Their solation–contraction hypothesis and model require little organization. The hypothesis is based upon the balance of forces that may exist in an actomyosin gel in which the force-producing interaction between actin and myosin can be stalled by the immobilization of actin by crosslinkers. Such a system could be mobilized if the network were partly disrupted by decreasing the number of crosslinks or by decreasing the length distribution of actin filaments so that the actin–myosin interaction would be released; the force would then produce motion in the form of contraction. The solation–contraction coupling hypothesis was devised as an explanation of the events that occur in the tail of a moving amoeba, but if the events occurred in the equatorial surface, the cell would be constricted.

The older hypothetical mechanism for explaining the activity at the base of the cleavage furrow follows more closely the sliding filament mechanism for striated muscle contraction. The demonstration that the circumferential filaments at the furrow base are in fact actin (Perry, John, and Thomas, 1971; Schroeder, 1973) stimulated investigations that resulted in the general acceptance of some kind of sliding filament mechanism for force production by the contractile ring. The conditions generally thought necessary for the operation of a sliding filament mechanism are present in the contractile ring. Actin filaments are roughly aligned and their polarity is about the same in both directions. The actin is connected to the surface and to the subequatorial cortical cytoskeleton, and myosin is present and apparently functional. Contractile ring, however, is not striated muscle, and many things about the furrow are not satisfactorily explained by a simple muscle model. Schroeder (1990) pointed out the complex association of the contractile ring with the plasma membrane (which must itself be changing as the circumference decreases), and the speed of its assembly and disassembly. It can also repair minor cytochalasin-induced breaks (Inoué, 1990), and its functional life can be prolonged by mechanically impeding its progress (Rappaport, 1977).

The schematic relation between molecular activity, arrangement, and force production proposed by Schroeder (1975) (Figure 6.14) established the pattern that has subsequently been embellished but not fundamentally changed as additional prospective components have been identified (Fujiwara, Porter and Pollard, 1978; Mabuchi, 1986; Pollard et al., 1990; Rappaport, 1991). The schematic relation contains the following sequence of events: (1) Filamentous actin is attached to the equatorial plasma membrane in a pattern that permits circumferential alignment of antiparallel microfilaments; (2) myosin becomes associated with antiparallel actin filaments and generates translational forces that bring their anchored ends closer together; (3) actin-crosslinking proteins spread the effect of translational forces in the equatorial region; (4) as the sites of actin attachment move closer, the diameter of the cylindrical equatorial region shrinks; and (5) as the myosin moves, the part of the actin filament that has already interacted with myosin is depolymerized so that the total volume of the contractile ring decreases with its diameter until it is completely disassembled when the sister cells are separated. This possible sequence of events nicely reconciles

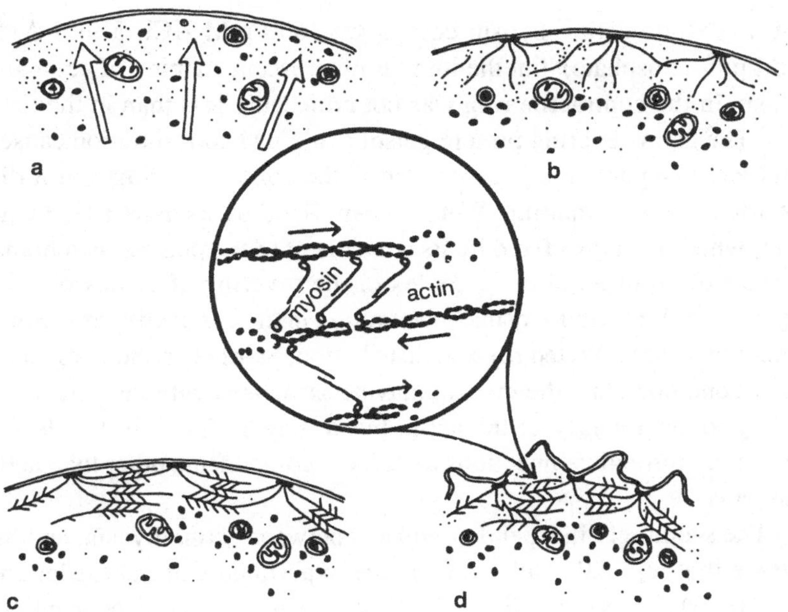

Figure 6.14 Possible sequence of events that relates molecular activity, arrangement, and force production in the cleavage furrow. (a) Mitotic apparatus triggers microfilament formation. (b) Microfilaments polymerize from the plasmalemma. Three attachment sites are shown. (c) Microfilaments of opposite polarity interact via myosin oligomeres acting as bridges; alignment results and filament sliding is initiated. (d) Sliding filaments cause contraction and the plasmalemma is thrown into folds. The inset shows bipolar myosin dimers bridging three actin microfilaments. Depolymerization of microfilaments into subunits (dots) is coupled to contraction. (From Schroeder, 1975, and with the permission of Raven Press)

structure, molecular properties, and function, but existing information permits some variation, and the proposed arrangement of myosin appears to be inconsistent with the results of experimental analysis. There is no ultrastructural evidence that myosin is directly attached to the plasma membrane, and in the model its position is fixed by its association with actin. Schroeder and Otto (1988) tested the assumption that the position of myosin is fixed by its association with actin by determining whether the normal distribution of myosin in the contractile ring requires the normal distribution of actin. The exposed, isolated cortex preparations containing the contractile ring were subjected to treatments that dissociate actin–myosin complexes (600 mM KCl, 5 mM ATP, 5mM pyrophosphate), but the immunofluorescent staining of actin and myosin in the contractile ring was not noticeably less than in the controls. The actin-severing protein gelsolin at 5 μM concentration caused complete disappearance of the actin in the contractile ring, but it did not affect myosin staining. Were myosin fixed by its association with actin, which is in turn fixed by its attachment to the plasma membrane, then the dissolution of the complex or the severing of actin would be expected to diminish myosin concentration in the contractile ring. Since neither measure affected myosin distribution, Schroeder and Otto tentatively concluded that the myosin may be associated with the cortex relatively independently of the actin, but it may be fixed to the plasma membrane through connections to cortical components other than actin filaments.

The seemingly independent linkage between actin, myosin, and the surface that logically follows from the experiments of Schroeder and Otto (1988) brings to mind how few experiments have been done on this important aspect of contractile ring biology, and how many other hypothetical constructions of the relation between structure, formation, and function currently may be possible. The following observations and conjectures are also consistent with existing information. Components of the cortical cytoskeleton can increase tension at the surface in the absence of demonstrable special organization or orientation. The tension fluctuations that normally occur during the cytoplasmic stiffness cycle in echinoderm eggs (Usui and Yoneda, 1982) and a general increase in tension associated with the nucleus and mitotic apparatus (Rappaport and Rappaport, 1988) are, if we accept cytochalasin sensitivity as a valid criterion, actin dependent. The existence of such fluctuations implies that at some phases of the cycle, the entire cortical cytoskeleton can contract, and it suggests that a significant part of the cortical actin and myosin is involved in isotropic contraction before the mitotic apparatus establishes the furrow position. According to K. Dan (1963, 1988) and Yoneda and Dan (1972), the earliest deformation requires the most energy, and since contractile ring structure is undeveloped at that time, its characteristic microfilamentous alignment may not be required when it exerts the maximum tension that is required of it. In this circumstance, it is desirable to know how the contractile ring differs from the rest of the cortex; that is, what qualities and components, if any, are unique to it. The actin–myosin interaction that appears to be ongoing in the entire cortex before furrow establishment has superimposed upon it a demonstrable shrinkage of the equatorial region that initially involves one-third to one-half of the entire cell surface (Scott, 1960b; Schroeder, 1972; Rappaport and Rappaport, 1976). This circumstance raises the possibility that some of the enhanced concentration of contraction-associated proteins may be due to the pulling together of the meshwork elements and their reorientation into a loosely circumferential pattern that also decreases the space between filaments. Both of these changes could result from the general configuration of more- and less-contractile regions (Scott, 1960b; Rappaport, 1975). It is, however, evident that some enrichment of actin and myosin takes place before meshwork reorientation and contractile ring formation (Yonemura and Kinoshita, 1986; Schroeder and Otto, 1988). The full extent of pre- and early-furrow shrinkage that normally occurs has not been satisfactorily documented. In *Arbacia punctulata,* Scott (1960b) described the shrinking region as 15 μm wide, and Schroeder (1972) found that the early contractile ring in eggs of the same species is 5 μm wide. It is not clear how much of the shrinking region becomes contractile ring, so that the actual amount of compaction is not calculable. In *Arbacia lixula,* the estimation of the amount of observable shrinkage is limited by the fact that the pigment granules are not displaceable after they contact each other. But during pigment stripe formation, the number of granules per unit area in the future stripe center more than doubles (Figure 6.5) (Rappaport and Rappaport, 1976). It follows that the number of meshwork elements per unit area is also probably doubled, so that the concentration of contractile proteins in the equatorial cortex could be enhanced independently of cortical flow from outside of the region.

The distribution of actin and myosin in the prefurrow cortex resembles that in an actomyosin thread more than that in a muscle, and in the cortex it changes from a punctate to a reticulate pattern as the first division time nears. Within the reticulum, the locations of actin and myosin become increasingly coincident, but they do not become identical (Schroeder and Otto, 1988). This nearly coincident distribution would be improbable if actin and myosin were independently linked to the plasma membrane but not also linked to each other. It might be expected if they were linked to each other, or to a third linear component of the cortical cytoskeleton. The degree of disorganization in the precleavage cortex, and the number of protein molecule species that are potentially present, suggest that surface and cortex may be linked haphazardly in a way that permits both generation of tension and force transmittal much as do the initially random components of an actomyosin thread.

The likelihood that the formation of the recognizable contractile ring is inextricably linked to its function adds a degree of simplification to the process. It may also be that its progressive reduction largely results from relatively simple consequences of its function (Chapter 8).

Energy requirements

Cytokinesis in animal cells requires energy, as does any other deformation, and determination of the necessary amount of energy was at one time considered a possible clue to the identity and efficiency of the division mechanism. Respiratory metabolism of cleaving eggs was studied by determining the effect of chemical agents on cell division and respiration, and by direct measurement of oxygen consumption or carbon dioxide production during cleavage (reviewed, Brachet, 1950; Boell, 1955). The techniques devised to determine whether cleavage is associated with rhythmic changes in oxygen consumption and carbon dioxide production were notably systematic and thorough (reviewed, Boell, 1955) but not readily confirmable. The difficulty with these approaches concerned the methods used, as they appeared to affect or measure respiratory metabolism as a whole, whereas the events of cell division require only a very small part of the energy available. Wolpert (1966) estimated the required energy from the amount of work done and also from the amount of energy necessary to stretch the surface 27%. Both approaches give values on the order of 10^{-5} ergs, which is very small compared to total energy production, because egg heat production during cleavage is 2 ergs. Boell had previously pointed out that such calculations do not take into account the efficiency of cleavage events. The problem of separating cleavage-related energy requirements from total energy production appears now to be overwhelming, and the subject is not now vigorously studied.

Formation of the division mechanism has been a logical and fruitful point of attack in the long history of attempts to analyze cytokinesis. From these investigations have come strong hints and definitive information concerning the location of the division mechanism, the sources of its components, the mechanisms of its assembly, and the biomolecular mechanisms that are employed in force production. Considering how long cytokinesis has been recognized as a major event in the life of the cell, the majority of the progress is relatively recent. Doubtless the recent introduction and perfection of new analytical methods have produced an acceleration, but it may also be that the amount of available information and its relatedness have focused investigations along lines that are more efficient and coherent than in the distant past. As attention to lower levels of organization increases, the old problems of artifact and active or passive roles will persist. As the number of potential participating proteins and their isoforms increases, visualization by itself becomes less satisfactory, and the temptation for untestable speculation will increase. Continued progress will require more ingenious combinations of experimentation and visualization.

7

The Stimulus–Response System

Mechanisms that divide cells by the action of a physically active surface require a causal link between the surface and the mitotic apparatus to account for the position and timing of furrow activity. The system requires a contraction promoter that originates from the mitotic apparatus and also requires a responsive surface. The specifics of this still-hypothetical system depend upon the nature of the proposed division mechanism. The contraction promoter is a form of stimulus, and it could have more than one possible mode of action (Greenspan, 1977). In a stable dynamic process, small causes produce small effects, and large effects are produced by an accumulation of small causes. In Bütschli's (1876) hypothesis, the additive effect of the astral rays dehydrated the equatorial surface to a greater extent than the rest of the surface and created a local increase in tension at the surface. Progressive furrowing required progressive increase in tension at the surface, which, in turn, required more dehydration. As Bütschli described the process, the continuing presence of the mitotic apparatus was required for completion, because each increment of tension increase required an increment of dehydration. In contrast, an unstable dynamic process requires only a relatively small perturbation to set in motion a system that is constituted in a way that amplifies the effect of the trigger. For example, the hypothetical cortical flow mechanism that might account for accumulation of mobile force-producing units at the equator requires only that the units move up a tension gradient and that such a gradient, no matter how slight, exists. After the flow begins, the local accumulation of the units increases the gradient, and the process is self-amplifying.

A stimulus could activate the division process by providing a missing component, by providing more contractile material, or by removing a component that blocks the process. Analysis of stimulus action has been hampered by several circumstances. The stimulus presently lacks an ultrastructural basis, and its immediate effect upon the surface that it designates as the future furrow is undetectable. Although likely possibilities may be imagined, we have no inkling of the actual chemical nature of the stimulus, and we must contend with the probability that it is not a substance that develops only at the time of division. It also appears probable that in most cells it is not restricted to any particular region, but it is merely concentrated at the equator. At present, a stimulus can only be judged effective on the basis of its effect on responsive surface, and the responsiveness of the surface can only be judged on the basis of its response to an effective stimulus. In the same way that it was possible to learn about the division mechanism by analyzing the geometrical requirements for setting it in motion, it has been possible to learn more about stimulation from studying the nature of the response, and to learn more of the nature of the response by studying the stimulus.

Modes of Furrow Establishment

Cells have two mechanisms for establishing furrows. In the direct mode, the furrow forms in direct response to the mitotic apparatus. Typically, cells that use only the direct mode have a relatively large, central mitotic apparatus and their furrows are symmetrical. Cells with a relatively small, eccentric mitotic apparatus typically also use a second, indirect mode in which the active tips of the directly established, relatively short furrow extend around the circumference independently of the mitotic apparatus. Yatsu's early (1912a) experiments not only demonstrated that the mitotic apparatus played no mechanical role in cleavage, but also showed that a part of the furrow was established independently of the mitotic apparatus. He cut the 1 mm *Beroë* egg roughly in half, perpendicular to the plane of the unilateral furrow, so that in some cases the furrow tip, or cleavage head, was isolated from the mitotic apparatus and, in other cases, it was included with the mitotic apparatus (Figure 3.2). When the cleavage head was included in the nucleated part, the unnucleated part did not divide, indicating that the mitotic apparatus had no significant effect on the vegetal region at the time it established the furrow directly in the animal region (Figure 3.2c–d). When the transverse cut was positioned so that the cleavage head was in the unnucleated part, furrowing continued and the part was divided permanently (Figure 3.2a–b). The first operation demonstrated that the vegetal part lacked any capacity for division when it was separated from the mitotic apparatus and the cleavage head. The second operation demonstrated that the cleavage head alone transferred the capacity to divide to the vegetal region. At present, this experiment is thought to have demonstrated that the ends of the advancing furrow converted nonfurrow surface to furrow as it progressed through a region that would otherwise have been unable to divide. Yatsu recognized that his results were inconsistent with Rhumbler's traction fiber hypothesis, but he was equally skeptical about Ziegler's (1903) suggestion that cleavage is accomplished by the contraction of a meridional ectoplasmic thickening. The amphibian egg furrow is also capable of extension by propagation at its tips, and results of recent investigations imply that microtubules are involved in the process. Sawai (1992) injected microtubule poisons (colchicine, vinblastine, nocodazole) around the small initial furrow or under the advancing furrow tip in the animal hemisphere of cleaving newt eggs. In both regions, furrow progress ceased, and either cleavage was incomplete or the furrow then regressed. Furrow-inducing cytoplasm from normal blastomeres was rendered ineffective by being mixed with microtubule poisons. Kubota and Sakamoto (1993) demonstrated similar susceptibility to colchicine in vegetal hemisphere furrows of *Xenopus* eggs.

All cells use the direct mode for establishing furrow; whether they also use the indirect mode appears to depend upon the geometrical circumstances, but all cells may have the capacity to use the indirect mode, whether or not it is normally necessary (Rappaport and Conrad, 1963). When a coelenterate egg that normally cleaves unilaterally is trimmed so that the mitotic apparatus is centered in a small volume of cytoplasm, the furrow is symmetrical (Figure 5.10). When the normally central mitotic apparatus of echinoderm eggs is pushed close to one part of the cell margin, cleavage is unilateral. And when Yatsu's experiments were repeated on a unilaterally cleaving sand dollar egg, the results were the same as those obtained in the studies of *Beroë* eggs: The unnucleated part did not cleave when it was separated from the furrow tip (Figure 7.1), but it cleaved when it included the tip (Figure 7.2). The ability of

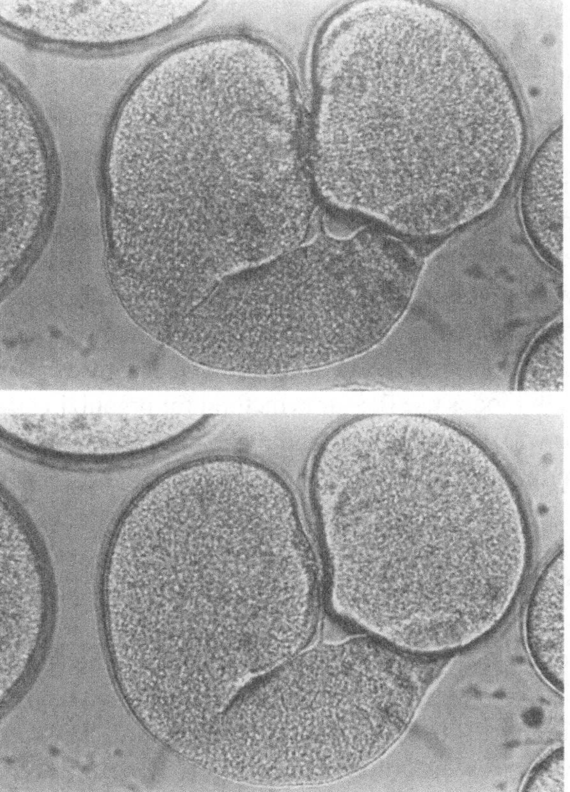

Figure 7.1 Sand dollar egg with unilateral furrow: absence of cleavage in area isolated from base of furrow. (From Rappaport and Conrad, 1963, and with the permission of Wiley-Liss)

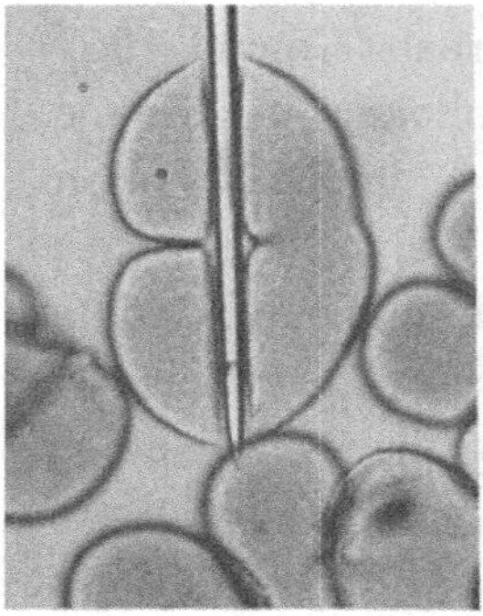 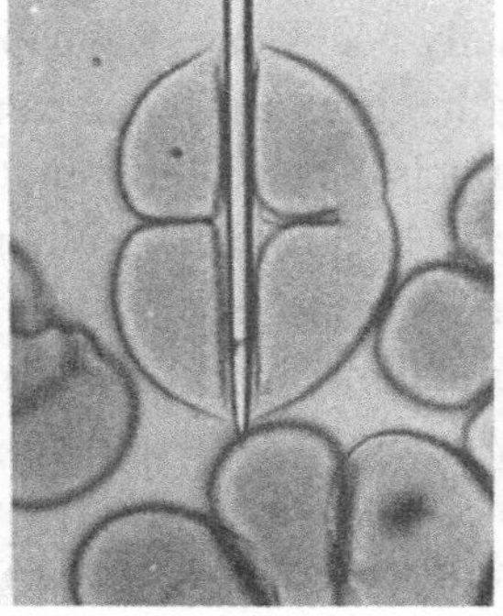 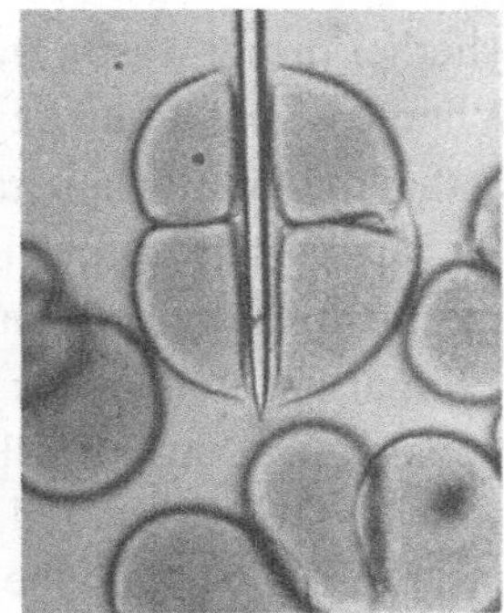

Figure 7.2 Sand dollar egg with unilateral furrow: progress of the furrow following separation of the base of the furrow from the telophase nuclei, which are isolated in the upper portions of the cell. (From Rappaport and Conrad, 1963, and with the permission of Wiley-Liss)

the coelenterate egg to form symmetrical furrows in altered circumstances is not surprising, because the alteration creates the same geometrical relation that normally exists later in development when cells no longer divide unilaterally. But normal sand dollar cells never encounter circumstances that result in unilateral furrows. It appears that both the tip of an advancing furrow and the mitotic apparatus have in common a capacity to convert ordinary surface to specialized surface with enhanced contractile activity. The furrowing activities resulting from the two modes appear similar, as do their ultrastructures (Arnold, 1969). Whether both modes use the same method for organizing the contractile material, or whether the cortex can make the same response to two different stimuli, remains to be determined.

The Rate of Stimulus Movement

The position of the mitotic apparatus relative to the surface is usually fixed long before the time of division; the furrow, however, is not established until shortly before it begins to function. The minimum time required for furrow establishment is brief. These facts imply that the mitotic apparatus is not specifically affecting the equatorial surface during the whole time that the apparatus is visible, and they suggest that at some point a signal that specifies the furrow position passes from the

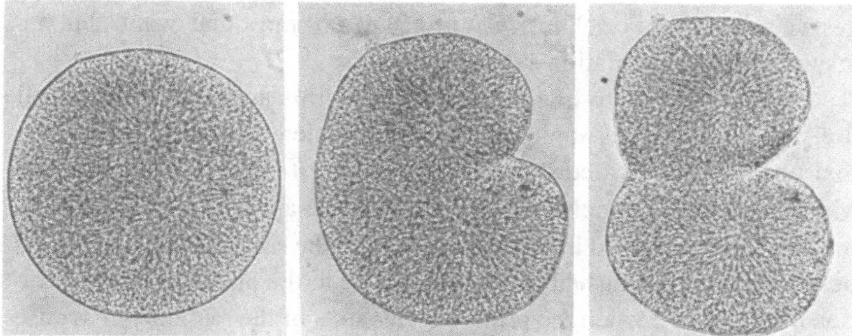

Figure 7.3 First cleavage of flattened sand dollar egg with eccentric mitotic apparatus. Left: Immediately before cleavage. Center: 5 min later, unilateral furrow in equatorial margin closest to the mitotic apparatus. Right: 2 min after center photograph, well-developed furrow in more distant margin. (From Rappaport, 1982, and with the permission of Wiley-Liss)

mitotic apparatus to the surface. The rate of signal movement has been estimated by flattening a spherical sand dollar egg in a way that holds the mitotic apparatus in an eccentric position and then comparing the differences between the distances from the mitotic axis to the two diametrically opposite parts of the equatorial cell margin with the differences between the times at which the furrow appears on the opposite margins. The furrow appears later in the margin more distant from the mitotic axis (Figure 7.3). The difference between the times when the furrow appears in the near and distant margins is directly proportional to the difference between the distances from the near and distant margins to the mitotic axis. The rate at which the stimulus traverses the additional distance is between 6 and 7 μm/minute (Rappaport, 1973, 1982). This rate approximates that of the elongation of microtubules (Swann, 1951; Aronson and Inoué, 1970; Cassimeris, Pryer, and Salmon, 1988).

How Long Does the Stimulus–Response System Remain Active?

When the furrow is established directly, the minimum period of interaction necessary to produce a complete furrow is relatively brief and its relation to the events of mitosis is regular. This circumstance suggests a sequence of brief episodes linked to chromosomal events, but the results of experimentation suggest otherwise.

It is instructive to begin an analysis of this aspect of animal cell division with a discussion of amphibian egg cleavage, because the normal process allows deductions about some of its important aspects. Morgan's (1899) conclusion that the mitotic apparatus could not be mechanically involved in furrowing because the second cleavage begins before the first is finished is also applicable to the question of the duration of the period when the principal elements of the stimulus–response system are competent to interact. The protracted period of furrow extension and function in amphibian cleavage means that, during the interval between the beginning of the first and second divisions, nonfurrow surface is converted to furrow, so the conclusion that the surface is responsive to some form of contraction-promoting factor during the entire interphase period becomes inescapable. As the first cleavage nears completion, the cortex at the vegetal end of the incipient blastomere forms new furrow in response to established active furrow, whereas at the animal end, the cortex forms new furrow in direct response to the mitotic apparatus or diastema. There is no time in the cycle when both contraction promotion and cortical response are not taking place. Direct furrow establishment appears to be confined to the mitotic period; indirect establishment, however, never stops, so that the surface must always be competent. According to Yatsu (1912a), *Beroë forskalii* can also begin the second cleavage before the first is finished.

Most evidence concerning direct furrow establishment was obtained by experimentation on dividing echinoderm eggs. The experiments described in Chapter 5 demonstrate that the mitotic apparatus and surface can interact to produce a furrow, both before and after the time when the event normally occurs. The total period when the interacting system can produce a furrow in sand dollars is about 30 minutes, which amounts to about two-thirds of the 48-minute cycle. During the period of competence in normal cells, there are changes in position and size of the spindle, asters, and chromosomes. These observations are inconsistent with the idea that the timing of cytokinesis is closely linked to one of a sequence of visible events in the mitotic apparatus.

The formation of a furrow requires a source of contraction promoter and a mechanism that will distribute it unequally under the surface. The achromatic part of the mitotic apparatus appears capable of doing both. Up to this time, precocious furrows have been produced only 7 minutes earlier than normal, when the mitotic apparatus is well developed (Rappaport, 1975; Rappaport and Rappaport, 1993). But a generalized cortical contraction that is dependent upon the mitotic apparatus has been demonstrated beginning about 30 minutes before the first cleavage (Rappaport and Rappaport, 1988). The generalized contraction, which often begins before the mitotic apparatus is distinct or large, extends continuously between the first and second cleavages regardless of the state of the mitotic apparatus (Rappaport, 1990). This increment of surface contractility requires the presence of nuclear material. These findings imply that the contraction promoter is produced during most or all of the cycle and that the normal tension at the cell surface between divisions is attributable partly to the effect of the mitotic apparatus and partly to the cytoplasmic cycle. During interphase, the surface reveals no pattern of unequal distribution of nuclear effect. The behavior of the surface suggests that contraction promoter production does not cease, but the possibility that it may fluctuate quantitatively remains. These observations may be combined in a parsimonious conjecture about the basis of the cycle of cytokinesis in sea urchin eggs:

1. Contraction promoter production is continuous but possibly varying in amount produced.
2. The cyclical development of the mitotic apparatus changes distribution of the promoter from uniform to nonuniform, so that its effect is most intense at the equator. In effect, this idea attributes the cyclical nature of division activity primarily to the cyclical elaboration of the visible mitotic apparatus; cortical responsiveness and promoter production need not fluctuate.

How Is the Nonuniform Distribution of the Cleavage Stimulus Accomplished?

A single aster centered in a spherical cell does not produce a furrow (Hiramoto, 1971a), but a single aster centered in a cylindrical cell produces an astral constriction that looks like a furrow and may completely and permanently divide the cell (Rappaport and Rappaport, 1985a). In these two cases, which have such different results, the only variable is the shape of the cell. Asters do not behave as though they are polarized; that is, the ability to induce furrows appears to be evenly distributed over their surfaces. In both circumstances, the asters can affect the surface, and in both cases surfaces respond, but only in the cylindrical cell does furrowlike activity appear. Furrow formation implies unequal surface contractility, which, in turn, implies nonuniform aster effect. The distance from the aster center to the surface in the spherical cell is uniform, but in the cylindrical cell it is not uniform, the closest surface being in the plane of the aster center, perpendicular to the cylinder axis. Because the aster or mitotic apparatus effect has been shown to decrease with distance, it is logical to attribute the furrow in the cylindrical cell to the nonuniform distribution of contraction promoter that results from the differing distances between the aster and the surface (Rappaport and Rappaport, 1985a). However, the same simple relationship between distance and intensity of aster effect cannot explain furrow formation in a spherical cell with two asters in a normal mitotic apparatus. It was pointed out in Chapter 5 that if distance were the sole determining factor, then artificial constriction in the equatorial plane should block cleavage, because it equalizes the distance from the surface to the two aster centers and, for the same reason that a spherical cell containing a single central aster fails to divide, an artificially constricted cell with two asters should also fail to divide. The fact that it does divide in that circumstance (Rappaport, 1964; Rappaport and Rappaport, 1984) is a powerful argument against the operation of a simple distance-related mechanism. The alternative is a mechanism in which nonuniformity results from additivity of the aster effect in the region between the aster centers. In this case, the distribution of effect is not uniform over the mitotic apparatus periphery, but greatest in the equatorial region where the effect of the asters somehow accumulates.

The idea that the effect of the asters may be additive appears originally to have been associated with the apparent convergence of astral rays and spindle sheath fibers at the equator. The way that additivity was used in formulating hypothetical division mechanisms depended upon one's convictions about the basis of the radial structure. Bütschli (1876) favored the idea that the rays represented currents of fluid flowing toward the surface, which could easily accumulate under the equator. He had observed that small differences in the composition of the medium could change the tension at the cell surface. He proposed that, as the accumulation progressed, the contraction increased. The idea implied that completed division required the continuing presence of the mitotic apparatus. The reasoning behind the idea that the nonuniformity of a physiological activity is associated with the nonuniform distribution or arrangement of visible cell components seems clear, but in this case, despite continuing effort, no patent, unarguable, likely structural arrangement has been demonstrated in the right place at the right time.

The form of the mitotic apparatus in living and fixed cells has been studied and described for about 150 years, but there is still controversy about apparently simple and basic aspects, such as the length of the astral rays, their relation to the cortex, and their cyclical changes (see Chapter 1). The rays extend from the centrosomal area, and in echinoderm eggs they appear at times to extend into the entire cytoplasm. The amount of precursor material available in cells is sufficient for many more than the two asters that are normally formed at one time. The length of the rays is affected by the space they occupy. When the normally central mitotic apparatus of an echinoderm egg is pushed into an eccentric position, the rays that extend toward the farthest equatorial margin are longer than they could possibly be when the mitotic apparatus is central (E. B. Harvey, 1935) (Figure 7.4). Light microscope studies have created the strong impression that some of the rays cross the equatorial plane, which implies that rays from two different asters can enter the same region. K. Dan (1943b) reviewed the older literature in which crossing rays were described, and more recently he published pictures of such rays in living cells (K. Dan, 1988). Recent investigations of fluorescently stained microtubules and microtubule-bound protein also show crossing rays (Henson et al., 1989; Hosoya et al., 1990). The impression of relative stability that may result from observation of fixed or even living cells is deceptive. Fry (1929) described the rays as "coarse" at some times during the mitotic cycle and "delicate" at other times. Although the microtubules that make up different regions of the mitotic apparatus appear to be indistinguishable, immunofluorescent studies have revealed heterogeneity. Oka, Arai, and Hamaguchi (1990)

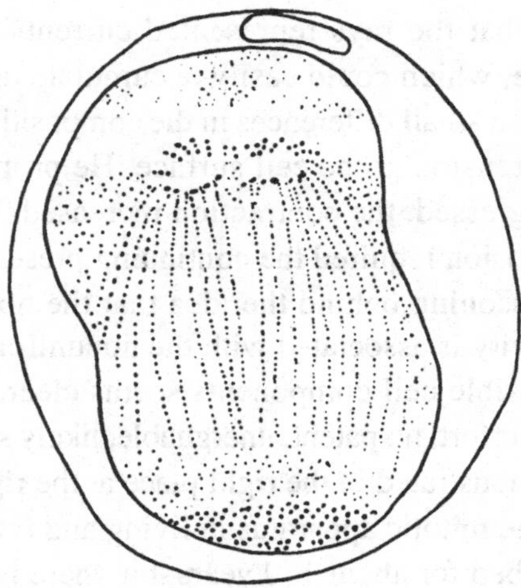

Figure 7.4 *Parechinus microtuberculatus*, mitotic figure shifted to centrifugal pole. Rays shortened toward centrifugal pole, elongate toward centripetal pole. (From E. B. Harvey, 1935)

demonstrated that, in sea urchin eggs, ∂-tubulin isotypes are localized mainly in spindle microtubules, whereas ß-tubulin isotypes are restricted to aster microtubules. The tubulin dimers that polymerize to form the microtubular component of the aster rays and spindle fibers are rapidly exchanged with those in the surrounding medium (Salmon et al., 1984), and microtubules rapidly collapse and reform (Cassimeris, Pryer, and Salmon, 1988).

Electron microscope studies have shown that the rays that can be seen with the light microscope are based on microtubules associated with aligned vesicles and endoplasmic reticulum (Rebhun and Sander, 1971; P. Harris, 1975). In a careful study of the mitotic apparatus microtubules in precleavage sea urchin eggs, Asnes and Schroeder (1979) concluded that the incidence of crossing rays was more limited than seemed apparent in the light microscope. They found that rays extend from the aster centers into the yolky cytoplasm at prophase and then shorten to occupy the yolk-free zone around the aster centers at metaphase. At early anaphase they extend into the yolky cytoplasm, and at late anaphase some microtubules radiate far into the cytoplasm, to within a few microns of the surface. At telophase they penetrate to the cortex of the poles and equator. Crossing rays are confined to the region close to the spindle axis at metaphase. At anaphase and early telophase, Asnes and Schroeder found no evidence of crossing any place and interpreted the astral configurations at that time as mutually exclusive hemispherical arrays. They compared the number of rays that enter the cortex in the polar and equatorial regions by counting microtubule profiles in sections cut 5 μm below the surface, parallel to a tangent to the two regions. They found no microtubules in the equatorial sections during the early period of cleavage, including the time when stimulation normally occurs, and only a few microtubules in the polar sections. Later, the number of profiles in sections made through both regions increased. The number at the poles always exceeded the number at the equator. At issue in this investigation was the question whether the demonstrable ultrastructural relation between the microtubules and the cortex is consistent with either equatorial stimulation or polar stimulation. If it is assumed that the intensity of mitotic apparatus–surface interaction is proportional to the number of microtubule profiles within 5 μm of the surface, the idea of equatorial stimulation is not supported and, as Asnes and Schroeder remark (1979, p. 338), neither is a mechanism of microtubule-mediated stimulation at the poles, "because significant penetrations of the polar cortex by microtubules occur only after the time of cleavage stimulation." In a sense, this approach to a better understanding of the stimulation phase of the process produced inconclusive results. It should also be pointed out that, although dynamic changes in microtubule length during mitosis encourage focusing attention on the circumstances that exist at the time that stimulation normally occurs, the cell's ability to divide is not restricted to either that particular set of geometrical relations or that particular time. The mitotic apparatus is effective over an extended period, during which the length of its rays and the incidence of their crossing varies. The cell can also divide when deformed in ways that would be expected to greatly modify the numbers of microtubules that approach within 5 μm of the polar and equatorial surfaces. Perhaps the best that can be said is that a mechanism that depends upon precise regional differences in the number of surface–ray contacts (Rappaport, 1965) does not now appear likely. Although the astral rays need not cross for their effects to be additive, it is interesting that attempts to verify and analyze the observations that originally prompted the hypotheses have been so inconclusive.

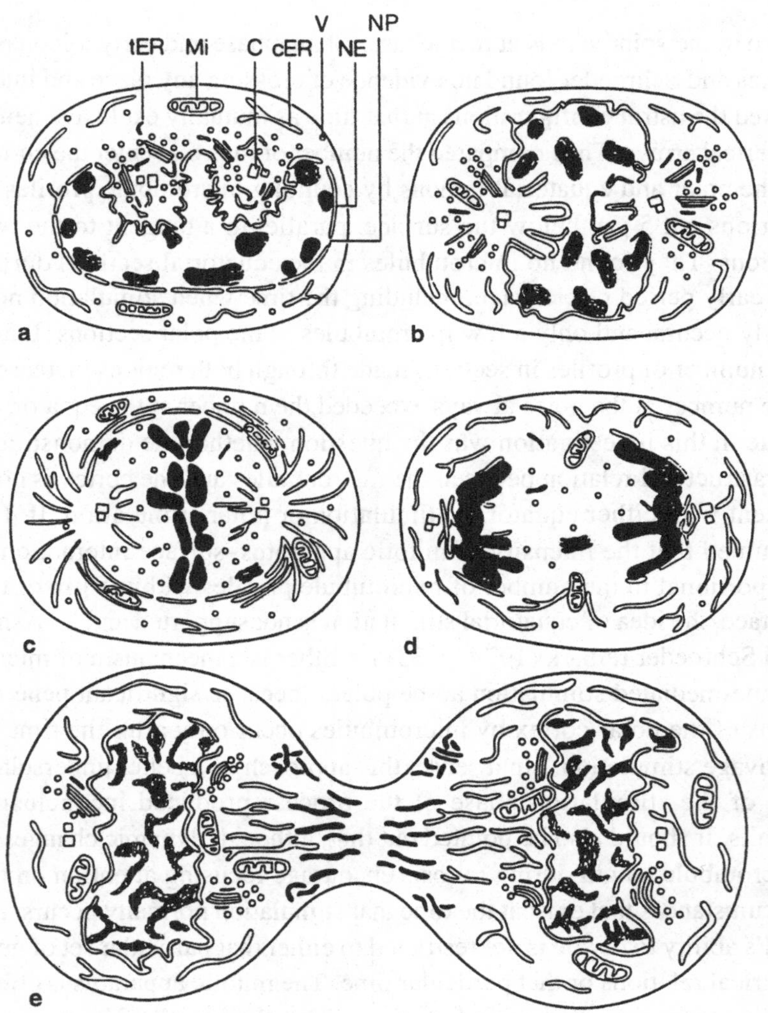

Figure 7.5 Schematic diagrams of the most prominent membranous structures of HeLa cells during some characteristic stages of mitosis. (a) Middle prophase. (b) Beginning of premetaphase. (c) Metaphase. (d) Middle or late anaphase. (e) Telophase. C = centrioles; cER = cisternae of the endoplasmic reticulum; tER = tubules of the endoplasmic reticulum; V = vesicles; Mi = mitochondria;.NE = nuclear envelope; NP = nuclear pores. (From Moll and Paweletz, 1980, and with the permission of the *European Journal of Cell Biology*)

The other component of the visible aster rays, the massive system of aligned membranes and vesicles, has been eclipsed until recently by the microtubules with their intriguing structure and chemistry. Even though protocols that preserve microtubules well are often less favorable to membranes and vesicles, most ultrastructural studies of the mitotic apparatus show the presence of the mitotic apparatus–associated vesicular elements variously called "endoplasmic reticulum," "smooth endoplasmic reticulum," or "Golgi bodies" (reviewed, P. Harris, 1975). The mitotic elongation of aster rays involves extension of both microtubules and vesicular arrays (P. Harris, 1975). As techniques for fixation and staining were modified, the magnitude of the membranous component in the ultrastructure of the mitotic apparatus was realized. During mitosis, the arrangement of the membranous component changes in parallel with the changes in the microtubular component (reviewed, Hepler and Wolniak, 1984). Lamellar, tubular, and vesicular elements of the endoplasmic reticulum have been described and followed in electron microscope studies of dividing HeLa cells (Moll and Paweletz, 1980; Paweletz, 1981). During prophase, the endoplasmic reticulum accumulates around the centrioles, as P. Harris (1962) had previously observed in sea urchin eggs. When the nuclear envelope is fragmented at the beginning of prometaphase, many of the pieces contribute to the population of endoplasmic reticulum membranes (Hepler and Wolniak, 1984) and may be included within the mitotic apparatus. In HeLa cells, Moll and Paweletz found flat cisternae near the asters, tubular elements near the spindle periphery, and vesicles inside the spindle. They estimated that the membranous structures increase from prophase to metaphase and then decrease. The endoplasmic reticulum forms various associations with the fibrous components of the mitotic apparatus (Paweletz and Finze, 1981) (Figure 7.5). The spindle is often ensheathed by concentric layers of endoplasmic reticulum, a condition that appears especially developed in the eggs of insects (Kubai, 1982). Within the asters, the membranes tend to parallel the radial microtubules. In anaphase and telophase, the concentric and radial arrangements dwindle, and in telophase, some of the endoplasmic reticulum associated with chromosomes is reconverted to nuclear envelope. After division of HeLa cells begins, the tubular elements and cisternae are oriented parallel to the mitotic axis and located between the separating chromosome groups. The tubules are in the equatorial plane, and the cisternae lie between them and the chromosomes (Moll and Paweletz, 1980).

Dynamic activity of the living endoplasmic reticulum in dividing sea urchin eggs was revealed by injecting dicarbocyanine, a fluorescent dye that spreads through the endoplasmic reticulum, but not through other organelles (Terasaki and Jaffe, 1991). Both the internal cisternae and the cortical tubular network were stained, and the dye diffusion pattern supported the idea that the endoplasmic reticulum is a continuous, interconnected compartment. The association between the cisternal endoplasmic reticulum and microtubular elements in the asters and the spindle was evident throughout the mitotic cycle. Most striking were the accumulations of fluorescent membranes around the aster centers, which, the authors point out, may be the basis of the "lake" characterized by a paucity of granular components described in the older literature. A calsequestrin-like protein located in the endoplasmic reticulum revealed in more detail the extension of endoplasmic reticulum radially along the astral rays, and confocal microscopy of antitubulin influorescence of the mitotic apparatus showed astral rays crossing in the equatorial region near the surface at anaphase (Figure 7.6).

The technique of stratifying the contents of living cells and eggs has been used for many years (reviewed, E. B. Harvey, 1956), and stratification provides a way to redistribute or remove the major part of the endoplasmic reticulum of sea urchin eggs after centrifugation at 3,000 x G for as little as 2 minutes. The visible contents of *Arbacia* eggs are rearranged into layers according to density (E. B. Harvey, 1956) (Figure 7.7). Harvey implied that the stratification sorted the contents to the extent that there was no mixture between layers, but Anderson (1970) showed that, except for oil and pigment, each zone or layer contained scattered representatives of adjacent zones. The "clear zone" of the stratified egg contains primarily endoplasmic reticulum, and when the distribution of the calsequestrin-like protein was determined by immunofluorescence, the protein was found primarily in the clear zone, with a sparsely distributed network in the rest of the centrifuged egg (Henson et al., 1989). Stratified eggs can be subdivided by additional centrifugation, or they can be subdivided with greater precision by cutting with the side of a needle. Rustad, Yuyama, and Rustad (1970) cut

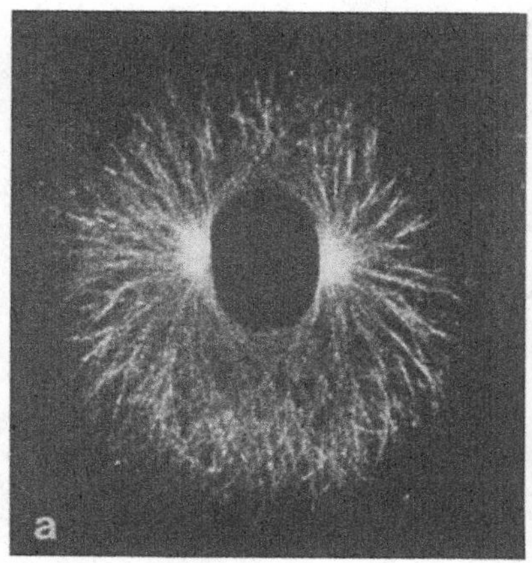

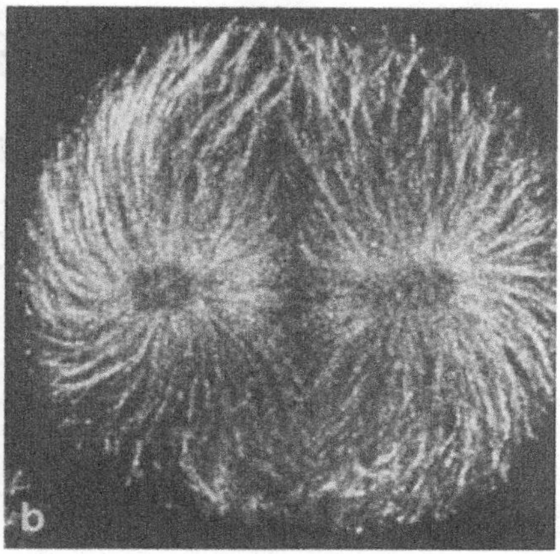

Figure 7.6 Confocal microscopy of antitubulin immunofluorescence of sea urchin embryos undergoing mitosis. (a) Prophase embryo showing microtubules concentrated in the early astral centers and extending out toward the cortex. (b) Anaphase embryo revealing the extension of astral ray microtubules from the astral centers to the submembranous region. Telophase embryo showing microtubules of the residual asters and the midbody. (From Henson et al., 1989, reproduced from the *Journal of Cell Biology*, 109:156, by copyright permission of the Rockefeller University Press)

the portion of the egg containing the clear zone, oil drop, and nucleus from the rest of the unfertilized stratified egg and then fertilized both halves (Figure 7.7). The ability of the egg portions to divide was not related to the presence or absence of the clear zone. Portions containing only yolk and pigment divided after some delay. When the mitochondrial layer was included, the timing was nearly normal. The important inference of the cutting experiments is that egg fragments can divide in the absence of the great majority of the endoplasmic reticulum.

In the cells commonly used in morphological and experimental studies of cytokinesis, there is no clear correlation between the distribution of visible subsurface components (other than the structural framework of the mitotic apparatus) and the position of the furrow. Although differences in appearance of the cytoplasm in the equatorial plane of these cells during the prefurrow period were described, the observations subsequently proved to be unrepeatable or artifactual. Redistribution or removal of the centrifugable cytoplasmic elements does not drastically affect the cells' ability to divide. The situation in amphibian and

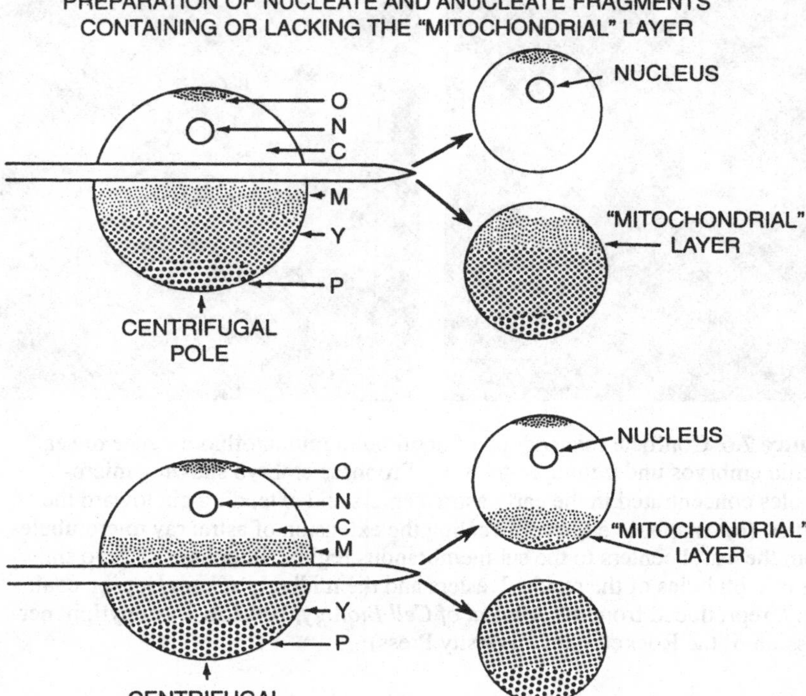

Figure 7.7 Distribution of visible cytoplasmic inclusions of an *Arbacia* egg after stratification by centrifugation. Nucleate fragments lacking or containing the "mitochondrial layer" are formed when the microneedle passes on the centripetal or centrifugal side of the layer, respectively. O = oil cap; N = nucleus; C = clear zone; M = "mitochondrial layer"; Y = yolk granules; P = pigment granules. (From Rustad et al., 1970, and with the permission of the *Biological Bulletin*)

sturgeon eggs appears to be different. The diastema described in the early studies of cleavage in these forms was recognizable because it contained fewer stainable yolk granules than the rest of the subcortical cytoplasm. Because the visible characteristics of the diastema are confined to the equatorial plane, its existence must result from a localized rearrangement of the cytoplasmic components so that stainable elements are either directly removed from the region or displaced by unstainable elements that are transported to the region. Both the timing

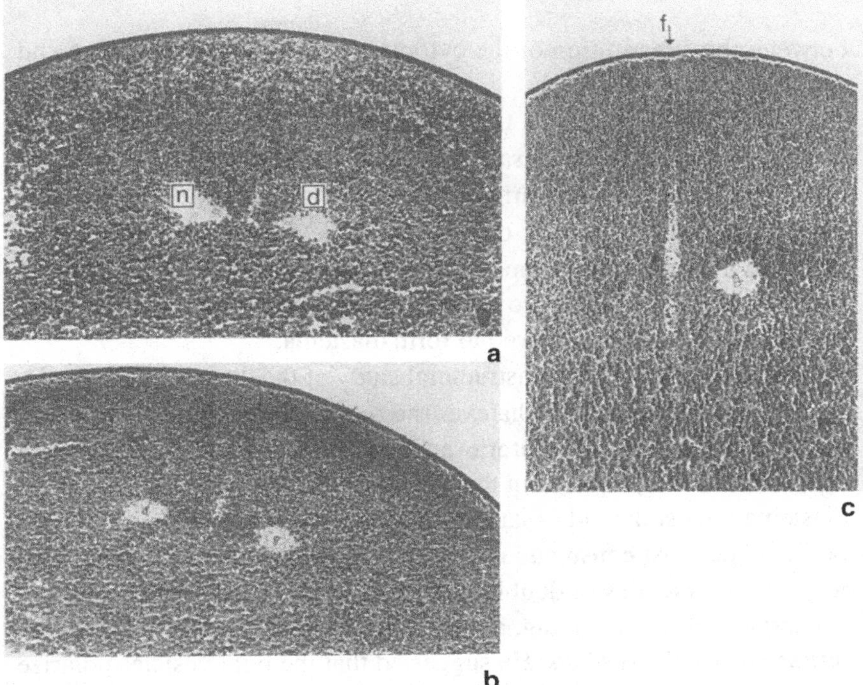

Figure 7.8 *Cynops* eggs (a) about 25 min and (b) 10 min before cleavage, and (c) just at cleavage. d = diastema; f = furrow; n = nucleus. (From Sawai and Yomota, 1990, and with the permission of the New York Academy of Sciences)

and the position of diastema formation are closely correlated with the timing and position of the mitotic apparatus events (Figure 7.8). The possibility of a physical role for the diastema in cytokinesis now appears unlikely, but the possibility that it may serve as a stimulatory intermediary between the mitotic apparatus and the cortex remains and has some experimental support (Zotin, 1964; Selman, 1982; Sawai and Yomota, 1990) (see Chapter 5). The diastema is usually described only in the first few cleavages and is not visible in later dividing blastomeres. The transition from cleavage mechanisms that require the diastema to somatic cell mechanisms that do not require the diastema is smoothly made, and there is cause to wonder whether the mechanisms are vastly different. It is possible that the density of the yolk granules and the great distance between the mitotic apparatus and the surface that exist in the earliest cleavages simply enhance the differences that normally exist between the constitution of the cytoplasm in the equatorial plane and elsewhere.

Zotin (1964) found that the fragmentation of the diastema that follows a brief immersion of salamander eggs in heavy water results in the formation of multiple furrowlike indentations in the animal surface, but the relation between the orientation of these furrows and the orientation of the diastema fragments was not described. He also reported formation of multiple furrows in similarly treated sea urchin eggs, which are not usually believed to form diastema.

Bluemink's (1970) ultrastructural study of the diastema in the salamander egg showed that whatever the role of the diastema may eventually prove to be, the mitotic apparatus can rearrange the visible elements of the cytoplasm in the equatorial plane. Bluemink found the diastema almost devoid of large inclusions like yolk granules and lipid bodies. The most conspicuous elements of the diastema territory were long, closed profiles of double-stranded membranes, which he termed "cisternae." Accompanying, and presumably originating from, the cisternae were light vesicles. He suggested that the light vesicles polarize and organize the filaments in the contractile ring and that they may serve as vehicles for calcium transport.

Degree of Structural Normality Required for Effective Operation

Normal activity and morphology usually coincide, but study of the normal state alone can fail to reveal which part of the normal configuration and activity is essential for a particular function. Imposition of abnormal conditions that do not prevent cleavage reveals that many of the circumstances that obtain in normal cells are not necessary for cytokinesis.

The surface

The tendency of detached cells to round up, especially before cleavage, gives the impression that their shape is determined solely by simple surface forces. But in cleaving sea urchin eggs, Isaeva and Presnov (1983) have shown that a kind of plastification takes place within 9 minutes of fertilization. When they reshaped eggs into discs during that period, they found that the flattening persisted after the eggs had been released and it continued through several divisions. Although the cleavage pattern was abnormal, development from gastrulation onward was normal. The egg's ability to retain the imposed shape during the cleavage period implies that some part of the cortical cytoskeleton became sufficiently less pliant during the 9-minute period so that the cyclically varying isotropic forces that exist within it were unable to reshape it into a least-surface figure. The anisotropic force exerted in the furrow must normally overcome not only the isotropic force in the rest of the cell surface, but also an added rigidity that tends to preserve the shape – whether spherical or other – that characterized the egg immediately after fertilization.

The shape of the surface envelope affects the geometrical relation between the mitotic apparatus and the surface and, thus, the place where the division mechanism forms, as well as other aspects of division mechanism function. The importance of cell shape in determining the position of the furrow led to the assumption that all cells, tissue cells as well as fertilized eggs, are spherical during the period when interaction between the surface and the mitotic apparatus occurs (Wolpert, 1960). Experiments have since demonstrated that almost any shape permits division, provided the equatorial part of the mitotic apparatus is close enough to the surface. The shape need not be fixed during the interaction period. Sand dollar eggs divided normally as they were repeatedly deformed by compression or extrusion during the precleavage period (Rappaport, 1961). During the experimental period, the shape was never fixed, because the eggs were either being deformed by physical manipulation or spontaneously returning to their original spherical shape by a phenomenon resembling elastic rebound. Because cytokinesis requires defined regional differences, these experiments and observations suggest that the effect of the mitotic apparatus upon the surface is not conveyed by a freely diffusible, easily displaceable intermediary.

In spherical cells, the prefurrow tension at the surface is uniform and isotropic, but it is not difficult to impose conditions that must alter the normal tension conditions. When sand dollar eggs are stretched by the weight of attached glass beads, they become cylinders with a diameter-to-length ratio of about 1:9 (Rappaport, 1960). Detachment of the bead allows the egg to rebound to its original spherical shape, so that the cylindrical cell must be under tension predominantly parallel to the direction of elongation. The cylindrical shape orients the mitotic apparatus parallel to the long axis, and the division plane is oriented normal to the mitotic axis. The tension at the cell surface is no longer isotropic, and linear elements in the cortex should be aligned parallel to the direction of stretch, as are the microfilaments in mechanically stretched epithelial cells (Kolega, 1986). Attachment of a glass bead has the effect of rapidly changing the pattern of tension in the weight-bearing cortex at the same time that the orientation of cytoskeletal elements changes; it may also affect the status of F-actin if the cortex behaves like slime mold plasmodium (Wohlfarth-Botterman and Fleischer, 1976). The cleavage furrow of the stretched cell forms in an anisotropic milieu, and the orientation of the microfilaments in the contractile ring must be perpendicular to the orientation of the cytoskeletal elements in the cortex in which it is located. These changes might be expected to interfere with division mechanism formation, but furrowing in stretched blastomeres and their companion control cells is synchronous (Rappaport, 1978).

The ability to respond to the mitotic apparatus is not confined to normal surfaces. A simple way to change the surface is to stretch it by compressing the cell. Sea urchin eggs that are flattened by mechanical compression frequently form small blebs that behave as though some of the endoplasm is leaking out into a blister or exovate that is covered by a small, weaker region of the original egg surface. With continued outflow, the distension of the exovate surface increases, and when the cortex normally contains pigment granules or vacuoles, it is possible to estimate the degree of stretching by comparing the number of vacuoles per unit area in the exovate and egg surfaces (Rappaport, 1976). When exovates were produced after the furrow for the second cleavage had developed, the mitotic apparatus could be forced into them with little damage, and a new furrow developed in the exovate surface within 4 minutes. Pigment vacuole counts revealed that between 1% and 2% of the exovate surface was formed from original blastomere surface, and the remainder was formed *de novo,* presumably by stretching. The exovate surface was formed in less than a minute; its response to manipulation suggested that it was thinner and more fragile than the normal surface. Yet the new furrow formed in a timely fashion and it functioned normally.

When invertebrate eggs are ruptured in ways that result in a massive outflow of the subcortical endoplasm, a new membrane boundary forms over the liberated material in a few seconds. The new membrane was termed the "protoplasmic surface film" by Chambers (1917a) and the "precipitation membrane" by Costello (1932). Without this new membrane, the endoplasm is rapidly dispersed by Brownian motion, but with it the endoplasm appears normal for hours. Membrane-bound endoplasm behaves more like a simple oil drop than an intact egg, and it is far more susceptible to rupture when prodded with a needle. At the time of the outflow, the original surface retracts and thickens, and the endoplasmic region frequently separates from the egg in the plane where the precipitation membrane and original surface meet (Figure 7.9). The

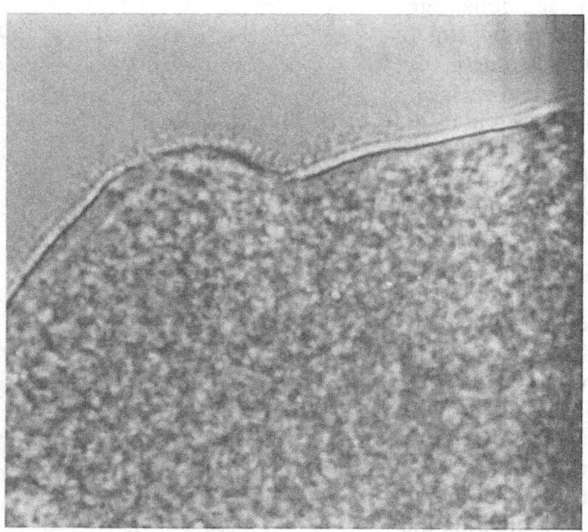

Figure 7.9 Conjunction of original (left) and new (right) surfaces after exovate formation. The radiating fibrous structures and the subsurface absence of granular material characteristic of shrunken original surface are lacking in the new surface. (From Rappaport, 1983, and with the permission of Wiley-Liss)

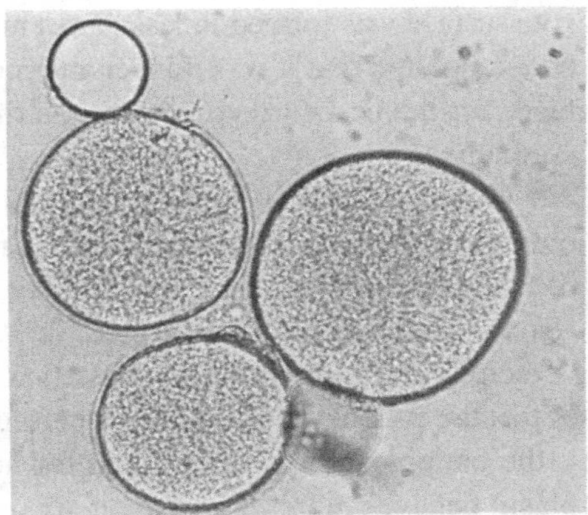

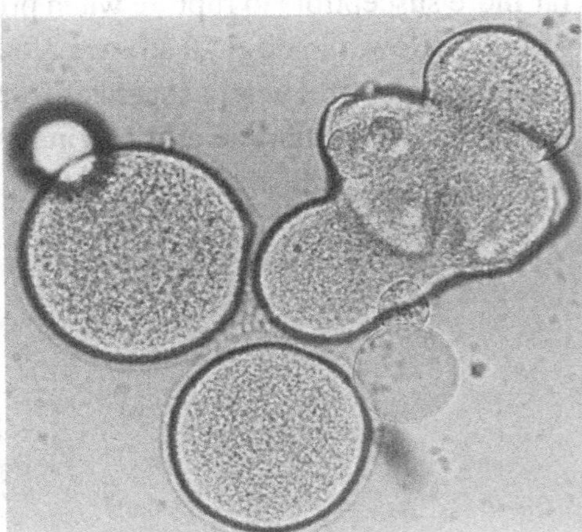

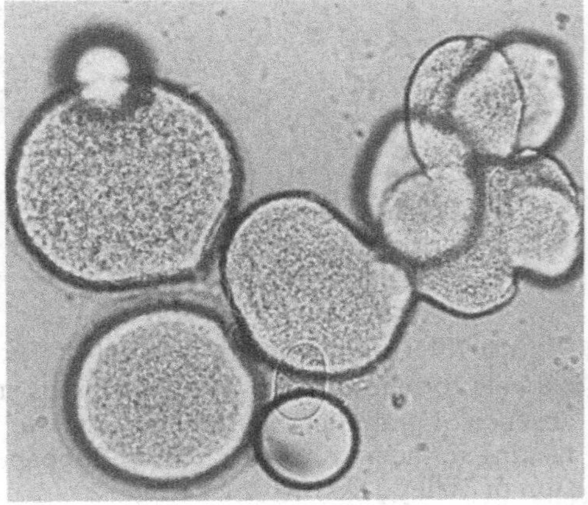

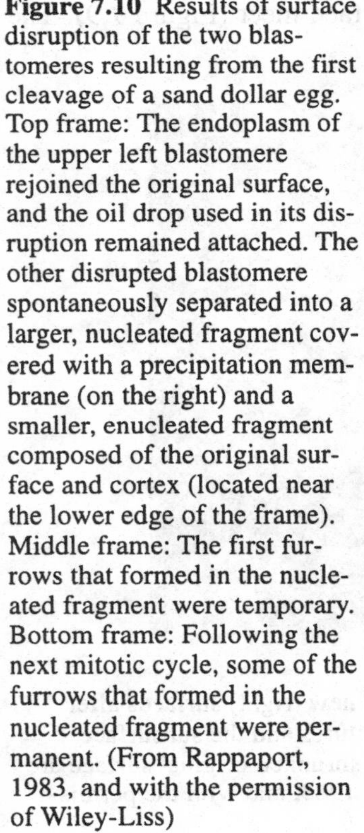

Figure 7.10 Results of surface disruption of the two blastomeres resulting from the first cleavage of a sand dollar egg. Top frame: The endoplasm of the upper left blastomere rejoined the original surface, and the oil drop used in its disruption remained attached. The other disrupted blastomere spontaneously separated into a larger, nucleated fragment covered with a precipitation membrane (on the right) and a smaller, enucleated fragment composed of the original surface and cortex (located near the lower edge of the frame). Middle frame: The first furrows that formed in the nucleated fragment were temporary. Bottom frame: Following the next mitotic cycle, some of the furrows that formed in the nucleated fragment were permanent. (From Rappaport, 1983, and with the permission of Wiley-Liss)

structure of a mitotic apparatus confined in an endoplasmic fragment appears normal. In contrast with the exovate covered by stretched original surface, the furrowing activity in those exovates immediately after the mitotic apparatus enters is abortive or absent. But in subsequent cycles, completed furrows are common (Figure 7.10). The surfaces of the exovates that formed in response to mechanical compression were certainly not normal, and their response to manipulation suggested that any cortical structure they may have had was reduced. The appearance of furrows in the precipitation membrane means that superficial *de novo* organization that may occur after rapid membrane formation is adequate for division mechanism formation. Further study of these preparations could reveal which structural features of the normal surface and cortex are not necessary for division mechanism formation.

The subsurface

The mitotic apparatus and the cytoplasm are considered separate compartments, but at the time of division they are interpenetrating in many animal cells and form a composite structure that fills the subcortical space. It is as accurate to describe the cytoplasm as extending wedgelike into the asters as it is to describe the aster rays as radiating into the cytoplasm. Morgan and Spooner (1909) showed that centrifugation that rearranges the visible cytoplasmic components without altering the mitotic apparatus does not prevent division. Chambers (1917b) found that mechanical agitation of the mitotic apparatus with a needle caused both the disappearance of the mitotic apparatus and the failure of cleavage. Other manipulations that do not completely obliterate the mitotic apparatus permit cleavage. The spindle and attached asters can be slid back and forth within the cell by compressing the polar regions alternately (Rappaport and Ebstein, 1965; Rappaport, 1975, 1978) or by pushing the polar surface of a cylindrical cell inward (Rappaport, 1985). Whether these manipulations are done continuously or intermittently, their cumulative effect is to reduce the apparent length of the astral rays so that they seem to end farther than normal from the surface. A constantly vibrating needle inserted between the mitotic apparatus and one part of the surface during furrow establishment does not interfere with formation of the furrow immediately above it (Rappaport, 1978). The

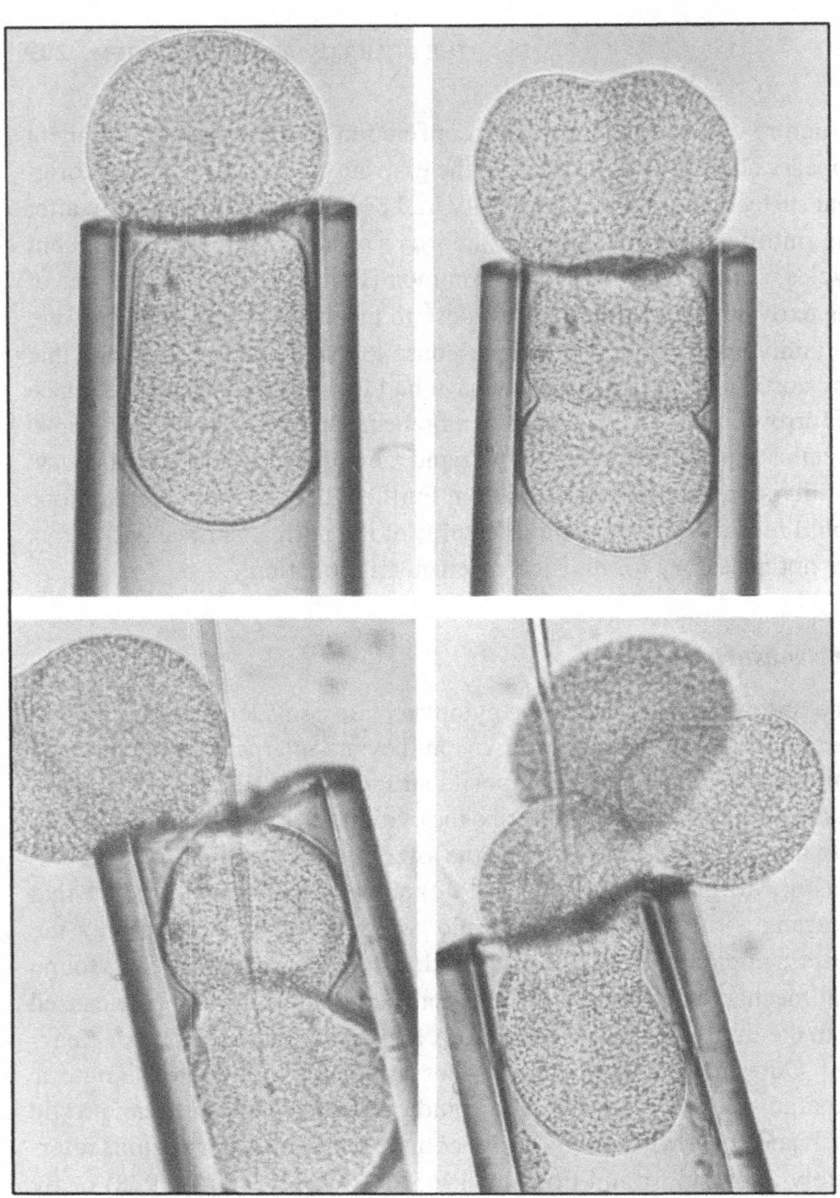

Figure 7.11 Furrowing following continual stirring of the zone between the mitotic apparatus and the surface, second cleavage. The unconfined cell attached to the pipet orifice is the companion blastomere that serves as a time control. Upper frames: Synchronous cleavage of confined and unconfined blastomeres, unoperated. Lower frames: Furrowing in confined blastomeres (two different operations) following continual stirring of the zone between the mitotic apparatus and the equatorial surface. (From Rappaport, 1978, and with the permission of Wiley-Liss).

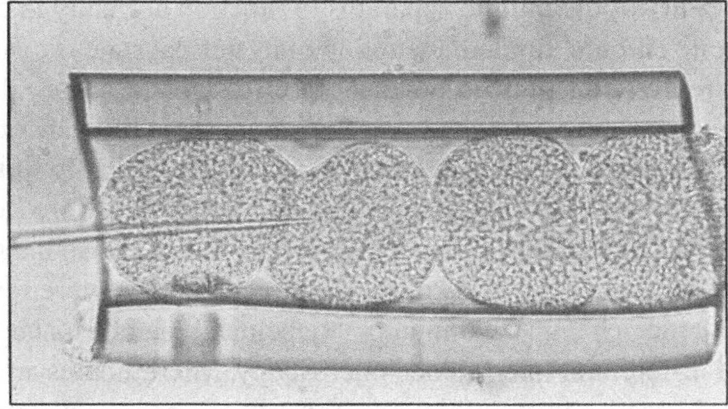

Figure 7.12 Second cleavage following repeated skewering of the mitotic apparatus in one blastomere. (From Rappaport, 1981, and with the permission of Wiley-Liss)

entire subsurface region in the equatorial and subequatorial areas can be disrupted by moving a needle oriented parallel to the mitotic axis around the perimeter of the spindle close to the cortex. To move a needle around the spindle perimeter required that a sand dollar egg be immobilized and its mitotic apparatus be simultaneously oriented by reshaping the egg into a cylinder in the tip of a pipet. A needle oriented parallel to the axis of the cylindrical cell and its mitotic axis was inserted through an exposed polar surface so that its point passed through the equatorial plane. As it swept through the subcortical region, the needle circumscribed the entire mitotic apparatus and sometimes caused it to rotate on its axis. When the operation was repeated continuously at rates of up to 1 revolution/minute or slower during the period when furrow establishment normally occurs, the furrow formed. Although it often appeared later and divided the egg more slowly than normal, the furrow was completed (Rappaport, 1978) (Figure 7.11). When, under similar circumstances, a needle was continuously thrust into and withdrawn from the cell along the mitotic axis to the extent that mitotic apparatus structure was visibly disrupted in the path of the needle, the furrow formed (Rappaport, 1981) (Figure 7.12). Unfortunately, the effect of these manipulations on the cell ultrastructure has not been studied, but the results suggest that the normal, precise number of, and configuration of, the microtubules and other ultrastructural elements are not necessary for division.

The ability of the mitotic apparatus to function normally in cytokinesis despite chronic, limited disruption of its normal state is consistent with the realization that its normal stability is in a sense illusory, because most of the microtubules that are the principal organizing elements of the spindle fibers and aster rays are dynamically unstable; they reversibly extend and retract. Older investigators were familiar with the cyclical changes in the appearance of fixed and stained astral rays. The dynamic properties of in vivo microtubules were revealed by fluorescence chemistry, immunocytochemical methods, and video-enhanced differential interference microscopy. Microtubules are structurally polarized because they are constructed from asymmetrical tubulin subunits. As a consequence, the rates of assembly at the two ends of the tubule are different. The end where more rapid assembly occurs is termed the "plus end," and the opposite end, where assembly is slower, is the "minus end." The microtubules of the mitotic apparatus are arranged so that the minus ends are at the centrosome or spindle pole and the plus ends are close to the cortex or equatorial plane. The spindle fibers are therefore antiparallel and overlap at the equator (Euteneuer and McIntosh, 1980). Microtubules are, as a population, in a steady-state exchange with a soluble pool of tubulin subunits (Inoué, 1981). The exchange is rapid; it is completed in 60–80 seconds in the sea urchin spindle (Salmon et al., 1984). In cultured mammalian epithelial cells, subunits are incorporated at the plus end of existing microtubules and in newly nucleated microtubules near the centrosomes (Soltys and Borisy, 1985). Each microtubule is either lengthening or shortening, the transition between the two states occurring rapidly and randomly in a population (T. J. Mitchison and Kirschner, 1984). In newt lung cells, microtubules elongate at 7 μm/minute and shorten at 17 μm/minute (Cassimeris, Pryer, and Salmon, 1988). During mitosis in cultured PtK_1 cells, the microtubules of the asters show greater dynamic instability than those of the interzone of the spindle between the separating chromosome groups. Within the interzone, incorporation increases from mid to late anaphase and decreases as the cells enter telophase (Shelden and Wadsworth, 1990). These findings suggest that the aster rays have the potential for rapid reextension toward the cortex after disruption of the plus ends of their constituent microtubules. The reduced length of the visible rays in the manipulated mitotic apparatus strongly suggests, however, that restoration in the disturbed area is incomplete.

Although dynamic instability appears to be a normal property of astral microtubules, there is cause to question whether it is essential to division mechanism establishment. Both hexylene glycol (Kane, 1962; Rebhun and Sawada, 1969) and taxol (Salmon and Wolniak, 1990) stabilize mitotic apparatus microtubules and increase the birefringence of the whole structure. Endo, Toriyama, and Sakai (1983) found that hexylene glycol increased the number of astral microtubules and strikingly increased the uniformity of their length. Treatment of starfish eggs during the period of meiotic divisions appeared to have similar effects upon the mitotic apparatus and also increased its size (Yamao and Miki-Noumura, 1988). In consequence, the polar body was larger than normal and, when the mitotic apparatus was centered, the egg divided into two cells of approximately equal volume in a way that resembled normal first cleavage. Salmon and Wolniak (1990) found that taxol increased mitotic apparatus birefringence in sand dollar eggs and increased the size of the spindles. Furrows formed, but cleavage was occasionally abortive. The asters that formed before the second cleavage contained an abnormally high concentration of microtubules, the boundary that separated them from the surrounding cytoplasm was unusually distinct, and their microtubules did not appear to elongate toward the surface to the normal extent. In treated flattened eggs, the mitotic apparatus induced furrowing, although its outer limit appeared to be farther than normal from the surface. These results imply that cytokinesis does not require the normal, dynamic assembly–disassembly activity of microtubules. They also suggest that the normal proportion of microtubule to nonmicrotubule aster components within the asters is not important.

Aster-Induced Relaxation or Aster-Induced Contraction?

The astral relaxation hypothesis and the astral contraction hypothesis described in Chapter 2 were both designed to explain how the division mechanism is established in the equatorial cortex. The hypotheses differ principally regarding the nature of the cortical response to the mitotic apparatus, and the region of the cortex that is subjected to maximum

Table 7.1 Summary of experiments involving alteration of normal geometrical relation between the mitotic apparatus and the surface

DESCRIPTION OF EXPERIMENT	PREDICTED OUTCOME		ACTUAL RESULT
	Astral Relaxation Hypothesis	*Astral Contraction Hypothesis*	
1. Interposition of blocks between the asters & polar & subpolar surfaces (K. Dan, 1943a; Rappaport, 1968; Rappaport & Rappaport, 1983)	Interference with furrow formation	Furrow	Furrow
2. Interposition of blocks between the asters & the equatorial surface (K. Dan, 1943a; Rappaport, 1968; Rappaport & Rappaport, 1983)	Furrow	Interference with furrow formation	Interference with furrow formation
3. Manipulation of polar & subpolar regions of asters, cytoplasm between surface & asters, & surface	Interference with furrow formation	No interference, furrow	Furrow
4. Relocation of asters farther from equator, closer to poles (Sluder & Begg, 1983; Rappaport & Rappaport, 1985a)	Furrow	No furrow	No furrow
5. Relocation in distant positions in (4) closer to equator (Sluder & Begg, 1983; Rappaport & Rappaport, 1985a)	Interference with furrow formation	Furrow	Furrow

6. Relocation of polar & subpolar surfaces away from aster centers (Rappaport, 1981; Rappaport & Rappaport, 1984)	Interference with furrow formation	Furrow	Furrow
7. Relocation of subpolar & polar surfaces closer to the aster centers (Rappaport, 1992)	Furrow	No furrow	No furrow
8. Artificial constriction at the equator & distention at poles that equalize or reverse distance relations to aster centers (Rappaport & Rappaport, 1984)	No furrow	Furrow	Furrow
9. Relocation of equatorial surface closer to mitotic axis in cells with asters too far apart to establish furrows (Rappaport, 1969b)	No furrow	Furrow	Furrow
10. Artificial constriction at the equator & distension at poles in urethane-blocked cells (Rappaport & Rappaport, 1984)	No furrow	Furrow	Furrow
11. Relocation of polar & subpolar surfaces closer to the astral centers in urethane-blocked cells (Rappaport & Rappaport, 1984)	Furrow	No furrow	No furrow

continued next page

Table 7.1 (continued)

DESCRIPTION OF EXPERIMENT	PREDICTED OUTCOME		ACTUAL RESULT
	Astral Relaxation Hypothesis	*Astral Contraction Hypothesis*	
12. Relocation of the mitotic apparatus after furrow forms (Rappaport & Ebstein, 1965; Rappaport, 1975; Rappaport & Rappaport, 1984)	No additional furrow after the first one	Possible multiple furrows	Multiple furrows
13. First cleavage of a torus-shaped cell (Rappaport, 1961)	Furrow at equator & between polar regions only of asters	Furrow at equatot only	Furrow at equator only
14. Location of an isolated aster close to the pole of an elongate cell (Rappaport & Rappaport, 1985a)	Furrow on side of aster away from pole (White & Borisy, 1983)	Constriction of polar & cylindrical surface, possibly no furrow	No furrow, constriction at polar & cylindrical surfaces (Rappaport & Rappaport, 1985b)
15. Location of isolated aster distant from poles in cylindrical cell (Rappaport & Rappaport, 1985a)	Possible constriction not in plane of aster	Constriction in plane of aster center	Constriction in plane of aster center

16. Location of equatorial surface next to spindle in circumstance where close relation to polar region of aster prevented (K. Dan, 1943a; Rappaport, 1968; Rappaport & Rappaport, 1974;Kawamura, 1977)	No furrow	Furrow	Furrow
17. Artificial constriction of subpolar surface close to plane of an aster center (Rappaport, 1988)	No furrow activity in artificial constriction	Furrow activity in artificial constriction	Furrow activity in artificial constriction
18. Artificial constriction so that mitotic apparatus confined to one side of constriction plane (Rappaport & Rappaport, 1988)	Cytoplasmic flow into nucleated part	Cytoplasmic flow out of nucleated part	Cytoplasmic flow out of nucleated part
19. Egg reshaped into a cone with mitotic & cone axis coincident (Rappaport & Rappaport, 1994)	Furrow at midpoint between asters where aster effect is least	Furrow closer to one aster where aster effect is greatest	Furrow closer to aster near cone apex

aster effect. Observations of the normal events that precede mechanism establishment have not produced decisive information, and direct measurement of the immediate effects of the asters on the surface has not been possible. However, the nature of the cortical response (whether the cortex relaxes or contracts) may be deduced in experimental circumstances, and the region that is maximally affected can be determined by experimental manipulation of the geometrical relation between the mitotic apparatus and the surface at the time when they interact. Many of the previously described experiments were designed to localize and characterize processes and events essential to division. But the results of a number of those experiments may also be used to evaluate the two hypotheses because they predict different outcomes for the same experiment. The experiments, which involved the imposition of unusual geometrical configurations at the time of division mechanism establishment, had to be performed on cleaving eggs and blastomeres, because these are incapable of autonomous shape change. Table 7.1 summarizes a number of such experiments and their outcomes. In science, one evaluates hypotheses, theories, and models on the basis of their predictive consequences. The table reveals that the astral relaxation hypothesis predicts none of the outcomes correctly, whereas the predictions of the astral contraction hypothesis are all confirmed.

Mitotic Apparatus Activity

Speculation about the way that the mitotic apparatus participates in division activity was closely related to speculation about the physical mechanism of the division process (see Chapter 2), but the elimination of all mechanical alternatives except equatorial contraction did not reveal much about mitotic apparatus activity. It is logical to assume that the activity causes localized changes in subsurface environment, and the idea that transport is involved is attractive. Fol (1873) proposed that the aster rays are converging streams of hyaline cytoplasm that accumulate centrally to form the relatively clear "lake" that is characteristic of echinoderm asters. Chambers (1917b) described the movement of 2–4 μm diameter oil drops from the aster periphery toward its center, and he repeatedly emphasized the importance of the phenomenon over the ensuing years. Other examples of accumulation at aster centers abound. Pigment granules in the cytoplasm of the vegetal blastomeres of cleaving amphibian eggs collect in aster centers and around interphase nuclei (Morgan, 1897; Kobayakawa, 1988). Stainable granules in invertebrate eggs behave in the same way (Fischel, 1906; Kojima, 1959; Rebhun, 1960) unless the microtubules are chemically disrupted. The same is true for injected plastic beads, which are transported regardless of charge (M. S. Hamaguchi, Hamaguchi, and Hiramoto, 1986) and remain near the nuclei (Wadsworth, 1987). Terasaki and Jaffe (1991) demonstrated similar behavior of internal cisternae of the endoplasmic reticulum, which also remain around the nucleus at interphase in cleaving sea urchin eggs. This transport system appears to have some selective properties; it does not move glass beads, diamond powder particles, paraffin oil droplets (M. S. Hamaguchi, Hamaguchi, and Hiramoto, 1986), or noncarboxylated plastic beads (Wadsworth, 1987). Whether the ability of a mitotic apparatus to establish furrows bears any direct relation to its ability to move small, visible cytoplasmic inclusions is unclear. The rearrangement of these inclusions by centrifugation does not seriously affect division (Morgan and Spooner, 1909) and, in the normal cell, their rearrangement is not cyclical, in that they remain near the aster centers or spindle poles. On the other hand, the machinery that moves the inclusions may have other effects that are integrated in furrow establishment.

The Cleavage Stimulus or Signal

The regional differences in cortical contractility that divide the cell arise from interaction between two major regions of the cytoskeleton: the prominently microtubular mitotic apparatus and the predominately microfilamentous cortex. The idea that the interaction is mediated by a stimulus or signal that moves between the interactants has a long history, and the proposed nature of the stimulus has varied with the mechanochemical basis of the hypothetical division mechanism. Current information concerning the structural relationship between the recognizable contractile ring and the periphery of the mitotic apparatus is intriguing. The 0.1–0.2 μm thick contractile ring lies immediately beneath the plasma membrane. The ring of cleaving echinoderm eggs is separated from the endoplasm and mitotic apparatus by 3–5 μm of cortical material. The linear elements of the mitotic apparatus do not now appear to penetrate the cortex. If it is assumed that the contractile ring is established by the direct effect of the stimulus, it is implied that the latter moves between the tips of the linear elements and the underside of the plasma membrane through the cortex. The ultrastructural effects of this event are confined to the very periphery of the cortex, and the structure of the deeper part of the cortex appears to be unaffected by the penetration of the stimulus.

The dimensions of the contractile ring and those of the furrow base suggest to the observer that the active force-producing region is relatively narrow and well defined. It is, however, possible that contractile activity exists beyond the edges of the region demarked by the contractile ring, so that the deforming force is distributed in a gradient in which the maximum is located at the deepest part of the furrow base. The dimensions of the mitotic apparatus and the whole cell, as well as the demonstrated ability of all regions of the aster periphery to induce furrow formation, suggest that the entire sea urchin egg surface may be affected when the division mechanism is established. The accumulated experimental evidence indicates that the effect is greater in the equatorial region, but the contours of the cortex and the periphery of the mitotic apparatus do not appear to be well suited to the formation of narrow, sharply defined interactive zones. Computer models suggest that the transition between the most and least affected zones is gradual (Devore, Conrad, and Rappaport, 1989; A. K. Harris and Gewalt, 1989). The cell's ability to form furrows when the region through which the signal passes is chronically disrupted by stirring (Rappaport, 1978) or is spread over a wider-than-normal area by the repeated sliding of the mitotic apparatus back and forth (Rappaport, unpublished results) also substantiates the idea that sharply defined interaction areas are not necessary.

The dimensions of the contractile ring may be determined either directly by the pattern of stimulus distribution or indirectly by a kind of self-organization that concentrates the maximally constricting cortex in a narrow band. The attractiveness of the latter alternative derives mainly from doubt about the suitability of the mitotic apparatus for the designation of a narrow, linear, sharply defined region in the cortex. It is interesting that the dimensions of the contractile ring are very similar in a wide variety of cells. Measurements suggest that its dimensions (other than its circumference) are not closely correlated with the size of the cell, the size of the mitotic apparatus as a whole (as well as the size of the spindle and the asters), the distance between the mitotic apparatus and the surface, and, in the amphibian egg, whether the contractile ring forms directly by action of the mitotic apparatus or of the diastema, or indirectly by extension of an active furrow (Schroeder, 1973). In other words, the width and thickness of the product of the interaction – the contractile ring – are more nearly constant than the dimensions of the source of the signal and other relations that might be expected to affect contractile ring assembly. In conical sand dollar eggs, the location of the furrow closer to the aster nearest the vertex (Rappaport and Rappaport, 1994) coincides with the region of maximum aster effect predicted by the Harris–Gewalt model. Experiments that involve sliding the mitotic apparatus back and forth during the interaction period indicate that a normal-looking furrow can form in a wider-than-normal interaction area. It may be that the contractile ring is centered in the region of maximum mitotic apparatus effect but that its dimensions result from the response of a subsurface region in which the events of reorganization are determined by the properties of the filamentous material that is reorganized, the nature of the matrix in which reorganization occurs, the nature of the mechanism that increases contractility, and the magnitude of local forces – all factors that could be similar in a wide variety of cells.

The problem of the molecular or chemical nature of the signal has been central in the study of cytokinesis since the general recognition of the role of the mitotic apparatus, and no clear solution is evident at this time. Early speculation about the nature of the cortical division mechanism appears to have assumed that the mechanism originated by modification or reorganization of locally present material and that a small amount of triggering influence could cause changes of relatively larger magnitude. Implicit in the terms "stimulus" and "signal" is the idea that the entity initiates the first in a possible sequence of events. But the potential for heavy two-way traffic between the equatorial subcortical region and the zone of the mitotic axis that involves a wide variety of cell components and is mediated by numerous types of motor molecules (Vallee, Shpetner, and Paschal, 1990) may complicate the certain identification of any single element with signal characteristics. There are many contraction-related regulatory molecules, and it will be necessary to discriminate between those that initiate the chain of events and those that play other essential roles in the chain.

Local fluctuation in concentration of any essential component of the contractile system could act as a signal (Rappaport, 1971b). Investigations involving studies of in vitro reactions and behavior of proteins that have been identified in the contractile ring or enriched in its vicinity, and in vivo studies of the distribution of these proteins, have clarified events during the period immediately before furrowing. Fluorescent staining and immunofluorescent methods have revealed that both in tissue cells (Sanger, Mittal, and Sanger, 1990) and in sea urchin eggs (Mabuchi, 1994), myosin, actin, and actin-associated proteins accumulate in the equatorial region before the furrow indents (Chapter 6). These findings raise the possibility that a stimulus may not be necessary, because the mitotic apparatus can transport the necessary components to the equatorial region where they could combine with those already present and reorganize. Mabuchi and Okuno (1977) found that injected myosin antibody blocks sea urchin egg cleavage when introduced at interphase, but it was less effective when injected immediately before cleavage. In one of their speculations concerning the basis of their finding, they proposed that the antibody, when injected during interphase, might have interfered with the signal and that later injection might have been ineffective because it occurred after signaling had been completed. Subsequently, Mabuchi (1986, 1990) pointed out that myosin has qualities that would be desirable in a signal molecule because, in addition to its role as a motor molecule that generates active shear, myosin in vitro can accelerate actin polymerization and induce polymerization of parallel actin filaments oriented in opposite directions. However, myosin, actin, and wheat germ agglutinin sites (which are distributed on the surface after anaphase in the same pattern as are actin filaments) appear in the contractile ring simultaneously (Mabuchi, 1994). Were myosin responsible for enhancement of local actin polymerization, it would be expected to aggregate at the equator first, but the appearance of simultaneity may be affected by the sensitivity of the staining methods and the rapidity of the events.

Chemistry of potential stimuli and signals

The results of biochemical investigations of in vitro activity of the structural and regulatory proteins having potential roles in cytokinesis, and information concerning fluctuations of their levels of activity that occur during the cell cycle, can be put together in ways that might explain how the events of cytokinesis may be linked at the biochemical level (Mabuchi, 1986, 1993; Pollard et al., 1990; Satterwhite and Pollard, 1992; Larochelle and Epel, 1993). The timing of cytokinesis has been ascribed to changes in the level of maturation-promoting factor, and the positioning of the division mechanism has been attributed to the localization of regions of high Ca^{2+} concentration. Maturation-promoting factor catalyzes the entry of eukaryotic cells into mitosis. It consists of a regulatory cyclin and the $p34^{cdc2}$ kinase; its level of activity is high at prophase and rapidly decreases at the transition point between metaphase and anaphase. For the purpose of this brief discussion, its ability to phosphorylate the regulatory light chain of myosin II is critical. Myosin regulatory light chains contain inhibitory and activating sites, and the maturation-promoting factor phosphorylates the inhibiting sites that are known to inhibit the actin-activated ATPase and to lower the affinity of myosin for actin filaments in vitro. The inhibitory sites are also phosphorylated by protein kinase C, which can inhibit the actin-activated ATPase of smooth muscle and promote its relaxation. High levels of maturation-promoting factor appear to be capable of inhibiting force-producing interaction between myosin and actin, which could, in turn, temporarily block cytokinesis. The benefit to the cell that has been proposed for the blockade is that it provides added assurance that cytokinesis does not occur until the transportation of the chromosomes to the spindle poles is less likely to be affected by constriction. The suggested effects of maturation-promoting factor are global, so that the cortex will not respond to the mitotic apparatus until the level of the factor's activity has decreased. Maturation-promoting factor has not been considered to play a role in determining the position of the furrow.

In recent thinking, furrow positioning has been closely associated with myosin regulation. Myosin filaments can organize parallel actin filaments of polarity opposite from that of G-actin, and they can also polymerize actin from the depactin–actin complex (Mabuchi, 1982). Myosin filament formation, in turn, requires phosphorylation of myosin light chains by a specific enzyme, myosin light chain kinase. Myosin light chain kinase has been considered likely to play an important role in initiating cleavage furrow contraction because cytoplasmic and smooth muscle myosin share the same regulatory light chain, and there is evidence that phosphorylation of the light chain is linked to smooth muscle contraction. Myosin phosphorylation also stabilizes myosin filaments in the presence of ATP and activates myosin ATPase activity by actin filaments. Inhibitors of myosin light chain kinase also inhibit cytokinesis in sea urchin eggs. Concentrations of inhibitors that block cleavage permit mitosis. In blocked eggs, fluorescent egg staining revealed no organization of actin (Mabuchi and Takano-Ohmuro, 1990).

Myosin light chain kinase is activated by calcium-calmodulin, and this relation raises the possibility that localized zones of high Ca^{2+} concentration could trigger the cascade of biochemical events previously described in brief. Because of its association with contractile activity, Ca^{2+} has often been proposed for a role in cell division. Difficulties in measuring the low concentrations that exist in single dividing cells have resulted in disagreements concerning the timing arrangement and the direction of transient changes (reviewed, Mabuchi, 1986; Hepler, 1989, 1992). Increases in intracellular Ca^{2+} concentration have been associated with furrowlike deformations (Schroeder and Strickland, 1974; G. W. Conrad and Davis, 1977, 1980), but detailed analysis of the phenomenon has not been possible. Schantz (1985) measured cytosolic, free Ca^{2+} concentrations in Medaka fish embryonic cells with calcium-selective electrodes and found one case in which Ca^{2+} concentration rose fourfold in two successive cleavages. He attributed the inconsistency of the results to the localization of the calcium transient and the small volume of cytoplasm monitored by the electrode. Fluck, Miller, and Jaffe (1991) also studied the relation of Ca^{2+} to the advancing furrow in Medaka eggs. By using aequorin, they could visualize a zone of high free calcium, which they estimated to be 5–8 μM at the peak. Early furrows of telolethical Medaka eggs resemble those of unilaterally cleaving eggs in that, after establishment by the mitotic apparatus, their tips extend by propagation across the large blastomere surface. The method used by Fluck and co-workers did not clearly show whether the calcium waves preceded or followed furrowing, so it was determined that the waves accompanied the advancing furrows. Fluck, Miller, and Jaffe proposed that the wave is propagated by stretch-sensitive calcium channels that release calcium and raise the local Ca^{2+} concentration. The resulting locally activated myosin light chain kinase could then initiate the sequence of events described previously. The researchers did not discuss the mechanism by which the mitotic apparatus initiates the beginning furrow. The hypothetical role of stretch-sensitive channels would appear to require a previously activated contractile mechanism, and the existence of the channels should be testable by micromanipulation experiments. In a subsequent investigation, Miller, Fluck, McLaughlin, and Jaffe (1993) found that in *Xenopus* eggs, injected calcium buffers could delay furrow formation and arrest elongation of active furrows. They concluded that the buffers suppressed formation of the localized zones of high calcium concentration that are essential for the formation, extension, and maintenance of the furrow.

A difficulty in proposing specific chains of events early in a period of biochemical exploration is the likelihood that important links may be missed because hypothetical schemes can be made to work without them. A rho-like protein investigated by Mabuchi et al. (1993) appears to be an example. Botulinum C3 exoenzyme specifically ADP-ribosylates and inactivates rho proteins. In cultured mammalian cells, C3 exoenzyme causes the disappearance of stress fibers and rounding up. When injected into sand dollar eggs during nuclear division, it interfered with furrow formation. Absence of furrows was accompanied by absence of actin filaments in the equatorial cortex. Injection during cytokinesis rapidly disrupted equatorial actin filaments and caused regression of active furrows. The only protein in isolated cleavage furrows that was ADP-ribosylated by C3 exoenzyme, comigrated with ADP-ribosylated rhoA derived from human blood platelets. Treatment did not affect nuclear structure or division, and treated cells became multinucleated after several rounds of mitosis. The results suggest that a rho-like protein is involved in the organization and maintenance of the contractile ring and that inactivation of the protein has effects as dramatic as cytochalasin. The rho-like protein appears to play an important role but has not yet been linked to other biochemical events.

It also appears that attempts to deduce an all-inclusive, hypothetical conserved mechanism on the basis of current information may be plagued by differences in response to experimental intervention that may be associated with species as well as with differences in method. Mabuchi and Takano-Ohmuro (1990) found that certain protein kinase inhibitors, which affect sea urchin egg myosin light chain kinase, block cleavage and actin filament formation in the equatorial cortex without affecting nuclear division. But injection of catalytic fragments of myosin light chain kinase into normal rat kidney cells, which would be expected to deregulate myosin light chain kinase activity and therefore cytokinesis, prolonged the period between nuclear envelope breakdown and anaphase onset, increased the motile surface activity during and after anaphase, but did not prevent the formation and function of furrows that progressed to completion at normal rates (Fishkind, Cao, and Wang, 1991). The sand dollar egg cortex responds to the mitotic apparatus by general contraction beginning about 30 minutes before the anticipated time of first cleavage, and it remains responsive until furrow formation (Rappaport and Rappaport, 1988). During the entire period between first and second cleavage, the nucleus or mitotic apparatus material causes a similar surface response (Rappaport, 1990). The cortex of sand dollar eggs can engage in furrowing activity for about half the cell cycle, and the activity is not correlated with the presence or absence of the nuclear envelope (Rappaport and Rappaport, 1993). But in *Xenopus* and other amphibia, successive early cleavages begin before the previous cleavage is completed. This normal circumstance implies that the cortex can form furrows throughout the cell cycle because, after direct establishment by the mitotic apparatus or the diastema, furrows are propagated at the furrow tips until completion of the furrows, and before the completion of the first furrow, the direct establishment of the second furrow has been completed and the second furrow has become functional.

Careful analysis by Satterwhite et al. (1992) revealed that inhibition sites on myosin II regulatory light chains are phosphorylated in a mitosis-specific pattern in activated *Xenopus* eggs. They propose that 34^{cdc2} phosphorylation inhibits actin-activated myosin ATPase during prophase and metaphase in a way that prevents premature cytokinesis.

Because at least one active furrow is present in the blastomere throughout the cycle, there may be some question whether there is a direct connection in *Xenopus* between phosphorylation and surface responsivity. It may also be relevant that the relatively long interval between the time that the position of the furrow is determined and the time that furrowing begins in frog, salamander, and sturgeon eggs (Selman, 1982) greatly reduces the need for a mechanism that prevents premature cytokinesis.

Few of the alternative signaling mechanisms that have been proposed have been subsequently eliminated by experimental testing, but aspects of all of them may be questioned. There is a spectrum of possibilities that ranges from transport of multiple components of the contractile ring to localized free-calcium transients. Numerous modulating and regulating proteins (Mabuchi, 1986, 1990) multiply the possibilities. The problem of testing the numerous alternatives in this aspect of cytokinesis is reminiscent of that confronted by investigators of a century ago when they tried to comprehend the process as a whole.

Signal transport

The mitotic apparatus–associated redistribution of cytoplasmic structures is undoubtedly linked to mobility phenomena associated with microtubules. A number of ingenious investigations have led to the general acceptance of the idea that movement is associated with two families of mechanochemical proteins – kinesin and cytoplasmic dynein – that produce force in opposite directions relative to the polarity of the microtubule. Kinesin moves structures toward the plus end; cytoplasmic dynein, toward the minus end (reviewed, McIntosh and Pfarr, 1991). In the mitotic apparatus, cytoplasmic dynein would mediate movement toward aster centers and spindle poles, and kinesin would mediate movement from these structures toward the cortex and equatorial plane. Movement requires energy, and vesicle movement requires accessory factors. This simple picture has great appeal. Microtubule-based transport systems provide a way to rapidly move intracellular compartments that are too large to be effectively carried by diffusion to specific locations (Schroer and Sheetz, 1991). The two families of proteins could account for the transport of particles in opposite directions on the same microtubule (Hayden, 1988), although the distinction between the families based upon direction of movement may be blurred (McIntosh and Pfarr, 1991). Kelly (1990) has emphasized the importance of microtubule-associated motility for both membrane traffic and organization. Vesicles can carry newly synthesized proteins from the Golgi apparatus to the surface; peripheral extension of the endoplasmic reticulum could result from kinesin-driven movement along the microtubules.

Clarification of the roles of motor proteins in furrow establishment may be complicated by their number and variety. Injection of kinesin heavy chain–specific antibodies into cleaving sea urchin eggs has no effect on mitosis even when the antibody is capable of blocking kinesin-driven motility in vitro and in mammalian cells (Wright, Terasaki, and Scholey, 1993). Some antibodies to kinesin-like proteins, however, interfere with normal mitotic apparatus formation, disrupt mitosis, and block cytokinesis. The authors suggest that some kinesin-like proteins may have essential mitotic function, but kinesin heavy chain does not.

Another form of mitotic apparatus–associated motion, termed "ejection force," was first described by Carlson (1938); Rieder et al. (1986) have tried to determine its cause. The force was proposed as a way of explaining the movement of large particles, inclusions, inclusions lacking membranes, and acentric chromosome fragments that drift away from the spindle poles or aster centers at about 2 μm/minute. Disruption of microtubules with nocodazole or colcemid abolishes ejection activity, whereas taxol, which promotes nonkinetochore microtubule polymerization, increases it. The existence of ejection force appears to be correlated with aster size, microtubule dynamics, and the relation between nonkinetochore microtubules and the chromosomes. The effects of the force are apparently maximal at prometaphase and metaphase, when microtubule turnover is greatest. Rieder et al. suggest that the force originates in the elongating microtubules, which collectively act as a "moving steric wall" that propels the visible objects before it.

The evidence suggests that the mitotic apparatus, by virtue of its general form and the transport capacity of its constituent radiate elements, is ideally suited to operate as a machine that sorts and rearranges cytoplasmic components. Vallee, Shpetner, and Paschal (1990) sug-

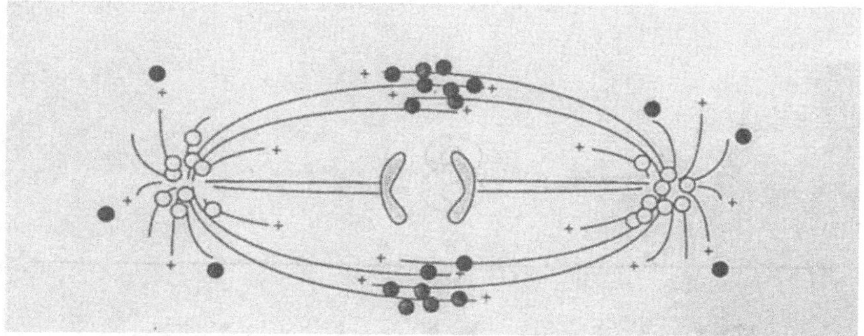

Figure 7.13 Suggested role of bidirectional organelle transport in specifying cleavage plane. Schematic diagram of a mitotic spindle showing predicted positions of minus-end–directed organelles (open circles) and plus-end–directed (+) organelles (shaded circles). The latter are envisioned to become concentrated at the zone of overlap of microtubules. This would occur if the plus-end–directed organelles, rather than diffusing away from the spindle, became trapped into a pattern of continuous back-and-forth movement along microtubules of opposite polarity. Organelles concentrated at the spindle mid-zone could release or take up diffusible factors involved in signaling the cell cortex to initiate furrow formation. (From Vallee et al., 1990, and with the permission of the New York Academy of Sciences)

gested that membranous organelles could act as triggers for cytokinesis by release or uptake of ions or other diffusible material. Because kinesin and dynein facilitate oppositely directed movement on microtubules, and because there is specificity in the attachments between motor proteins and organelles (Schnapp, Reese, and Bechtold, 1992), it is hypothetically possible for organelles linked to different motor proteins to accumulate in different regions (Figure 7.13). Within the spindle, plus-end–directed organelles would tend to accumulate in the equatorial zone, and within the asters, they would move to the general periphery as well as to the equatorial plane. This idea is consistent with the observation that two asters without an intervening spindle can establish a furrow. The possibility that motor proteins are also involved in mitotic chromosome movement has excited great interest (McIntosh and Pfarr, 1991). Should it eventually prove that motor-protein–mediated transport along microtubules is the central phenomenon in both chromosome movement and cleavage furrow establishment, the expectation of early students of cell division that the role of the mitotic apparatus in both mitosis and cytokinesis is based upon the same physical activity would be realized, although not in the way that may have been expected.

Division Mechanism Function and Its Consequences

The data obtained by experimental studies on both cleavage and tissue cells indicate that the forces that deform cells during division originate at the surface and cortex and that they are contractile in nature. Much of the energy and ingenuity that has been expended in the studies of dividing cells was directed toward the identification of the physical mechanism. Inability to distinguish between cause and effect permitted divergent interpretations, but the observations generally have been reasonably accurate and useful. Qualitative data excite the imagination, but quantitative data are often necessary to make a point.

Duration of the Functional Period

The period from the first indentation at the equatorial surface to the complete separation of the daughter cells is about 8 minutes in normal cleaving sand dollar (*E. parma*) eggs. The period between cleavages is 48 minutes, so that active furrowing comprises about 17% of the cycle. On the other hand, in some large eggs with relatively small, highly eccentric mitotic apparatuses and, consequently, unilateral furrows – such as the eggs of amphibians and some ctenophores – the second cleavage begins before the first is completed. In these cases, active furrowing is in progress throughout the entire cycle. In order to understand whether the relatively brief functional life of the sand dollar furrow results from intrinsic factors, or from the milieu in which it operates, or from the consequences of its activity, it is necessary to understand the factors that begin and end division activity. The formation of premature furrows following confinement of the mitotic apparatus in a narrow cytoplasmic cylinder was described in Chapter 5. Relative to this discussion, these experiments imply that the mitotic apparatus and the surface are competent to interact about 5 minutes before they normally do and that their interaction is prevented by intervening distance, which is reduced in the normal cell by the lengthening of the astral rays, and in experimental cells by moving the surface. Cleavage normally ends when the cells are completely separated and nothing remains to be divided. The cleavage phase can be prolonged by increasing the dimensions of the part of the cell to be divided. Sand dollar eggs form unilateral furrows after the mitotic apparatus is pushed toward the cell margin (Rappaport and Conrad, 1963). The length of the region through which the furrow extends can be increased by stretching the region with the side of a needle (Figure 8.1). In this circumstance, the furrow deepens for about 22 minutes (Rappaport and Rappaport, 1993). If it reaches the cell margin during that period, division is complete and permanent. If it fails to reach the cell margin, it stops and usually regresses. In the proper geometrical circumstances, the operated cell and its control begin the subsequent cleavage nearly synchronously. These simple experiments indicate that the surface can engage in furrowing activity about 5 minutes earlier and 14 minutes later than normal, or over about 56% of the cycle. The normal beginning and end of the period when the division mechanism is active are determined by the proximity of the mitotic apparatus and by the physical separation of the cells. The stimulus mechanism and the responding surface are competent to carry out their roles before and after they normally do.

These data and others from previous chapters can be represented in a summary chart (Figure 8.2). The combined information emphasizes the extended period during which the cell can carry out division-related

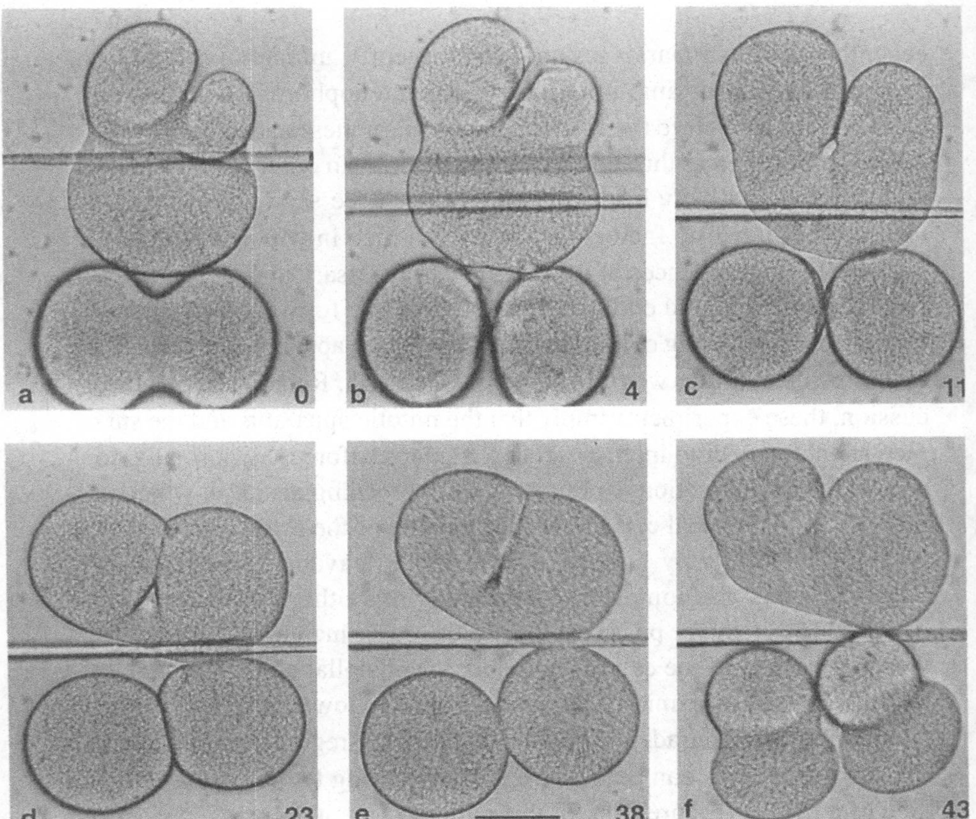

Figure 8.1 Prolongation of furrowing in a flattened elongate blastomere. (a) Unilateral furrow in upper, operated cell and symmetrical furrow in lower control. (b) Control division completed. (c) Circular opening above the base of the furrow. (d) Maximum extent of unilateral furrow. (e) Unilateral furrow partly receded, asters in operated and controls. (f) Early furrows in operated and controls. Numbers show minutes elapsed since (a). Bar = 50 μm. (From Rappaport and Rappaport, 1993, and with the permission of Academic Press)

events. The minimum period between the beginning of stimulation and the completion of furrowing is normally 13 minutes, or 27% of the 48-minute cycle, but under experimental conditions, the same activities can be extended to 32 minutes, or 67% of the cycle. It is also apparent that division activity is possible whether the nuclear membrane is present or broken down, whether the cortical stiffness is maximal or near minimal, and apparently without a fixed relation to a specific mitotic stage.

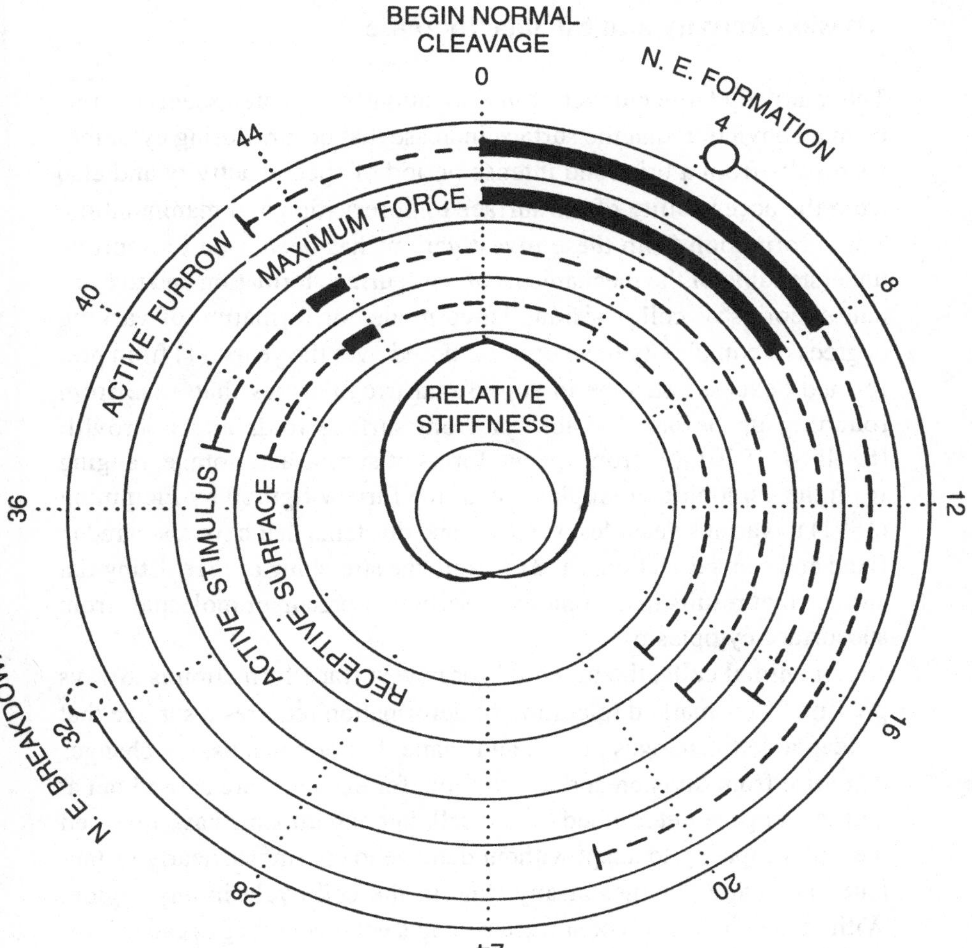

Figure 8.2 Summary of time relations among division-coordinated events in the second and third cleavages of sand dollar eggs. Data for maximum force were taken from Rappaport (1977); data for relative stiffness were taken from Yoneda and Schroeder (1984). The complete cycle requires 48 minutes. Dark bar = normal duration or minimum time necessary for activity. Dashed line = time during which event can occur. (From Rappaport and Rappaport, 1993, and with the permission of Academic Press)

Division Activity and Surface Increase

The relatively large number of investigations in this area seems to arise from a conviction that the surface increase that occurs during cytokinesis results from a brief and intense period of special activity and also from the accessibility of the surface to observation and manipulation. The information from these investigations has been used to improve understanding of the mechanisms of new surface formation and to evaluate theories of cell division. Three modes of formation of varying degrees of complexity have been proposed over the years: (1) from preformed "excess" surface in superficial irregularities that arise from rounding up before division, or from surface folds and microvilli (Szollosi, 1970); (2) from special forms of subsurface storage, ranging from the osmophilic cytoplasm near the furrow figured by Flemming (1891) to various vacuoles, vesicles, and diastema that have appeared in fixed and stained sections; and (3) from the stretching of preexisting flat surface that results in the random insertion of precursor molecules from subsurface cytoplasm.

In animal cells, the potential for new surface formation is always present. Their marked toleration of deformation requires a surface that is adaptable to changes in area and shape. In normal cells, the changes that arise from division and locomotion, for example, are carried out at a time and pace determined by the cell, but experiments have revealed the cell's capacity to adapt without damage to enormous, nearly instantaneous, shape changes at any time in the cell cycle in any region. Without any indication of surface damage, echinoderm eggs can be artificially constricted until they appear to have cleaved (Horstadius, 1973; Rappaport and Rappaport, 1988), or they can be cut, subjected to multiple punctures (K. Dan, 1943a; Rappaport, 1968), stretched into ribbons (Scott, 1964) and elongate cylinders (Rappaport, 1960), folded (Rappaport and Conrad, 1963), repeatedly extruded (Rappaport, 1961), or locally distended by aspiration (Rappaport, 1992) or by inward mechanical pressure (Rappaport, 1985). These manipulations and others can be imposed at any time in the cell cycle without damage and without interfering with cytokinesis. In the interests of simplicity, there is cause to wonder whether the surface formation that must accompany normal division requires any special cyclical activity.

Documentation of normal surface changes

The methods used for studies of normal surface changes require cells with regular geometry and an inability to change shape except as part of the division process. For these reasons, cleaving eggs have furnished the most detailed and reliable information. Quantitative descriptions of the simple and easily seen events of cytokinesis are not easy to make. As Hiramoto (1958) pointed out, even among sea urchin eggs from the same female there are size differences and differences in sphericity. It is also true that the mitotic apparatus may not be precisely centered, the furrow may therefore be somewhat unsymmetrical, and it may not exactly coincide with the equatorial plane of the cytoplasmic sphere. Pooling of the data from measurements on a number of cells requires that the data from each cell be normalized as for an egg having a radius of unity. It is not surprising that Hiramoto's (1958) study of protoplasmic movement in a sea urchin egg stands alone and that comparable studies of dividing tissue cells do not exist and may not be possible.

SMALL EGGS. The pattern of new surface formation in sea urchin eggs was the subject of a controversy between Motomura and K. Dan (see Chapter 2) that appears to have produced the first attempts at careful documentation. Motomura believed that all the new surface is formed in the furrow, whereas Dan proposed that it could form elsewhere, especially at the poles. Dan and his collaborators (K. Dan, Yanagita, and Sugiyama, 1937; K. Dan, Dan, and Yanagita, 1938; K. Dan and Dan, 1940, 1947; K. Dan and Ono, 1954; K. Dan, 1943b) undertook a series of investigations in which they tracked the movement of clay particles attached to the surface. They found that surface increase is not confined to the furrow, and they described a wave of expansion that begins at the poles and progresses toward the furrow region. The necessary new surface appeared to arise from the stretching of the preexisting surface. Their results were qualitative and were limited to particles that lodged by chance on the part of the egg margin that was clearly visible when the microscope was focused on the principal diameter. Calculation of area changes from linear particle movement on the spherical surface was difficult, and the relation between movements on the surface and the underlying endoplasm was not clearly described. Ishizaka (1958), using similar methods, showed that two narrow, annular surface regions that straddled the equator neither changed

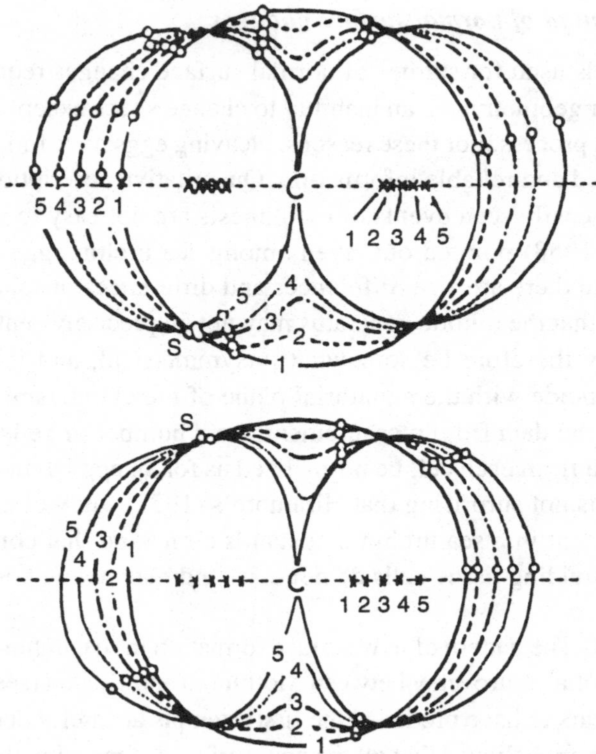

Figure 8.3 Series of sketches, subsequently superimposed, of carbon particles applied to egg surfaces to visualize displacement of the cortex. Top, *Mespilia globulus;* below, *Hemicentrotus pulcherimmus.* S indicates position of stationary surface ring. (From Ishizaka, 1958, and with the permission of the *Journal of Experimental Biology*)

in diameter nor moved relative to the geometrical center of the sphere of the uncleaved egg during division (Figure 8.3). He named the regions "stationary surface rings."

Hiramoto (1958) refined the methods of K. Dan and his collaborators and emphasized the quantitative aspects of the investigation. His results indicate that the stationary rings demarcate two somewhat differently behaving surfaces. Wolpert (1960) pointed out that about 80% of the new surface is formed between the stationary rings and the poles. The region between the rings and the equator is mostly contractile, but it also contributes the remaining 20%. The data are summarized in

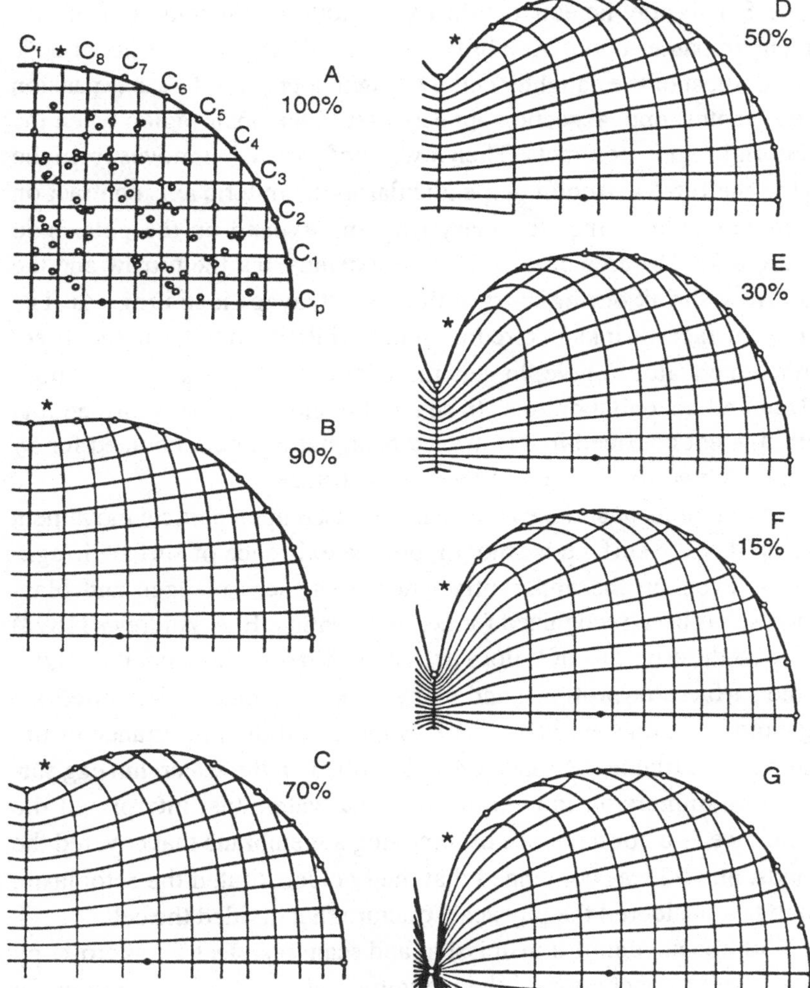

Figure 8.4 Diagram indicating protoplasmic movement during cleavage of *Clypeaster* egg. A quarter of the egg is shown. (A) Stage 100%; (B) stage 90%; (C) stage 70%; (D) stage 50%; (E) stage 30%; (F) stage 15%; (G) stage 0%. Cross lines within the egg indicate different strata of the endoplasm in the largest optical section of the egg. Movements of white circles on the cell surface in (A) – C_p, C_1, C_2, C_3, C_4, C_5, C_6, C_7, C_8, C_f – can be followed in the other figures. Black circles indicate positions of the astral center. (From Hiramoto, 1958, and with permission of the *Journal of Experimental Biology*)

Figure 8.4. While the equator shrinks to nothing, the length of the arc that encompasses about 50% of the uncleaved egg stretches about 60% to cover most of the daughter cell. During that process, linear expansion in the greatest optical section appears first in the polar region when the equatorial region contracts. Then a wave of expansion moves from the pole to the furrow. Annuli perpendicular to the mitotic axis contract on the furrow side of the stationary ring and expand on the polar side (Figure 8.5). The area of most regions expands, but the furrow and the subfurrow shrink during most of the time cleavage is in progress. The furrow surface shrinks in circumference while it stretches in the direction from equator to pole. In general, Hiramoto's results confirm those of Dan and his collaborators. In sea urchin cleavage, new surface formation is not confined in a restricted area, but can be accounted for by varying degrees of stretch of preexisting surface.

Calculations of surface area changes based upon particle movement assume that the surface is smooth, but the existence of surface irregularities, even on the apparently smooth surfaces of eggs and blastomeres, has been recognized for nearly a century. E. A. Andrews (1898) reviewed descriptions of "filose activity of finer pseudopodial threads" on the surfaces of protozoa and many living chordate and invertebrate eggs and blastomeres. Just (1939) emphasized the importance of the "ectoplasmic filaments" that Herbst described on the sea urchin egg surface after immersion in calcium-free sea water. Just interpreted the appearance as a consequence of removing a membrane that covered the threads, and he was convinced that the two constituted the ectoplasm, which he considered the physical structure that divided the cell.

With the advent of transmission and scanning electron microscopy came a better conception of the magnitude of surface irregularities. In sea urchin eggs, the number and length of microvilli change with time (summarized, Schroeder, 1981b). The microvilli elongate at fertilization and at a second time before the first cleavage. According to Schroeder, the aggregate of the surfaces of the microvilli increases the total membrane area to four times what it would be were the egg surface a smooth sphere. This would mean that the surface gain at cleavage constitutes an increase of 6% rather than 26%. The microvilli have been proposed as an immediate source of membrane at division, but there is no marked difference in the concentration of microvilli during cleavage, as would

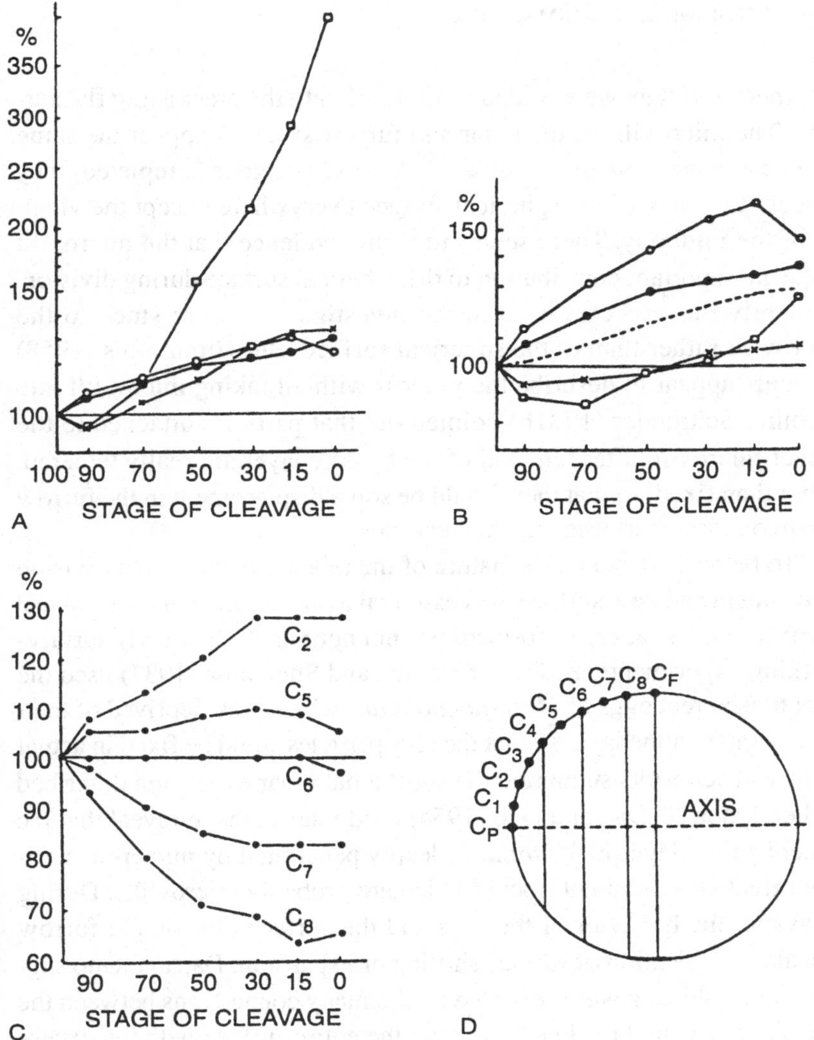

Figure 8.5 Linear and area changes in the cleaving sea urchin egg. (A) Linear changes of the surface along the line of the greatest optical section. ○ = length between C_p and C_4 (polar region); ● = length between C_4 and C_6 (subpolar region); X = length between C_6 and C_8 (subfurrow region); □ = length between C_8 and C_f. The stages of cleavage refer to the ratio of the diameter of the furrow at C_f to the initial diameter, given in percentages in Figure 8.4. (B) Changes in regional surface. The regions are those defined by the lengths of the surface in (A). The broken line represents the total area of the egg. (C) Linear changes in the perpendicular distance from designated points on the surface to the line of the mitotic axis during cleavage. (D) Positions of points on the surface before cleavage. (After Hiramoto, 1958, with the permission of the *Journal of Experimental Biology,* and Wolpert, 1960, with the permission of Academic Press)

be expected if they were folded or flattened into the preexisting flat surface. The microvilli of the polar and furrow surfaces appear the same when cleavage is nearly completed. After cleavage is completed, they obscure the view of the spherical surface everywhere except the vicinity of the midbody. There seems to be no evidence that the microvilli make an important contribution to the spherical surface during division. Evidently the particles used in the investigations were stuck to the microvilli rather than to the spherical surface. But Hiramoto's (1958) numbers appear to describe the process without taking microvilli into account. Schroeder (1981b) pointed out that particles attached to the longer microvilli after removal of the hyaline layer are really tethered, rather than fixed, so that there could be some displacement in the furrow where opposed surfaces may be very close.

To better understand the nature of the relation between the division mechanism and new surface, it is essential to know whether division can occur when the pattern of formation is not normal. In their early surface-marking experiments, K. Dan, Yanagita, and Sugiyama (1937) used the very transparent eggs of *Mespilia globulus*, which they deprived of their thick, rigid hyaline layer so that the clay particles could be fixed in actual surface. The results supported the substantial polar expansion described earlier. Later, K. Dan and Ono (1954) used vital stains to reveal that the intact hyaline layer in *Mespilia* is deeply penetrated by numerous radially oriented attachment fibers (which are probably microvilli). During cleavage, the behavior of the rays and the surface outside the furrow reveals no indication of surface shifting or expansion. Dan and Ono suggested that shifting was prevented by the many connections between the fibers and the rigid hyaline layer over the entire surface with the exception of the equatorial zone, where they appear stretched and broken. This circumstance led them to conclude that the newly formed furrow surfaces of the normal egg with an intact hyaline layer must be formed by the stretching of a narrow strip of surface material that girdles the precleavage equator. This conclusion was supported by the results of experiments in which extensive areas of the surface of sand dollar eggs were physically restrained (Rappaport and Ratner, 1967). Their experiments showed that completed cleavage was prevented only when the movement of surface close to the furrow was prevented. Surface formation in *Barnea candida* cleavage appears to resemble that in normal *Mespilia* eggs (Pasteels and de Harven, 1962). *Barnea* is a bivalve mollusk whose 45 μm diameter eggs undergo no visible changes at fertilization. The surface is invested with a regular arrangement of microvilli up to 1 μm long that are inserted for about four-fifths of their length into an extracellular layer of chorionic material that superficially resembles the hyaline layer of sea urchin eggs. The space between the chorionic material and the egg surface is about 0.2 μm wide and does not change during cleavage. A population of distinctive subcortical granules is distributed within about 1 μm of the surface. Because the microvilli are firmly fixed in the chorion, the normal *Barnea* egg is an ideal object for the study of surface behavior during division. Pasteels and de Harven found that the formation and deepening of the furrow did not greatly deform the chorion; the microvilli remained attached to the surface, so that they were stretched to several times their original length before they parted (Figure 8.6). In some cases, bits of detached cytoplasm remained with the chorion as the furrow surface, with its associated subcortical granules, progressed through the deeper regions of the egg. Pasteels and de Harven detected no specialized structures in the cytoplasm in advance of the furrow. The microvilli revealed that the region of the original egg surface that gave rise to the interblastomeric surface was an annulus 3.5 μm wide that was stretched to five to six times its original area by the time cleavage had been completed. Because the plasmalemma thickness did not change during the process, the researchers reasoned that it was augmented by molecular reorganization. These observations and experiments suggest that the pattern of new surface formation in small eggs may be facultative and dependent upon physical circumstances associated with the periphery.

The early meticulous studies of division-related surface behavior in echinoderm eggs appear to have been stimulated primarily by a desire to know more about the physical nature of the division mechanism, but, in fact, surface changes continue after division is completed in these cells. K. Dan, Dan, and Yanagita (1938) observed that attached kaolin particles moved into the furrow during cleavage then, after the completion of cleavage, moved away from the point of last connection between the blastomeres toward the poles. K. Dan and Dan (1940) found that the movements of endogenous cortical pigment vacuoles in *Strongylocentrotus pulcherimmus* coincided with the movements of the

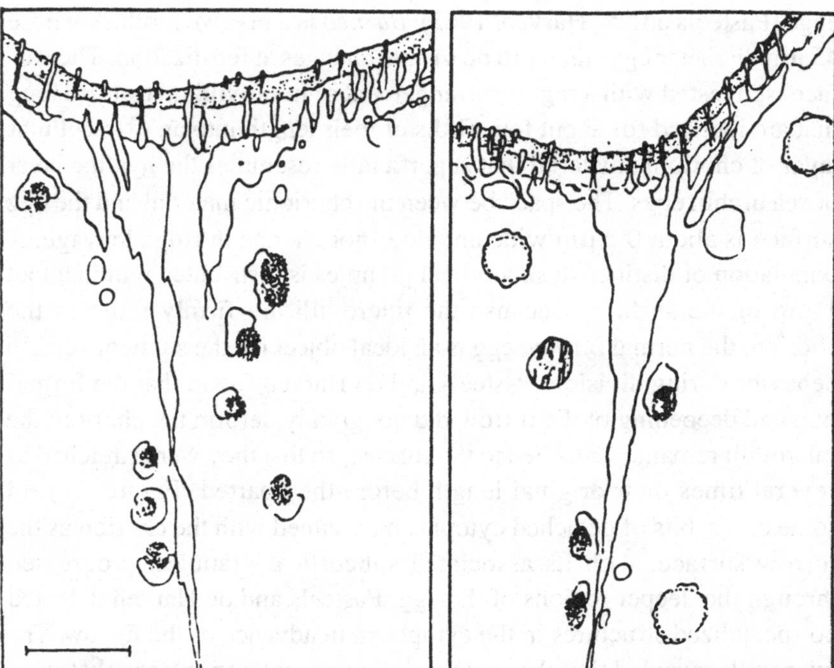

Figure 8.6 Tracings of electron microscope photographs of *Barnea candida* furrows. Furrow almost completed (left) and complete (right). In the two cases, subcortical granules are dragged downward. Microvilli, normally at the periphery, are drawn downward in the zone of the furrow. They appear to be firmly attached to the chorion. Note, in the center of the left-hand figure, a protoplasmic lobe remaining adherent to the chorion. Bar = 1 μm. (From Pasteels and de Harven, 1962, and with the permission of *Archives de Biologie*)

attached kaolin particles. They further observed that pigment accumulated in the furrow during cleavage and formed a dark region where it concentrated around the site of the last interblastomeric connection. After cleavage, the concentration of pigment dispersed so that the cortical distribution of pigment was at first uniform; then the pigment apparently disappeared in the interblastomeric region (Figure 8.7). K. Dan and Dan attributed the disappearance to a combination of optical illusion, vacuole movement toward the poles, and new surface formation. The unpigmented area they equated with regional new surface formation, and they estimated that it covered about 35% of the blastomere surface. They speculated that the specific site of new surface formation was the furrow tip. Division-related

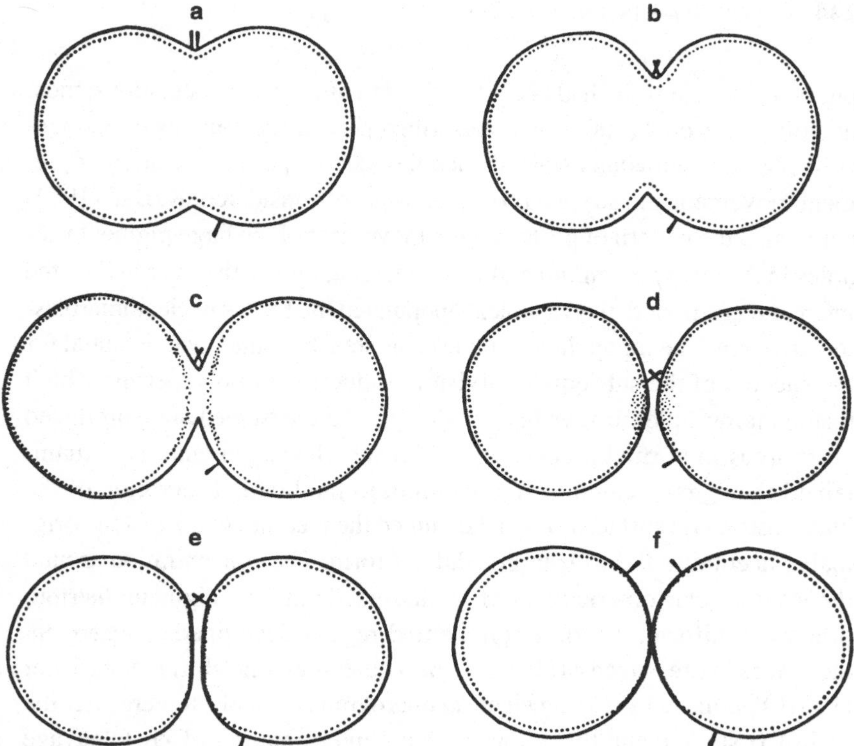

Figure 8.7 Diagram of typical *Strongylocentrotus* experiment to show, simultaneously, behavior of kaolin particles adhering to surface, extragranular zone, and granular layer. (Changing outline of cleaving egg and positions of adhering particles are traced from photographs. Positions of granules are indicated by dots.)

(a) Cleaving egg; kaolin particles are situated in furrow region; extragranular zone has narrowed at poles, presumably because of stretching.
(b) Kaolin particles are carried inward by deepening of furrow.
(c) Extragranular zone in furrow region widens; at poles, it becomes extremely narrow.
(d) Extragranular zone regains uniform width around blastomeres; granules are accumulated at point of last connection between blastomeres.
(e) Formation in furrow region of new surface, lacking in extragranular zone. Kaolin particles have been pushed out of the furrow; accumulated granules have dispersed: They are now lined up in close proximity to newly formed surface.
(f) Late interkinetic stage; extragranular zone over polar two-thirds of blastomeres has widened as though released from tension; upper particles are now well outside of furrow, at boundary between old and new surfaces. (From K. Dan and Dan, 1940)

pigment vacuole shifts had been described in previous investigations, most notably by Fischel (1906), who described similar movements of endogenous pigment and vitally stained vacuoles, but the possible relation of pigment movement and surface formation was not considered. K. Dan (1954) confirmed the pattern of postcleavage movement of the large pigment vacuoles in *Arbacia punctulata* and observed that, during the later half of the intercleavage period, the vacuoles repopulated the formerly clear interblastomeric area. The repopulation of the clear area, he found, was essential for the success of the subsequent division. Induction of polyspermy, which results in multiple furrows in a single egg at the first cleavage, produced clear areas in normal association with each cleavage plane. Dan maintained the eggs in calcium-free medium to facilitate observation of the interblastomeric surfaces. When he diluted the medium to 80% of the original concentration, the clear area did not form. The concentration around the last interblastomeric connection dispersed until the pigment distribution was uniform, but no unpigmented region developed. Subsequent cleavages were successful, with the usual pigment shifts. Schroeder (1981b) published scanning electron micrographs of a blastomere after the end of first cleavage that support Dan's interpretation of postcleavage events. The micrographs show more widely-spaced microvilli around the remnant of the furrow surface, which is consistent with the idea of regional surface stretching, but perhaps may not be consistent with the idea of entirely *de novo* surface formation. Investigation of changes in microvillar spacing provides an opportunity to study the relation between the capacity for cleavage and both the cortex and the cell surface. The unessential nature of the normal relation is strongly suggested by the egg's ability to cleave, despite failure of local pigment reduction in dilute medium. K. Dan (1954) proposed that localized pigment depletion results from polar contraction, which implies that, after division, the cortex is divided into two large regions of different functional states separated by a boundary located in a region that can be recognized with a simple light microscope. The two regions behave identically during the second cleavage. Dan (1954) also found that blastomeres in which pigment remained segregated failed to cleave, suggesting that cortical reorganization is normally required. Although experiments in which cells fail to cleave must be designed and interpreted cautiously, comparison of the ultrastructure of competent and noncompetent cortex could be interesting.

LARGE EGGS. A special pattern of new surface formation is implied by a paling of the surface that develops near the cleavage furrow in some, but not all, amphibian eggs in which the animal surface region is pigmented. The phenomenon has suggested that the localized intrusion of new unpigmented surface dilutes the concentration of surface pigment that was deposited during egg formation. In contrast with most invertebrate and eutherian mammal eggs, amphibian eggs are measured in terms of millimeters rather than scores of microns. Their size and their physical consistency make them relatively flaccid, so their normal dimensions are determined in many species by the surrounding vitelline membrane. When viewed from an angle that permits observation of both the animal and vegetal hemispheres, the animal pole appears flattened so that a fluid-filled space exists between the animal polar surface and the closest part of the spherical vitelline membrane. The space may be especially large in salamander eggs, and its existence enabled early amphibian embryologists to remove the membrane with sharpened forceps without damaging the egg. The dimensional changes that follow membrane removal clearly show its importance. The demembranated egg, when it comes to rest against the flat bottom of a container, flattens into a disc with dimensions determined by the elasticity of the cortex and by gravitational forces. As a result, membrane removal has the effect of rapidly increasing the tension at the animal surface at the same time that total egg surface increases.

Selman and Waddington (1955) investigated cleavage of the European newt egg in detail using time-lapse photography, vital staining, and elastimeter measurements, as well as sectioned and stained fixed eggs. They described an initial movement of the margin of the pigmented zone toward the animal pole in the cleavage plane. Wrinkles that are oriented perpendicular to the cleavage plane then form in the surface within a short distance of the very early furrow. The pigmented cortex on either side of the furrow then moves poleward, and lighter cortex appears adjacent to the cleavage plane. In the last phase of the process, unpigmented cortex moves back toward the cleavage plane and is subsequently confined to the region where the resulting blastomeres are in contact. Overall, Selman and Waddington observed no net change in the surface area of the pigmented cortex, which would imply that the necessary new surface originated in the pale furrow walls. They rejected the possibility that simple, local surface stretching might be responsible, because they calculated that it would require a 10-fold increase, and their elastimeter measurements failed to reveal any stiffness decrease in the supposedly stretched region. These findings led them to adopt the diastema splitting theory described in Chapter 2, in which the mechanisms that divide the cell and form the new surface are identical.

K. Dan and Kojima (1963) repeated some earlier experiments of Waddington (1952) that involved making cuts through the surface and into the amphibian egg cytoplasm in various relations to the established furrow. They were able to show that isolated cell fragments containing a part of the active furrow could complete division. Their results led them to suggest that propagation of a contractile furrow over the surface of the amphibian egg is followed by the centripetal growth of a septum that separates the blastomeres.

In their transmission electron microscope studies of newt cleavage, Selman and Perry (1970) observed that the surface of the early, shallow furrow is pigmented and that accumulations of pigment granules and yolk platelets extend in a sheet about 15 μm wide and 70 μm deep in the cytoplasm of the cleavage plane. In position and size, the accumulation suggests the diastema. The higher resolution afforded by electron microscopy also revealed that a good portion of the region that had previously been considered to be diastema was in reality the closely apposed walls of the already formed furrow. Microvilli first appear in the bottom of the groove over the contractile filaments and up to 30 μm on either side before unpigmented surface appears in the furrow walls. Selman and Perry point out that surface irregularities were unaltered throughout cleavage except in a strip less than 20 μm wide, in which the irregularities become flattened in the early stages. They estimated that the flattening could provide less than 1% of the necessary new surface. They observed few instances of vesicles fusing with the expanding, unpigmented furrow surface and rarely observed vesicles in the cleavage plane ahead of the furrow.

Kalt (1971a, 1971b) recognized in cleaving *Xenopus* eggs an ectoplasm that was divisible into cortical and subcortical cytoplasm. The ectoplasm varied in thickness from about 30 μm at the animal pole to 5–15 μm at the vegetal pole and was distinguishable from the underlying endoplasm. As the time for cleavage neared, the subcortical region of the ectoplasm thickened and expanded toward the zygote center in the future cleavage plane. This region appears to be the diastema described by other investigators. Kalt found that furrows, whether they were in the animal or vegetal region, were always bounded by ectoplasm that contained pigment granules, vacuoles, and some small yolk granules. As furrows deepened, they were surrounded by cytoplasmic vacuoles. Areas of glycogen underlying the furrow walls were released into the furrow space by a process that at the same time added new membrane to the furrow surface. This process was considered to contribute to both cytokinesis and blastocoel formation.

Denis-Donini, Baccetti, and Monroy (1976) emphasized the possible role of microvilli in new surface formation during *Xenopus* cleavage. Their scanning electron microscope studies revealed more-abundant and thickened microvilli in the presumptive furrow area, which flattened and gave the impression of being incorporated into the walls of the deepening furrow. In *Rana pipiens*, the early furrows do not usually develop pale surfaces (Rugh, 1951), and bands of microvilli form in the furrow grooves on both sides of the base (Beams and Kessel, 1976). Although the relation between the bands and the furrow base is the same through several cleavages, the width of the bands varies. The microvilli are restricted to the bands and to a region ahead of the advancing furrow tip before it extends around the entire cell circumference. Outside these regions, the nearby surface may bear a few stubby projections, and farther away it appears even smoother. Beams and Kessel described no ridges or lamellipodia.

Bluemink and deLaat (1977) extensively reviewed ultrastructural and physiological information concerning amphibian cleavage and concluded that the preexisting membrane does not contribute new surface and that membrane formation is restricted to the furrow region. They reasoned that free-living eggs require a permeability barrier and that insertion of new membrane over the whole surface would reduce the effectiveness of the barrier. New surface, they proposed, would have less effect if confined to the intercellular region. They suggested that new surface may be assembled by fusion of Golgi vesicles either with existing membrane or with a more complex assembly of glycoprotein in the membrane, or by molecular insertion of phospholipids, neutral lipids, and lipoproteins. Scanning electron microscope studies of the surfaces of fertilized eggs and two-cell, four-cell, and eight-cell embryos of three species of newts led Semik and Kilarski (1986) to the opinion that the calculated constant total surface they found during that period resulted from decreased microvillar and microfold surface that compensated for the increased new surface formed in the furrows.

T. J. Byers and Armstrong (1986) radioiodinated surface proteins of demembranated, precleavage *Xenopus* eggs and followed protein redistribution by fixing and sectioning eggs at different stages and then preparing radioautographs. Near the end of cleavage, the outer, pigmented surface and a narrow band at the furrow base (or "cleavage head") were heavily labeled. The lightly pigmented area on the furrow walls was also lightly labeled. Byers and Armstrong concluded that the leading edge of the furrow is formed mostly from original animal cap surface, whereas the furrow walls form from mostly unlabeled sources. They suggest that vesicles that underlie the surface near the junction between the pigmented and pale areas may be a source of new surface that is inserted among preexisting surface elements. It might be instructive to use their methods to compare the normal furrow wall with that which forms in an artificial constriction made by the pressure of a glass fiber or the tightening of a hair loop or silk fiber, using the techniques of the older amphibian embryologists. The artificial constriction could be made at any time in the cycle, including the time when an active furrow is present.

Bieliavsky and Geuskens (1990) compared by electron microscopy the organelles isolated by differential centrifugation and *in situ* organelles of intact, cleaving *Xenopus* eggs and, based upon their observations, proposed that formation of interblastomeric cell membranes results from sequential participation of vesicles derived from mitochondria, Golgi apparatus, and smooth endoplasmic reticulum. They proposed that the bulges that form on mitochondria in the cleavage plane become separate and enlarge after they assimilate lipid droplets. These vesicles move toward the surface in the upper furrow region, where their membranes are inserted into the plasma membrane. They also reported that vacuoles containing Golgi vesicles are present, and they suggest that these vesicles are inserted into the precursor membranes, along with other vesicles derived from the smooth endoplasmic reticulum. Other large vesicles that may have been formed by the coalescence of vesicles originating from the smooth endoplasmic reticulum contain flocculent material and become associated with the cortical plasma membrane at some distance from the upper border of the furrow. Bieliavsky and Geuskens suggest that the contents of those vesicles could stimulate the surface to form the shallow surface groove that appears at the beginning of division and could contribute material used in the formation of the contractile ring.

Study of static preparations of fixed cleaving eggs does not clearly reveal the pattern of new surface formation. Investigators who adopted this method conveyed the general impression that the majority of new surface appears close to the region where the pigmented and unpigmented regions join, but surface-marking studies of living newt eggs suggest otherwise. Sawai (1987) and Ohshima and Kubota (1985) stripped the vitelline membranes from newt eggs and marked the surface in the region of the early furrow with carbon particles. Timed photographs of the particle positions during cleavage revealed that although the pigmented margins on opposite sides of the furrow move apart, the distance from the particles to the nearest margin is relatively unchanged

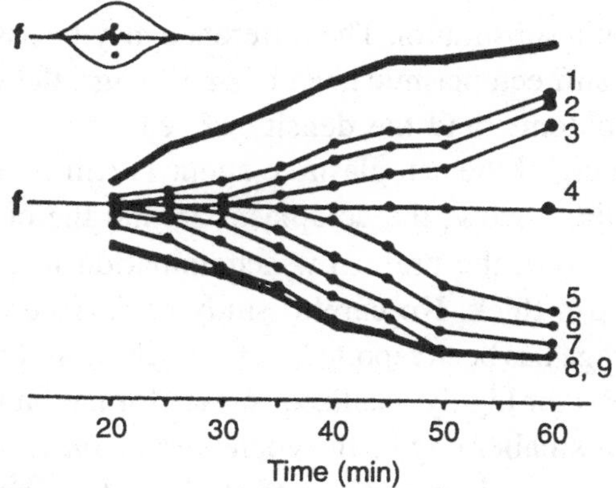

Figure 8.8 Graphic representation of particle movement on the pale surface of the cleaving newt egg in the middle stage, in the direction perpendicular to the cleavage plane. The top figure shows the rough positions of a cluster of particles on the pale surface at the time when the study was started. The movements of nine particles (1–9) are traced. All the particles except no. 4 shifted away from the furrow bottom (line f); particle no. 4 remained at the bottom. The two thick outer curves show the border between the pigmented and unpigmented surfaces. Time = minutes after the appearance of the furrow. (Sawai, 1987, and with permission of the Zoological Society of Japan)

and their distance from the cleavage plane changes greatly (Figure 8.8). The distance between particles scarcely changes. The results imply that new surface formation is concentrated in narrow strips close to, and on either side of, the cleavage plane. When the egg is spherical and confined within the vitelline membrane, division into two similar hemispheres requires a surface gain of about 50% (Wolpert, 1960). Were this amount of new surface so rapidly formed in such a restricted area by participation of ultrastructurally demonstrable organelles, one might expect visible evidence of the process within a few micrometers of the base of the furrow, but numerous studies have thus far failed to reveal it.

Most cleaving amphibian eggs are larger than most cleaving invertebrate eggs, but the diameter of the egg of *Beroë ovata* (1.0–1.2 mm) is roughly the same as that of many frogs (Rugh, 1948), including *Xenopus* (0.9–1.0 mm). Not only are they both spheres of about the same diameter, but their mitotic apparatuses are also eccentric, with *Beroë's* appearing to be closer to the surface. It follows that their furrows are unilateral, with little deformation of the vegetal surface during cleavage. The inability of *Xenopus* eggs to remain spherical after removal of the membrane was previously described, but *Beroë* eggs retain their spherical shape when they rest on a substratum. The difference may be associated with the 15 μm thick, stiff ectoplasmic layer of *Beroë* eggs, the specific gravity of the egg contents, and the density of sea water. Yatsu (1912a) described a thinning of the ectoplasm to about 11 μm when it is drawn downward into the furrow; the ectoplasm around the cleavage head, which is formed from the prefurrow accumulation in the equatorial plane, is 33–35 μm thick. No careful study of surface movement in ctenophore cleavage has been reported, but the behavior of the ectoplasm suggests little need for highly localized surface formation in the furrow. In the similar but smaller egg of the coelenterate *Spirocodon*, K. Dan (1958) used attached carbon particles to study surface movements, or, more precisely, stationary surface ring behavior in the presence of a unilateral furrow. He found that the stationary surface rings inclined toward each other in the animal region as the furrow deepened (Figure 8.9). This behavior would be expected if preexisting animal-region surface were drawn into the deepening furrow.

The pronounced difference in volume between eggs and tissue cells is attributable mainly to the amount of yolky material stored in the for-

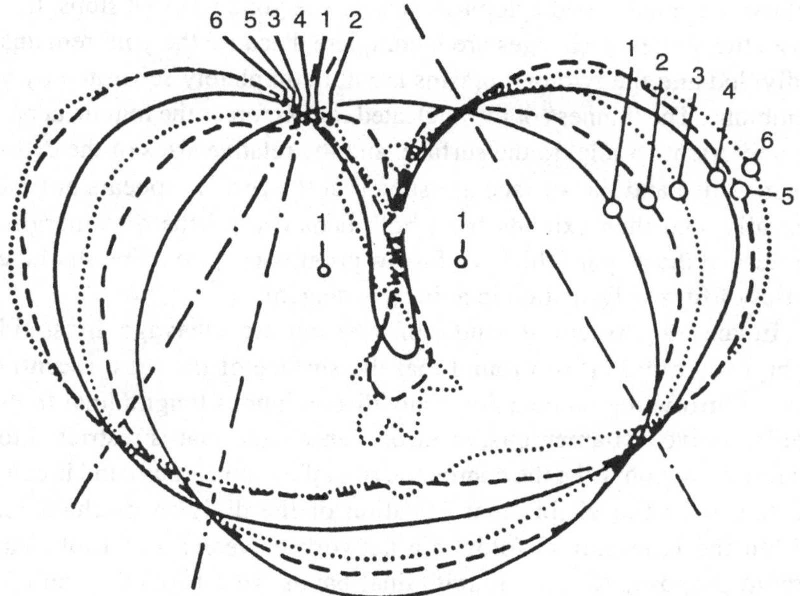

Figure 8.9 Superposition of contour sketches of cleaving egg of the medusa, *Spirocodon saltatrix*, seen perpendicular to the egg axis. In spite of indefiniteness of vegetal intersection points, it is clear that the stationary surface rings lean toward each other, embracing a very obtuse angle. (From K. Dan, 1958, and with the permission of the *Journal of Experimental Biology*)

mer. Yolk is metabolically inactive, and its inert qualities are such that, at one time, the rate of division of a cell or of a part of an egg was considered to be inversely proportional to its yolk content (Balfour's Rule), but Wilson (1928) pointed out many exceptions to the generalization. From species to species, the degree of mixture of yolk and cytoplasm is variable. At one extreme of the spectrum, the two are almost completely separated, although at the beginning of development, they are contained within one membrane. The term "telolethical" is applied to eggs in which the yolk is accumulated at one pole. When the nucleated, active, less yolky cytoplasm is gathered in a thin, superficial disc at the opposite pole, it undergoes a cleavage pattern that has been termed "meroblastic." Typically, birds, reptiles, and bony fish embryos cleave meroblastically. In this pattern, the furrow cuts downward from the free surface of the nucleated cytoplasm toward the yolk, where it stops. In a sense, the earliest cleavages are incomplete, because the yolk remains undivided and the two cytoplasms are not completely separated by a membrane. The thinness of the nucleated region forces the mitotic apparatus to orient parallel to the surface, and the relative sizes of the cytoplasmic disc and the surface are such that the furrow appears at first centrally, and then extends from both ends toward the disc margin. Because of the way in which the furrow progresses, it contains all of the phases of furrow formation in a linear arrangement.

In an ultrastructural study of meroblastic cleavage in squid embryos, Arnold (1969) found that the surface of the most recently formed furrow region contains many discontinuous longitudinal folds that lie above a filamentous, electron-dense band that is restricted to the furrow region. Like the contractile ring, the filamentous band is considered to be the visible manifestation of the division mechanism. Within the longitudinal folds is a network of vesicles and tubules. Arnold proposed that the longitudinal bands were formed when the width of the filamentous band shrank during the early stages of its formation. The deepening of the furrow is associated with a reduction in the number of longitudinal folds, and he suggests that they are flattened and assimilated in the walls. Subsequent studies of the surface with scanning electron microscopy revealed that microvilli formed on the surface in advance of the furrow tip. In time, the microvilli fused into longitudinal folds. Arnold (1974) proposed that the folds were the principal source of plasma membrane that lies between the blastomeres. The complete separation of the central part of the population of dividing cells results from the furrow bases of the first and second cleavages. These extensions cut parallel to the upper boundary of the yolky region in the absence of preformed microvilli and folds, and Arnold (1976) presumed that any necessary new surface was formed from internal sources, like cytoplasmic vesicles that fuse with the plasma membrane. But he concluded from measurements of membrane length in thin sections that about 75% of the required additional surface came from the flattening of the longitudinal folds.

Meroblastic cleavage takes place in bird eggs while the internally fertilized egg moves down the oviduct toward the outside. An electron microscope study of cleaving chicken eggs revealed complex furrow-associated structures that may be possible sources of new surface (Gipson, 1974). Glycogen-filled folds were present along the upper edge of shallow furrows, as were irregularly arranged, sometimes elaborate, protrusions that projected into the furrow space. Areas of concentrated, convoluted membranes and vesicles were present in the adjacent cytoplasm. In deeper furrows, the wall protrusions were reduced in the upper regions, but present close to the furrow base. Additional complex furrow base bodies, with complex arrangements of sinuous rays and sheets of surface containing yolk spheres and glycogen, extended into the cavity of the furrow bases.

TISSUE CELLS. In principle, surface contour irregularities are a potential immediate source of material for rapid surface expansion. The rounding up of tissue cells that often precedes division appears to reduce the surface area, and the possibility that such a cyclically formed excess could be briefly stored in some form and then be rapidly retrieved and redistributed during cytokinesis has been recognized for about a century (Rhumbler, 1897). Ultrastructural techniques have revealed superficial contour irregularities that Wolpert (1963) estimated would permit the apparently smooth sea urchin egg to cleave without adding new surface. Scanning electron, phase contrast, and differential interference microscopy are especially suited to the observation of cultured tissue cells. These techniques have focused attention on events and structures that appear on cell surfaces and margins. In a study by Porter, Prescott, and Frye (1973), scanning electron microscopy of confluent cultures of Chinese hamster ovary cells (CHO) revealed microvilli, blebs, and ruffles that varied in number during the cell cycle . They were numerous during the G1 period, and then, except for the ruffles, they tended to diminish during S when the cells spread over the substratum. During G2, when the cells thickened before they rounded up, the number of microvilli and ruffles increased, and in late G2, the long filopodia that characterize the surface of the mitotic cell appeared. The cyclical formation and loss of surface irregularities may be associated with the overall shape changes that characterize the cycle in confluent CHO cultures. When Rubin and Everhart (1973) used the same technique to study CHO cells in low-density cultures where cell-to-cell contact was rare, they found that the cells remained rounded up and covered with blebs throughout the cycle. These observations confirmed those of Follett and Goldman (1970), who found that cultured fibroblast cells bore more microvilli and blebs when they were rounded up than when they were flattened and spread upon the substratum, whether or not the cells were dividing. The surface projections, they proposed, are a means by which surface is conserved and subsequently used to cover the cell in its normally flattened interphase state. The idea that microvilli are rapidly formed and depleted by events that affect total cell surface area is supported by the dehydration experiments of Albrecht-Buehler and Bushnell (1982). They found that following the removal of 50–75% of the water content by immersion in hyperosmolal sorbitol, cultured tissue cells formed numerous microvilli whether they were in rounded division stages or flattened. Upon return of cells to the normal medium, the surface reverted to its pretreatment form.

In order to avoid the complications that arise from the study of cultured flattened cells, Knutton, Sumner, and Pasternak (1975) examined the cyclical changes in surface morphology of mouse mastocytoma cells, which remain spherical in suspension culture because they are unable to adhere to a substratum. They found that cells, as they progressed through the cycle, increased in size and their surfaces were increasingly elaborated with microvilli. Calculations showed that the doubling of volume was accompanied by a doubling of surface area by reason of microvillar development, so that the ratio of volume to surface area remained constant throughout the cycle. In general form, the mastocytoma cells looked like cleaving sea urchin eggs without the hyaline layer. The polar surfaces were relatively smooth, and the microvilli were concentrated in the furrow region. The regional differences in microvillar number and size may reflect surface redistribution during which time there is a flow of membrane toward the furrow region (Pasternak, 1987). The cell appeared to produce surface at a nearly constant rate and store the excess until it was required at cytokinesis. Erickson and Trinkaus (1976) made similar observations with spread and rounded baby hamster kidney cells and found that when the previously smooth, spread cells rounded up before division, they became richly invested with microvilli and blebs. Erickson and Trinkaus also speculated that the surface irregularities represented membrane reserves that were drawn upon in division and subsequent spreading. Their figures of dividing cells show none of the equatorial accumulation and polar diminution of microvilli that appear to characterize dividing mastocytoma cells. Measurements of the amount of major plasma membrane components (proteins, phospholipids, and cholesterol) doubled between the start of the new cycle (early G_1) and the beginning of cytokinesis (late G_2) (reviewed, Pasternak, 1987). The increase in surface components appeared to be accounted for by the increase in the number in microvilli.

Osmotic swelling of mastocytoma cells was accompanied by the reduction of microvilli in a study by Knutton et al. (1976). These results complement those of Albrecht-Buehler and Bushnell (1982), which revealed that microvilli form on flattened cells in response to osmotic shrinkage. Together, results of the two studies suggest the possibility that the distribution of preformed surface between the protruding and the nonprotruding regions may be determined by physical events in the superficial regions. But Schroeder (1981b) has cautioned against attempts to implicate the phenomenon of surface storage in superficial irregularities as a universal and important event in animal cell cytokinesis. His reservations are based upon the variety in microvillar distribution and behavior among different cultured cells and the absence of information indicating a necessity for microvilli in cleavage divisions. Schroeder also pointed out that, whatever may prove to be the roles of microvilli, blebs, and folds, their existence means that the amount of new surface that might be required during division, when expressed as the percent of original surface per minute of division activity, is far less than would be needed were the surface of the dividing cells smooth.

Animal cells appear to have several ways in which they could increase their surface area at division, but despite extensive study and speculation, we know little beyond the possibilities. All animal cells appear to use a division mechanism based upon active equatorial constriction, and in all cases, the position of the mechanism is directly or indirectly established by the mitotic apparatus; however, there is no evidence of any single mechanism for surface increase, and as yet we cannot be certain that division requires any special surface properties or activity. The idea that surface irregularities result from the cyclical gathering of temporarily excess material to prepare for division seems attractive for those cases where the accumulation exists, but the phenomenon is hardly universal. The microvilli of PtK_2 cells are roughly uniformly distributed over the surface at the beginning of cleavage and remain so until completion. The number of microvilli in the furrow does not differ from the number over nonfurrow areas (Sanger, Reingold, and Sanger, 1984). There are, in fact, examples in which microvilli do not appear until the time of cleavage. The surface of cleaving *Halocynthia roretzi* (an ascidian) egg is smooth until immediately before the first cleavage. At that time, microvilli appear over the entire surface and increase in number during division in each cleavage cycle up to at least the 32-cell stage (N. Satoh and Deno, 1984). The findings of N. Satoh and Deno suggest that not only are microvilli not stretched into smooth surface during division, but also their formation probably increases the amount of surface that must be formed in the brief period of division activity. Among dividing cultured tissue cells, some show a reduction of microvilli at the poles (Knutton, Sumner, and Pasternak, 1975), others at the equator (Schroeder, 1981b), and others show no regional change (Erickson and Trinkaus, 1976). The cyclical formation of microvilli and of other forms of "stored surface" in cultured cells is correlated with the rounded state (Follett and Goldman, 1970; Rubin and Everhart, 1973), but Sit, Bay, and Wong (1993) found that the mitotic rounding up of human Chang liver cells is accompanied by a reduction in surface area, which may be caused by plasma membrane internalization. It is also true that many types of cultured cells can divide in the flattened state (reviewed, Rieder and Hard, 1990) when surface irregularities are minimal.

Because tissue cells are relatively small, are often covered with surface irregularities, and tend to adhere to the substratum, studies of surface behavior during division are understandably scarce. Wang, Silverman, and Cao (1994) traced the movement of surface receptors on dividing, cultured, normal rat kidney cells with attached fluorescent latex beads. They found that movement began after anaphase onset and was most pronounced in the interzone region between the separating chromosome groups. Movement at the poles was random, which may be associated with the local attachments to the substratum in these flattened cells. Beads initially located on both sides of the equatorial plane tended to converge upon it, but no evidence of local bead accumulation was presented. The authors concluded that the forces responsible for bead movement are not restricted to the equatorial plane. Scott (1960b) and

Rappaport and Rappaport (1976) reached a similar conclusion from observations of pigment granule movement in *Arbacia* eggs. The nature of the forces responsible for bead movement is not made evident by simple observations of the event. Wang, Silverman, and Cao favor cortical flow as the mechanism that most likely underlies surface movement, but an increasing distance between beads that develops in the direction of their movement can be directly measured on some of the figures. The increased distance between the beads would not be expected if their movement were driven by simple flow of cortical material.The pattern of bead movement suggests that subequatorial surface stretching analogous to that occurring in sea urchin cleavage may also take place.

It seems unlikely that any animal cell is limited to a single mode of new surface formation, and it is not unreasonable to suggest that the principal mode employed may depend upon the immediate circumstances. The phenomenon has prompted much observation, but little experimentation, at the cellular level. Approaches that combine simple physical manipulation with ultrastructural analysis appear likely to produce interesting results.

Bulk Cytoplasmic Movement

In animal cells, cytokinesis is the consequence of shifting cytoplasm out of the equatorial plane. The movements that are associated with division have been observed in living eggs with prominent cytoplasmic inclusions for about a century. The movements were at first thought to be consistent with the then prevalent opinion that the appearance of the mitotic apparatus was based upon radiating streams of cytoplasmic components that were directed toward or away from the aster centers; the fibrous structures that appeared in fixed and stained preparations were considered to be artifacts. The understanding of cytoplasmic movement, and its possible role in the division process, was plagued by the now familiar problem of discriminating between cause and effect. The phenomenon was at times thought to be involved in furrow establishment, or to be the physical mechanism of division, or to be a simple passive response to the immediate intracellular physical environment.

Von Erlanger (1897) studied cleavage in several nematode eggs in which rapid division activity is accompanied by striking cytoplasmic currents. He observed subsurface movement from the poles that converged at the equator, then turned downward toward the mitotic axis in the plane of the early furrow, and along the mitotic axis back toward the poles. Conklin (1902) deduced the existence of cytoplasmic currents in studies of fixed and stained, whole and sectioned Crepidula embryos. From positional changes of the mitotic apparatus and yolk spheres, he concluded that there is a vortical flow similar to that described by von Erlanger, although he did not see movement of yolk particles through the spindle. Spek (1918) restudied nematode eggs and found that the movement from the poles toward the equator coincided with the time when the spindle and asters were growing maximally (Figure 8.10a), and the movement in the equatorial plane paralleled nearby surface movement (Figure 8.10b). The equatorial flow appeared to begin and end at the same time that furrowing began and finished. When furrows were unilateral, the movement was restricted to the cleaving side (Figure 8.10c). Although attention was focused upon symmetrical flow patterns that seemed to be associated with the mitotic apparatus, other types of cleavage-related movement were noted by early investigators. Ziegler (1898b) described a reciprocating flow that crossed the equatorial plane of Beroë eggs during cleavage, and Chambers (1938) reported a similar phenomenon in dividing tissue cells.

There have been many assumptions and speculations about the phys-

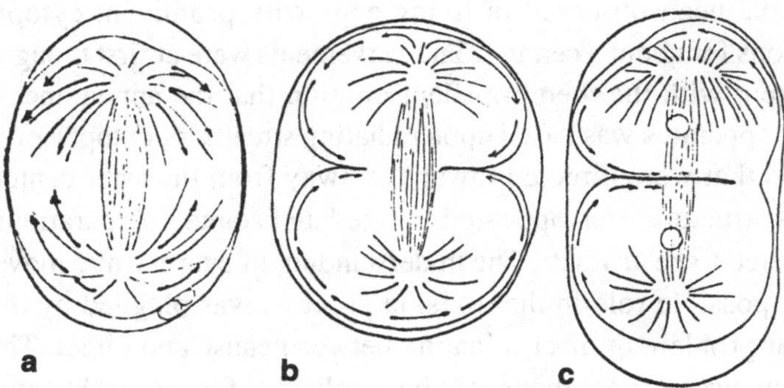

Figure 8.10 Patterns of cytoplasmic movement in *Rhabditis*. (a) Movement from the poles along the surface toward the equator during spindle elongation. (b) Movement in the equatorial plane parallel to the furrow surface. (c) Unilateral furrowing accompanied by unilateral flow in the equatorial region. (From Spek, 1918)

ical basis of the division-associated flow. Conklin (1902) and von Erlanger (1897) implied that it was active and autonomous. Henley and Costello (1965) proposed that it resulted from aster transport. Spek (1918) pointed out the similarity between the pattern in dividing nematode eggs and that which takes place in suspended oil drops when the surface tension of diametrically opposed surface regions is reduced (Chapter 2). He and others considered the pattern to strongly support the surface tension theory of cytokinesis. The early movement of cytoplasm out of the region between the spindle pole and the polar surface could be attributed to physical displacement by the concurrent spindle elongation and aster growth. The data of Hiramoto (1958, 1971b) generally confirm Spek's description and strongly suggest that after cleavage begins, endoplasmic movement can be accounted for by mechanical displacement caused by the advancing furrow and a kind of viscous drag exerted by the spreading surface.

The functional significance of the flow pattern remains enigmatic, although it has been the subject of speculation. Conklin (1902) proposed that the flow creates tension differences at the surface that subsequently divide the cell. In contrast, Spek (1918) appeared to be convinced that the movement was caused by preexisting surface tension differences and, by implication, that the movement itself played no physical role in division. Loeb (1895, cited in Spek, 1918), on the other hand, proposed that the vortical flows could exert enough force to bisect the cell. Cornman and Cornman (1951) and Henley and Costello (1965) proposed that the pattern of flow distributes a "furrow organizer" or cleavage stimulus that originates near the aster centers and accumulates at the equator where the streams converge.

Were the pattern of cytoplasmic flow causally connected with an essential, active component of the establishment or the operation of the division mechanism, disruption of the flow's normal pattern would be expected to affect cleavage, but no such effect has been demonstrated. Before and during furrow establishment, the cell shape and the pattern of movement appear to be closely related, but various types of cell deformation do not interfere with the process. If the visible flow were involved in furrow establishment, then its essential role would be completed before furrowing; however, in nematode eggs, furrowing and flow begin simultaneously (Spek, 1918). Changing the cell into a cylinder (Rappaport, 1960) or a torus (Rappaport, 1961, 1970) greatly changes the relation between the aster and the polar surface without interfering with division. Deep cuts and notches located near the equatorial plane also disrupt the normal pattern, but they have little or no effect (Scott, 1960a; Rappaport, 1970); neither does repeated extrusion (Rappaport, 1961). After the furrow is established, it appears to function regardless of internal rearrangements. Removal of the mitotic apparatus and its replacement with oil or sea water have no effect (Hiramoto, 1956, 1965); neither do slit punctures in the furrow region (Scott, 1960a). Also without effect are massive, reciprocal cytoplasmic flows induced by alternate mechanical compression of the polar regions and constant stirring of the cytoplasm in the division plane (Rappaport, 1966). And, finally, furrowing continues in fragments of the equatorial region that have been nearly (Scott, 1960a) or completely (Rappaport, 1969a) isolated from normal underlying cytoplasmic events.

In summary, it now appears that the bulk cytoplasmic movements that may be associated with cleavage are passive responses to physical events initiated by nearby structures. They play no demonstrably important role in cytokinesis.

Disappearance of the Contractile Ring

Effects of component interaction

The disappearance of the contractile ring must be counted among the consequences of cytokinesis, and this event appears to have stimulated as much commentary as the ring itself. The absence of contractile ring material during interphase (at least in cleavage divisions) implies cyclical dissolution of contractile ring material. Schroeder's (1972) documentation of the dimensional changes in the ring during cleavage indicates that, during the second half of the process, the volume of the ring decreases as the equatorial circumference shrinks, until the ring is completely absent shortly before the connection between the blastomeres is severed. The assumption of a sliding filament mechanism naturally led to the use of the more stable striated muscle as a model, and from that ensued a certain amount of wonder about the way in which the ring undergoes its unique dissolution. The hypothesis of a biochemical mechanism that depolymerizes actin filament material after interaction with myosin (Schroeder, 1975; Mabuchi, 1986) has been frequently considered, but there is no experimental evidence for the existence of such a biochemical mechanism. Persisting rings of actin, myosin, and associated proteins that constrict the distal ends of the midbody in PtK_2 cells (Sanger, Mittal, and Sanger, 1990; Sanger et al., 1994) suggest that depolymerization after interaction is not inevitable. Other division- and contraction-related events of a biophysical nature that involve component interaction and cell surface deformation that could contribute significantly to a decrease in ring volume must also be considered.

The distribution of actin and myosin at the time equatorial constriction begins resembles an actomyosin thread in its degree of disorganization more than it resembles a muscle. During contraction, a muscle changes from one highly organized state to another without significant volume change, but a contracting actomyosin thread may undergo considerable volume change as it changes from a disorganized state to one involving more order. Comparisons between the actomyosin threads' behavior during contraction and the contractile ring are edifying.

In actomyosin threads prepared from various sources, ranging from myxomycete plasmodia to vertebrate striated muscle, the actin and myosin in the uncontracted state form a random network (D'Haese and Komnick, 1972a). F-actin filaments are decorated with myosin in the arrowhead configuration, and the actin filaments are cross-connected by small myosin aggregates. The size of the myosin aggregates suggests that they may be dimers and tetrameres, and they range in size up to one-third the size of a myosin filament. When relaxed threads are passively stretched, the actin filaments align, but no myosin filaments develop. In the prefurrow equatorial cortex, actin and myosin form a meshwork of randomly distributed pore size (Yonemura and Kinoshita, 1986; Schroeder and Otto, 1988). The correspondence between the distribution of actin and that of myosin in the network is close but not complete. Both thread and equatorial cortex are characterized by close association of actin and myosin. The near correspondence of their localization in the cell would be improbable if they were independently linked to the plasma membrane, but they would be expected to be perfectly colocalized if they were linked only to each other. The distribution might be accounted for by linkage to other filamentous cortical elements, as well as by linkage to each other.

The effects of ATP-induced contraction on actomyosin threads depend upon the initial organization of the threads' internal elements. The unstretched thread, in which the filaments are randomly oriented, contracts to 12% of its original volume as its original length and diameter decrease by 50% (D'Haese and Komnick, 1972a). The filamentous network is condensed while fluid is apparently expressed. When threads are passively stretched before contraction or when they contract isometrically, the filaments line up parallel to the direction of stress. Upon contraction, thick filaments appear to be associated with actin filaments; the amount of overlap among the filaments increases, and there is filament condensation. Stretched threads contract almost exclusively in the longitudinal direction, with little or no change in diameter. They may decrease to 25–30% of their original length as, in contrast with striated muscle, they decrease to 25% of their original volume. The contraction of stretched threads is more rapid than that of unstretched (D'Haese and Komnick, 1972a, 1972b).

The dimensions of the contractile ring are also reduced in the longitudinal (circumferential) direction. The measurable thickness does not change, but the width decreases somewhat (Schroeder, 1972). The transition from a random network to a circumferential array could change the equatorial properties during contractile ring activity from those of an unstretched thread to those of a prestretched thread. The rearrangement of ring components would enhance contractility and tend to stabilize ring width and thickness. The structure of the contractile ring and the presence of nonfilamentous material do not favor careful analysis of the amount of filament condensation that may occur as the ring functions (Schroeder, 1972). Actomyosin threads demonstrate that an elongate mass of chaotically arranged components can hold together, self-organize into a nonrandom configuration, and actively shorten in one dimension in the absence of an enveloping membrane or of any other potential anchoring site other than its constituent actin and myosin (D'Haese and Komnick, 1972a), and that those changes can be accompanied by a significant decrease in volume.

Effects of cell surface deformation

Hiramoto's (1958) study of protoplasmic movement during sea urchin egg cleavage revealed that there is no significant slippage between the surface and the immediately adjacent endoplasm. The shifting is made possible by linear surface stretching, which varies in amount in different regions and is maximal at 370% in the furrow region by the time cleavage is completed (Figure 8.4). In normal eggs, furrow surface stretching is accompanied by a regional surface increase of about 25%, but local surface increase is not necessary, because colchicine-treated eggs complete division, although the furrow surface area decreases slightly while its surface is linearly stretched to about 350% of its original length (Hiramoto, 1958). The linear stretching appears to result from the local surface's yielding to the localized tension generated in the contractile ring; for the purpose of this discussion, linear stretching is more important than an increase in area. In his analysis of surface changes in *Arbacia* cleavage, Scott (1960a) photographed the process through a polar surface and likened the furrow surface to an iris with a decreasing aperture. Most of the iris-like region originated from Hiramoto's stretched furrow surface region, and the pigment granules are clearly condensed on the pupil margin.

The effects of stretching upon the contractile ring can be estimated from Hiramoto's (1958; see Figure 8.4) data. *Clypeaster japonicus* eggs are 110 μm in diameter (Osanai, 1975) and the intervals between the cross lines in the grid that indicate strata of the endoplasm are 5.5 μm. In Figure 8.4A, the line next to the equator, indicated by an asterisk, lies immediately outside the outer limit of the region that will lie over the contractile ring, which, according to Schroeder (1972), extends about 4 μm from the equatorial plane. When the contractile ring first appears at the 90% stage (Figure 8.4B), it occupies 87% of the zone between the equator and the asterisk. During cleavage, that zone is stretched to a greater extent than any other part of the surface, and much of it loses the underlying contractile ring ultrastructure. Between the 90% stage and the 15% stage (Figure 8.4F), the proportion of surface between the asterisk and the equator underlain by the contractile ring is reduced to 10%, assuming that its 4 μm width persists. The stretching continues until completion. When the equator is reduced to about 6% of its original diameter, the contractile ring structure disappears (Schroeder, 1972). The pronounced linear stretching raises the probability that during cleavage, a large proportion of contractile ring material is carried away from the equator, where its characteristic structure is lost as it reverts to an open meshwork. Absence of contractile ring organization does not exclude the possibility of contractility, for example, during the earliest cleavage deformation. Eggs in which constriction of the furrow base is blocked show deformation in the subfurrow region that would not be possible if all circumferential contraction were confined to the zone underlain by the contractile ring (Figure 4.1b).

These observations and deductions suggest that several aspects of the progressive changes in the form and function of the contractile ring may be simple consequences of its initial structure, dimensions, location in the cell, and ability to contract. The early contraction of the belt-shaped region can decrease the cell diameter and reorient the underlying cortical filamentous network of the belt region. The contraction of the initially random network can also substantially reduce the network volume. The subsequent parallel arrangement of the contractile ring's constituent filaments can restrict its dimensional change to a decrease in length. The constriction of the equatorial ring can strain the contiguous surface and attached cortex so that the greatest degree of cortical stretch is located at the ring perimeter, and during division, material that was associated with the contractile ring periphery can be moved away from the equator. The contractile ring undergoes changes that could simultaneously increase and decrease its cross-sectional area. The decrease in circumference could cause an increase in cross-sectional area; at the same time, the shrinkage that accompanies contraction in structures of similar organization (for example, an actomyosin thread), as well as the attenuation of contractile ring material caused by stretching at the ring periphery, could cause cross-sectional area to decrease. The ring may retain its cross-sectional dimension to an approximate degree by a balance among the factors that tend to increase or decrease this dimension and by the strength of the linkages among the ring's anisotropically arranged elements. These possible modes of ring volume control do not require special dynamic activities or unknown mechanisms. The existence of these modes does not preclude the existence of active actin depolymerization, but the modes raise questions concerning the necessity of depolymerization. Neither do these modes of ring volume control explain the final disappearance of contractile ring organization. A hypothetical exercise such as the foregoing also illustrates the ease of framing alternative explanations when the available information is descriptive rather than decisive.

9

Informative Variations on the Normal Process

Unequal Division

The usually orderly and predictable pattern of division that occurs during plant and animal development led to generalizations like Sachs's rules and Hertwig's modification of Sachs's rules (discussed in Wilson, 1928). These generalizations reflected the idea that the pattern of divisions is affected by the size and proportions of the protoplasmic mass that is undergoing division. Wilson (1928, p. 982) translated Sachs's rules as follows: "(1) Cells typically tend to divide into equal parts. (2) Each new plane of division tends to intersect the preceding one at right angles." O. Hertwig's modifications related the statements more closely to the circumstances in animal egg cleavage patterns (Wilson, 1928, p. 984). "(1) The typical position of the nucleus (and hence of the mitotic figure) tends toward the center of its sphere of influence, i. e., of the protoplasmic mass in which it lies. (2) The axis of the spindle typically lies in the longest axis of the protoplasmic mass, and division therefore tends to cut this axis transversely." The general subject interested embryologists at the turn of the century because they had many opportunities to observe the relationships under ideal conditions, and because the cleavage pattern appeared to play an important role in the normal distribution of "formative substances" in the early development of some major animal groups. But the use of the term "rules" in combination with necessarily vague terms like "tends to," "typically," and "protoplasmic mass" led to some confusion and dissatisfaction among those who sought to understand the basis of the generalizations. In 1927, Morgan wrote (p. 487), "It has been suggested by Oscar Hertwig that the spindles develop in the direction of the greatest protoplasmic mass. This statement is sometimes spoken of as a law, but it should not pass unnoticed that it is little more than a restatement of what is found in certain cases, and it does not seem to apply in other cases. There are also other results that appear to contradict such a 'law'."

The basis of the near constant relation between the mitotic apparatus and the division plane was discussed extensively in previous chapters. The normal relation also obtains in cases of unequal division; that is, unequal division follows when the mitotic apparatus is eccentric and its axis coincides with a cell radius. An understanding of the process begins with an analysis of the mechanisms that position the mitotic apparatus when the position of the furrow is determined and then hold it in place during cleavage. The simple expedient of forcing normally spherical eggs to cleave while flattened between parallel flat glass plates was devised by Pflüger in 1884 (described in Morgan, 1927, and Wilson, 1928). He found that the division planes of flattened frog eggs were altered so that cytoplasmic regions of blastomeres formed under compression were forced to associate with nuclei they would not normally have contacted. After release, the population of blastomeres rounded up and developed into a normal embryo. This finding had important implications for theories of development at the time, and Pflüger postulated that the developmental potential of egg cytoplasm was uniform and that the way the cytoplasm was divided up and packaged with the nuclei was unimportant. Others repeated the experiment on eggs of different animal groups and, although the effect on development differed in some groups, the effect on the cleavage pattern was the same.

Most of those who repeated the compression experiment were interested in its developmental effect, but Ziegler (1894) appeared to concentrate his effort on understanding how flattening changed the division planes. Others had pointed out that compression put the longest protoplasmic axis in one plane and that subsequent furrows developed in the normal relation to the mitotic apparatus, but the cause of mitotic apparatus reorientation was unknown. In the first experiments, done with amphibian eggs, the mitotic apparatus was not visible, but Ziegler could clearly see the internal structures and events in sea urchin eggs (Figure 9.1a). He accurately described the association between elongation of

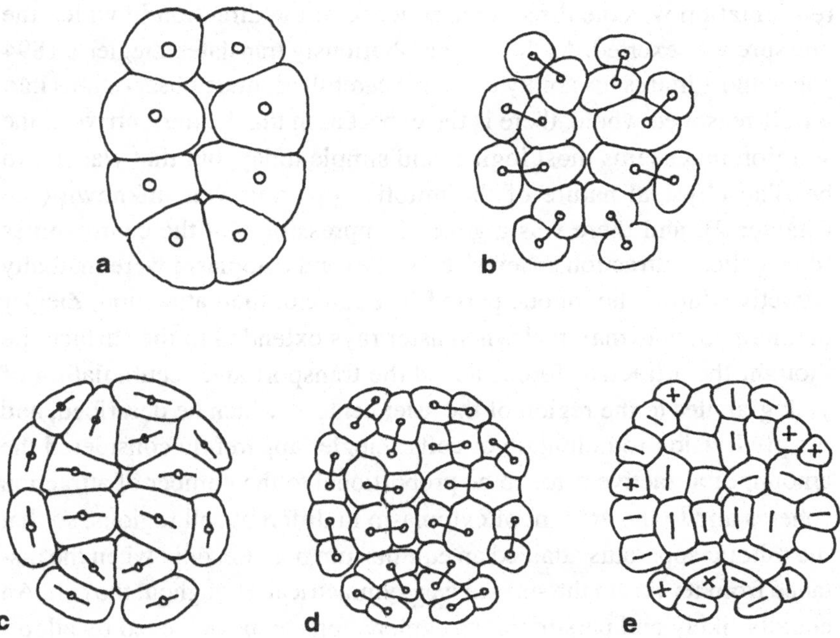

Figure 9.1 Cleavage of a flattened sea urchin embryo. At the 8-cell stage (a) all the blastomeres are in the same plane and the nuclei are central. After the next cleavage, the 16 rounded blastomeres are approximately the same size (b) and remain in the same plane. The straight lines joining the nuclei indicate the pairs of cells that originated from a single blastomere. Several minutes later, the blastomeres have flattened against each other, and subsequently the spindles orient parallel to the longest cell dimension (c). After cleavage, all 32 cells lie in the same plane (d), but before the next cleavage (e) some of the spindles are oriented parallel (-) to the plane of flattening and some are perpendicular (+), so that at the 64-cell stage, all the blastomeres will not lie in the same plane. (From Ziegler, 1894)

the aster rays and rotation of the mitotic apparatus until it lay parallel to the flattening glass surfaces. He recognized that reorientation and ray elongation often were not completed until shortly before the furrow appeared. He described similar events in an experimentally produced, elongate ellipsoid cell in which an initially small mitotic apparatus that lay perpendicular to the cell's long axis shifted to a central position and oriented parallel to the long axis as the asters expanded. His experiments and observations led Ziegler to propose that the important factor in the reorientation of the mitotic apparatus was the shapes that were imposed upon the compressed cells. Others before him had vaguely proposed that reorientation was the direct consequence of the direction in which the pressure was exerted. As the reader laboriously translates Ziegler's 1894 paper and admires his ability to weave careful, accurate observations into a well-reasoned whole, there is the expectation that he must arrive at the solution that seems most logical and simple today, but that was not to be. The physical nature of the mitotic apparatus was unknown (see Chapter 2), and there was a general impression that the centrosomes (then called "attraction spheres" by some) and cytoplasm were mutually attractive during the mitotic period. The force of their attraction, Ziegler maintained, was maximal when aster rays extended to the surface; he thought the attractive force caused the transport and accumulation of yolk granules to the region of the aster centers, which he described, and the predivision rounding up of cells. Ziegler apparently considered the amount of attractive force to be proportional to the number of attraction spheres and to the amount of cytoplasm in different cell regions, so that the mitotic apparatus attained an equilibrium position only when the distance from its axis to the surface was symmetrical throughout the cell. An equally likely and parsimonious explanation can be achieved by adopting Teichmann's (1903) assumption that visible aster growth is the consequence of an expanding, spherical zone of solidification located in more liquid surroundings. In this circumstance, the asters could push against the underside of the cortex and reach an equilibrium point when the repulsion forces were symmetrical. Chambers (1917b) substantiated Teichmann's speculation concerning the nature of the asters by microdissection experiments, but regardless of the mechanism that orients the mitotic apparatus, the results of the pressure experiments of the 1890s made unlikely Conklin's later (1902) contention that the plane of cleavage is predetermined in the cytoplasm and that the mitotic apparatus shifts to coincide with it.

It follows that unequal daughter cells are associated with unsymmetrical placement of the mitotic apparatus, but simple eccentricity is insufficient to cause inequality. When an eccentric mitotic apparatus is oriented parallel to a tangent, as in coelenterate and ctenophore egg cleavage, the subsequent unilateral furrow cuts the egg into equal blastomeres. Given the propensity of an eccentric mitotic apparatus to align itself parallel to a tangent with the guidance of the concave inner surface of the cortex, there is some question how the mitotic apparatus is maintained in the basically radial orientation necessary for formation of unequal blastomeres. Cleavages that result in strikingly unequal daughter cells are not unusual in invertebrate development, and early embryologists linked unequal daughter cells with unequal asters (reviewed, Conklin, 1902). Conklin was convinced that the asters were of equal size at the beginning of division and that the size differences arose after the beginning of cytokinesis as the aster and centrosome size accommodated to the volume of the enveloping cytoplasm as the furrow deepened. More recent investigations have shown that unequal cleavage may be associated with the migration of the nucleus or mitotic apparatus, asters of unequal size, and the attachment of the mitotic apparatus to the cortex.

Micromere formation

Unequal cleavage in invertebrate eggs frequently results in the sequestering of material with special developmental properties in one of the two blastomeres. Such a division occurs in the fourth cleavage of sea urchin eggs, when the four cells nearest to the vegetal pole divide unequally to form four smaller micromeres at the pole and four larger cells with dimensions that approximate those of the other eight cells that make up the embryo (Figure 9.2). Micromere cells are instrumental in the formation of the calcified, needlelike larval skeleton, and when unequal division is prevented by treatment with detergent, micromere formation fails and no skeleton develops, although the cytoplasm that would have been contained by the micromeres is still present in the embryo (Tanaka, 1976). K. Dan's (1979) observations of living cells convinced him that unequal division in this case is normally preceded in

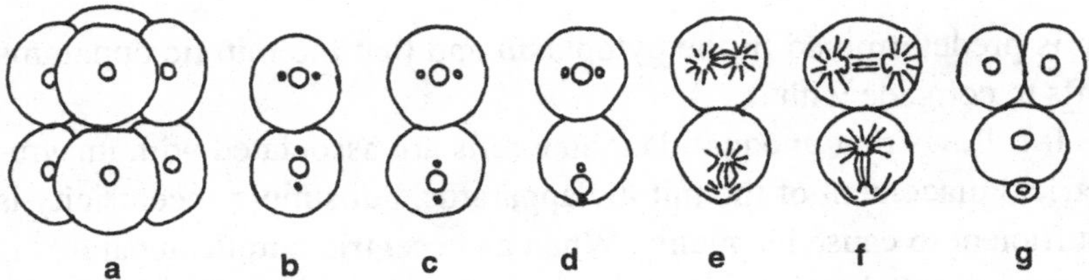

Figure 9.2 Nuclear behavior at the 8-cell stage in normal sea urchin development according to K. Dan. In the intact embryo (a), four animal cells rest upon four vegetal cells. In a single animal–vegetal cell pair (b), the vegetal nucleus rotates so that one of the associated centrosomes (dots) faces the vegetal pole. Then the vegetal nucleus and centrosomes move toward the vegetal pole (c, d). After nuclear membrane breakdown (e), the vegetal aster is confined between the spindle pole and the vegetal surface so that its rays are deflected (f). In the ensuing division, the cleavage planes of the animal and vegetal blastomeres are perpendicular (g) and the progeny of the vegetal blastomere are highly unequal. (From K. Dan, 1978, and with the permission of Academic Press)

each of the four vegetal cells by positioning of the centrosome near the region of the nuclear membrane that faces the vegetal pole and by the migration of nucleus and centrosome from an initially central position toward the vegetal region of the blastomere. The mitotic apparatus subsequently forms and, by mid metaphase, the aster closer to the surface appears smaller and flattened against the cortex. The micromere spindle pole also contains less centrosomal material than the macromere spindle pole (Holy and Schatten, 1991). Schroeder (1987), on the other hand, reinvestigated the process using an immunofluorescent method for the localization of tubulin and concluded that early nuclear migration precedes both centrosomal realignment and the development of distinct differences between the asters. Precocious micromeres can be elicited at the third cleavage by treatment with sodium sulfite, and the structure of the mitotic apparatus associated with the unequal division is unsymmetrical in the same way as that associated with normal micromere formation. Micromeres are not formed unless the asters are unequal in size, and the asters are not unequal unless the mitotic apparatus is radially positioned. When micromere formation is suppressed by detergent treatment, both mitotic apparatus movement and the normal size disparity are prevented (Tanaka, 1976; K. Dan, 1979).

Ziegler's (1894) compression experiment demonstrated that flattening can prevent micromere formation (Figure 9.1b), although he did not discuss it. K. Dan (1987) proposed that micromere formation fails in this circumstance because nuclear contact with the flattened cell surfaces prevents migration. He also showed that micromere formation could occur by chance when compression brought the mitotic apparatus close to the cell margin. The correlation between mitotic apparatus position and disparity in aster size suggests that the disparity is not intrinsic to the mitotic apparatus but imposed by its immediate surroundings. The result of the size disparity is that the geometric center of the spherical cell coincides with the center of the large aster, rather than with the midpoint between the aster centers. Czihak (1973) speculated that this circumstance arises because the vegetal cortex of the cell can cause nearby aster rays to disaggregate. Because the aster farthest from the vegetal surface would be unaffected, its extending rays would push against other regions of the surface, and it would center itself; at the same time, it would force the smaller aster to a position closer to the vegetal region. Czihak determined the effect of the removal of different amounts of uncleaved sea urchin egg vegetal cortex upon subsequent micromere formation. He found it necessary to remove nearly the entire vegetal half to suppress the micromeres, and he concluded that the special property of the vegetal cortex that affected peripheral aster formation originates when the animal–vegetal axis is established. Schroeder (1987) questioned whether differential aster growth can explain nuclear migration, because the asters are not very extensive at the time of nuclear migration. The importance of microtubules in mitotic apparatus relocation was substantiated by the findings that when mitotic microtubules are disassembled with colcemid, relocation fails and the asters and resulting blastomeres are nearly equal. Cytochalasin D has no effect on nuclear movement (Lutz and Inoué, 1982). The idea that the vegetal cortex is instrumental was supported by Morgan and Spooner's (1909) observation that micromere formation occurred at the vegetal pole of the embryo when the cytoplasm normally located in that region had been displaced by centrifugation. This special property of the cortex appears to develop three hours after fertilization, regardless of the number of intervening cleavages, because ultraviolet-induced cleavage delay results in micromere formation at the third cleavage simultaneously with micromere formation in controls at the fourth cleavage (K. Dan and Ikeda, 1971). A similar clock mechanism appears to be involved in the specification of cleavage time and form in *Ilyanassa* (Cather, Render, and Freeman, 1986).

Another aspect of micromere formation that must be accounted for is the precision that characterizes mitotic apparatus relocation. The smaller aster approaches the vegetal cortex whether or not the blastomere is normally integrated into the embryo (K. Dan, 1987). K. Dan, Endo, and Uemura (1983) discovered that the region of cortex that abuts the smaller aster is distinguished by a sparser population of subcortical vesicles than are the other free surfaces of the blastomere, and they speculate that the mitotic apparatus is attracted to that part of the cortex where the gaps between the vesicles are greatest. In support of their hypothesis, they pointed out that detergent treatment that interferes with mitotic apparatus relocation and micromere formation also disrupts the vesicular pattern (Tanaka, 1979). For this mechanism to operate, one would expect the vesicle-free region to form before the beginning of nuclear migration, but the data are not clear on that point. It should also be pointed out that the cortex in the regions where the blastomere surfaces abut is also vesicle free, but it does not appear to attract asters.

After the polar portion of the smaller aster has reached its closest proximity to the vegetal surface, its radiate structure appears to be lost or deflected toward the future division plane, and a fusion between the cortex and the truncated aster appears possible (K. Dan, Endo, and Uemura, 1983). An attachment in the vegetal area seems useful because the intruding furrow would tend to propel an unattached mitotic apparatus toward the opposite end of the cell, and the original furrow would probably regress as another furrow formed in the normal relation to the new position of the mitotic apparatus (Rappaport, 1985). More positive evidence for the attachment is desirable. K. Dan and Tanaka (1990), however, have shown that the mitotic apparatus remains close to the vegetal pole when the embryo is rapidly flattened and when the vegetal pole is pushed toward the cell center.

Unequal cleavage

In several major invertebrate groups, the early cleavages are patterned so that the substance of the vegetal pole region is restricted to a small number of blastomeres. In some it is accomplished by sequestering vegetal pole material in a temporary lobe at cleavage times (which will be described later), and in others the cleavages are unequal and the vegetal region is passed undivided to the larger daughter cell through several successive divisions. The 19th-century studies of this form of cleavage (reviewed, Conklin, 1902) emphasized the correlation between aster size and blastomere size, but Conklin maintained that the disparity was the effect, rather than a contributing cause, of unequal cleavage. A commonly used example of unequal cleavage is the egg of the clam *Spisula*, in which the first cleavage plane is tilted relative to the animal–vegetal axis and the resulting blastomeres are unequal (Allen, 1953). The refractility of the yolk particles obscures internal events, but Rebhun (1959) found that vital staining of cytoplasmic particles improved observation of aster-related activity because the particles tended to accumulate around the aster centers. The mitotic apparatus formed in the center of the egg and then slid along its axis so that one aster was close to the egg cortex and the other was roughly positioned in the geometrical center of the spherical egg. The relocation required about 45 seconds, and shortly after its completion, the aster closest to the surface began an oscillatory motion that in some cases did not cease until after the beginning of cytokinesis. (The oscillatory motion had been observed and drawn by earlier observers, and its significance is still unknown.) Shortly before cleavage, the egg elongated parallel to the mitotic axis and was divided by the furrow into unequal blastomeres, the larger of which contained the central aster. Rebhun's preparations were not well suited to observations of aster morphology, but he noted that during the oscillatory or rocking motion, a conical, relatively granule-free region occupied the space between the peripheral aster and the surface.

In a study of cleavage spindles isolated from *Spisula*, K. Dan and Ito (1984) found that the metaphase asters are initially about equal in size and then become unequal, but not, they felt, to an extent that would completely account for the relocation of the mitotic apparatus. As the smaller aster approaches the surface, the rays between its center and the surface are redirected into a roughly tangential orientation, and in some preparations, surface material remains associated with the smaller aster of the isolated mitotic apparatus. With improved viewing and recording methods, K. Dan and Inoué (1987) studied spindle migration in living cells and concluded that migration is brought about by the centering movement of the larger aster, which is a consequence of its large size. At that time, the position of the large aster is fixed and the position of the small aster subsequently becomes fixed by attachment to adjacent cortex.

Ganglion cell formation

Formation of certain tissues and organs during later development of invertebrate embryos may be based upon small populations of cells that are the source for a particular cell type. These cells divide into daughters of different developmental capacity in that the development of one is fixed so that it will give rise to, for example, nerve or muscle, and the other retains the qualities of the mother cell in that it can subsequently divide into unlike daughter cells; the result is that the population of basic or source cells remains relatively constant in number while the tissue cells accumulate and form bands or columns in the embryonic body.

The mode of formation of the ventral nerve cord in the insect midline is a typical example of this process (Snodgrass, 1935). The superficial ventral ectoderm forms bilateral neural ridges separated by a median neural groove (Figure 9.3a). Beneath the outermost layer of ectoderm that covers the ridges and the groove is a population of larger neuroblast cells that are separated from the superficial ectoderm by small cap cells (Figure 9.3b). The neuroblasts are polarized and are oriented so that the division plane is perpendicular to the embryonic dorso–ventral axis. Successive divisions produce a configuration in which the large neuroblast is positioned at the base of the column or strand of the smaller, more-dorsal ganglion cells that it had produced (Figure 9.3b). The ganglion cells become rearranged into the ventral nerve cord. Before they differentiate into neurons, the ganglion cells can be distinguished from neuroblast cells by the shape and size of the nucleus, by mitotic activity, and by the equality or inequality of their subsequent cytokinesis (Yamashiki and Kawamura, 1986). The neuroblasts are mitotically active when nerve cord formation is in progress. The region of neural development can be conveniently dissected from grasshopper embryos and maintained for some hours in relatively simple solutions and mounted in a chamber for microdissection experi-

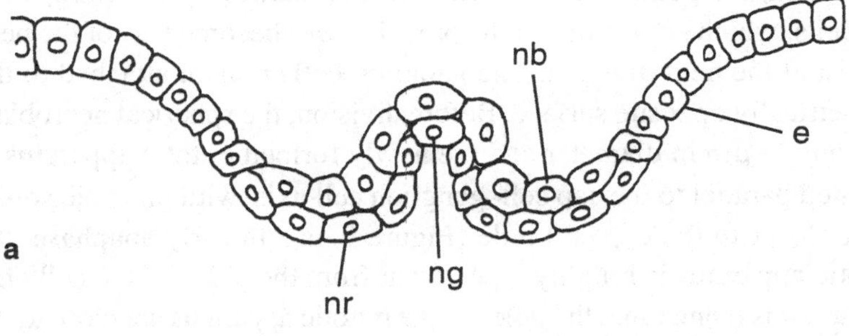

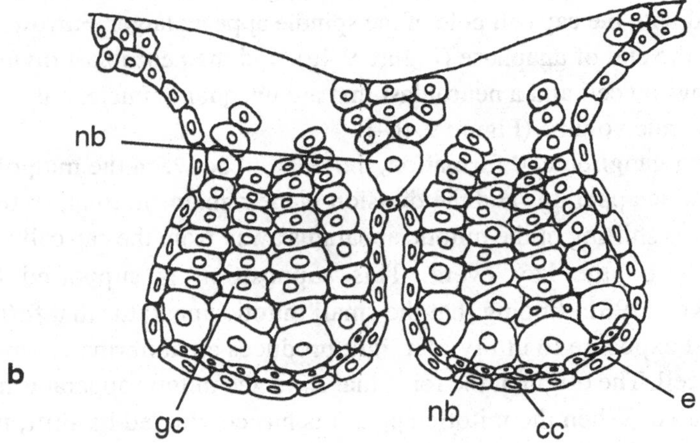

Figure 9.3 Schematic sections of a developing insect ventral nervous system. In the early stage (a) the ventral area neural ridges (nr) form on either side of the median neural groove (ng). The neuroblasts (nb) lie under the superficial ectoderm. Later (b), ganglion cells (gc) form columns in association with the neuroblasts, which are separated from the superficial ectoderm (e) by small cap cells (cc). (After Snodgrass, 1935)

ments (Carlson, 1952; Roberts, 1955). The large chromosomes allow precise determination of mitotic stage and, although the cells do not tolerate puncturing well, the elasticity and deformability of the surface permit rearrangement of the visible organelles without penetration of the interior (Carlson, 1952).

In situ, the polarity of the neuroblast is marked by the contiguous cap cells, which are located at the pole that will become part of the neuroblast at the next division. The ganglion cells remain attached to the diametrically opposite surface. Before division, the spherical neuroblast is about 25 μm in diameter and the newly formed mitotic apparatus is oriented parallel to the cap cell–ganglion cell axis, with one pole somewhat closer to the cap cell pole (Figure 9.4a). In early anaphase, the mitotic apparatus is roughly equidistant from the poles of the cell and, because it is elongating, the poles of the mitotic apparatus are close to the surface (Figure 9.4b–c). As the spindle enlarges, the cell elongates so that it becomes prolate. At the same time, the spindle shifts toward the ganglion cell pole and appears to attach there (Figure 9.4c–e), and the aster associated with the cap cell pole of the spindle appears larger. Furrowing begins at the end of anaphase (Figure 9.4g), and the neuroblast divides into a ganglion cell and a neuroblast that are unequal in nuclear as well as cytoplasmic volume (Figure 9.4h–j).

During ganglion cell formation, the relation between the midpoint of the mitotic apparatus and the division plane remains normal, so that the previous shifting of the mitotic apparatus away from the cap cell end appears to be the key event. This supposition is supported by Yamashiki's (1981) finding that the small mitotic apparatus that forms after local exposure to ultraviolet light produces an abnormally small ganglion cell. The origin of the force that shifts the mitotic apparatus has been debated. When the mitotic apparatus has completed its shift, the aster at the cap cell end is larger. Because the asters appear approximately equal in size when the mitotic apparatus is central, it is natural to assume that the simultaneous growth of one aster and shrinkage of the other within the confines of the cortex could propel the mitotic apparatus toward the end of the ellipsoid most distant from the larger aster (Kawamura, 1960, 1977). But immunofluorescence microscopy studies with antitubulin antibodies have revealed that the size of the asters in dividing neuroblasts is equal until after the shifting of the mitotic apparatus is completed (Yamashiki, 1990). A second phenomenon that may move the mitotic apparatus is a contraction wave that moves simultaneously from the cap cell pole to the ganglion cell pole (Kawamura, 1988; Kawamura and Yamashiki, 1990). The wave takes the form of a

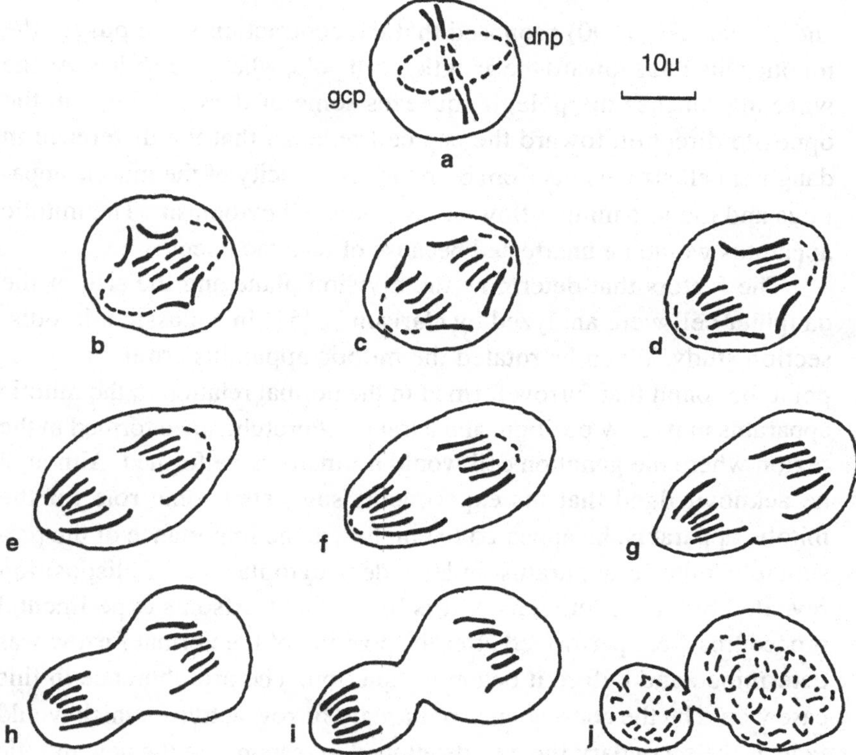

Figure 9.4 Division of an isolated grasshopper neuroblast. The chromosomes are shown diagrammatically and in reduced number for clarity. At metaphase (a), the spindle pole is closer to the daughter neuroblast pole (dnp) adjacent to the cap cells. The mitotic apparatus is equidistant from the poles in early (b) and mid (c) anaphase. As the cell elongates in the mitotic axis (c–f), the spindle pole remains close to the ganglion cell pole (gcp) and separates from the daughter neuroblast pole (dnp). The furrow develops in the plane of the spindle midpoint (g), and the daughter cells are unequal in nuclear and cytoplasmic volume (h–j). (From Roberts, 1955)

narrow constriction zone that moves at about 5 μm per minute and takes 5.7–10.6 minutes to traverse the cell. Furrowing begins when the wave has progressed to a point about halfway to the ganglion cell pole. Interestingly, cytochalasin B, which blocks cleavage, does not block the contraction wave but rather produces repeated waves of normal amplitude and direction that pass one at a time over the surface. Kawamura and Yamashiki (1990) proposed that the contraction wave pushes the mitotic apparatus toward the ganglion cell pole, where it attaches. As the wave approaches the pole, it squeezes some of the cytoplasm in the opposite direction toward the cap cell pole, so that the difference in daughter cell size results from both the eccentricity of the mitotic apparatus and the last-minute flow of ganglion cell cytoplasm. The mitotic apparatus would be unaffected because of its attachment.

The factors that determine the division plane and the size of the daughter cell were analyzed by Carlson (1952) in a classic microdissection study. When he rotated the mitotic apparatus around its midpoint, he found that furrow formed in the normal relation to the mitotic apparatus in its new position, and a transient protuberance formed in the region where the ganglion cell would normally have formed. Although he acknowledged that his experiments suggested some role for the mitotic apparatus, he appeared to emphasize the importance of the presumably mitotic apparatus–independent cytoplasmic predisposition revealed by the protuberance. It is likely that Carlson's experimental procedures were performed after the position of the normal furrow was determined and before it began to function. The protuberance in this case would be the consequence of primary furrow activity, which would fade as the secondary furrows developed in response to the new mitotic orientation (Kawamura, 1960, 1977; Yamashiki and Kawamura, 1990). Other factors, both internal and external, have been shown to affect daughter cell size. Mechanical removal of the cap cells and ganglion cells before early anaphase produces equal-sized daughter cells, but when the cells are removed by trypsin treatment, normal unequal division ensues (Kawamura and Carlson, 1962). On the other hand, neuroblasts divide with normal inequality after cap cell removal if the ganglion cells are not disturbed (Kawamura, 1965). Intracellular mechanical connections have also been implicated in the orientation of the mitotic apparatus and, thus, of the cleavage plane. Carlson (1952) deduced the existence of fibrous connections between the spindle poles and the cortex from the shifting of the spindle axis that occurs when a needle is passed between them. He also reported the presence of fibers in the same region of fixed preparations. When spindles are rotated by micromanipulation, they tend to return to their original orientation after the needle is removed, but when colcemid-treated cells are subjected to the same operation, the spindle does not return. Cytochalasin B has no effect (Kawamura, 1982). An electron-dense layer of material that appears at metaphase in the neuroblast cortex adjacent to the cap cells has also been described. The material's disappearance after mid anaphase coincides with the time when the mitotic apparatus loses its ability to return autonomously to its original orientation after rotation, and Kawamura and Yamashiki (1990, 1992) have postulated an important affinity between the dense layer and the polar cytoplasm.

Before neuron differentiation, neuroblasts and ganglion cells can be distinguished on the basis of nuclear morphology, mitotic activity, and equality or inequality of division at the subsequent cytokinesis (Yamashiki and Kawamura, 1986). These rapidly developing characteristics have enabled analysis of some of the factors that might determine the developmental fate of the daughter cells. When the neuroblast divides into equal daughter cells, both have neuroblast characteristics. There is a critical range of volumes that seems to determine cell fate: Cells smaller than 2,200–2,600 μm^3 become ganglion cells, and those that are larger become neuroblasts. This relation holds without regard to the original polarity of the nuclei and the cortex.

The structure of the furrow in dividing neuroblasts resembles that of other animal cells, and Yamashiki and Kawamura (1990) were able to accomplish a welcome combination of micromanipulation and electron microscopy. They found a circumferentially oriented microfilamentous layer about 100 nm thick in the cortex near the spindle midpoint that appeared 0.5–1.0 minute before the beginning of furrowing activity. Furrowing began at late anaphase and, as the furrow deepened, the microfilamentous band decreased in width but not thickness. The mitotic apparatus was rotated 90° and fixed 1.0 minute later. At that time, the protuberance that appeared at the original site of the ganglion cell developed and microfilaments oriented parallel to the original division plane appeared at its base. Four minutes after mitotic apparatus rotation, the microfilaments in the original division plane were disorganized and diminished. A new contractile ring appeared in the cortex near the middle of the rotated mitotic apparatus. The ring was incomplete, but the cell showed signs of furrowing activity (Yamashiki and Kawamura, 1990).

Polar body formation

In the late 19th century, when careful observation and reasoning led to fruitful insights concerning the relation between genetics and development, on one hand, and cytological detail on the other hand, it became apparent that meiosis in the egg is a process in which a specialized form of cell division creates a necessarily haploid gamete. At the same time, meiosis conserves and maximizes the amount of stored cytoplasmic material available for subsequent development of the zygote (historical review, Wilson, 1928). The extremely unequal form of cell division that accomplishes the process can easily be observed shortly following insemination in marine invertebrate eggs because, in most species, meiosis is in a state of arrest that is broken at the time of egg activation. The point in the succession of meiotic events at which arrest occurs varies among animal groups and, in a relatively small number, including coelenterates and sea urchins, the process is completed before fertilization (John, 1990). In brief, at the time of the meiotic divisions, the nuclear material moves from the central region of the oocyte toward the periphery and, sooner or later, develops into an apparatus that resembles that of mitosis, but is usually smaller. The mitotic apparatus usually orients so that its axis is parallel to a radius of the spherical egg, and in most oocytes, a bulge appears on the surface region at the point where one of its asters contacts the cortex (Figure 9.5). The mitotic apparatus shifts part way into the bulge as it

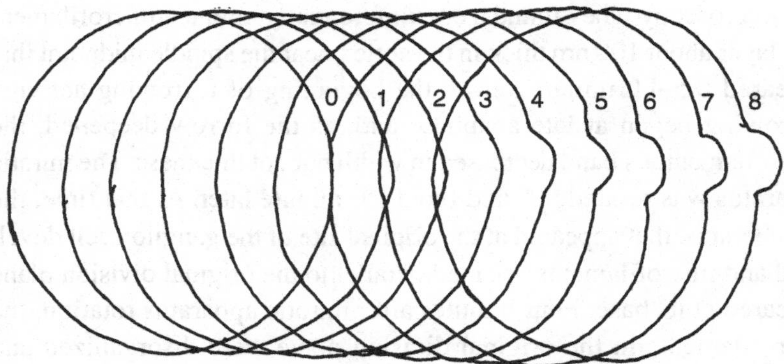

Figure 9.5 Tracings of changes in animal region marginal contour during polar body formation in *Pisaster gigantia* over a 10-minute period. The numbers next to each tracing indicate the time in minutes elapsed since the first observation. The undercutting of the protuberance began in the fifth minute.

enlarges. The connection between the nucleated bulge and the oocyte is severed, and the polar body, or polocyte, thus formed consists mostly of nuclear material. The ratio between the volumes of the polar body and its nucleus is variable, but in most cases, the polar body is relatively minute. Meiosis is essential to sexual reproduction, and the similarity in the appearance of the oocyte and the polar body among animal groups and the apparent participation of similar organelles in the formation process encourage the expectation that a single conserved mechanism is responsible, but significant differences do exist.

To the observer of the living oocyte, the formation of the bulge looks like a small protrusion in a restricted part of the egg perimeter, so the process was often called "polar body extrusion." It was apparent to early investigators that polar body formation could be considered a two-phase process in which bulge formation and subsequent separation involved different mechanisms (Yatsu, 1909). When the polar body separates from the oocyte, the relation between the mitotic apparatus and the plane of separation approximates that in normal cytokinesis, but the events preceding separation – such as the migration of the mitotic apparatus, its change in symmetry, its special orientation relative to the oocyte surface, its adhesion to the surface, and the formation of the bulge – are not usually encountered in other forms of cytokinesis. Each of these special events has been investigated.

Study of the movement of the maternal diploid nucleus, or germinal vesicle, toward the animal pole can be most conveniently done on eggs that are normally fertilized at the germinal vesicle stage. In the egg of the gastropod *Crepidula*, Conklin (1899) described the appearance of a yolk-free region around the spindle as it formed. The initial orientation of the spindle is roughly perpendicular to the animal–vegetal axis, but the spindle rotates until it is parallel to the axis and then, at metaphase, it moves toward the animal pole until one of its asters contacts the cortex. Conklin attributed the movement to a decrease in spindle length and to a general shifting of the egg contents, which he did not directly observe in the relatively opaque egg but deduced from the study of fixed and sectioned specimens. The arrival of the mitotic apparatus at the periphery is accompanied by a difference in the appearance of the asters. The peripheral aster that contacts the animal surface appears to be flattened or truncated in the area of contact and appears to be smaller than the internal aster, which continues to expand. Others who have described these events assumed that the difference arose as the mitotic apparatus moved, but Conklin (1902) maintained that it developed after relocation and that the size of an aster depended upon its cytoplasmic surroundings. In their ultrastructural study of these events in *Spisula*, Longo and Anderson (1970) found that germinal vesicle breakdown and aster formation begin minutes after fertilization. As the asters enlarge, they separate and establish the poles of the mitotic apparatus; large cytoplasmic structures are excluded from the centriolar regions; and the small bundles of microtubules that constitute the aster rays radiate among endoplasmic reticulum and various organelles. As the mitotic apparatus and its surrounding area of less-particulate cytoplasm move along a radius toward the periphery, inclusions initially located under the animal pole are displaced as the peripheral aster approaches. After the peripheral aster stabilizes, it appears smaller than the central aster. Mitotic apparatus translocation in *Spisula* meiosis is blocked by taxol but not by cytochalasin B (Kuriyama, 1986). Taxol prevents normal aster (but not spindle) development, and Kuriyama suggests that it may prevent translocation by interfering with a mechanism that moves the mitotic apparatus by the simultaneous growth of the inner aster and shrinkage of the peripheral aster. Lutz, Hamaguchi, and Inoué (1988) studied the meiotic spindles of *Chaetopterus* eggs using video-enhanced polarized light microscopy and concluded that the numbers and lengths of the microtubules in both asters are similar, and they interpreted the form of the peripheral aster as flattened rather than diminished. In the late metaphase of *Ilyanassa,* microtubules of the peripheral aster were seen to radiate at all angles, so that those between the aster center and the pole were short and straight and ended in the cortex (Burgess, 1977).

In starfish oocytes, attachment of the mitotic apparatus to the oocyte surface at metaphase results in asymmetric distribution of microtubules in the spindle as well as in the asters (S. K. Satoh, Oka, and Hamaguchi, 1994). At that time, microtubules in the peripheral half-spindle become more numerous than those in the inner half-spindle, and the difference increases until late anaphase. When attachment of the mitotic apparatus to the surface is prevented by centrifugation, the distribution of microtubules in both the asters and the spindle halves is symmetrical.

In amphibia (most observations have been made on *Xenopus*), typical centrioles and centrosomes are absent, and structures and events that develop between the breakdown of the germinal vesicle membrane and the radial orientation of the mitotic apparatus are strikingly different from those in invertebrate oocytes. Typically, the large germinal vesicle lies in the animal half of the oocyte and, after hormonal initiation of maturation, a fibrous mass develops close to the germinal vesicle vegetal surface (Brachet, Hanocq, and Van Gansen, 1970). The breakdown of the germinal vesicle membrane begins near the fibrous mass, and the contents of the germinal vesicle combine with the mass and move to the subcortex of the animal pole. The mitotic apparatus forms from fibrous-mass material, and the condensed chromosomes become associated with it. Gard (1992) reexamined the process using fluorescent antitubulin antibodies in fixed and living *Xenopus* oocytes. He found that the fibrous mass is a transient microtubule array that is assembled in association with a discoidal, microtubule-organizing center that is located near the vegetal surface of the germinal vesicle membrane when membrane dissolution begins. The microtubule mass moves from its point of origin toward the animal surface, where its microtubules form a compact aggregation in which the linear elements radiate toward the surface from the discoidal base. The mitotic apparatus is resolved from the microtubular mass. Initially, it is oriented parallel to the surface, and the aster microtubules are sparse. During the one-hour period after its formation, the spindle reorients. The process is completed shortly before, or during, cytokinesis. The mitotic apparatus for the second meiotic division goes through a similar reorientation. The chromosomes are first visible embedded among the microtubules of the transient array during the array's migration toward the animal pole. In the transversely oriented spindle, the chromosomes are in the prometaphase configuration and they progress through the subsequent stages during and after the shifting to the radial orientation. In some cases, formation of the second polar body occurs 50 μm from the site of the formation of the first polar body following movement of the mitotic apparatus. The state of the microtubular material during the migration period is distinctive, and it is inconsistent with any generalization about an essential role for the asters. Absence of centrioles in later meiotic stages is also characteristic of mice and other mammals (Szollosi, Calarco, and Donahue, 1972).

In rats and mice, also, the mitotic apparatus is oriented with its axis parallel or chordal to the surface when its migration toward the surface is complete (Kirkham, 1907). In the rat, the furrow develops at the spindle midpoint and at first progresses on a plane that intersects with the oocyte center. As the furrow impinges upon the mitotic apparatus, the latter is rotated to a radial position, and the line of the furrow is bent into an arc (Selman, 1966). The curving furrow extends and, when it is completed, the outer margins of the polar body and the oocyte are flush in flattened preparations. Stefanini, Õura, and Zamboni (1969) found that one of the spindle poles in the mouse egg is more intimately associated with the surface than the other and, in its vicinity, the cortical area is modified. This pole becomes the pivot point around which the mitotic apparatus subsequently rotates. The furrow develops in the plane of the spindle equator, and the 90° rotation of the spindle and the redirection of the furrow follow as in the rat.

The movement of the mouse meiotic spindle to the cortex does not occur in the presence of cytochalasin B. Associated with the failure of relocation is an absence of the changes in the surface and cortex that normally occur after the mitotic apparatus takes up its position subadjacent to the cortex. The normal regional enrichment of cortical filamentous actin fails to appear, as does the conoidal elevation that develops at the animal pole of denuded eggs (Longo and Chen, 1985). Colchicine treatment prevents meiotic spindle development, but the chromosomes move to the cortical region, which in their presence progresses through the normal structural and form changes.

The strength and appearance of the region where the peripheral aster attaches to the cortex are remarkable. Conklin (1917) centrifuged *Crepidula* eggs and found that before the formation of the first polar body, while the mitotic apparatus was distant from the surface, the mitotic apparatus was easily displaced. At metaphase, however, after the aster was positioned, its connection to the surface could not be broken. The spindle could be shifted to the interior, but its persistent attachment pulled the surface into a dimple. In the second meiotic division, Conklin found that at metaphase the attachment was weaker, so that the mitotic apparatus could be shifted without deforming the surface; however, the mitotic apparatus returned to its original position after centrifugation had stopped. When the mitotic apparatus was held in its displaced position by prolonged centrifugation, the polar body was formed wherever the spindle happened to be. At anaphase, the attachment could not be broken, and Conklin assumed that the connection consisted of aster fibers. In the same year, Chambers (1917b) probed the mitotic apparatus of *Cerebratulus* eggs with glass needles. While the mitotic apparatus was in the central region, it could be pulled and pushed around to the extent that the astral rays formed curves and spirals, but after the peripheral aster rays contacted the surface, it could not be dislodged without extensive damage. In a more recent micromanipulation investigation of the phenomenon, Lutz, Hamaguchi, and Inoué (1988) found that in nucleated fragments of *Chaetopterus,* the surface was also pulled inward when the mitotic apparatus was pulled inward. After the connection broke, the surface assumed its original contour, and the peripheral aster expanded to the dimensions of the inner aster. When the mitotic apparatus was released while its connection to the surface was intact, it returned to its original site and, because the yolk content of the fragment was reduced by a centrifugation procedure, aster fibers connecting the aster to the cortex could be observed.

Observations and descriptions of polar body formation have emphasized the events that occur in the immediate vicinity of the animal pole, but M. S. Hamaguchi and Hiramoto (1978) found that, in starfish eggs, synchronized changes occur in more distant regions (Figure 9.6). Before

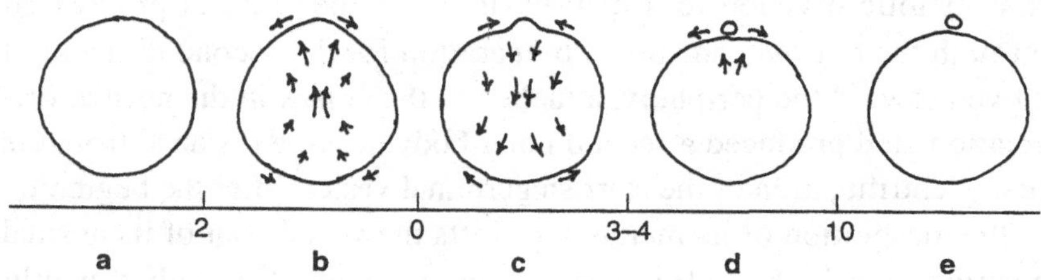

Figure 9.6 Relation between polar body formation, cell shape, and cytoplasmic movement in starfish oocytes. The change from spherical (a) to ovate (b) form is accompanied by animal surface expansion, vegetal surface contraction, and cytoplasmic movement from the vegetal to the animal region. During bulge formation, the direction of surface expansion and cytoplasmic movement is reversed (c). As the bulge is undercut (d), the directions of nearby cytoplasmic flow and surface movement are again reversed. After polar body formation is completed, movements cease (e). (From M. S. Hamaguchi and Hiramoto, 1978, and with the permission of Academic Press)

the bulge develops, the animal hemisphere surface expands, the vegetal surface contracts, and the overall shape changes from a sphere to a shape resembling a hen's egg, with the animal pole located at the narrow end. Internally, cytoplasm shifts from the vegetal region toward the animal region, predominantly in the central region. When the bulge has been formed, the directions of surface contraction and expansion are reversed, as is the direction of cytoplasmic movement. The final separation of the polar body from the oocyte is accompanied by surface expansion and animal-ward cytoplasmic movement in the polar region.

The factors that determine precisely where the polar body will develop are presently unclear. Conklin (1902) was convinced that the spot is predetermined and that it controls mitotic apparatus behavior. Schroeder (1985b) described cytological observations indicating the presence of animal–vegetal polarity in the arrested premeiotic stage and immediately after stimulation with a meiosis-inducing hormone in starfish oocytes. Lutz, Hamaguchi, and Inoué (1988) demonstrated a form of behavior control by showing that the detached mitotic apparatus moved back to the original site, and only to the original site, of attachment when it was positioned less than 35 μm from it.

It is alternatively possible that the site is determined by the mitotic apparatus. Yatsu (1909) speculated that bulge formation might result from the action of the centriole upon the surface. Chambers (1917b) described experiments in which he dragged the daughter nucleus of the first meiotic division to a new position; as the nucleus progressed through the normal changes in preparation for the second division, it moved toward the periphery, attached to the cortex in the normal orientation, and produced a second polar body at some distance from the first. Centrifugation of the starfish germinal vesicle after the beginning of the dissolution of its membrane shifts the vesicle out of its normal position. A polar body forms at the new position of the subsequently formed mitotic apparatus, but the original polarity of the developed zygote is undisturbed (Shirai and Kanatani, 1980). Bulge formation appears to be associated with the proximity of asters to the surface. When the mitotic apparatus is oriented parallel, rather than perpendicular, to the surface during maturation divisions of *Asterias*, it produces two bulges rather than one (Rappaport and Rappaport, 1985b) (Figure 9.7). When one pulls outward on the tip of the bulge that develops

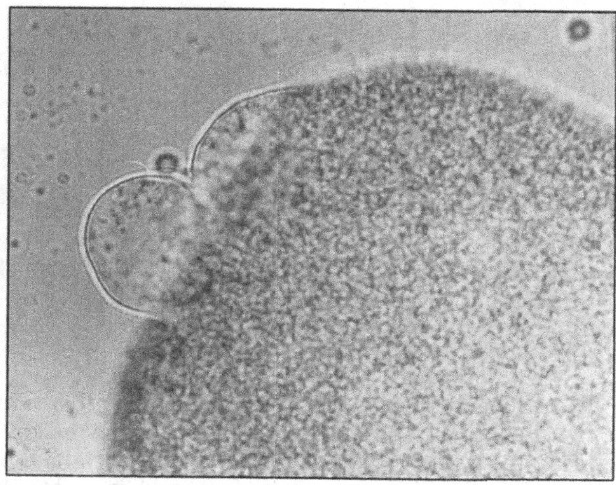

Figure 9.7 Formation of two bulges when the mitotic apparatus for the first polar body in an unoperated starfish oocyte is oriented nearly parallel to the surface. (From Rappaport and Rappaport, 1985b, and with the permission of Wiley-Liss)

before formation of the first polar body, the inner aster is brought to the surface, and a second bulge forms around it (Figure 9.8). When all or part of the mitotic apparatus is shifted to a new position close to the oocyte surface, a bulge forms nearby in an average of 1.74 minutes (Rappaport and Rappaport, 1985b). The large asters that form in starfish oocytes in response to hexylene glycol treatments are associated with large bulges that, in turn, become large polar bodies (Yamao and Miki-Noumura, 1988). Shimizu (1990) has found an exception in the eggs of *Tubifex,* an annelid worm. He centrifuged the oocyte nucleus toward the vegetal pole after the formation of the first polar body. At the time second polar bodies appeared in normal controls, a bulge appeared at the animal pole of the centrifuged cell. It contained no nuclear material and the repositioned mitotic apparatus elicited no bulge. This result suggests a different mechanism at work than in *Cerebratulus* and *Asterias,* where relocation of the oocyte nucleus prevents bulge formation in the normal location at the subsequent division cycle (Chambers, 1917b; Rappaport and Rappaport, 1985b).

The physical mechanism that produces the bulge has long been a

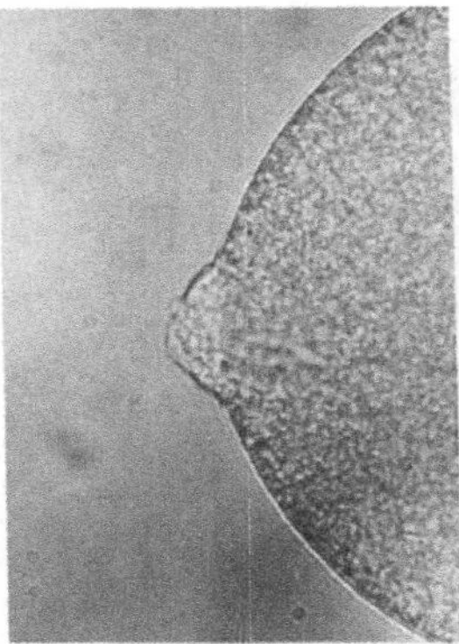
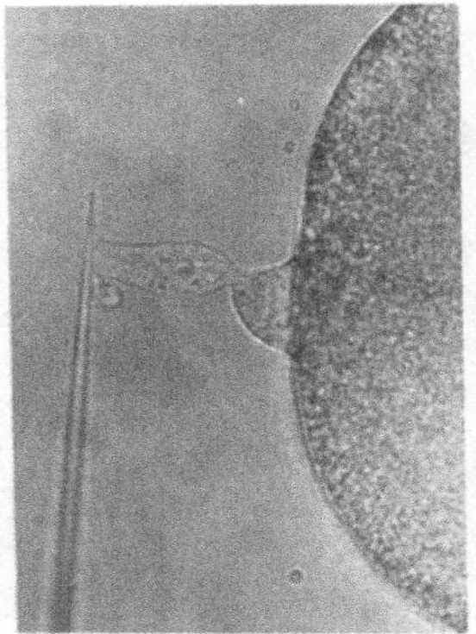
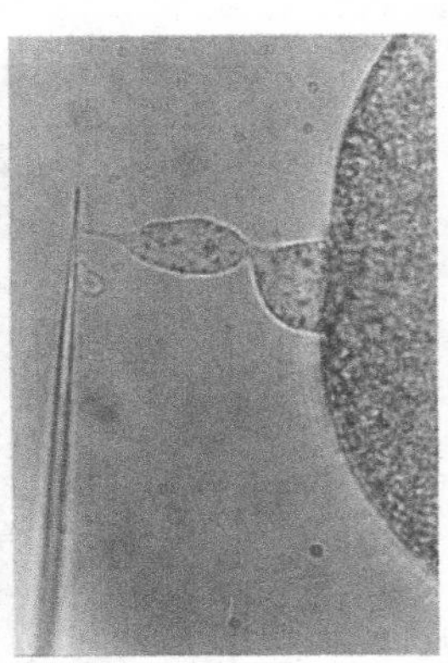

Figure 9.8 Consequences of pulling the first polar body protuberance outward (left). The attached mitotic apparatus moves with the tip, and the furrow develops (middle). A secondary bulge forms around the central aster of the mitotic apparatus as it shifts closer to the surface (middle, right). (From Rappaport and Rappaport, 1985b, and with the permission of Wiley-Liss)

subject of speculation. The idea that it resulted from a localized decrease in surface tension or from tension at the surface appeared early and then reappeared frequently (Conklin, 1902; Yatsu, 1909; Chambers, 1917b; Spek, 1918; Wolpert, 1960). K. Dan (1943b), on the other hand, suggested that it might arise from the force generated by an autonomously elongating spindle anchored to the cortex in a radial orientation. Whatever the mechanism may prove to be, the site of polar body formation does not appear to be restricted to a small part of the oocyte surface in starfish oocytes. Because neither the attachment nor the orientation of the mitotic apparatus material that produced accessory bulges was necessary, bulge formation does not appear to require a precise geometrical relation between the interactants. The ability of a mitotic apparatus to induce two bulges when it lies parallel, and close, to the surface (Rappaport and Rappaport, 1985b) suggests that this effect of the mitotic apparatus is not restricted to its polar regions (Figure 9.6). Even the normal size of the mitotic apparatus is not an important factor. There are other instances in which relatively small asters in the cleaving *Beroë* egg produce transient bumps or bulges in relatively wide expanses of surface. Ziegler (1898b) described and figured the small "noselike" processes that form adjacent to the asters on either side of the forming furrow. Fullilove and Jacobson (1971) also described bumps that arise close to the asters in the cleaving *Drosophila* blastoderm. The possibility that this type of change in shape may be a common response in a large region of cortex to a relatively small and restricted region of astral proximity may merit further investigation.

Although the bulges that form during polar body formation bear a superficial resemblance to the blebs that form on the same eggs after treatment with agents that damage surface integrity, polar body bulges and surface blebs respond differently to mechanical treatment (Rappaport and Rappaport, 1985b). Blebs about the same size as polar body bulges form on the surface of *Asterias* oocytes after treatment with cytochalasin B. When the internal hydrostatic pressure is raised by flattening the oocyte or by immersing the oocyte in dilute sea water, the diameter of the bleb increases rapidly, but when the same measures are used to increase the hydrostatic pressure in an untreated oocyte, the height of a preformed, normal bulge rapidly diminishes, and the bulge appears to spread into the adjacent region (Figure 9.9).

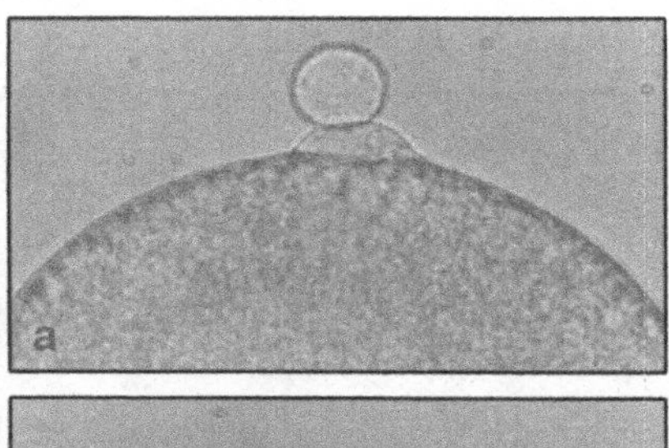

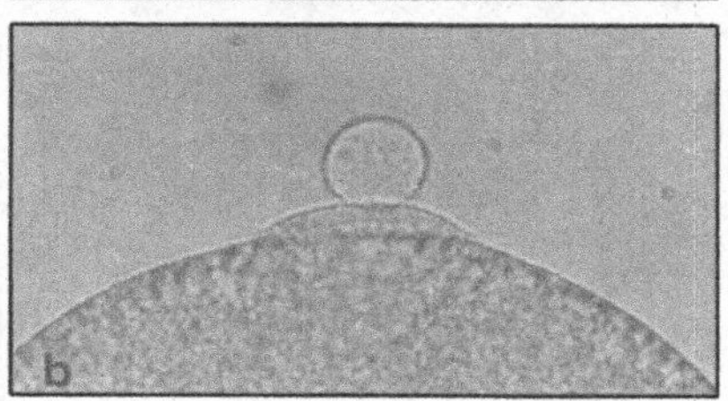

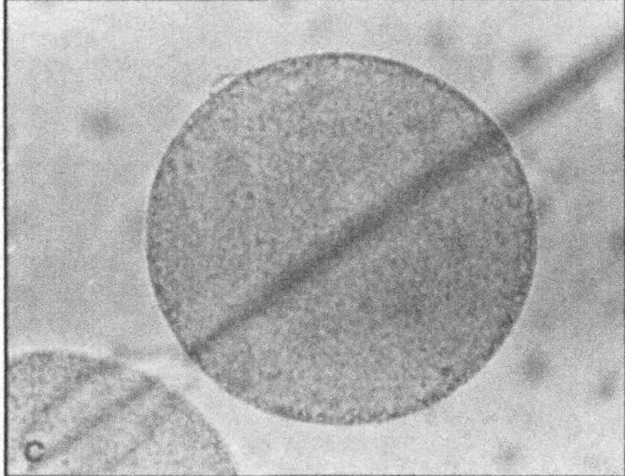

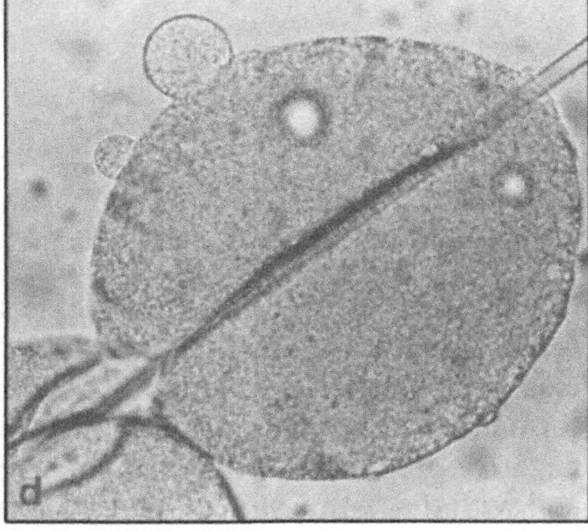

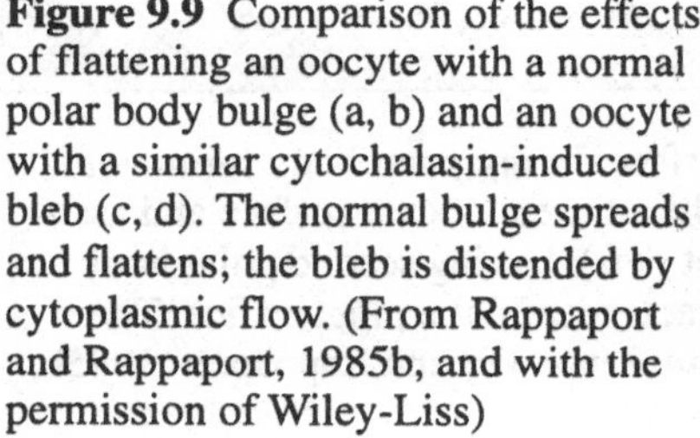

Figure 9.9 Comparison of the effects of flattening an oocyte with a normal polar body bulge (a, b) and an oocyte with a similar cytochalasin-induced bleb (c, d). The normal bulge spreads and flattens; the bleb is distended by cytoplasmic flow. (From Rappaport and Rappaport, 1985b, and with the permission of Wiley-Liss)

Bulge formation and subsequent polar body separation, like the usual form of cytokinesis that results in daughter cells of equal volume, are preceded and accompanied by biophysical and structural changes. The general resistance to deformation or stiffness of starfish eggs increases before the contour of the animal surface changes, and decreases during and after completion of both meiotic divisions (Hiramoto, 1976; Nakamura and Hiramoto, 1978). Local stiffness changes were visualized by observing changes in the height of a bulge formed when the surface of the polar body region was sucked into a wide-mouth pipet and held at constant negative pressure. The experiments revealed increasing stiffness during bulge formation (Ohtsubo and Hiramoto, 1985; Rappaport and Rappaport, 1985b) (Figure 9.10). Yamamoto and Yoneda (1983) bisected starfish eggs 10 minutes after germinal vesicle breakdown and found that the enucleated fragments underwent two cycles of increased resistance to deformation, with a temporal pattern similar to that of nucleated fragments. Cycling of enucleate fragments occurred only when the eggs were bisected after ger-

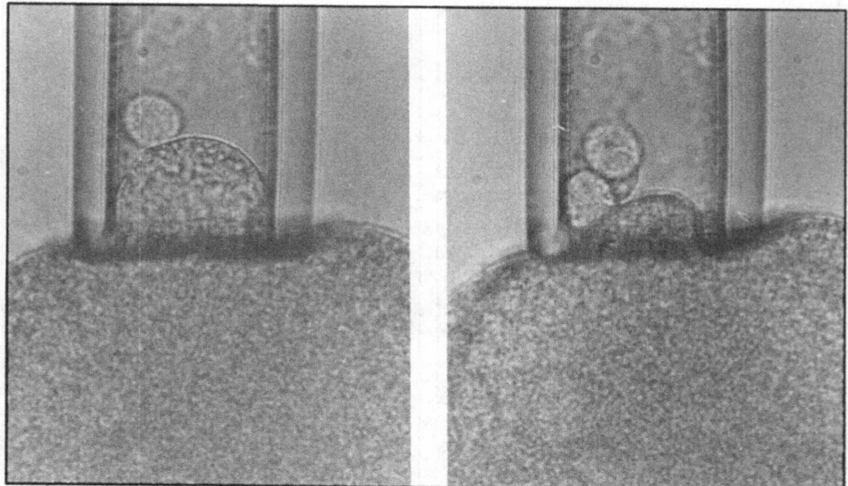

Figure 9.10 Retraction of the oocyte surface during polar body formation. After formation of the first polar body, the adjacent oocyte surface and associated nucleus were sucked into the pipet (left). During second polar body formation (right), the oocyte surface retracts toward the pipet orifice. The negative pressure was not changed. (From Rappaport and Rappaport, 1985b, and with the permission of Wiley-Liss)

minal vesicle breakdown, and Yamamoto and Yoneda concluded that cycling is associated with cytoplasmic changes that take place in the presence of germinal vesicle material. In *Tubifex*, there is a consistent relationship between mitotic stage, animal pole surface contour, and F-actin distribution in rh-phalloidin–treated, isolated cortex preparations (Shimizu, 1990). At metaphase, before the bulge forms, there is a dimple in the center of the prospective bulge region that is underlain by a dotlike F-actin thickening. At early anaphase, when the bulge is low and somewhat flattened at its apex, the actin dot persists and actin distribution in the immediately surrounding cortex appears about the same as that in the remainder of the animal cortex. At mid anaphase, the bulge is near maximum height, the actin dot has disappeared, and the actin mesh in the apical region of the bulge is much diminished. The actin bundles appear to accumulate at the base of the bulge and in the immediately surrounding cortex, and in both areas they are oriented circumferentially (Figure 9.11). Both the rate of decrease in apical bundles and the rate of increase in bundles at the base increase as the bulge is undercut by the constriction that severs the polar body from the oocyte.

M. S. Hamaguchi and Hiramoto (1978) proposed that the animal pole surface is relaxed by the contraction of a ring of surrounding cor-

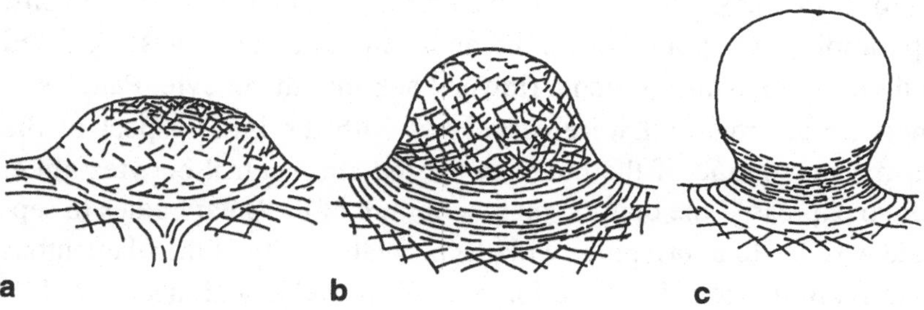

Figure 9.11 Diagrammatic summary of the changes in organization of actin bundles in the animal pole cortex during polar body bulge formation in *Tubifex*. A small cap of actin underlies the prospective bulge surface region at metaphase. The cap persists at the beginning of bulge formation in early anaphase (a). At mid anaphase, the cap is not present (b), the concentration of actin at the tip is diminished, and circumferentially oriented actin accumulates around the base. When the furrow begins to undercut the bulge at early telophase (c), actin is concentrated at the base and is not detectable at the tip and the lateral surface of the forming polar body. (From Shimizu, 1990, and with the permission of the New York Academy of Sciences)

tical material, and that this restricted, relaxed area is distended by inflowing cytoplasm when the internal pressure rises as a consequence of the cyclical increases in tension at the surface that develop in conjunction with the division cycle. Shimizu (1990) pointed out that the ring of actin filaments he found around the base of the bulge could restrict the lateral expansion of the bulge as it grows, and the reduction of F-actin at the apex would enhance its pliancy and tendency to yield as internal pressure increases. Microtubules are also present in the premeiotic cortex, but the possibility that they may participate in polar body formation appears to be obviated by the discovery of Schroeder and Otto (1984) that the presence of cortical microtubules is correlated with the presence of an intact nucleus and the absence of a division apparatus. The virtual absence of cortical microtubules during the period of polar body formation can be reconciled with the absence of nuclear membrane re-formation between meiotic divisions. The results of experiments in which cytochalasin B or D was administered to cells during the period of polar body formation are not easy to weave into broad generalizations. In *Spisula*, fertilization occurs during germinal vesicle breakdown, before the formation of the first polar body (Costello et al., 1957). When Longo (1972) began cytochalasin B treatment 3 minutes after fertilization, he found that although bulge formation occurred at the time of both the first and second meiotic divisions, separation of the polar bodies did not occur, and the bulges regressed as their cytoplasmic contents flowed back into the oocyte. Patches of fine-textured material were associated with the inner aspect of the plasma membrane of the bulge in treated and control oocytes alike. Longo expressed the opinion that the failure of the polar bodies to separate was due to interference with the normal activity of the filamentous material in the cortex around the base of the bulge and that bulge formation could result from the action of the peripheral aster upon the animal pole cortex. Tsukahara and Ishikawa (1980) found that cytochalasin B, when applied for 10 minutes to starfish oocytes at metaphase of the first meiotic division, reduced the size of the bulge and suppressed polar body formation. When oocytes were treated for 10 minutes beginning 15 minutes after the formation of the first polar body, the second polar body was suppressed, and no bulge formation was described. The sensitivity of two other starfish, *Pisaster* and *Asterias*, to cytochalasin B appeared to differ, but effective treatment had the same consequences. In *Pisaster*, treatment begun immediately before first polar body formation permitted development of the first polar body but suppressed development of the second polar body. In *Asterias*, all polar body formation activity was blocked (Rappaport and Rappaport, 1985b). Treatment of *Tubifex* eggs shortly after first polar body formation completely inhibited extrusion of the second polar body without producing an effect on the ultrastructure of the filamentous cortical layer or on the development of the second mitotic apparatus (Shimizu, 1981a). When the treatment was begun earlier, extrusion of the first polar body was normal, a cortical layer of filamentous material that normally lay under the animal pole was missing, and the peripheral aster and spindle of the mitotic apparatus for the second meiotic division were missing. The coincidence between diminution of the filamentous material and defective formation of the mitotic apparatus led Shimizu to suggest that the filamentous cortical material may play a role in the formation of the aster as well as in its attachment to the cortex.

Interpretation of these experiments is complicated by the possibility that the effect of cytochalasin B is not uniform when applied to oocytes of different major animal groups and, perhaps, when applied to different actin-rich or actin-actuated structures within the same oocyte. In the early period of cytochalasin use, its beauty derived from its apparent specificity and reversibility upon removal, and from the correlation between its physiological and ultrastructural effects. The bulge formation and cytokinesis that comprise polar body formation require forces that can originate from a general or a regional increase in tension at the surface. We have come to expect that tension increase is actin related. In all cases, cytochalasin B inhibits the furrowing part of polar body formation given adequate time. In starfish, both bulge formation and separation of the polar body are blocked (Tsukahara and Ishikawa, 1980; Rappaport and Rappaport, 1985b). In *Spisula*, bulge formation is not blocked (Longo, 1972), suggesting different susceptibility. In *Tubifex*, extrusion as well as separation is blocked, with no apparent effect on the ultrastructure of the filamentous cortical layer (Shimizu, 1981a). But if the inhibitor is washed out of the eggs, a low bulge develops on the surface within 40 minutes (Shimizu,1981b). There is a good correlation between mitotic apparatus position and bulge formation (Chambers,

1917b; Conklin, 1917; Rappaport and Rappaport, 1985b; Dan and Tanaka, 1990), which is consistent with a more general relation between mitotic apparatus position and actin redistribution as, for instance, in contractile ring formation. Usually, the repositioning of the entire mitotic apparatus to a subcortical location at a distance from the animal pole results in bulge formation and subsequent polar body separation in the new location with no vestiges of activity in the original location (Chambers, 1917b; Conklin, 1917; Dan and Tanaka, 1990). *Tubifex* appears to be exceptional, because centrifugation does not abolish bulge formation at the animal pole (Shimizu, 1990), although the events that lead to bulge formation are apparently cytochalasin B sensitive (Shimizu, 1981a). These intriguing experimental facts and near paradoxes suggest that further investigation is required.

There has been a persistent confidence that the mechanism that finally separates the polar body from the oocyte is identical to that which separates the daughter cells in mitosis. Not only are the consequences of the two events the same, but the same cell organelles appear to be involved. The differences between cleavage and polar body separation have usually been considered superficial. Wilson (1900, p. 375), while commenting on the smaller size of the peripheral aster, remarked that "the size of the aster, in other words, depends upon the extent of the cytoplasmic area that falls within the sphere of the centrosome. . . . If, therefore, the polar amphiaster could be artificially prevented from moving to its peripheral position, the egg would probably divide equally." The commonality, if not identity, of mechanisms was implied in the results of early experiments that suggested interconvertibility of the two division types. Conklin's (1917) centrifugation experiments demonstrated that the form of the mitotic apparatus during meiosis is a consequence of its location and that its simple relocation results in cell division that is indistinguishable from cleavage. Morgan's (1937) similar centrifugation of *Ilyanassa* oocytes revealed that giant polar bodies were formed only when cells were drawn into cylindrical form, which demonstrated that the effect of the mitotic apparatus upon the surface decreased with distance. A similar relation had previously been demonstrated by Wilson (1901b) using ether-treated, cleaving sea urchin eggs. The ether reduced the aster size and blocked cytokinesis but not mitosis.

After several mitotic cycles, when the reduced asters were crowded closer to the surface, furrows developed. The similarity of mechanical activity in mitosis and meiosis was also implied by the bundling together of the spindle fibers that occurs in the later stages of both kinds of division. But most convincing has been the similarity of the ultrastructures. The resemblance of the structures at the furrow base is convincing (Longo and Anderson, 1970), and the likeness was even apparent when preparative methods produced artifactual sheets of vesicles in the cleavage plane (Humphreys, 1964). In the period when many hypothetical cell division mechanisms stood undisproven, the ability of the mechanism to explain polar body formation was sometimes applied as a test. K. Dan (1943b) questioned the constriction ring hypothesis because it required that, during polar body formation, the constricting region be nearly tangential to the surface. He appeared to assume that the same mechanism produced both the bulge and the separation. Later, Dan (1963) pointed out the difficulty of explaining polar body formation using Wolpert's (1960) aster relaxation hypothesis because, if the tension at the surface on both sides of the forming constriction were equal, then the contents of the polar body should flow into the oocyte because its internal pressure would be higher in the polar body by reason of its smaller radius. Although the cell structures that actively participate in polar body formation and cleavage are similar, the size difference between the polar body and the oocyte in the usually studied species is greater than what can be accounted for by the simple displacement of a normal mitotic apparatus out of its usual central position.

Several special circumstances appear to enhance the chances that the polar body will be smaller than the oocyte. Perhaps the most important is the movement of the mitotic apparatus toward the periphery. The same hypothetical mechanisms that have been used to explain the similar event in unequal cell division have been applied to polar body formation, and in some discussions the two events are almost inextricably enmeshed. Polar body formation occurs at the animal pole, and the relation between the animal pole and the early cleavage planes is predictable. In the development of eggs in which normality is dependent upon a relation between the cleavage planes and the timely separation of cytoplasmic developmental domains, the underlying importance of the relationship is apparent, but it has not been possible to demonstrate that polar body formation is restricted to the animal pole region. Transport of the unattached mitotic apparatus to cortical regions distant from the animal pole by centrifugation (Conklin, 1917) and by manipulation (Chambers, 1917b) result in polar body formation in any region, regardless of which aster was near the cortex. Repositioning the mitotic apparatus after it attaches reveals a fibrous connection between the peripheral aster and the cortex (Conklin, 1917; Lutz, Hamaguchi, and Inoué, 1988). The connection is strong enough to deform the surface. Lutz, Hamaguchi, and Inoué (1988) studied the return of the mitotic apparatus to the site of its original attachment in fragments of *Chaetopterus* eggs and found that return is blocked by colcemid but not by cytochalasin. If the mitotic apparatus is inverted so that the inner aster is closer to the attachment site, the inner aster leads the mitotic apparatus back to the attachment site. Shifting the mitotic apparatus far enough to break the fibrous connection prevents its return unless it is pushed to a region within 35 μm of the site, whereupon it moves as though it had not previously been detached. Lutz, Hamaguchi, and Inoué found the attachment site to be uniquely capable of holding the mitotic apparatus firmly to the cortex. They argued that the return of the mitotic apparatus to its attachment site, although microtubule dependent, is not due to persistent connections to the surface. They tend to equate the return phenomenon with the normal movement of the mitotic apparatus toward the surface, and, in this connection, further experiments involving manipulations of the previously unattached mitotic apparatus in the absence of an active attachment site would be useful. Longo and Chen (1985) found that the initial movement of the mitotic apparatus in mouse oocytes does not occur in the presence of cytochalasin B.

In addition to movement of the mitotic apparatus as a whole, several changes in the mitotic apparatus itself would also tend to put the midpoint of the spindle closer to the cortex and thereby reduce the size of the polar body. Conklin (1902) found that in *Crepidula* the mitotic apparatus attains, at its maximum, a length equal to the oocyte radius and, by late anaphase, it shrinks to half that length. He cites similar observations made by others in other species, and Lutz, Hamaguchi, and Inoué (1988) also observed shrinkage in the *Chaetopterus* spindle. The asymmetry of the asters has been variously described and explained in the preceding pages, but, whatever the cause, the diminution or flattening of the peripheral aster brings the midpoint of the spindle closer to the surface than it would be if the polar part of the aster retained the usual hemispherical form. Saiki and Hamaguchi (1993) explored the causes of the difference in size between starfish cleavage asters and the asters that form during the meiotic divisions by transplanting polar body and cleavage centrosomes. After stabilization with hexylene glycol, normal cleavage asters and meiotic asters had diameters of, respectively, 40.8 μm and 28.8 μm. When the cleavage centrosome was transplanted to an oocyte, the aster that formed around it was the same size as the endogenous asters, which were of normal diameter. This result raises the possibility that the normally diminished aster size in oocytes may result from an oocyte cytoplasmic effect, because the aster rays that develop after centrosomal transplantation derive principally from host cell material. Another aspect of meiosis that may minimize the size of the polar body is the tendency, at least in some forms, of the division plane to be closer to the peripheral, rather than to the inner, aster center. This circumstance was described by Wolpert (1960) and has been illustrated in some of the older (Conklin, 1902), as well as the more recent (Shimizu, 1981b), literature. Because the intruding furrow can move the mitotic apparatus (Rappaport and Rappaport, 1994), the relation between them probably is determined in the earliest stage of cytokinesis. In polar body formation, the distance from the peripheral aster center to the surface is less than the distance from the inner aster center to the surface, and the geometrical relations resemble those in conical cells, in which the furrow characteristically appears closer to the aster nearest the vertex (Rappaport and Rappaport, 1994).

Unequal cell division occurs in different cells in a variety of organisms at different developmental times. The similarities of the phenomenon of unequal cell division in different circumstances appear to have led to a degree of conflation of the observations, which in turn may lead to the masking of important differences. The phenomenon is usually characterized by a shifting of the mitotic apparatus into a radial position and by the diminution or reconfiguration of the peripheral aster. The shifting of the mitotic apparatus invariably occurs, but it is uncertain whether it is in all instances driven by the same mechanism. Reduction of peripheral aster size usually occurs, but accounts of the peripheral aster's morphology and its relation to the surface are not all the same. Associated with, and perhaps a contributing cause of, the reduction in size is the apparent attachment of the aster to the cortex. Attachment could be necessary to stabilize the aster position in circumstances that might result in propulsion of the mitotic apparatus toward the center of the cell as the furrow deepens. K. Dan and Tanaka (1990) pointed out that, at present, the attachment only of the meiotic spindle in *Crepidula* and the spindle of dividing grasshopper neuroblast cells (Kawamura, 1977) has been unequivocally established, but they adduce persuasive indirect evidence of attachment in other forms of unequal division. It is interesting to note that the mitotic apparatus of cleavage cells shows no evidence of attachment or truncation when crowded against the cortex (Rappaport, unpublished results).

Polar Lobe Formation

Some molluscan and annelidan eggs form a temporary, anucleate protuberance at the vegetal pole at the time of meiotic divisions and the first two cleavages. These polar lobes (sometimes called "antipolar lobes" or "yolk lobes" in the old literature) contain morphogenetic determinants that, during the earliest cleavages, become associated with one of the blastomeres, so that their volumes are subsequently unequal. Although unequal distribution of the determinants is necessary for normal development, some mollusks can accomplish the unequal distribution, and

Figure 9.12 Shape changes of an *Ilyanassa* egg during cleavage. The spherical egg constricts in the region of the dashed line, causing elongation of the animal–vegetal axis. The cleavage furrow and the constriction of the polar lobe subsequently deepen simultaneously. (From Schmidt et al., 1980, and with the permission of Academic Press)

also the inequality of blastomere size, without forming polar lobes (Kumé and Dan, 1968). Eggs that produce polar lobes are initially spherical and their mitotic apparatus is positioned near the animal pole. The first indication of lobe formation is a shrinkage of the egg diameter perpendicular to the animal–vegetal axis; this causes elongation of the axis and reshapes the vegetal region into a roughly conical form in which the rounded, vegetal pole is the vertex (Figure 9.12). The shrinkage occurs in the same location during both the meiotic divisions (when the mitotic apparatus is parallel to the animal–vegetal axis) and the early cleavages (when the mitotic apparatus is perpendicular to the animal–vegetal axis), so that the relation between the mitotic apparatus and the plane of constriction is not so simple as in mitosis. During the cleav-

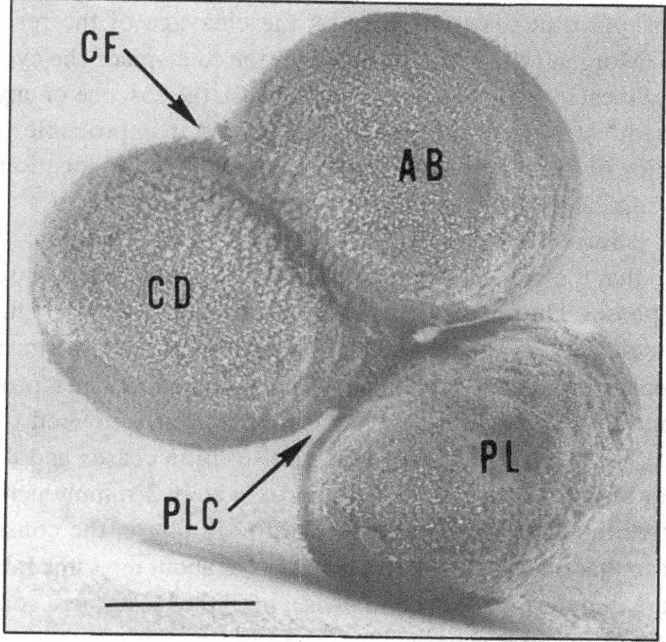

Figure 9.13 Scanning electron micrograph of an *Ilyanassa* egg during formation of the first cleavage furrow (CF), when the polar lobe constriction (PLC) is nearly maximal. When the process is completed, the AB and CD blastomeres will be completely separated and the polar lobe (PL) will be incorporated by the CD blastomere after the polar lobe constriction relaxes. Bar = 50 µm. (From G. W. Conrad et al., reproduced from the *Journal of Cell Biology*, 1973, 59:229)

age period, polar lobe formation begins during prophase; the degree of constriction ultimately attained varies among species. The furrow first appears next to the mitotic apparatus, and it advances in the direction of the vegetal pole. The furrow and the constriction that demarcates the polar lobe deepen simultaneously, and, in some species, for a brief period the egg appears to be dividing into three nearly equal masses (Figure 9.13). But following the regression of polar lobe constriction and the completion of cleavage, the only outward evidence of polar lobe activity is the inequality of the blastomeres.

At the end of the 19th century, simple amputation experiments revealed that the absence of polar lobe material resulted in deficiency in several larval structures, and exploration of the developmental role of the polar lobe began. Wilson (1904) isolated the vegetal region of a mollusk egg and found that changes in form resembling polar lobe formation took place at the same time as the cleavage of the rest of the embryo. Morgan (1935) used centrifugation to displace the cytoplasm normally located in the vegetal region and, in the absence of any effect on subsequent lobe formation, he considered it improbable that the mechanics of the process had anything to do with the location of the mitotic spindle or the contents of the lobe region.

The careful observations of G. W. Conrad and Williams (1974a) revealed that the process of polar lobe formation in *Ilyanassa* occurs in several phases. During the first phase, constriction is slow and linear for a 40-minute period. The second phase, which begins at the time the furrow appears, is characterized by more-rapid constriction that progresses for about 15 minutes, until the polar lobe is nearly severed from the embryo (Figure 9.14). At that time, constriction ceases and the neck diameter remains constant for the approximately 3-minute period that comprises the third phase. During the fourth phase, the constriction relaxes so that the neck diameter increases at about the same rate that it had decreased during the second phase, until the blastomere containing polar lobe material is spherical.

During phase two, both the cleavage furrow and the polar lobe constriction are associated with microfilamentous actin (Schmidt et al., 1980), which may disappear in phase three. Application of cytochalasin B to spherical eggs prevented subsequent formation of both furrows and polar lobe constrictions. By manipulating the concentration of the drug, it was possible to demonstrate that the first cleavage furrow was less susceptible to cytochalasin B than the accompanying polar lobe constriction. Concentrations of cytochalasin that caused regression of furrows and constrictions in the second phase also resulted in the absence of microfilaments and the disappearance of the normal cortical structure (G. W. Conrad and Williams, 1974a).

In the presence of colchicine, *Ilyanassa* eggs enter phase one but fail to undergo the rapid constriction of phase two, although phase one persisted for long periods (G. W. Conrad and Williams, 1974a). When colchicine was applied late in phase one, normal polar lobe activity continued during that cycle but failed in the subsequent cycle. No microtubules were found in colchicine-treated eggs. With their experiments, G. W. Conrad and Williams (1974a) demonstrated that microfilaments and microtubules are active participants in both cleavage furrows and polar lobe constrictions, but the mechanisms are not identical. Phase one appears to be microfilament dependent, but microtubule independent, and the older experiments suggest that the characteristic limited contraction is based upon inherent constitutional differences in the lobe cortex and upon cyclic cytoplasmic activity (G. W. Conrad and Rappaport, 1981). Phase two, on the other hand, is microtubule dependent until late in phase one, in a manner that is reminiscent of the fact that the cleavage furrow in sea urchin eggs is colchicine sensitive until mid anaphase (Beams and Evans, 1940). Further support for an active role for microtubules is provided by observations indicating that microtubule-enhancing substances enhance polar lobe constriction. Silver ions cause the complete separation of the polar lobe from the rest of the embryo (A. H. Conrad et al., 1994); hexylene glycol blocks phase four relaxation and may also cause complete separation (A. H. Conrad, Stephens, and Conrad, 1994).

Polar lobes appear to be relatively restricted in occurrence, although their formation involves elements common to all dividing cells and there is some justification for wondering why they are not more common. The furrows that appear conjointly with polar lobe formation usually begin unilaterally, as would be expected when the mitotic apparatus is eccentric, but many eggs that cleave unilaterally do not produce polar lobes. Unusual are the shape of the egg during phase one and the orientation of the mitotic apparatus perpendicular to the reshaped cell's

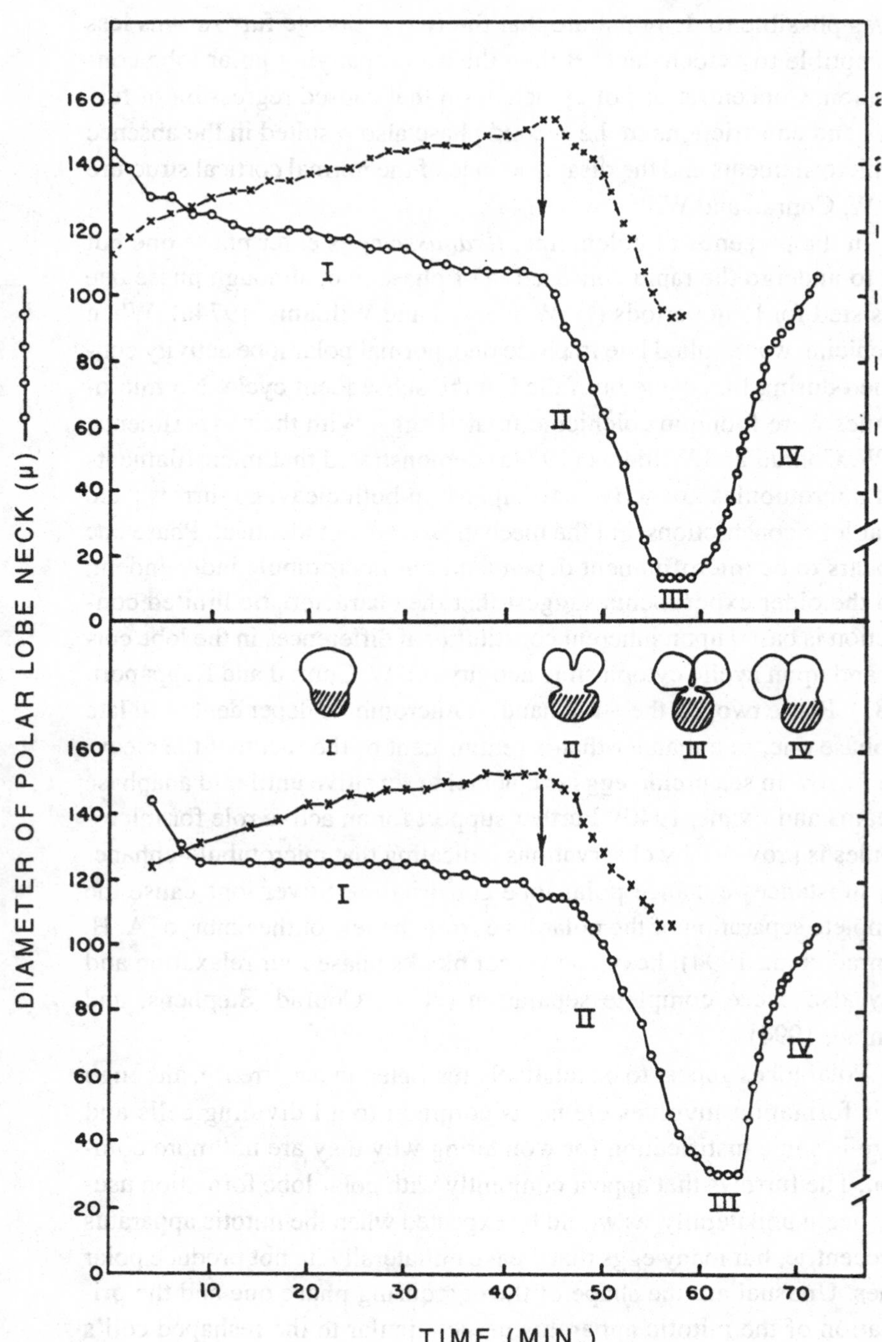
DIAMETER OF POLAR LOBE NECK (μ)
160
140
120
100
80
60
40
20
0
I
II
III
IV
I
II
III
IV
I
II
III
IV
10
20
30
40
50
60
70
TIME (MIN.)

long axis (Figure 9.12). In this circumstance, the mitotic apparatus could simultaneously interact with a closer, relatively hemispherical animal surface and with a more distant vegetal surface that bears some resemblance to a truncated cone. It was possible to assess the importance of this special geometrical configuration by imposing it upon an egg that does not normally form polar lobes. A method for reshaping spherical sand dollar eggs into cones was described previously, and all that is required is the additional step of inserting the egg into the cone so that the mitotic apparatus is oriented at right angles to the cone axis (Rappaport and Rappaport, 1994). In this circumstance, constrictions permanently or temporarily isolate the vertex region in a way that closely resembles the constrictions associated with polar lobe formation (Figure 9.15). Isolation of the vertex region by a circumferential constriction can be produced at both the first and the second cleavages, and it occurs only on a conical surface (Figure 9.16). As the constricting ring isolates the vertex, the furrow advances down the length of the egg, veering as it nears the constriction so that the lobular region usually remains associated with only one of the daughter cells. These results suggest that conical sand dollar eggs can duplicate the basic features of polar lobe formation. They also imply that the ability to carry out phase one is the minimum essential feature that separates eggs with eccentric nuclei that can form polar lobes from other eggs with eccentric nuclei that cannot. This simple, parsimonious hypothesis redirects attention back to the way in which the mitotic apparatus creates localized, linear

Figure 9.14 Changes in the diameter of the polar lobe neck, the length of the egg, and the distance to the base of the cleavage furrow during the third and fourth polar lobe formation and resorption. Roman numerals denote the four phases of polar lobe activity described in the text. The time of cleavage furrow formation is indicated by arrows. Open circles represent the diameter of the polar lobe neck. The crosses represent the length of the egg from animal pole to vegetal pole before cleavage furrow formation, and the distance between the base of the furrow and the vegetal pole after cleavage furrow formation. Upper graph shows changes that occur during formation of the third polar lobe and first cleavage furrow. The lower graph shows changes during formation of the fourth polar lobe and second cleavage furrow; in this case, the AB and CD blastomeres were separated after the first cleavage, and the measurements were made on the isolated CD blastomere. (From G. W. Conrad and Williams, 1974a, and with the permission of Academic Press)

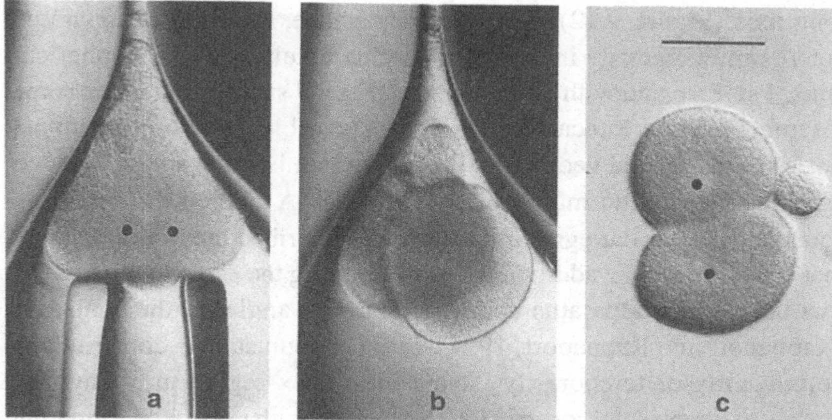

Figure 9.15 Pattern of constriction zones in a conical sand dollar egg in which the mitotic apparatus is oriented perpendicular to the cone axis (a). The cleavage furrow and a perpendicular constriction separate the egg into three regions (b). The association between the lobe and one of the blastomeres is evident when they are freed from the confining pipet (c). The dots indicate the positions of the aster centers. Bar = 50 μm. (From Rappaport and Rappaport, 1994, and with the permission of Academic Press)

contractile areas in the cortex that are essential for phase two. The importance of microtubules has been clearly documented (G. W. Conrad and Williams, 1974a; A. H. Conrad, Stephens, and Conrad, 1994; A. H. Conrad et al., 1994). Aster rays are visible near the vertex in conical sand dollar eggs, but the way in which they can establish a contractile ring oriented parallel to the mitotic axis is unresolved. There is, however, some basis for speculation. The effect of the asters in furrow establishment appears to be directed in straight lines (K. Dan, 1943a). It diminishes with distance (Rappaport, 1982) and with methods that discourage additive aster effect (Rappaport, 1969b). In conical cells, diminution of effect could result from both distance and the absence of additive effects in different regions. In the region between the planes of the aster centers, the vertex may be beyond the effective range of the asters, and in conical surfaces closer to the mitotic apparatus the likelihood of additive effect is diminished (Figure 9.17). When the cell is conical, there is a restricted, circular band of cortex situated within the effective range of the mitotic apparatus near the vertex. More-distant surface may be beyond the effective range, and less-distant surface would be outside of the region of effective additivity. Failure of spherical and cylindrical cells to engage in lobe formation does not appear to be attributable to deficiencies in either the mitotic apparatus or the cortex, and the angle between the surface and the aster rays may eventually prove to be the limiting factor.

Although this simple hypothesis may satisfactorily explain experimentally induced polar lobe activities in sand dollar eggs, it does not entirely account for some aspects of true polar lobe formation. In sand dollar eggs, lobe formation takes place only when the mitotic apparatus is perpendicular to the cell axis (Rappaport and Rappaport, 1994). But in *Ilyanassa*, both phase one and phase two occur during the meiotic divisions of polar body formation, when the mitotic apparatus is oriented parallel to the cell's long axis, which resulted from the phase one contraction (G. W. Conrad and Williams, 1974a). On the other hand, Costello et al. (1957) do not describe any polar lobe formation during maturation divisions in the annelid *Sabellaria*; in another species of the same genus, Render (1983) described a "broad" polar lobe at the second meiotic division that may resemble a phase one polar lobe in *Ilyanassa*. Both sabellid species form deeply constricted polar lobe stalks during the early cleavage divisions.

The plane of polar lobe constriction is affected by the dimensions and volume of the egg. When unfertilized *Dentalium* eggs are cut in half perpendicular to the animal–vegetal axis and then fertilized, the animal half forms no polar lobe and cleaves into equal blastomeres. The fertilized, vegetal half forms a polar lobe proportional in size to the half-egg and subsequently cleaves into unequal blastomeres (Wilson, 1904; Render and Guerrier, 1984). Render and Guerrier's diagrams (1984) indicate that the egg halves are spherical after bisection. The inability of isolated animal halves to form polar lobes indicates that the process requires more than a mitotic apparatus in the proper orientation. The proportional size of the polar lobe in the fertilized, vegetal half implies that the constriction plane is not predetermined in the vegetal cortex. The results admit the possibility that the position of the constriction plane is determined by the mitotic apparatus, the size of which may be determined by the volume and dimensions of the egg fragment.

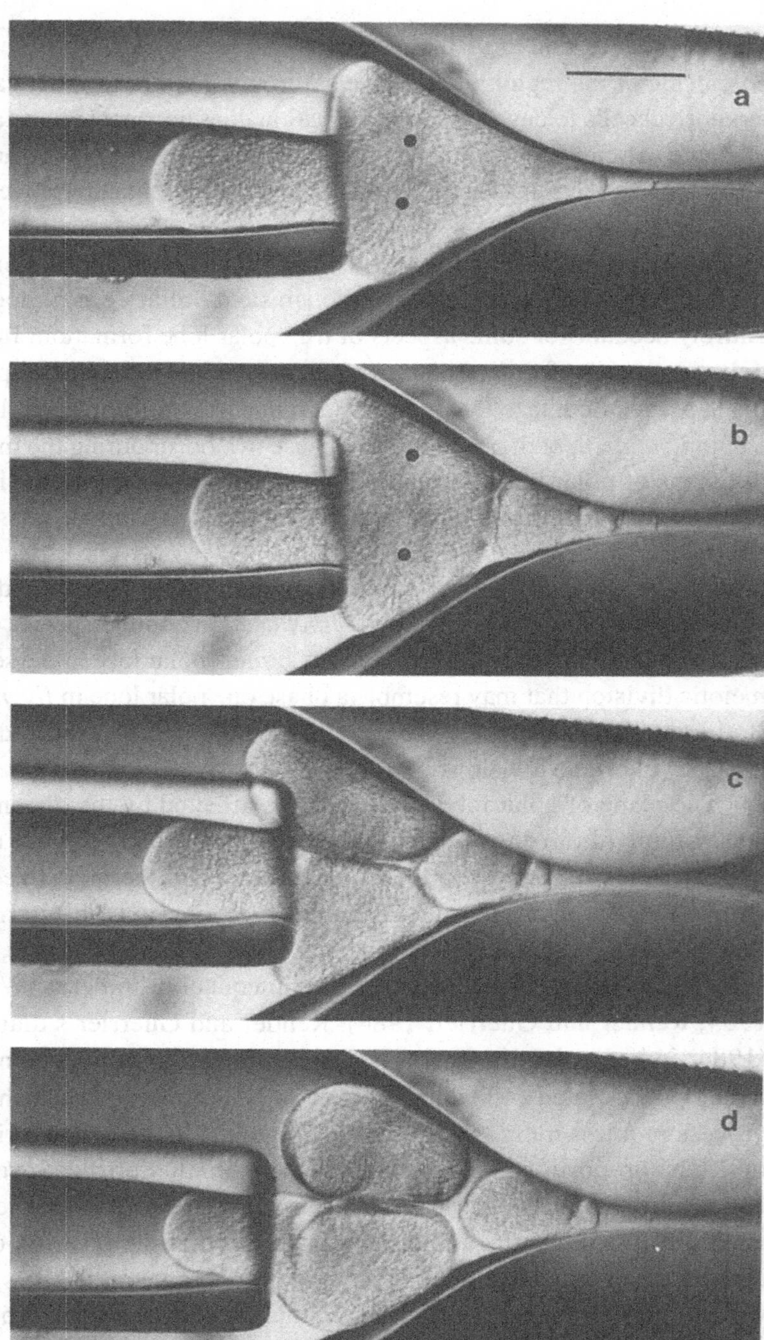
a
b
c
d

Delayed Cytokinesis

Early sectioning studies of developing invertebrate embryos revealed instances (especially among the arthropods) in which temporary or regional failure of furrow formation resulted in the formation of syncytia (summarized in Wilson, 1883). Typically, the failure occurred in a region of concentrated yolk; sometimes, the nuclei were later separated by multiple furrows and the cells subsequently formed were assimilated or contributed to the embryonic body form. The failure occurred in a regular predictable fashion, and, although it was recognized as an exception to the usual relation between nuclear and cytoplasmic division, it was not often cited in the cell division literature, perhaps in part because the opacity of the yolk in which it occurred, and the tough chorion that in many cases surrounded the egg, made observation and experimentation nearly impossible.

In 1883, Wilson described the cleavage and early development of the alcyonarian coelenterate *Renilla reniformis,* in which the large, opaque, membraneless egg undergoes a dissociation of mitosis and cytokinesis with an unusual degree of unpredictability. The variability of cleavage is not associated with developmental variability, and in his description of this phase of development, Wilson emphasized the implications concerning the cell division mechanism. The first cleavage, he found, could divide the egg directly into 2, 4, 8, 16, or 32 cells, and the interval between fertilization and cleavage varied between 1.5 and 3.0–4.0 hours (Figure 9.18). The blastomeres were often unequal in size and, in some zygotes, cleavage at first was confined to one pole and then spread in subsequent cleavages. After the first cleavage, the divisions of the embryonic cells were usually irregular, but always synchronous. The

Figure 9.16 Comparison of lobe-forming activity in conical and cylindrical surfaces. Within a single cell, the mitotic apparatus is positioned to interact simultaneously with cylindrical and conical surfaces, both oriented perpendicular to the mitotic axis (a). A constriction forms in the conical surface (b) and completely isolates the vertex (c, d). There is no significant deformation in the cylindrical surface. The distances from the mitotic axis to the conical and cylindrical surfaces are the same, and the central axis of the cone and the cylinder intersect the midpoint of the spindle (a). The dots indicate aster centers. Bar = 50 μm. (From Rappaport and Rappaport, 1994, and with the permission of Academic Press)

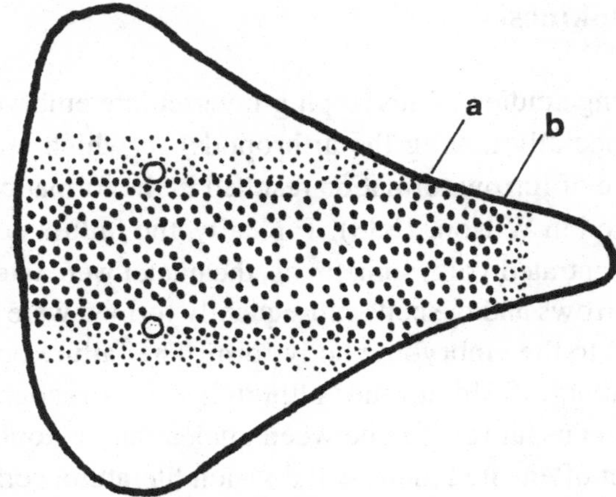

Figure 9.17 Hypothetical relationships that might account for simultaneous furrowing and lobe formation in conical sand dollar eggs. The region of maximum aster effect is three dimensional and lies between the planes of the aster centers (open circles), oriented perpendicular to the mitotic axis (large stipples). The relation between the mitotic apparatus and the surface at the base of the cone duplicates that in an egg that normally forms a unilateral furrow. Near the vertex, the region of maximum aster effect is associated with a band of conical surface that is bounded by areas of diminished aster effect. The boundary of the region of maximum aster effect closest to the base (a) is determined by the width of the region in which the effect of the asters is additive. The boundary closest to the vertex (b) is determined by the limited range of effectiveness of the mitotic apparatus. The existence of the band of affected surface bounded by areas of diminished mitotic apparatus effect is dependent upon the vertex angle of the conical surface. Were the surface in that region cylindrical rather than conical, the same geometrical relation could not be established.

first cleavage was often preceded by the formation of multiple surface mounds that Wilson associated with the mitotic activity that preceded cytokinesis.

The large size (350 μm diameter) and opacity of *Renilla* eggs required that their internal events be studied in microscopic sections. Wilson (1883) found that cytokinesis often fails to follow mitosis. After several mitotic cycles, the nuclei may not divide simultaneously, and nuclei in interphase and nuclei with well-developed amphiasters may appear in the same section. At first, the nuclei tend to be near the egg center, but as they multiply and the time of cleavage nears, more appear

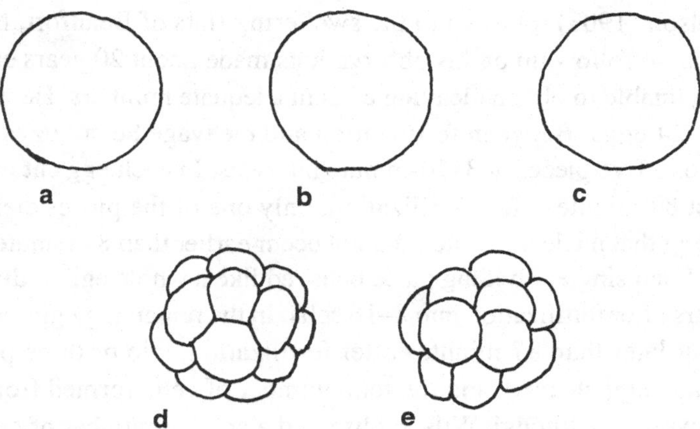

Figure 9.18 Early division activity in *Renilla*: successive outline drawings of the same egg. The perfectly spherical form at fertilization (a) becomes irregular as a consequence of the formation of multiple low prominences (b, c). It later divides directly into 16 cells (d); (e) is a different view of (d), and it shows that the entire egg has divided into cells. Wilson considered this the usual type of cleavage. No times were given. (From Wilson, 1883)

in the periphery. When the superficial mounds appear, each has a single nucleus at its center, suggesting that mitosis precedes mound formation. When the furrows subsequently appear, they progress from the surface toward a central, initially undivided, yolky mass, which is eventually apportioned among the cells. Wilson speculated that the irregularity in the timing and location of divisions was associated with an unhomogeneous distribution of yolk. Nuclear divisions, he felt, were fundamentally regular, but the uneven distribution of yolk and "division energy" created the appearance of chaos. C. W. Hargitt (1900) described similar irregular cleavage in the eggs of *Pennaria*, a sessile, branched, subtidal, relatively common coelenterate of colder Atlantic waters. His first observations of cleavage in the 400–500 μm yolky eggs led him to believe that they were abnormal, but he accidentally discovered that the irregularly cleaving eggs developed into normal larvae. The rates of cleavage among the eggs were variable, as were the patterns, and he concluded that nuclear division outran cytoplasmic division as a result of a fundamental disturbance of a rhythmic relation that normally exists between nucleus and cytoplasm.

Wilson (1903) returned to the sweltering flats of Beaufort, North Carolina, to follow up on his observations made about 20 years earlier but was unable to obtain cleaving eggs in adequate numbers. He used a total of 24 eggs. Between fertilization and cleavage he cut eggs each into two to five pieces at 3–10-minute intervals. In each egg cut during the first 87 minutes after fertilization, only one of the pieces cleaved, indicating that nuclear division did not occur earlier than 87 minutes (18 cases). Each single, dividing piece behaved like a whole egg. It divided 2.5 hours after fertilization into 8–16 cells. In the remaining eggs, which were cut later than 87 minutes after fertilization, two or three pieces from each egg cleaved, and the total number of cells formed from the pieces was 16. Although Wilson observed a smaller number of eggs in 1903, his results were more uniform than those of his 1883 investigation. The number of blastomeres formed was 8 or 16, but it had earlier varied between 2 and 32; the time of cleavage was restricted to 2.5 hours, but it had earlier varied between 1.5 and 3.0–4.0 hours.

C. W. Hargitt (1904) pointed out the similarities between the normal cleavage of *Pennaria* and that of the sea urchin eggs that Wilson (1901b) had treated with ether. In both cases, mitosis proceeded without cytokinesis, the nuclei moved from the central region to the periphery before division, the surface became lumpy, and multiple furrows subsequently cut the egg into more than two cells. A systematic cytological study of cleaving *Pennaria* and *Tubularia* eggs (G. T. Hargitt, 1909) supported the idea that cytokinesis lags behind mitosis. In G. T. Hargitt's study of *Pennaria,* the spindle for the second cleavage formed when the first cleavage was barely started, and the third mitotic division of the nucleus began before the first cleavage was completed. In *Tubularia*, the irregularity of cleavage appeared to be enhanced by an irregularity of egg form caused by close confinement in the gonophore, in contact with other eggs in haphazard arrangements.

Other examples of similarly irregular cleavage have been described in other animal groups, but careful scrutiny and experimental analysis of the phenomenon appears to be limited to these relatively few studies. Irregular cleavage is found among relatively large, yolky eggs in which the nucleus is initially located well below the surface. Because large egg size is usually a consequence of massive yolk accumulation and the optical properties of yolk obscure internal events, experimental analysis is hampered. Wilson's (1901b) investigation of the effect of ether on sea urchin egg cleavage suggests that the irregular behavior that he observed in Renilla could have arisen by chance associations between the mitotic apparatus and the surface, but he did not discuss that possibility in his 1903 paper. Ether treatment reduced the size of the asters and mitotic apparatus and also deferred cytokinesis. As persistent mitosis increased the number of nuclei, they became spaced so that some neared the surface. Wilson (1901b) noted that single asters could organize surface mounds around themselves and that deeper furrows formed between asters that were closer to each other and to the surface. These observations may indicate that the tendency toward multiple furrows in the first cytokinesis of Renilla results from the initially great distance between a normally centered mitotic apparatus and the surface. If, by chance, the nucleus was not centered or the egg was not spherical, then the necessary interaction between the mitotic apparatus and the surface could occur during the first, second, or third mitotic cycle, but if the zygote nucleus was centered, then the furrowing was delayed until the fourth or fifth cycle, when the egg contained many more asters. The superficial similarity that C. W. Hargitt (1904) pointed out between the cleavage of Renilla eggs and that of etherized sea urchin eggs might be attributed to a similar disparity in the relative sizes of the mitotic apparatus and the cytoplasm. Were that the case, then we could expect that the Renilla eggs that cleaved into fewer blastomeres would do so earlier than those that cleaved into more blastomeres, because more time would be necessary for more mitotic cycles, but Wilson did not describe the time relations in the necessary detail. We might also expect that the results of Wilson's 1903 experiments would have been different. In the nucleated fragments that are about one-fourth the volume of the whole egg, the distance between the mitotic apparatus and the surface is smaller and, were the normal delay a simple consequence of geometry, we would expect them to divide earlier than the whole egg into a smaller number of cells, but, according to Wilson, the timing and number did not differ from the intact egg. He concluded that cleavage depends upon a progressive change in the cytoplasm that reaches a critical point at a given time whatever the cytoplasmic volume may be, and that the number of nuclei is unimportant (Wilson, 1903). The question of nuclear versus cytoplasmic responsi-

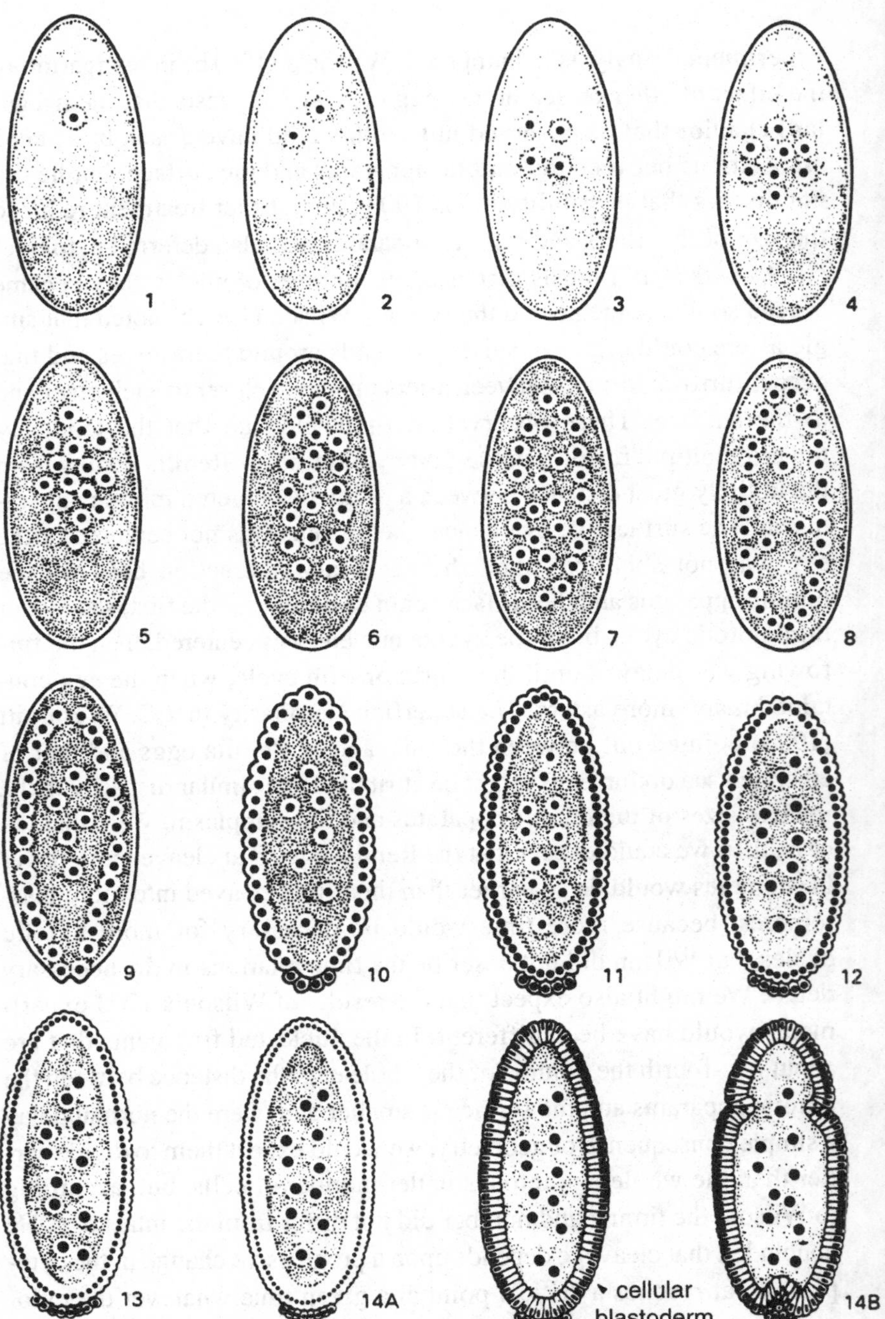
1
2
3
4
5
6
7
8
9
10
11
12
13
14A
cellular
blastoderm
14B

bility for delayed cytokinesis requires further analysis. The difficulties that arise from availability of experimental material and egg opacity remain, but advances are possible.

Early insect development is also characterized by a period in which mitosis is unaccompanied by cytokinesis but, in contrast with the previously described examples, the process is carried out with remarkable precision in both timing and morphology. In *Collembola* eggs, cytokinesis begins after two mitoses, when the nuclei move toward the periphery (Jura, 1975b). But the more complex events of early development in several higher insects have been more extensively investigated. Well-established methods for the rearing of *Drosophila* and for the management of their embryos, and the introduction of immunofluorescent methods for the observation and identification of proteins, have permitted thorough studies of early divisions despite some inherent inconveniences (reviewed, Warn et al., 1990). The eggs are banana shaped and average less than 0.5 mm long. They are covered by a chorion, which can be mechanically removed, and, beneath that, a relatively impervious vitelline membrane. The sperm enters through a micropyle, and pronuclear fusion occurs near the egg center (Doane, 1967). The mitoses that follow are roughly synchronous and take place at 10-minute intervals. Each nucleus is surrounded by a yolk-free zone, and, as the nuclei increase in number, they become spaced equidistantly in a roughly spherical array (Figure 9.19).

During this early period, the nuclei are involved in intermittent periods of division and movement, and the underlying causes of the movement have been the subject of several investigations. The gall midge *Wachtiella persicariae* follows a developmental pattern similar to that of *Drosophila,* and Wolf (1973) concluded that nuclear movement is

Figure 9.19 Schematic drawings of *Drosophila* developmental stages up to gastrulation. The number beside each embryo corresponds to the total number of nuclear divisions undergone by the nuclei beginning with stage 1, the fertilized zygote. The embryos are shown in sagittal section with the anterior end up. Solid black circles represent nuclei; yolky regions are indicated by stippling. Up to stage 5, all the nuclei are shown, although they are not normally all in the same plane. See the text for an account of division-related events. (From Foe and Alberts, 1983, and with the permission of the *Journal of Cell Science*)

associated with both active and passive processes. The passive movement results from general cytoplasmic flow, whereas the active movement appears to be microtubule based. The microtubules are assembled in a "migration aster" that is derived from an aster positioned at the pole of the spindle of the previous mitotic cycle. The microtubules are oriented radially and extend as far as 80 μm. They were thought by Wolf to function by attaching to peripheral structures and by moving the attached nucleus toward the egg surface by traction. In Wolf's 1978 study, colchicine injection disrupted the microtubules, stopped transport of yolk particles toward the aster centers, and blocked active nuclear transport. In the presence of cytochalasin B, active nuclear transport continued but ooplasmic flow ceased. Active nuclear transport occurred only while the elongate microtubules of the migration cytaster were present, from the onset of anaphase to the early prophase of the next mitotic cycle (Wolf, 1980).

During the period between the 4th and 6th cycles in *Drosophila*, the array of nuclei changes to a hollow ellipsoid as some of the nuclei move toward the egg poles at a rate of about 50 μm per cycle (Figure 9.19). The rate of movement is variable and reaches 23 μm per minute during prophase. This axial expansion of the nuclei is not affected by colcemid, but it is blocked by cytochalasin and by cyclohexamide, so it is assumed to require actin activity (Baker, Theurkauf, and Schubiger, 1993) (Figure 9.19). After the formation of the ellipsoidal configuration, from the 7th through the 10th cycles, most of the nuclei move at 7–8 μm per minute toward the egg cortex, which is yolk-free to a depth of about 3 μm below the plasma membrane. The cortical migration is also blocked by colcemid and is assumed to be microtubule dependent. The centrosome-associated microtubules that are directed toward the egg center are longer than those that are directed toward the cortex, and Baker, Theurkauf, and Schubiger (1993) proposed that the center-directed microtubules may form antiparallel associations with microtubules associated with the nuclei that remain at the egg center axis and those on the opposite side. This arrangement could result in propulsion of the nuclei away from the center by force generated by plus-end–directed motor molecules. The 50 or so yolk nuclei that do not migrate, divide and become polyploid. The migrating nuclei do not reach the cortex simultaneously. During the 9th cycle, the nuclei of the future pole cells reach the posterior egg surface; later in development, they become the source of the gametes and of some cells of the gut wall. During the next cycle, the now 400 remaining nuclei, as they near the surface, orient so that their centrosomes lie between the nuclear membrane and the cortex (Fullilove and Jacobson, 1971). During the first 8 cycles, the cortex contains both actin and tubulin, but during the 9th cycle the cortical microtubule density decreases near the approaching nuclei (Karr and Alberts, 1986). During mitoses of the 10th cycle, the pole cells pinch off and become uninucleated several cycles earlier than the similar event that takes place over the remainder of the embryonic periphery. At mid interphase of the 10th cycle, the migration ends with the nucleus about 2 μm from the plasma membrane. Then occurs the formation of a "cap" of filamentous actin, microtubules, and intermediate filament proteins over each centrosome. The characteristic cap domain is contained within a restricted cortical domain lying over each aster (Karr and Alberts, 1986) that cyclically bulges, flattens, and divides.

The four additional mitotic divisions that follow the cortical migration that ends during the 9th division populate the egg surface with about 6,500 nuclei (Turner and Mahowald, 1976). Before each division, the centrosome splits and migrates to opposite poles of the nucleus; the spindle lies parallel to the egg surface and the egg is as yet undivided. The superficial complex of nuclei and yolk-deficient cytoplasm is termed the syncytial blastoderm. Mitotic activity of the syncytial blastoderm nuclei is closely synchronized with behavior of the caps or bulges to the extent that a causal relation between the two is assumed. With each successive mitotic cycle, the number of bulges increases and their diameter and intervening distance decrease (Turner and Mahowald, 1976). As the nuclei divide, so do the caps, and the mutual displacement that accompanies the increasing number of syncytial nuclei is also reflected in the repositioning of the caps (Warn, Magrath, and Webb, 1984). During prophase of cycles 11, 12 and 13, pseudo-cleavage furrows pull down around each nucleus so that each spindle is temporarily surrounded by a shallow indentation (Stafstrom and Staehelin, 1984). Aphidicolin slows, but does not block, centrosomal replication, microtubule formation, and cap formation. Raff and Glover

(1988) suggest that these results may reflect a role for the centrosome in cap formation. They confirm the observation that caps subside or fail to form when nuclei move to the interior after exposure to antitubulin antibody (Warn, Flegg, and Warn, 1987).

Immediately before furrowing, the nuclei enlarge by increasing in length in the axis perpendicular to the surface. At the same time, microtubules, also oriented perpendicular to the surface, appear between the nuclei (Fullilove and Jacobson, 1971). Each cap is surrounded by contiguous rings or polygons of flattened surface so that, in superficial view, they resemble a net with a uniform mesh size. The meshlike furrow complex intrudes into the egg substance, accompanied by filamentous material (Fullilove and Jacobson, 1971) that has been identified as F-actin (Warn and Robert-Nicoud, 1990). The tip of the furrow expands into a space, the furrow canal, that appears vesicle-like in cross section and resembles the cleavage head of ctenophore eggs and flattened echinoderm eggs. The rate of furrow intrusion is not uniform. Intrusion progresses from the surface to the level of the bottom of the nucleus about half as fast as it progresses from the bottom of the nucleus to the yolk. During the rapid phase of furrow intrusion, the furrow canals expand parallel to the plane of the blastoderm surface so that the connection between the cell and the underlying yolky region is reduced to a narrow gap surrounded by F-actin (Warn and Robert-Nicoud, 1990). During furrowing, myosin colocalizes with actin (Kiehart et al., 1990), and there is a general assumption that the biomechanical activity that simultaneously divides the syncytium into thousands of cells is the same as that which appears to function in more conventional cytokinesis. There are, however, important differences in the way the two types of division occur.

Conventional cytokinesis is often compared to a purse string that divides the cell by constricting its middle, but in the formation of the cellular blastoderm, the actin–myosin ring maintains a relatively constant diameter as it slides along the periphery of the elongate nucleus toward the yolk. Its progress resembles that of the furrow in the squid blastoderm (Arnold, 1969), and it appears to have a similar requirement for anchorage. The only anchorages that appear to be available in the insect embryo are the contiguous rings, and this arrangement could result in a complex interplay of forces during the half-hour process.

Warn and Robert-Nicoud (1990) have suggested the necessity for a push mechanism in the expanding furrow walls.

The sources of the additional surface membrane required for the transition from syncytial to cellular blastoderm are no better known than in conventional cytokinesis. The large number of nuclei leads to the initial expectation of a period of very intense new surface formation, but the relative slowness of the process results in a calculated surface increase of 5% per minute (Warn et al., 1990), which is similar to that of the sea urchin egg (Schroeder, 1981b). Various cytoplasmic structures have been suggested as surface precursors. Sanders (1975) could identify no contribution from endoplasmic reticulum, Golgi bodies, or nuclear envelope, but he described lamellar bodies 4–5 μm below the surface that became sparser as cleavage progressed. Another possible source lies in an array of microvilli and surface folds that normally cover the nucleus-associated caps or bulges on the syncytial blastoderm surface. By the time the furrows have extended almost to the level of the bases of the nuclei, the apical surfaces of the forming cells are flat (Fullilove and Jacobson, 1971; Turner and Mahowald, 1976).

The timing and location of the furrowing that occurs during cellularization and subsequent divisions are precise and, once begun, furrowing is completed in a brief period. The process in insects is more regular than in *Renilla* and other large yolky eggs, but the factors that determine the point when syncytial divisions give way to cellularization and cytokinesis are not well understood. Fullilove and Jacobson (1971) showed that both the timing and the position of furrows could result from the changes in the geometrical relations among the mitotic apparatuses and between the mitotic apparatus and the surface that develop as the number of nuclei in the periphery increases from about four hundred to more than six thousand. Conversion of the syncytium into a blastoderm in which all cells are uninucleate requires that furrows form between asters that are not connected to a common spindle. Their measurements revealed that at the time of cellular blastoderm formation, the synchronously formed mitotic apparatuses are crowded to the extent that the distances between asters in the same mitotic apparatus and between asters of adjacent mitotic apparatuses are similar, and Fullilove and Jacobson (1971) compared this circumstance with one that can be experimentally created in sea urchin eggs (Rappaport, 1969b). In the same way, the failure of permanent nuclear separation during the first 13 cycles could result from the distance between the mitotic apparatus and the surface, and from the distances between the individual mitotic apparatuses. The similarity of the organelles involved, their geometrical relations, and the apparent similarity of the ensuing physical activity suggest that cellularization and cytokinesis share common mechanisms, but direct comparisons are hampered by the unsuitability of *Drosophila* embryos for micromanipulation.

Drosophila is, however, useful for investigations based upon the effect of cell division mutants that affect the early embryo, and it has been possible to demonstrate dependency relationships among division-associated structures and activities. The maternal-effect mutations often used in such investigations depend entirely upon the genotype of the mother, and they involve genes that are essential only during oogenesis. Because the zygote genome is not actively involved in control during the earliest stages of development, the phenotypic deficiencies of the embryo are attributed to alterations in, or absence of, essential gene products that are normally provided in the egg. Three such mutations that affect the transition from syncytial to cellular blastoderm were investigated by Rice and Garen (1975) and were found to result in different, specific deficiencies. All three mutants form pole cells but show different degrees of blastoderm cellularization. In *mat (3) 6,* cellularization occurs only at the anterior and posterior ends, and 70% of the embryo surface is noncellular. The nuclei of the noncellular area remain spherical and move away from the surface. No blastoderm cells are formed by *mat (3) 1.* The nuclei remain spherical and also move away from the surface. Rice and Garen compared *mat (3) 1* with *grandchildless* in *D. subobscura,* which forms no pole cells, but the normal blastoderm cells develop into a sterile adult. They concluded that not only must there be separate controls for pole cell and blastoderm cell formation, but there must also be regionally specific controls for the blastoderm. With the introduction of immunofluorescent techniques, more detailed information about the nature of the deficiencies has become available.

Sullivan, Minden, and Alberts (1990) investigated the basis of the abnormalities that arise in embryos bearing the *daughterless-abo-like (dal)* mutation, which affects the number of surface nuclei and internal nuclei at the time of cellularization. They found that events proceeded normally to cycle 10, but that at cycle 11 some irregularities of nuclear shape and size appeared. The abnormalities, which were associated with the failure of normal centrosome separation and with the improper formation of pseudocleavage furrows, often resulted in partial nuclear fusions. In cycles 12 and 13, the abnormal nuclei moved into the deeper cytoplasm, leaving a reduced number of apparently normal nuclei in the normal relation to the surface. At cycle 14, cellularization occurred when about half the normal number of nuclei were at the surface. The nuclei and cells were larger than normal. The results suggest that cellularization is not directly linked to the absolute number of nuclei in the surface. Embryos that develop from eggs laid by females homozygous for a mutation of the *sponge* locus lack F-actin caps and the accompanying surface bulges, as well as the pseudofurrows that normally appear between the dividing syncytial nuclei (Postner, Miller, and Wieschaus, 1992). Other F-actin is present in the cortical layer at cycle 10, but it is depleted over the nuclei and, during division, over the position of the spindle. Nuclear position and division are nearly unaffected during cycle 10, despite the absence of caps. During cycle 11, the absence of pseudofurrows is associated with the formation of spindlelike structures between the centrosomes of different nuclei and with interference with normal chromosome separation. During cycles 12 and 13, the number of aberrations increases. By cycle 14, however, most of the abnormal nuclei have sunk below the surface, and the actin in the surface is continuous where the nuclei were previously located, but depleted above the nuclei that remain in the normal location. Where the nuclei that remain near the surface are clustered, hexagonal arrays of actin filaments form around the nuclei, which remain spherical. This actin network supports cellularization, although the cells are not as long as in the wild type. Postner, Miller, and Wieschaus (1992) concluded that the *sponge* gene products are essential during the syncytial stage, but not during cellularization. Although the actin that normally forms the linked hexagons during cycle 14 is at least in part recruited from the caps, the ability of *sponge* embryos to form hexagonal actin arrays demonstrates that the caps are not essential, and the *sponge* mutation does not affect the ability of actin filaments to form ordered structures. Sullivan, Fogarty, and Theurkauf (1993) compared three mutations that affect previously uncharacterized genes and that produce embryos with mitotic division anomalies that appear only after the nuclei have migrated to the surface. Each mutation appears to affect a different aspect of pseudocleavage and cellularization, but they are all considered to affect the cytoskeleton. The previously described *dal* mutation results in abnormal pseudocleavage furrows, which are in turn associated with incomplete centrosome separation and frequently defective spindle formation. Normal spindles in *dal* embryos are surrounded by normal furrows. In *nuf*^{1} embryos, large gaps in the actin-based furrows are associated with nuclear cycle 12. The spacing and structure of the spindles appear normal, as do the actin caps during interphase. During later cycles, fused and unequally spaced spindles appear, and cellularization is abnormal. In *sced* embryos, pseudofurrow formation is deficient and spindle defects follow. Actin cap formation occurs at interphase. During cellularization, the nuclei fail to elongate normally and the configuration of microtubules that normally surrounds them is missing. Cells with nuclear abnormalities detach from the surface and sink and, as in *dal* embryos, the remaining blastoderm nuclei are larger than normal, as are the cells that contain them after cellularization. To this list of mutations may be added *pebbles,* in which pseudocleavage furrowing and cellularization occur, but subsequent cytokinesis fails (Hime and Saint, 1992). Sullivan, Fogarty, and Theurkauf (1993) pointed out that pseudocleavage formation, cellularization, and cytokinesis are all actin-mediated events, yet each requires a unique gene product.

By the period of cellularization, the transcription rates from the zygotic nuclei have increased, so embryonic dependence on maternal gene products is decreased. At that time, zygotic mutations can change the normal hexagonal actin–myosin pattern in the blastoderm. Failure of normal furrow formation between adjacent nuclei subsequently results in the formation of multinucleate cells in the blastoderm. Schejter and Wieschaus (1993) described three genes associated with cellularization, and expression of these genes is limited to the period of the syncytial and cellularizing blastoderm. Immunolocalization techniques have revealed that the protein products of the three genes are restricted to the membranes that separate the nuclei during cellularization. *Peanut (pnut)* is another *Drosophila* gene required for cytokinesis (Neufeld and Rubin, 1994). Its period of expression is less restricted, and the *pnut* protein may be maternally supplied. Homozygous mutants survive on the maternal protein until the larval stages. The larvae, which are characterized by severely reduced imaginal discs, die shortly after pupation. Within the discs, failure of cytokinesis results in large, multinucleate cells. Immunolocalization staining reveals localization of *pnut* protein in the furrow in dividing cells, and also at the leading edges of the membranes that separate the nuclei during cellularization and in the rings that separate the blastoderm cell cytoplasm from the underlying yolk.

Combining the application of molecular and genetic methods in studies of the early *Drosophila* embryo has produced wonderfully detailed analyses of the distribution of specific gene products and has also facilitated establishing the correlation between the presence and normal distribution of the products within the embryonic cells and their ability to engage in normal cytokinesis. The data reveal, for instance, that cellularization and cytokinesis are vulnerable to mutation-induced deficiencies of different gene products, and they suggest that the two processes may employ different control and functional mechanisms, which are not detectable by simple observation. Understanding the basis of functional deficiencies in cytoskeletal activity that follow modification or elimination of gene activity may require the use of locally introduced reagents that act in a well-defined manner to produce effects that are not obtainable by mutation.

Division of Amoebae

The prefurrow events of *Amoeba proteus* division were described and figured (Figure 2.3) in Chapter 2. Telophase elongation reshapes the contour of the surface between the planes of the nuclei into an approximately cylindrical form. At the beginning of cytokinesis, the short pseudopodia in the middle of the cylindrical region retract, so the early furrow surface is smooth and shaped like a hyperboloid (Figure 9.20). As the wide furrow rapidly deepens, the cytoplasm surges back and forth across the division plane. Normally, the pseudopod-covered prospective daughter cells attach to the substratum and they begin to move apart, parallel to the mitotic axis, before the connection between them is actively severed, so that it becomes stretched and then snaps. This final event of the process led Chalkley (1951) to believe that

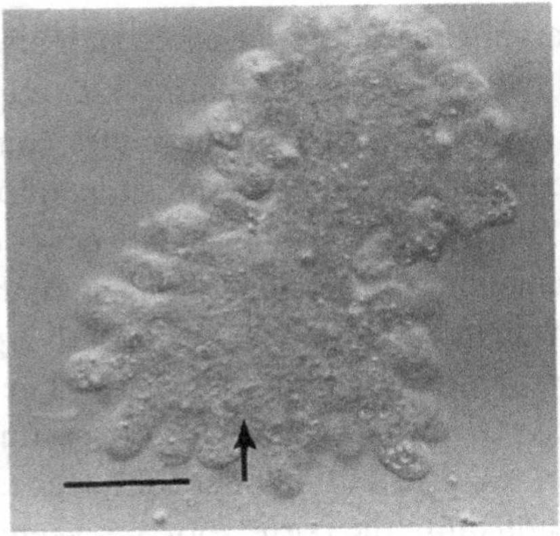

Figure 9.20 Telophase elongation and early furrowing in *Amoeba proteus*. The distinctive furrow region lacks pseudopodia. The arrow indicates one of the telophase nuclei. Bar = 50 μm. (From Rappaport and Rappaport, 1986, and with the permission of Wiley-Liss)

cytokinesis is entirely due to coordinated locomotion (Chapter 2). Such a hypothetical mechanism makes cytokinesis dependent upon attachment to a substratum, and the idea has been tested by forcing the amoeba to divide in circumstances where one or both poles cannot attach. Chalkley (1935) found that immersion in distilled water prevents adhesion but not division, which, in this circumstance, he attributed to the pushing of the prospective daughter-cell pseudopodia against each other. His observations could not be confirmed, and other methods of preventing effective attachment have also failed to block division (Rappaport and Rappaport, 1986). Cytokinesis is not blocked by repeated detachment of pseudopodia with a blunt probe, suspension of an amoeba from a pipet by one pole, or placement of an amoeba on a liquid interface. The locomotion-associated stretching and final parting of the last connection between daughter cells occurs in cultured tissue cells as well as in amoeba. It does not appear to be the only way that severing can occur, and the elongation does not require the active process proposed by B. Byers and Abramson (1968). Bisection and cytoplasmic amputation experiments revealed that important division-related events occur before the daughter nuclei form and separate. When late prophase amoebae were bisected, only the nucleated halves divided or showed division activity, but when the same operation was done at metaphase, 21 of 23 enucleated halves developed transient furrows and 2 completed the process. The establishment and the function of the division mechanism in *Amoeba proteus* resemble those in tissue cells (Rappaport and Rappaport, 1986). Contrary to Chalkley's theory, the position of the furrow is determined by the mitotic apparatus between prophase and telophase, and the equatorial surface actively constricts. The final separation of the daughter cells can be accomplished by the furrow, but it normally occurs by traction. These findings confirm the predictions made by W. H. Lewis (1942).

The amoeboid stage of the life cycle of the cellular slime mold *Dictyostelium discoideum* offers the investigator several advantages over *A. proteus*, and it is now one of the best-studied nonmuscle contractile systems (reviewed, Fukui, 1993). The cytoplasmic events of *Dictyostelium* division closely resemble those of the larger *A. proteus*, but the nuclear events are very different. Moens (1976) made an ultrastructural study of spindle and kinetochore morphology and described a spindle pole body adjacent to the nuclear membrane that appears to behave like the centrosome. At prophase, the chromosomes condense and the kinetochores become more distinct. The spindle pole body comes to lie in an indentation in the nuclear membrane near the kinetochores. The spindle pole body divides into two discs, which separate but remain attached by microtubules. During mitosis, the nuclear membrane persists and the spindle pole bodies take up diametrically opposite positions near its outer surface. Between the spindle pole bodies, microtubules pass through openings in the nuclear membrane and form a spindle that is about 2 μm long at metaphase. The chromosomes are located near the spindle center. About 20 microtubules extend from pole to pole. At anaphase, the spindle elongates to 4–5 μm and the nuclear membrane becomes constricted near the spindle midpoint, where the microtubules overlap before cytoplasmic division begins. Spindle elongation continues during telophase, and the nuclear membrane division is completed when the spindle is 7–8 μm long. During mitosis and cytokinesis, the spindle pole bodies are associated with both spindle and cytoplasmic microtubules. The microtubules extend in radiate patterns that expand so that during telophase they resemble asters with sparse, sometimes curving rays. Similar microtubule organization is present in migrating interphase cells (Rubino et al., 1984). A small bundle of microtubules that extends between the spindle pole bodies persists until cytokinesis is completed, when the spindle is 10 μm long. The spindle aperture in the nuclear membrane closes around the remaining microtubules.

Although their small size appears to have discouraged experimentation by micromanipulation, *Dictyostelium* amoebae possess other qualities that make them well suited for investigations based upon molecular techniques. They grow rapidly on either bacterial or axenic media. They are normally haploid, but diploidy may be induced, and they tolerate a degree of flattening that facilitates high-resolution immunofluorescent studies of cytoskeletal proteins. Their genetic constitution can be manipulated by molecular methods such as homologous recombination (DeLozanne and Spudich, 1987) and antisense RNA inactivation (Knecht and Loomis, 1987), and their general morphology resembles animal tissue cells.

Fukui and colleagues have studied changes in the distribution of cytokinesis-related cytoskeletal molecules by immunofluorescent methods, which are apparently less disruptive to nonmuscle myosin II (conventional myosin) and other proteins than conventional ultrastructural treatment (Yumura and Fukui, 1985; reviewed, Fukui, 1993). In vegetative *Dictyostelium* amoebae, cortical myosin II takes the form of 0.2 μm by 0.7 μm rods, which at the time of division disappear except in the furrow region, where they orient circumferentially (Yumura and Fukui, 1985). Before furrowing, at late prophase, cortical myosin decreases and endoplasmic myosin increases. In anaphase, before the furrow appears, endoplasmic myosin decreases as cortical myosin at the equator increases. In early telophase, during elongation, the apparent translocation continues and, as the furrow progresses, the endoplasmic myosin filaments almost disappear (Kitanishi-Yumura and Fukui, 1989). During the division period, actin filaments are colocalized with myosin until late anaphase, at which time the actin is distributed over the entire cortex and myosin is restricted to the furrow, suggesting the possibility of independent organizing mechanisms. Actin filaments in the furrow are circumferentially aligned (Yumura and Fukui, 1985). In contrast with myosin II, nonfilamentous myosin I is concentrated in the polar lamellipodia rather than in the equatorial cortex (Fukui et al., 1989), and ∂-actinin is excluded from the furrow region but present in the remainder of the cortex (Fukui, 1993). High-extinction polarized light microscopy and differential interference contrast video microscopy

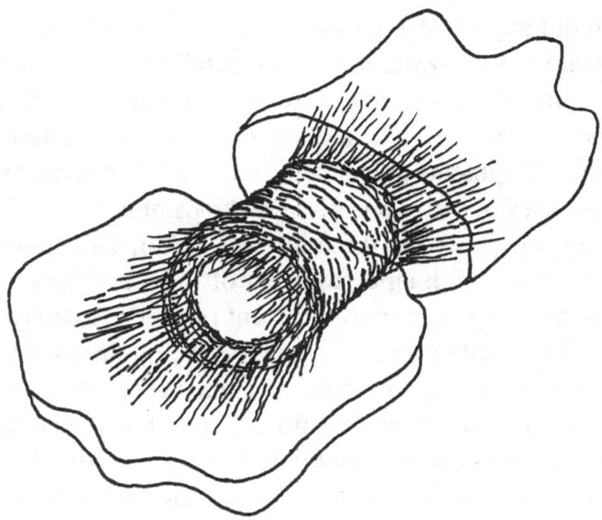

Figure 9.21 Diagram of the organization of actin and myosin II during cytokinesis of attached *Dictyostelium.* Some F-actin in the furrow region is oriented circumferentially parallel to, and some is oriented perpendicular to, the division plane. The circumferentially oriented actin is associated with myosin II, and it is restricted to the equatorial region. The actin oriented perpendicular to the division plane is not associated with myosin II, and it extends into the daughter cytoplasm. (Redrawn from Fukui, 1990, and with the permission of the New York Academy of Sciences)

enhanced with digital image processing reveal more details of the dynamic activity of the cytoskeleton during cytokinesis (Fukui and Inoué, 1991). The formation of the furrow is accompanied by an obvious, but weak, birefringence parallel to the equator and by a second birefringent array, oriented perpendicular to the equatorial plane and parallel to the mitotic axis (Figure 9.21). The second array is actin-based and takes the form of a hollow cylinder that extends into the daughter cell cytoplasm; the array is not associated with myosin II. Both configurations persist throughout cytokinesis. After cytokinesis, the structure is partitioned equally in the posterior cortex of the daughter cells.

Molecular methods that make possible the disruption of selected genes permit more detailed study of the force production mechanisms that operate during cytokinesis in *Dictyostelium* amoebae. Utilizing homologous recombination, DeLozanne and Spudich (1987) inserted the myosin heavy chain gene, so the mutants failed to express native myosin II but did express the myosin fragment. Mutant cells varied in size and nuclear number, which suggests that failure of cytokinesis was common. The altered cells were not, however, totally devoid of motility and shape-change activity, and some underwent fission – after which, nucleated fragments presumably survived. The results indicate that the forces necessary to express several kinds of normal cell behavior are present when myosin II is absent. Fukui, DeLozanne, and Spudich (1990) undertook an assessment of the role of myosin II in *Dictyostelium* by using immunofluorescent methods to determine the organization of the mutant heavy meromyosin protein, as well as the organization of actin and microtubules in mutant cells. They found that the truncated myosin was always diffusely distributed, and they reasoned that it cannot assemble into thick filaments. Actin distribution was essentially normal in vegetative cells, in that it was rich in pseudopodia and lamellipodia. Actin was also demonstrable where cells were attenuated to form pseudocleavage furrows. The microtubules of mutant cells penetrated farther than normal into the region of the pseudopodia and lamellipodia; in multinucleate cells, a microtubule network was often associated with each nucleus. The information concerning pseudocleavage furrows permits some speculation concerning the forces that create them. Their appearance does not coincide with mitotic activity or the presence of spindles. Their formation requires that the cells be attached and that the regions of attachment be outside of the area where the pseudofurrows form. Cells that cannot form myosin II do not increase in number when they are grown in suspension, although the mitotic cycle continues (Knecht and Loomis, 1987). Myosin I in mutants is concentrated in the polar pseudopodia and only diffusely distributed in the remainder of the cell (Fukui et al., 1989).

The descriptive information is consistent with a mechanism that, near the regions of attachment, exerts traction forces that are cumulatively sufficient to deform the cell and to break it occasionally in the region where the cross-sectional area is initially greatest. The close resemblance between normal dividing amoebae and those that form single, centrally located pseudocleavage furrows implies that many of the visible characteristics of a partially divided normal amoeba are not mitotic apparatus–mediated. Because normal amoebae divide in the attached state and detached amoebae do not form an elongate hyperboloid furrow region, it appears likely that in normal division the equatorial region is simultaneously stretched parallel to the mitotic axis and compressed perpendicular to it. Traction force sufficient to break the cell would also be sufficient to realign the microfilamentous cytoskeleton parallel to the direction of tensile stress (Kolega, 1986). These relationships suggest that the axial array of F-actin that appears with the circumferentially oriented contractile ring of normal amoebae (Fukui, 1990; Fukui and Inoué, 1991), and the modest amount of actin in the pseudocleavage furrow of myosin-deficient amoebae (Fukui, DeLozanne, and Spudich,1990), may result from passive realignment of F-actin in response to traction exerted in the region of the polar pseudopodia. The axial array of F-actin may, perhaps, also be augmented by the effects of stretch-activation of isometric contraction (Wohlfarth-Bottermann and Fleischer, 1976).

When we consider that attached *Dictyostelium* amoebae can exert traction sufficient to bisect themselves despite myosin II deficiency, it is unexpected to find that the final rupture of the thin, persistent, connecting cytoplasmic thread may be dependent upon the level of calmodulin in the midbody region. Liu, Williams, and Clarke (1992) constructed a *Dictyostelium* strain in which calmodulin expression could be conditionally suppressed by introduction of antisense RNA. Reduction of calmodulin expression was accompanied by the development of giant multinucleated cells, which appeared to have resulted from repeated failure of cytokinesis. The early steps of cytokinesis were normal. The contractile ring formed and the equatorial region was reduced to an attenuated thread, but the thread did not part at the midbody. Eventually, the daughter cells merged in the region of the persistent connection to form a single, binucleate mass. In normal dividing *Dictyostelium,* calmodulin is enriched in the midbody region, and the authors suggest that its suppression may affect cytoskeletal behavior or membrane fusion.

The unevenness in the intensity of investigations of the various kinds of cytokinesis activity makes comparisons of the results difficult. When studies begin to involve a wider variety of cell types and to involve more and new methods of analysis, we can expect to find an increasing number of special mechanisms and different behaviors and physical properties. The most common features among the examples discussed are the relation between the position of the mitotic apparatus and the position of the division plane, and the structure, composition, and physical activity of the division mechanism.

Echinoderm eggs can be made to reproduce features of unequal division, polar lobe formation, and delayed cytokinesis by methods that involve different kinds of geometrical manipulation, but any assumption that such simple causes are entirely responsible for a particular type of variation on the normal must be tested.

There is a likelihood that different cell types may use somewhat different mechanisms to accomplish superficially similar ends, as is apparently the case in the division mechanism establishment that is accomplished by the widest part of the spindle in ganglion and tissue cells and that accomplished by the enlarged asters in cleavage cells. The relocation of the mitotic apparatus to its definitive predivision position (where it interacts with the cortex) that takes place in polar body formation, micromere formation, and unequal cleavage, may well be based upon mechanisms that differ from one another. The delayed cytokinesis that results in temporarily syncytial zygotes in different major animal groups could result from early failure to meet certain geometrical requirements, but temporary inability of either the mitotic apparatus or the cortex to play its normal role in furrow establishment would produce the same results.

Most intriguing are instances in which a part of the division process is accomplished in a completely different fashion, like the formation of the mitotic apparatus for the first meiotic division in frog egg, and the formation of the bulge in *Tubifex* polar body formation independently of the mitotic apparatus. The strong attachment between the aster and the cortex that characterizes polar body and micromere formation has not been replicated by simple relocation of the mitotic apparatus during other cleavages. The similarity of aster size and form that follows when close association between an aster and the cortex is prevented suggests that the special conditions that facilitate the attachment arise in the cortex.

The search for functional heterogeneity at the molecular level has barely begun. The principal molecular participants and potential participants, such as actin, myosin, and other motor molecules, exist in multiple forms (isoforms) that can be arranged in families and superfamilies. The roles of the different isoforms in the different kinds of division activity await further analysis.

10

Conclusion

The "slender but consistent" series of investigators responsible for early progress in the understanding of cytokinesis produced an appreciable body of thought and observation that seems to stand to a great extent by itself. The importance of understanding contraction, deformation, surface formation, and other manifestations of cell activity that seemed especially important during cytokinesis was recognized, but fundamental information concerning the nature of these activities was lacking. It was not usually possible to explain by adducing relevant supporting data from other fields because they did not exist. W. H. Lewis's (1942) insight concerning the similarities among cytokinesis, amoeboid locomotion, and cell shape change drew attention to the probable importance of surface contractile activity in cell biology, but it did not at that time offer a way to understand the underlying mechanisms.

The 50-year efflorescence of comment and conjecture about cytokinesis that was produced by some of the best-known minds in cell and developmental biology was not accompanied by marked advances in an understanding of the details of the process or by improvements in general methods of analysis of dynamic cell activity. Attempts to understand cytokinesis by reasoning from descriptive information were usually unsuccessful because of the nature and scale of the process. The actions and interactions occur in a broad spectrum of levels of organization, and there are serious problems in determining chronology, which is a basic way to differentiate between cause and effect. The inability to discriminate clearly between cause and effect contributed to the ease with which competing cell division theories were created. Early interests focused upon understanding the immediate bases of visible events: the relation between mitosis and cytokinesis, the general nature of the division mechanism, its location in the cell, and the roles played by the different structures of the cell, for example. Points of view were clearly articulated, and rational schemes were coherently presented, but, inevitably, most of them eventually failed the experimental test. A case can be made for the idea that the relatively late acquisition of basic information can be related to a delay in the application of thoughtful, systematic experimentation. Few who proposed theories tested them. In retrospect, most of the measurements and experiments that were done appear to have been searches for supporting data rather than evaluations of predictive accuracy. The delay may also have been based upon a reluctance to use any form of intervention in the normal process that might possibly introduce an artifact, and a faith that measurements of different aspects of a normal process would eventually provide a basis for understanding, as they had in physics and chemistry.

Simple physical experimentation that employed micromanipulation devices provided answers to many of the general questions that were central at the turn of the century. Experimentation by physical manipulation has the virtue of introducing a simpler form of disruption than do chemicals, and the extreme sensitivity of the living dividing cell is an effective indicator of unacceptable artifact. The wonder is that the tools that eventually facilitated physical experimentation were not used for that purpose until long after they had become available. The Barber three-movement pipet holder was described around 1907, and it was originally used to isolate single microorganisms. It was in effect a stage-mounted rack-and-pinion manipulator having the same general characteristics as those that are in production today. In use, the object was suspended in a hanging drop mounted in a moist chamber that could be positioned with a separate, conventional mechanical stage. Kite (1913) and then Chambers (1915) used it to explore the physical properties of protoplasm. In practice, the needle was positioned in the field with the pipet holder, and rapid movements were obtained by adjusting the mechanical stage. Similar devices appeared in Europe, such as the Peterfi instrument, which was based upon three screw-driven, tilting movements, and in 1921 Chambers described his own manipulator based upon hinged, rigid bars that were moved apart by screws against springs. By the mid-1920s, the devices were developed to a degree that they could have been used to perform most of the manipulation experiments that were done in the following 60 years, but they were not systematically used in cytokinesis experiments for about 25 years after that time. This gap is puzzling because Chambers spent most summers working at the Marine Biological Laboratory at Woods Hole, where there was a tradition of lively interest in cytokinesis. Although his life's work was focused upon the physical properties of protoplasm, he had worked with dividing cells and intermittently published results of experiments and speculation about mechanisms of animal cell division.

Between 1955 and 1975, the experimental information necessary to answer the basic 19th-century questions was available, if not universally accepted. Mitosis and cytokinesis are closely associated in time and place, because the short-lived structure in which the former occurs induces the activity and the formation of the structure that is responsible for the latter. The division mechanism is a restricted band of contractile material located at the interface between the surface and the cortex. The general behavior of animal cells indicates that contractile bands can exert forces sufficient for surface deformation without developing special ultrastructural qualities, but well-developed division furrows are associated with an underlying contractile ring. The contractile ring is characterized by the orientation and concentration of its filamentous components. Actin and myosin appear to be its active force-producing elements, but other proteins are also locally enriched. As the contractile ring functions, the volume of its ultrastructurally recognizable volume decreases because its circumference shortens. The cross-sectional dimension and the tension exerted by the ring are not diminished. The duration of the functional period of the contractile ring is not constitutionally restricted to the time period when it normally operates.

The circumferential shrinkage of the contractile ring appears to be responsible for many of the visible consequences of deformation. The subsurface equatorial cytoplasm is displaced, with consequent distension at the poles. An ability to form additional surface during cell deformation can be demonstrated throughout the cell cycle, and the normal pattern of new surface is not essential to division.

The division mechanism is established by the mitotic apparatus. During establishment, linear radial elements of the mitotic apparatus are essential, and the position of the mechanism is based upon a simple geometrical relation between the mitotic apparatus (or an aster) and the receptive surface. Successful interaction requires no stable morphological connection between the interacting regions, and the period when establishment can occur may be experimentally lengthened. Some induced abnormalities of the mitotic apparatus and the surface do not prevent division.

The immediate effect of the mitotic apparatus upon the surface is to enhance its contractility. General surface contractility can be enhanced by an incompletely developed mitotic apparatus, but a normal, permanent furrow requires a factor associated with the nucleus and with well-developed aster rays. The zone of enhanced cortical contractility is initially wider than the contractile ring that subsequently forms. The dimensions of the contractile ring are similar in a wide variety of cells. They do not appear to be scaled to the size of the mitotic apparatus or (excepting the circumference) to the size of the cell. It may be that the visible properties of the ring are not directly determined by the mitotic apparatus, but rather by the nature, arrangement, and interplay of the responding cortical elements.

At the organelle level, cytokinesis has a pleasing simplicity, but clearer understanding of the nature and importance of events that take place at lower levels of organization will involve a new set of challenges. Although physical experimentation has identified basic events, it has limited applicability. The techniques are most useful for the determination of necessity. If removal of a cell part does not prevent subsequent division, the part is not essential for the division process. If division fails, determination of the time when the part is necessary can reveal whether the part is important for the function of the division mechanism or for its creation. By a process of elimination, it has been possible to identify some of the essential elements, but this approach does not reveal a relation in which a part assists the whole but is not essential. It follows that this type of experimentation can identify minimal mechanisms and that the demonstration of contributing, but not essential, roles depends upon the use of more sensitive experimental end points.

The events of cytokinesis and the preliminary steps that set it in motion involve all the levels of cellular organization. Concerning the many visible and measurable structures and events that normally precede and accompany division, there are still questions about the identity of those that are essential. At a subtler level, we could soon encounter some of the same problems that 19th-century investigators faced when the amount of accumulated descriptive data permitted so many different interpretations, and it will be necessary to concentrate attention on division-related phenomena that are essential to the process. Some of the investigations described in previous chapters involved imposition of conditions in ways that allowed a judgement whether they affected a normal state, structure, or activity required for division. Because the data were obtained by experimentation, they involved studies on cleavage divisions. The results are relatively consistent, and they show that cytokinesis can proceed despite measures that disrupt or extensively modify many, if not most, of the normal characteristics of dividing cells (Table 10.1).

The table shows that a number of the common features of cytokinesis in animal cells are not required for successful permanent division. The basic necessities are a restricted source of contraction promoter and a responsive boundary membrane with a closely associated complex of potentially contractile material. The experimental results favor the idea that formation of a linear furrow requires that the more-contractile region be bounded by less-contractile regions. The cell has a remarkable ability to divide despite the disruption of cell regions that appear to play important roles. The boundary membrane that forms in a few seconds on the surface of exposed endoplasm, for instance, can form contractile regions in response to the mitotic apparatus in less than an hour. Continuous mechanical agitation of the periphery of the mitotic apparatus and the subcortical region of the cytoplasm does not prevent furrow establishment. These and similar findings suggest that the levels

Table 10.1 Effects of experimental perturbation on dividing cells

NORMAL STATE, STRUCTURE, OR ACTIVITY	EXPERIMENTALLY INDUCED CONDITION	EFFECT ON CYTOKINESIS
A. *Precleavage*		
Orientation of cortical microfilaments & other linear elements isotropic	Realignment perpendicular to future division plane by attached weight (Rappaport, 1960)	Normal furrow
Isotropic tension at the surface	Imposed anisotropic tensile stress perpendicular to future cleavage plane (Rappaport, 1960, 1978)	Normal furrow
Normal cortex & surface	a. Cortex & surface stretched up to 100-fold (Rappaport, 1976)	a. Normal furrow
	b. Cortex & original surface removed, new boundary membrane formed by interaction between endoplasm & medium (Rappaport, 1983)	b. Furrow delayed until next mitotic cycle
Mitotic apparatus present	Mitotic apparatus removed or dissolved (Hiramoto, 1956,1958; Y. Hamaguchi 1975; Rappaport, 1981)	
	a. More than 4 min before division	a. No furrow
	b. Less than 4 min before division	b. Normal furrow

Spindle & chromosomes present	No spindle & no chromosomes between asters (Rappaport, 1961; Hiramoto, 1976)	Normal furrow
Aster centers closer to polar surfaces than to equatorial surface	Asters closer to equatorial surface than to polar surfaces; artificial constriction in equatorial plane (Rappaport & Rappaport, 1984)	Normal furrow
Aster pair present	Single aster present a. In spherical cell (Hiramoto, 1971a)	a. No furrow
	b. In cylindrical cell near midpoint (Rappaport & Rappaport, 1985)	b. Delayed furrow
	c. In cylindrical cell near pole (Rappaport & Rappaport, 1985)	c. No furrow; shrinkage of polar surface
Asters formed in the presence of nuclear material	Nucleus separated from asters before nuclear membrane breakdown (Rappaport, 1991)	Furrow forms, then usually regresses
Relation between mitotic apparatus & equatorial cortex stable, no motion	Zone between mitotic apparatus & cortex chronically disrupted (Rappaport, 1978)	Delayed furrow

continued next page

Table 10.1 (continued)

NORMAL STATE, STRUCTURE OR, ACTIVITY	EXPERIMENTALLY INDUCED CONDITION	EFFECT ON CYTOKINESIS
No barriers between mitotic apparatus & polar or equatorial surface	Barriers introduced (Rappaport, 1968; Rappaport & Rappaport, 1983) a. Between aster & polar surfaces b. Between aster & equatorial surfaces	a. Normal furrow b. Interference with furrow formation
B. *Cleavage*		
Furrow appears when cortical stiffness maximal (Yoneda & Schroeder, 1984)	Time of furrow appearance protracted by repeated shifting of mitotic apparatus (Rappaport & Rappaport, 1993)	Furrow appeared when cortical stiffness near minimal
Furrow appears near midpoint between 2 asters in spherical cell	a. 2 asters in cylindrical cell (Rappaport, 1981) b. 1 aster in cylindrical cell (Rappaport & Rappaport, 1985a)	a. Furrow appears near midpoint between aster centers b. Furrows appear in all cells, about half in plane of aster centers

	c. Cell reshaped into a cone, mitotic apparatus and cone axis parallel (Rappaport & Rappaport, 1994)	c. Furrow nearer aster closest to vertex
Asters, spindle & chromosomes present during division	Dissolution, disruption, displacement, removal (Beams & Evans, 1940; Hiramoto, 1956, 1958, 1965; Y. Hamaguchi, 1975; Rappaport, 1981)	Cell divides
Subequatorial cytoplasm stable during division	Subequatorial cytoplasm disrupted & displaced (Rappaport, 1966, 1978; Hiramoto, 1965)	Cell divides
The majority of the new surface appears near the poles	a. Polar surfaces immobilized (Rappaport & Ratner, 1967)	a. Cell divides
	b. Furrows isolated (Rappaport, 1969a)	b. Fragment completes division

of order and complexity that characterize normal dividing cells are greater than are necessary to carry out the basic events of cytokinesis. It also follows that the study of such experimentally modified but division-capable cells may provide more decisive information concerning the basic mechanisms of cytokinesis than the study of normal cells.

Establishment of the division mechanism appears to be a consequence of interaction between two major discrete regions of the cytoskeleton. The principal role of the mitotic apparatus in cytokinesis is thought to be the transport of contraction promoter from the region of the mitotic axis. All of the achromatic parts (the spindle and both asters) of the mitotic apparatus can establish furrows, and the dependence of cleavage cells on the integrity of the mitotic apparatus is based on geometrical rather than physiological requirements. Agents that disrupt the normal radiate structure of the mitotic apparatus also interfere with division mechanism establishment, and the idea of an association between the linear structure and the movement of cytoplasmic elements remains as attractive as it has been for the past century.

The microtubules of the mitotic apparatus do not appear to penetrate the microfilamentous region of the cortex, and experimental evidence indicates that neither the division mechanism nor its function is dependent upon the integrity of any connection between the cortical and subcortical cytoskeletons. The lack of any evidence that the dimensions of the contractile ring are scaled to the dimensions of the cell or to the dimensions of the mitotic apparatus suggests that the dimensions of the division mechanism may be determined by events that occur in the cortex after it has been affected by the mitotic apparatus.

A kind of resiliency revealed by the cell's ability to divide successfully in other than normal circumstances is also characteristic of the cell's response to manipulations that alter the timing of division-related events. The time when the mitotic apparatus can interact with the cortex to produce a furrow can be retarded or advanced, and the results demonstrate that the components of the division system are competent to act and interact over almost 60% of the cycle. During the period of competence, the mitotic apparatus assumes its normal sequence of forms, the nuclear envelope is present about half the time, and the relative stiffness varies from maximal to near minimal. Nevertheless, interaction between the mitotic apparatus and the cortex at any time during the period results in normal furrowing. There is no evidence of a well-defined sequence of regulatory events after the cell becomes competent to divide. Rather, it appears that the physiological capacity of different cell regions to act and react is acquired in a relatively brief period, and it persists over a substantial proportion of the cycle. The cell's capacity to ignore, or to compensate for, variations in the arrangement of its organelles and the timing of their division-related activities would be expected to ensure successful division activity despite minor abnormalities.

With the invention of very sensitive and specific immunofluorescent methods that permit identification of molecular species in living as well as fixed cells, we enter a new era of observational biology, which presents some of the interpretational challenges that faced the older era. Attempts to deduce function solely from appearances have not always been successful. Again, discrimination between active and passive may be problematical. The relation between function and regional differences in concentration is not always clear. The controversies that waxed and waned around the distribution of actin in the cleavage furrow are illustrative. The question whether actin enrichment actually occurs in the furrow was debated for some time (reviewed, Rappaport, 1986b). There is some question whether local augmentation of actin is essential for local cortical contraction and, conversely, it is questionable whether localized accumulation of actin is a reliable indicator of localized cortical contraction. Is augmentation of any entity in the cleavage furrow a reliable indication of its active participation or necessity in cytokinesis? Concanavalin A receptors accumulate in the cleavage furrows of sea urchin eggs (McCaig and Robinson, 1982) and mouse macrophages (Koppel, Oliver, and Berlin, 1982), but their role in cytokinesis is unclear. Do all circumferentially oriented actin arrays originate from the same set of causes? They occur at the base of large blebs that form on *Ilyanassa* eggs following immersion in $CaCl_2$ solutions (G. W. Conrad and Williams, 1974b), and beneath peristaltic constrictions that move from the animal pole to the vegetal pole during the maturation period of barnacle eggs (C. A. Lewis, Chia, and Schroeder, 1973). Investigations of the responses of model and actual cortical actin meshworks to physical and geometrical manipulation might aid in discriminating between changes caused by dynamic biological activity and changes caused by more passive response to changes in the physical milieu.

Studies of in vitro effects and activity are essential to cell biology, and they have revealed proteins that could function in vivo in ways that might conceivably be important to cytokinesis. In vivo importance has been assessed by blocking interactions specifically and by studying mutant cells that are deficient in production of the molecule whose role is to be evaluated. The effectiveness of the general method depends upon the absence of alternative pathways and overlapping mechanisms. The cell's capacity to divide following elimination of a protein with strong in vitro activity does not necessarily mean that its normal role in cytokinesis is insignificant (de Hostos et al., 1993). This fact encourages skepticism of hypothetical control systems in which a complex activity is controlled by the state or concentration of a single protein species. The interpretational complications illustrate the need for new methods and for an experimental design better suited to an analysis of events that can be driven by multiple causes and control mechanisms. The complications also cause one to wonder whether it is reasonable to expect intellectually satisfying answers for every question that can be framed by the ingenious mind.

Although it is true that many of the old questions have been answered at a certain level of comprehension, the resolution of the underlying causes of cytokinesis requires further evaluation of multiple hypotheses and of potentially overlapping mechanisms. One can only hope that the rate of real progress will be more rapid than it was during the past century.

References

Aimar, C. (1988). Control of cell cycle length in amphibian egg: evidence for a temporal relationship between the nucleus and egg cytoplasm. *Development*, **104**, 415–422.

Akkas, N. & Engin, A. E. (1981). Ultrastructure of animal cells and its role during cytokinesis from a structural mechanics viewpoint. In *Mechanics of Structured Media*, ed. A. P. S. Selvaduri, pp. 187–207. *Proceedings of International Symposium on the Mechanical Behaviour of Structured Media.* Amsterdam & New York: Elsevier.

Albrecht-Buehler, G. & Bushnell, A. (1982). Reversible compression of cytoplasm. *Experimental Cell Research*, **140**, 173–189.

Allen, R. D. (1953). Fertilization and artificial activation in the egg of the surf-clam, *Spisula solidissima. The Biological Bulletin*, **105**, 213–239.

Allen, R. D. (1954). The fertilization reaction in isolated cortical material from sea urchin eggs. *Experimental Cell Research,* **8**, 397–399.

Anderson, E. (1970). A cytological study of the centrifuged whole, half, and quarter eggs of the sea urchin *Arbacia punctulata. The Journal of Cell Biology*, **47**, 711–733.

Andreassen, P. R., Palmer, D. K., Wener, M. H. & Margolis, R. L. (1991). Telophase disc: a new mammalian mitotic organelle that bisects telophase cells with a possible function in cytokinesis. *Journal of Cell Science*, **99**, 523–534.

Andrews, E. A. (1898). Filose activity in metazoan eggs. *Zoological Bulletin* **2**, 1–13.

Andrews, G. F. (1897). *The Living Substance As Such: and As Organism.* Boston: Ginn & Co.

Arnold, J. M. (1968). Formation of the first cleavage furrow in a telolecithal egg (*Loligo pealii*). *The Biological Bulletin*, **135**, 408–409.

Arnold, J. M. (1969). Cleavage furrow formation in a telolecithal egg (*Loligo pealii*). I. Filaments in early furrow formation. *The Journal of Cell Biology*, **41**, 894–904.

Arnold, J. M. (1974). Cleavage furrow formation in a telolecithal egg (*Loligo pealii*). III. Cell surface changes during cytokinesis as observed by scanning microscopy. *Developmental Biology*, **40**, 225–232.

Arnold, J. M. (1976). Cytokinesis in animal cells: new answers to old questions. In *The Cell Surface in Animal Embryogenesis and Development*, ed. G. Poste & G. L. Nicolson, pp. 55–80. Amsterdam: Elsevier/North-Holland Biomedical Press.

Aronson, J. & Inoué, S. (1970). Reversal by light of the action of N-Methyl N-Desacetyl colchicine on mitosis. *The Journal of Cell Biology*, **45**, 470–477.

Asnes, C. F. & Schroeder, T. E. (1979). Cell cleavage: ultrastructural evidence against equatorial stimulation by aster microtubules. *Experimental Cell Research*, **122**, 327–338.

Baker, J., Theurkauf, W. E. & Schubiger, G. (1993). Dynamic changes in microtubule configuration correlate with nuclear migration in the preblastoderm *Drosophila* embryo. *The Journal of Cell Biology*, **122**, 113–121.

Beams, H. W. & Evans, T. C. (1940). Some effects of colchicine upon the first cleavage in *Arbacia punctulata*. *The Biological Bulletin*, **79**, 188–198.

Beams, H. W. & Kessel, R. G. (1976). Cytokinesis: a comparative study of cytoplasmic division in animal cells. *American Scientist*, **63**, 279–920.

Begg, D. A., Rodewald, R. & Rebhun, L. I. (1978). The visualization of actin filament polarity in thin sections. *The Journal of Cell Biology* **79**, 846–852.

Bell, L. G. E. (1962). Some mechanisms involved in cell division. *Nature*, **193**, 190–191.

Bergh, R. S. (1879). Studien über die erste Entwicklung des Eies von *Gonothyraea loveni* (Allm.). *Morphol. Jahrbuch* **1**, 1–42.

Bieliavsky, N. & Geuskens, M. (1990). Interblastomeric plasma membrane formation during cleavage of *Xenopus laevis* embryos. *Journal of Submicroscopic Cytology & Pathology* **22**, 445–457.

Bjerknes, M. (1986). Physical theory of the orientation of astral mitotic spindles. *Science*, **234**, 1413–1416.

Bluemink, J. G. (1970). The first cleavage of the amphibian egg. An electron microscope study of the onset of cytokinesis in the egg of *Ambystoma mexicanum*. *Journal of Ultrastructure Research*, **32**, 142–166.

Bluemink, J. G. (1972). Cortical wound healing in the amphibian egg: an electron microscopical study. *Journal of Ultrastructure Research*, **41**, 95–114.

Bluemink, J. G. & deLaat, S. (1977). Plasma membrane assembly as related to cell division. In *The Synthesis, Assembly and Turnover of Cell Surface Components*, ed. G. Poste and G. L. Nicolson. *Cell Surface Reviews*, **4**, 403–461. Amsterdam: Elsevier / North-Holland Biomedical Press.

Boell, E. J. (1955). Energy exchange and enzyme development during embryogenesis. In *Analysis of Development*, ed. B. H. Willier, P. A.

Weiss, and V. Hamburger, pp. 520–555. Philadelphia & London: W. B. Saunders Co.

Bonnevie, K. (1906). Untersuchungen über Keimzellen: Beobachtungen an den Keimzellen von *Enteroxenos östergreni*. *Jenaische Zeitschrift für Naturwissenschaft*, **41**, 315–421.

Boveri, T. (1897). Zur Physiologie der Kern- und Zellteilung. Sitzungs-Berichte der Physikalisch-Medicinisch Gesellschaft zu Würzburg.

Brachet, J. (1950). *Chemical Embryology*. New York: Interscience Publishing Co.

Brachet, J., Hanocq, F. & Van Gansen, P. (1970). A cytochemical and ultrastructural analysis of *in vitro* maturation in amphibian oocytes. *Developmental Biology*, **21**, 157–195.

Bray, D. & White, J. G. (1988). Cortical flow in animal cells. *Science*, **239**, 883–888.

Brown, D. E. S. (1934). The pressure coefficient of "viscosity" in the eggs of *Arbacia punctulata*. *Journal of Cellular & Comparative Physiology*, **5**, 335–346.

Buck, R. C. & Tisdale, J. M. (1962). An electron microscopic study of the development of the cleavage furrow in mammalian cells. *The Journal of Cell Biology*, **13**, 117–125.

Buckley, I. K. & Porter, K. R. (1967). Cytoplasmic fibrils in living cultured cells. *Protoplasma*, **64**, 349–380.

Burgess, D. R. (1977). Ultrastructure of meiosis and polar body formation in the egg of the mud snail, *Ilyanassa obsoleta*. In *Cell Shape and Surface Architecture*, ed. J. P. Revel, U. Henning, and C. F. Fox, pp. 569–579. New York: Alan R. Liss, Inc.

Burgess, D. R. & Schroeder, T. E. (1977). Polarized bundles of actin filaments within microvilli of fertilized sea urchin eggs. *The Journal of Cell Biology*, **74**, 1032–1037.

Burridge, K., Nuckolls, G., Otely, C., Pavalko, F., Simon, K. & Turner, C. (1990). Actin–membrane interaction in focal adhesions. *Cell Differentiation & Development* **32**, 337–342.

Burrows, M. T. (1927). The mechanism of cell division. *American Journal of Anatomy*, **39**, 83–134.

Bütschli, O. (1876). Studien über die ersten Entwicklungsvorgange der Eizelle, die Zelltheilung und die Conjugation der Infusorien. *Abhandlungen, Herausgegeben von der Senckenbergischen Naturforschenden Gesellschaft*, **10**, 213–464.

Byers, B. & Abramson, D. H. (1968). Cytokinesis in HeLa: post-telophase delay and microtubule-associated motility. *Protoplasma*, **66**, 413–435.

Byers, T. J. & Armstrong, P. B. (1986). Membrane protein redistribution during *Xenopus* first cleavage. *The Journal of Cell Biology*, **102**, 2176–2184.

Cao, L.-G. & Wang, Y.-L. (1990a). Mechanism of the formation of contractile ring in dividing cultured animal cells. I. Recruitment of preexisting actin filaments into the cleavage furrow. *The Journal of Cell Biology*, **110**, 1089–1095.

Cao, L.-G. & Wang, Y.-L. (1990b). Mechanism of the formation of contractile ring in dividing cultured animal cells. II. Cortical movement of microinjected actin filaments. *The Journal of Cell Biology*, **111**, 1905–1911.

Carlson, J. G. (1938). Mitotic behavior of induced chromosomal fragments lacking spindle attachments in the neuroblasts of the grasshopper. *Proceedings of the National Academy of Sciences*, **24**, 500–507.

Carlson, J. G. (1952). Microdissection studies of the dividing neuroblast of the grasshopper, *Chortophaga viridifasciata* (De Geer). *Chromosoma*, **5**, 199–220.

Carré, D. & Sardet, C. (1984). Fertilization and early development in *Beroë ovata. Developmental Biology*, **105**, 188–195.

Cassimeris, L. (1993). Regulation of microtubule dynamic instability. *Cell Motility and the Cytoskeleton*, **26**, 275–281.

Cassimeris, L., Pryer, N. K. & Salmon, E. D. (1988). Real-time observations of microtubule dynamic instability in living cells. *The Journal of Cell Biology*, **107**, 2223–2231.

Cather, J. N., Render, J. A. & Freeman, G. (1986). The relation of time to direction and equality of cleavage in *Ilyanassa* embryos. *International Journal of Invertebrate Reproduction and Development*, **9**, 179–194.

Chalkley, H. W. (1935). The mechanism of cytoplasmic fission in *Amoeba proteus. Protoplasma,* **24**, 607–621.

Chalkley, H. W. (1951). Control of fission in *Amoeba proteus* as related to the mechanism of cell division. In *The Mechanisms of Cell Division*, ed. M. J. Kopac. *Annals of the New York Academy of Sciences*, **51**, 1303–1310.

Chambers, R. (1915). Microdissection studies on the physical properties of protoplasm. *The Lancet-Clinic*, 1–8 (read before the Cincinnati Research Society, March 4, 1915).

Chambers, R. (1917a). Microdissection studies. I. The visible structure of cell protoplasm and death changes. *American Journal of Physiology*, **43**, 1–12.

Chambers, R. (1917b). Microdissection studies. II. The cell aster: a reversible gelation phenomenon. *The Journal of Experimental Zoology*, **23**, 483–505.

Chambers, R. (1919). Changes in protoplasmic consistency and their relation to cell division. *Journal of General Physiology,* **2**, 49–68.

Chambers, R. (1921). The formation of the aster in artificial parthenogenesis. *Journal of General Physiology*, **4**, 33–39.

Chambers, R. (1924). The physical structure of protoplasm as determined by micro-dissection and injection. In *General Cytology*, ed. E. V. Cowdry, pp. 237–309. Chicago: The University of Chicago Press.

Chambers, R. (1937). The physical state of the wall of the furrow in a dividing cell. *The Biological Bulletin,* **73**, 367–368.

Chambers, R. (1938). Structural and kinetic aspects of cell division. *Journal of Cellular & Comparative Physiology*, **12**, 149–165.

Chambers, R. & Chambers, E. L. (1961). *Explorations into the Nature of the Living Cell.* Cambridge: Harvard University Press.

Chambers, R. & Fell, H. B. (1931). Micro-operations on cells in tissue cultures. *Proceedings of the Royal Society B,* **109**, 380–403.

Chambers, R. & Kopac, M. J. (1937). The coalescence of sea urchin eggs with oil drops. *Carnegie Institution of Washington Year Book*, **36**, 88–89.

Cole, K. S. (1932). Surface forces of the *Arbacia* egg. *Journal of Cellular and Comparative Physiology*, **1**, 1–9.

Cole, K. S. & Michaelis, E. M. (1932). Surface forces of fertilized *Arbacia* eggs. *Journal of Cellular and Comparative Physiology*, **2**, 121–126.

Conklin, E. G. (1899). Protoplasmic movement as a factor of differentiation. In *Biological Lectures Delivered at the Marine Biological Laboratory, 1898*, pp. 69–92. Boston: Ginn & Co., The Athenaeum Press.

Conklin, E. G. (1902). Karyokinesis and cytokinesis in the maturation, fertilization and cleavage of *Crepidula* and other Gasteropoda. *Journal of the Academy of Natural Science of Philadelphia*, Series II, **12** (Part 1), 1–121.

Conklin, E. G. (1908). The habits and early development of *Linerges mercurius*. *Carnegie Institution of Washington*, publication **103** (Papers from the Tortugas Laboratory of the Carnegie Institution of Washington, Vol. II), 155–170.

Conklin, E. G. (1917). Effects of centrifugal force on the structure and development of the eggs of *Crepidula*. *Journal of Experimental Zoology*, **22**, 311–419.

Conrad, A. H., Clark, W. A. & Conrad, G. W. (1991). Subcellular compartmentalization of myosin isoforms in embryonic chick heart ventricle myocytes during cytokinesis. *Cell Motility & the Cytoskeleton*, **19**, 189–206.

Conrad, A. H., Stephens, A. P. & Conrad, G. W. (1994). The effect of hexylene glycol–altered microtubule distributions on cytokinesis and polar lobe formation in fertilized eggs of *Ilyanassa obsoleta*. *Journal of Experimental Zoology*, **269**, 188–209.

Conrad, A. H., Stephens, A. P., Paulsen, A. Q., Schwarting, S. S. & Conrad, G. W. (1994). Effects of silver ions (Ag^+) on contractile ring function and microtubule dynamics during first cleavage in *Ilyanassa obsoleta*. *Cell Motility & the Cytoskeleton*, **27**, 117–132.

Conrad, G. W. & Davis, S. E. (1977). Microiontophoretic injection of calcium ions or of cyclic AMP causes rapid shape changes in fertilized eggs of *Ilyanassa obsoleta*. *Developmental Biology*, **61**, 184–201.

Conrad, G. W. & Davis, S. E. (1980). Polar lobe formation and cytokinesis in fertilized eggs of *Ilyanassa obsoleta*. III. Large bleb formation caused by Sr^{2+}, ionophores X537A and compound 48/80. *Developmental Biology*, **74**, 152–172.

Conrad, G. W. & Rappaport, R. (1981). Mechanisms of cytokinesis in animal cells. In *Mitosis/Cytokinesis*, ed. A. M. Zimmerman and A. Forer, pp. 365–396. New York: Academic Press.

Conrad, G. W. & Williams, D. C. (1974a). Polar lobe formation and cytokinesis in fertilized eggs of *Ilyanassa obsoleta*. I. Ultrastructure and effects of cytochalasin B and colchicine. *Developmental Biology*, **36**, 363–378.

Conrad, G. W. & Williams, D. C. (1974 b). Polar lobe formation and cytokinesis in fertilized eggs of *Ilyanassa obsoleta*. II. Large bleb formation caused by high concentrations of exogenous calcium ions. *Developmental Biology*, **37**, 280–294.

Conrad, G. W., Williams, D. C., Turner, F. R., Newrock, K. M. & Raff, R. A. (1973). Microfilaments in the polar lobe constriction of fertilized eggs of *Ilyanassa obsoleta. The Journal of Cell Biology*, **59**, 228–233.

Cooke, C. A., Heck, M. M. S. & Earnshaw, W. C. (1987). The inner centromere protein (INCENP) antigens: movement from inner centromere to midbody during mitosis. *The Journal of Cell Biology*, **105**, 2053–2067.

Cornman, I. & Cornman, M. E. (1951). The action of podophyllin and its fractions on marine eggs. In *The Mechanisms of Cell Division*, ed. M. J. Kopac. *Annals of the New York Academy of Sciences*, **51**, 1443–1481.

Costello, D. P. (1932). The surface precipitation reaction in marine eggs. *Protoplasma*, **17**, 239–257.

Costello, D. P., Davidson, M. E., Eggers, A., Fox, M. H. & Henley, C. (1957). Annelida (Polychaeta) *Sabellaria vulgaris.* In *Methods for Obtaining and Handling Marine Eggs and Embryos*, pp. 93–97. Lancaster, PA: Lancaster Press.

Coue, M., Lombillo, V. A. & McIntosh, J. R. (1991). Microtubule depolymerization promotes particle and chromosome movement in vitro. *The Journal of Cell Biology*, **112**, 1165–1175.

Czihak, G. (1973). The role of astral rays in early cleavage of sea urchin eggs. *Experimental Cell Research*, **83**, 424–426.

Czihak, G., Kojima, M., Linhart, J. & Vogel, H. (1991). Multipolar mitosis in procaine-treated polyspermic sea urchin eggs and in eggs fertilized with UV-irradiated spermatozoa. *European Journal of Cell Biology*, **55**, 255–261.

Dan, J. C. (1948). On the mechanism of astral cleavage. *Physiological Zoology,* **21**, 191–218.

Dan, K. (1943a). Behavior of the cell surface during cleavage. V. Perforation experiment. *Journal of Faculty of Science, Tokyo Imperial University*, Series IV, **6**, 297–321.

Dan, K. (1943b). Behavior of the cell surface during cleavage. VI. On the mechanism of cell division. *Journal of Faculty of Science, Tokyo Imperial University,* Series IV, **6**, 323–368.

Dan, K. (1954). The cortical movement in *Arbacia punctulata* eggs through cleavage cycles. *Embryologia,* **2**, 115–122.

Dan, K. (1958). On 'the stationary surface ring' in heart-shaped cleavage. *Journal of Experimental Biology,* **35**, 400–406.

Dan, K. (1963). Force of cleavage of the dividing sea urchin egg. In *Cell Growth and Cell Division*, ed. R. J. C. Harris. *Symposia International Society for Cell Biology*, **2**, pp. 261–276. New York: Academic Press.

Dan, K. (1978). Unequal division: its cause and significance. In *Cell Reproduction: In Honor of Daniel Mazia*, ed. E. R. Dirksen, D. M. Prescott, and C. F. Fox, pp. 557–561. ICN-UCLA Symposium on Molecular and Cellular Biology, **12**. New York: Academic Press.

Dan, K. (1979). Studies on unequal cleavage in sea urchins. I. Migration of the nuclei to the vegetal pole. *Development, Growth & Differentiation*, **21**, 527–535.

Dan, K. (1987). Studies on unequal cleavage in sea urchins. III. Micromere formation under compression. *Development, Growth & Differentiation*, **29**, 503–515.

Dan, K. (1988). Mechanism of equal cleavage of sea urchin egg: transposition from astral mechanism to constricting mechanism. *Zoological Science*, **5**, 507–517.

Dan, K. & Dan, J. C. (1940). Behavior of the cell surface during cleavage. III. On the formation of new surface in the eggs of *Strongylocentrotus pulcherrimus*. *The Biological Bulletin*, **78**, 486–501.

Dan, K. & Dan, J. C. (1947). Behavior of the cell surface during cleavage. VII. On the division mechanism of cells with excentric nuclei. *The Biological Bulletin*, **93**, 139–162.

Dan, K., Dan, J. C. & Yanagita, T. (1938). Behaviour of the cell surface during cleavage. II. *Cytologia*, **8**, 521–531.

Dan, K., Endo, S. & Uemura, I. (1983). Studies on unequal cleavage in sea urchins. II. Surface differentiation and the direction of nuclear migration. *Development, Growth & Differentiation*, **25**, 227–237.

Dan, K. & Ikeda, M. (1971). On the system controlling the time of micromere formation in sea urchin embryos. *Development, Growth and Differentiation*, **13**, 285–301.

Dan, K. & Inoué, S. (1987). Studies of unequal cleavage in molluscs. II. Asymmetric nature of the two asters. *International Journal of Invertebrate Reproduction & Development*, **11**, 335–354.

Dan, K. & Ito, S. (1984). Studies of unequal cleavage in molluscs. I. Nuclear behavior and anchorage of a spindle pole to cortex as revealed by isolation technique. *Development, Growth & Differentiation*, **26**, 249–262.

Dan, K. & Kojima, M. K. (1963). A study on the mechanism of cleavage in the amphibian egg. *Journal of Experimental Biology*, **40**, 7–14.

Dan, K. & Ono, T. (1954). A method of computation of the surface area of the cell. *Embryologia*, **2**, 87–89.

Dan, K. & Tanaka, Y. (1990). Attachment of one spindle pole to the cortex in unequal cleavage. In *Cytokinesis*, ed. G. W. Conrad and T. E. Schroeder. *Annals of the New York Academy of Sciences*, **582**, 108–119.

Dan, K., Yanagita, T. & Sugiyama, M. (1937). Behavior of the cell surface during cleavage. I. *Protoplasma*, **28**, 66–81.

Danchakoff, V. (1916). Studies on cell division and cell differentiation. I. Development of the cell organs during the first cleavage of the sea urchin egg. *Journal of Morphology*, **27**, 559–603.

Danielli, J. F. (1952). Division of the flattened egg. *Nature*, **170**, 496.

Danielli, J. F. & Harvey, E. N. (1935). The tension at the surface of mackerel egg oil, with remarks on the nature of the cell surface. *Journal of Cellular & Comparative Physiology*, **5**, 483–494.

de Hostos, E. L., Rehfueß, C., Bradtke, B., Waddell, D. R., Albrecht, R., Murphy, J. & Gerisch, G. (1993). *Dictyostelium* mutants lacking the cytoskeletal protein Coronin are defective in cytokinesis and cell motility. *The Journal of Cell Biology*, **120**, 163–173.

DeLozanne, A. & Spudich, J. A. (1987). Disruption of the *Dictyostelium* myosin heavy chain gene by homologous recombination. *Science*, **236**, 1086–1091.

Denis-Donini, S., Baccetti, B. & Monroy, A. (1976). Morphological changes of the surface of the eggs of *Xenopus laevis* in the course of development. *Journal of Ultrastructure Research*, **57**, 104–112.

Detlaff, T. A. & Detlaff, A. A. (1961). On relative dimensionless characteristics of the development in embryology. *Archives de Biologie (Liège & Paris)*, **71**, 1–16.

Devore, J. J., Conrad, G. W. & Rappaport, R. (1989). A model for astral stimulation of cytokinesis in animal cells. *The Journal of Cell Biology*, **109**, 2225–2232.

D'Haese, J. & Komnick, H. (1972a). Fine structure and contraction of isolated muscle actomyosin. 1. Evidence for a sliding mechanism by means of oligomeric myosin. *Zeitschrift Zellforschung*, **134**, 411–426.

D'Haese, J. & Komnick, H. (1972b). Fine structure and contraction of isolated muscle actomyosin. 2. Formation of myosin filaments and their effect on contraction. *Zeitschrift Zellforschung*, **134**, 427–434.

Doane, W. W. (1967). *Drosophila.* In *Methods in Developmental Biology*, ed. F. H. Wilt and N. K. Wessells, pp. 219–244. New York: Thomas Y. Crowell Co.

Driesch, H. & Morgan, T. H. (1895). Zur Analysis der ersten Entwicklungsstadien des Ctenophoreneies. I. Von der Entwicklung einzelner Ctenophorenblastomeren. II. Von der Entwicklung ungefurchter Eier mit Protoplasmadefekten. *Archiv für Entwicklungsmechanik*, **2**, 204–226.

Drüner, L. (1894). Studien über den Mechanismus der Zellteilung. *Jenaische Zeitschrift für Naturwissenschaft*, **29**, 271–344.

Durham, A. C. H. (1974). A unified theory of the control of actin and myosin in nonmuscle movements. *Cell*, **2**, 123–135.

Earnshaw, W. C. & Cooke, C. A. (1991). Analysis of the distribution of the INCENPS throughout mitosis reveals the existence of a pathway of structural changes in the chromosomes during metaphase and early events in cleavage furrow formation. *Journal of Cell Science*, **98**, 443–461.

Edds, K. T. (1993). Effects of cytochalasin and colcemid on cortical flow in coelomocytes. *Cell Motility and the Cytoskeleton*, **26**, 262–273.

Endo, S., Toriyama, M. & Sakai, H. (1983). The mitotic apparatus with unusually many microtubules from sea urchin eggs treated by hexyleneglycol. *Development, Growth & Differentiation*, **25**, 307–314.

Erickson, C. A. & Trinkaus, J. P. (1976). Microvilli and blebs as sources of reserve surface membrane during cell spreading. *Experimental Cell Research*, **99**, 375–384.

Erlanger, R. von (1897). Beobachtungen über die Befruchtung und ersten zwei Teilungen an den lebenden Eiern kleiner Nematoden. *Biologisches Zentralblatt*, **17**, I – 152–159; II – 339–346.

Euteneuer, U. & McIntosh, J. R. (1980). Polarity of midbody and phragmoplast microtubules. *The Journal of Cell Biology*, **87**, 509–515.

Fischel, A. (1906). Zur Entwicklungsgeschichte der Echinodermen. I. Zur

Mechanik der Zellteilung. *Archiv für Entwicklungsmechanik*, **22**, 526–534.

Fishkind, D. J., Cao, L.-G. & Wang, Y.-L. (1991). Microinjection of the catalytic fragment of myosin light chain kinase into dividing cells: effects on mitosis and cytokinesis. *The Journal of Cell Biology*, **114**, 967–975.

Flemming, W. (1891). Neue Beiträge zur Kenntniss der Zelle. *Archiv für Mikroscopische Anatomie*, **37**, 685–751.

Flemming, W. (1895). Zur Mechanik der Zelltheilung. *Archiv für Mikroscopische Anatomie und Entwicklungsmechanik*, **46**, 696–701.

Fluck, R. A., Miller, A. L. & Jaffe, L. F. (1991). Slow calcium waves accompany cytokinesis in Medaka fish eggs. *The Journal of Cell Biology*, **115**, 1259–1265.

Foe, V. E. & Alberts, B. M. (1983). Studies of nuclear and cytoplasmic behaviour during the five mitotic cycles that precede gastrulation in *Drosophila* embryogenesis. *Journal of Cell Science*, **61**, 31–70.

Fol, H. (1873). Die erste Entwicklung des Geryonideies. *Jenaische Zeitschrift für Medicin und Naturwissenschaft*, **7**.

Follett, E. A. C. & Goldman, R. D. (1970). The occurrence of microvilli during spreading and growth of BHK 21/C13 fibroblasts. *Experimental Cell Research*, **59**, 124–136.

Forer, A. & Behnke, O. (1972). An actin-like component in spermatocytes of a crane fly (*Nephrotoma suturalis* Loew). II. The cell cortex. *Chromosoma*, **39**, 175–190.

Fry, H. J. (1925). Asters in artificial parthenogenesis. II. Asters in nucleated and enucleated eggs of *Echinarachnius parma* and the role of the chromatin. *The Journal of Experimental Zoology*, **43**, 49–81.

Fry, H. J. (1929). The so-called central bodies in fertilized *Echinarachnius* eggs. I. The relationship between central bodies and astral structure as modified by various mitotic phases. *The Biological Bulletin*, **56**, 101–128.

Fujiwara, K. & Pollard, T. D. (1976). Fluorescent antibody localization of myosin in the cytoplasm, cleavage furrow, and mitotic spindle of human cells. *The Journal of Cell Biology*, **71**, 848–875.

Fujiwara, K., Porter, M. E. & Pollard, T. D. (1978). Alpha-actinin localization in the cleavage furrow during cytokinesis. *The Journal of Cell Biology*, **79**, 268–275.

Fukui, Y. (1990). Actomyosin organization in mitotic *Dictyostelium* amoebae. In *Cytokinesis*, ed. G. W. Conrad and T. E. Schroeder. *Annals of the New York Academy of Sciences*, **582**, 156–165.

Fukui, Y. (1993). Toward a new concept of cell motility: cytoskeletal dynamics in amoeboid movement and cell division. *International Review of Cytology*, **144**, 85–127.

Fukui, Y., DeLozanne, A. & Spudich, J. (1990). Structure and function of the cytoskeleton of a *Dictyostelium* myosin-defective mutant. *The Journal of Cell Biology*, **110**, 367–378.

Fukui, Y. & Inoué, S. (1991). Cell division in *Dictyostelium* with special emphasis on actomyosin organization in cytokinesis. *Cell Motility and the Cytoskeleton*, **18**, 41–54.

Fukui, Y., Lynch, T. J., Brzeska, H. & Korn, E. D. (1989). Myosin I is located at the leading edges of locomoting *Dictyostelium* amoebae. *Nature*, **341**, 328–331.

Fullilove, S. L. & Jacobson, A. G. (1971). Nuclear elongation and cytokinesis in *Drosophila montana*. *Developmental Biology*, **26**, 560–577.

Gallardo, A. (1902). Interpretación Dinámica de la División Celular. *Teses Universidad Nacional de Buenos Aires*, Casa Editora de Coni Hermanos.

Gard, D. L. (1992). Microtubule organization during maturation of *Xenopus* oocytes: assembly and rotation of the meiotic spindles. *Developmental Biology*, **151**, 516–530.

Gardiner, E. G. (1895). Early development of *Polychoerus caudatus*, Mark. *Journal of Morphology*, **11**, 155–176.

Ghosh, S. & Paweletz, N. (1984). Synchronous DNA synthesis and mitosis in multinucleate cells with one chromosome in each nucleus. *Chromosoma*, **89**, 197–200.

Gipson, I. (1974). Electron microscopy of early cleavage furrows in the chick blastodisc. *Journal Ultrastructure Research*, **49**, 331–347.

Gittes, F., Mickey, B., Nettleton, J. & Howard, J. (1993). Flexural rigidity of microtubules and actin filaments measured from thermal fluctuations in shape. *The Journal of Cell Biology*, **120**, 923–934.

Goodenough, D. A., Ito, S. & Revel, J.-P. (1968). Electron microscopy of early cleavage stages in *Arbacia punctulata*. *The Biological Bulletin*, **135**, 420–421.

Gray, J. (1931). *A Textbook of Experimental Cytology*. London: Cambridge University Press.

Greenspan, H. P. (1977). On the dynamics of cell cleavage. *Journal of Theoretical Biology*, **65**, 79–99.

Greenspan, H. P. (1978). On fluid-mechanical simulations of cell division and movement. *Journal of Theoretical Biology*, **70**, 125–134.

Gurwitsch, A. (1904). Streitfragen und Theorien der Zellteilung. In *Morphologie und Biologie der Zelle*, pp. 268–334. Jena: Verlag von Gustav Fischer.

Hamaguchi, M. S., Hamaguchi, Y. & Hiramoto, Y. (1986). Microinjected polystyrene beads move along astral rays in sand dollar eggs. *Development, Growth & Differentiation*, **28**, 461–470.

Hamaguchi, M. S. & Hiramoto, Y. (1978). Protoplasmic movement during polar body formation in starfish oocytes. *Experimental Cell Research*, **112**, 55–62.

Hamaguchi, Y. (1975). Microinjection of colchicine into sea urchin eggs. *Development, Growth & Differentiation*, **17**, 111–117.

Hamaguchi, Y. & Mabuchi, I. (1982). Effects of phalloidin microinjection and localization of fluorescein labeled phalloidin in living sand dollar eggs. *Cell Motility*, **2**, 103–113.

Hargitt, C. W. (1900). A contribution to the natural history and development of *Pennaria tiarella* McCr. *American Naturalist*, **34**, 387–415.

Hargitt, C. W. (1904). The early development of *Pennaria tiarella* McCr. *Archiv für Entwicklungsmechanik der Organismen*, **18**, 453–488.

Hargitt, G. T. (1909). Maturation, fertilization, and segmentation of *Pennaria tiarella* (Ayres) and of *Tubularia crocea* (Ag.). *Bulletin of the Museum of Comparative Zoology at Harvard College*, **53**, 161–212.

Harris, A. K. (1990). Testing cleavage mechanisms by comparing computer simulations to actual experimental results. In *Cytokinesis*, ed. G. W. Conrad and T. E. Schroeder. *Annals of the New York Academy of Sciences*, **582**, 60–77.

Harris, A. K. & Gewalt, S. L. (1989). Simulation testing of mechanisms for inducing the formation of the contractile ring in cytokinesis. *The Journal of Cell Biology*, **109**, 2215–2223.

Harris, P. (1962). Some structural and functional aspects of the mitotic apparatus in sea urchin embryos. *The Journal of Cell Biology*, **14**, 475–487.

Harris, P. (1968). Cortical fibers in fertilized eggs of the sea urchin *Strongylocentrotus purpuratus*. *Experimental Cell Research*, **52**, 677–681.

Harris, P. (1975). The role of membranes in the organization of the mitotic apparatus. *Experimental Cell Research*, **94**, 409–425.

Harris, P., Osborn, M. & Weber, K. (1980). Distribution of tubulin-containing structures in the egg of the sea urchin *Strongylocentrotus purpuratus* from fertilization through first cleavage. *The Journal of Cell Biology*, **84**, 668–679.

Harrison, R. G. (1907). Observations on the living developing nerve fiber. *Anatomical Record*, **1**, 116–118.

Harrison, R. G. (1908). Regeneration of nerves. *Anatomical Record*, **1**, 209.

Harrison, R. G. (1969). *Organization and Development of the Embryo*. New Haven: Yale University Press.

Harvey, E. B. (1935). The mitotic figure and cleavage plane in the egg of *Parechinus microtuberculatus* as influenced by centrifugal force. *The Biological Bulletin*, **69**, 287–297.

Harvey, E. B. (1936). Parthenogenetic merogony or cleavage without nuclei in *Arbacia punctulata*. *The Biological Bulletin*, **71**, 101–121.

Harvey, E. B. (1956). *The American Arbacia and Other Sea Urchins*. Princeton: Princeton University Press.

Harvey, E. N. (1954). Tension at the cell surface. In *Protoplasmatologia*, ed. F. Weber and L. V. Heilbrunn, pp. 1–30. Vienna: Springer Verlag.

Harvey, E. N. & Shapiro, H. (1934). The interfacial tension between oil and protoplasm within the living cells. *Journal of Cellular & Comparative Physiology*, **5**, 255–267.

Hayden, J. H. (1988). Microtubule-associated organelle and vesicle transport in fibroblasts. *Cell Motility & the Cytoskeleton*, **10**, 255–262.

Heidenhain, M. (1897). Neue Erläuterungen zum Spannungsgesetz der centrirten Systeme. *Morphologische Arbeiten*, **7**, 281–365.

Heilbrunn, L. V. (1928). *The Colloid Chemistry of Protoplasm*. Berlin: Borntrager.

Henley, C. & Costello, D. P. (1965). The cytological effects of podophyllin and podophyllotoxin on the fertilized eggs of *Chaetopterus*. *The Biological Bulletin*, **128**, 369–391.

Henson, J. H., Begg, D. A., Beaulieu, S. M., Fishkind, D. F., Bonder, E. M., Terasaki, M., Lebeche, D. & Kaminer, B. (1989). A calsequestrin-like protein in the endoplasmic reticulum of the sea urchin: localization and dynamics in the egg and first cell cycle embryo. *The Journal of Cell Biology*, **109**, 149–161.

Hepler, P. K. (1989). Calcium transients during mitosis: observations in flux. *The Journal of Cell Biology*, **109**, 2567–2573.

Hepler, P. K. (1992). Calcium and mitosis. *International Review of Cytology*, **138**, 239–268.

Hepler, P. K. & Wolniak, S. M. (1984). Membranes in the mitotic apparatus: their structure and function. In *International Review of Cytology*, **90**, ed. G. H. Bourne and J. F. Danielli, pp. 169–238. New York and London: Academic Press.

Higashi, A. (1972). The thickness of the cortex and some analytical experiments of thixotropy in sea urchin eggs. *Annotationes Zoologicae Japonensis*, **45**, 119–144.

Hime, G. & Saint, R. (1992). Zygotic expression of the *pebble* locus is required for cytokinesis during the postblastoderm mitoses of *Drosophila. Development*, **114**, 165–171.

Hiramoto, Y. (1956). Cell division without mitotic apparatus in sea urchin eggs. *Experimental Cell Research*, **11**, 630–636.

Hiramoto, Y. (1957). The thickness of the cortex and the refractive index of the protoplasm in sea urchin eggs. *Embryologia*, **3**, 361–374.

Hiramoto, Y. (1958). A quantitative description of protoplasmic movement during cleavage in the sea urchin egg. *Journal of Experimental Biology*, **35**, 407–424.

Hiramoto, Y. (1963). Mechanical properties of sea urchin eggs. I. Surface force and elastic modulus of the cell membrane. *Experimental Cell Research*, **32**, 59–75.

Hiramoto, Y. (1965). Further studies on cell division without mitotic apparatus in sea urchin eggs. *The Journal of Cell Biology*, **25**, 161–167.

Hiramoto, Y. (1968). The mechanics and mechanism of cleavage in the sea urchin egg. In *Aspects of Cell Motility, 22nd Symposium for the Society of Experimental Biology*, pp. 311–327. Cambridge: Cambridge University Press.

Hiramoto, Y. (1970). Rheological properties of sea urchin eggs. *Biorheology*, **6**, 201–234.

Hiramoto, Y. (1971a). Analysis of cleavage stimulus by means of micromanipulation of sea urchin eggs. *Experimental Cell Research*, **68**, 291–298.

Hiramoto, Y. (1971b). A photographic analysis of protoplasmic movement during cleavage in the sea urchin egg. *Development, Growth & Differentiation*, **13**, 191–200.

Hiramoto, Y. (1975). Force exerted by the cleavage furrow of sea urchin eggs. *Development, Growth & Differentiation*, **17**, 27–38.

Hiramoto, Y. (1976). Mechanical properties of starfish oocytes. *Development, Growth & Differentiation*, **18**, 205–209.

Hiramoto, Y. (1979). Mechanical properties of the dividing sea urchin egg. In *Cell Motility: Molecules and Organization*, ed. S. Hatano, H. Ishikawa, and H. Sato, pp. 653–663. Tokyo: University of Tokyo Press.

Hiramoto, Y. (1990). Mechanical properties of the cortex before and during cleavage. In *Cytokinesis*, ed. G. W. Conrad and T. E. Schroeder. *Annals of the New York Academy of Sciences*, **582**, 22–30.

Hird, S. N. & White, J. G. (1993). Cortical and cytoplasmic flow polarity in early embryonic cells of *Caenorhabditis elegans. The Journal of Cell Biology*, **121**, 1343–1355.

Holy, J. & Schatten, G. (1991). Differential behavior of centrosomes in unequally dividing blastomeres during fourth cleavage of sea urchin embryos. *Journal of Cell Science*, **98**, 423–431.

Horstadius, S. (1973). *Experimental Embryology of Echinoderms*. London: Oxford University Press.

Hosoya, N., Hosoya, H., Mohri, T. & Mohri, H. (1990). A 70 kD microtubule-binding protein from starfish eggs: purification, characterization, and localization during meiosis and mitosis. *Cell Motility & the Cytoskeleton*, **15**, 168–180.

Humphreys, W. J. (1964). Electron microscope studies of the fertilized egg and the two-cell stage of *Mytilus edulis*. *Journal of Ultrastructure Research*, **10**, 244–262.

Huxley, H. E. (1969). The mechanism of muscular contraction. *Science*, **164**, 1356–1366.

Huxley, H. E. (1972). Molecular basis of contraction in cross-striated muscles. In *The Structure and Function of Muscle*, ed. G. H. Bourne, pp. 301–387. New York and London: Academic Press.

Inoué, S. (1981). Cell division and the mitotic spindle. *The Journal of Cell Biology*, **91**, 131s–147s.

Inoué, S. (1990). Dynamics of mitosis and cleavage. In *Cytokinesis*, ed. G. W. Conrad and T. E. Schroeder. *Annals of the New York Academy of Sciences*, **582**, 1–14.

Isaeva, V. V. & Presnov, E. V. (1983). Stabilization following fertilization and maintenance of experimentally altered egg shape during cleavage in the sea urchin. *Cytology*, **25**, 200–203.

Ishizaka, S. (1958). Surface characters of dividing cells. I. Stationary surface rings. *Journal of Experimental Biology*, **35**, 396–399.

Janson, L. W. & Taylor, D. L. (1993). *In Vitro* models of tail contraction and cytoplasmic streaming in amoeboid cells. *The Journal of Cell Biology*, **123**, 345–356.

John, B. (1990). *Meiosis*. Cambridge: Cambridge University Press.

Jordan, M. A., Thrower, D. & Wilson, L. (1992). Effects of vinblastine, podophyllotoxin and nocodazole on mitotic spindles. *Journal of Cell Science*, **102**, 401–416.

Jura, C. (1975a). Analysis of cytokinesis by means of UV microbeam during first cleavage division in *Succinea putris* L. (Mollusca) egg. *Acta Biologica Cracoviensia,* Series Zoologica, **18**, 67–77.

Jura, C. (1975b). Cleavage without cytokinesis of the eggs of

Tetrodontophora bielanensis (Waga) (*Collembola)* after periplasm damage with thermocauter. *Acta Biologica Cracoviensia*, Series Zoologia, **18**, 97–102.

Just, E. E. (1922). Studies on cell division. I. The effect of dilute sea-water on the fertilized egg of *Echinarachnius parma* during the cleavage cycle. *American Journal of Physiology*, **61**, 505–515.

Just, E. E. (1939). *The Biology of the Cell Surface.* Philadelphia: P. Blakiston's Son & Co.

Kallenbach, E. (1963). A staining reaction associated with the cell membrane of dividing cells in the cleavage furrow. *Experimental Cell Research*, **33**, 581–583.

Kalt, M. R. (1971a). The relationship between cleavage and blastocoel formation in *Xenopus laevis*. I. Light microsopic observations. *Journal of Embryology & Experimental Morphology*, **26**, 37–49.

Kalt, M. R. (1971b). The relationship between cleavage and blastocoel formation in *Xenopus laevis*. II. Electron microscopic observations. *Journal of Embryology & Experimental Morphology*, **26**, 51–66.

Kane, R. (1962). The mitotic apparatus: isolation by controlled pH. *The Journal of Cell Biology*, **12**, 47–55.

Karr, T. L. & Alberts, B. M. (1986). Organization of the cytoskeleton in early *Drosophila* embryos. *The Journal of Cell Biology*, **102**, 1494–1509.

Kawamura, K. (1957). Studies on mitotic apparatus and cytokinesis of spermatocytes of nine species of grasshopper. *Cytologia*, **22**, 337–346.

Kawamura, K. (1960). Studies on cytokinesis in neuroblasts of the grasshopper, *Chortophaga viridifasciata* (De Geer). I. Formation and behavior of the mitotic apparatus. II. The role of the mitotic apparatus in cytokinesis. *Experimental Cell Research*, **21**, 1–18.

Kawamura, K. (1965). Studies on cytokinesis in neuroblasts of the grasshopper, *Chortophaga viridifasciata* (De Geer). IV. Further experiments on factors determining neuroblast polarity. *Journal of the College of Dairy Agriculture*, **2**, 120–124.

Kawamura, K. (1977). Microdissection studies on the dividing neuroblast of the grasshopper, with special reference to the mechanism of unequal cytokinesis. *Experimental Cell Research*, **106**, 127–137.

Kawamura, K. (1982). Determinant factors of the spindle axis in grasshopper neuroblast. *Development, Growth & Differentiation*, **24**, 408.

Kawamura, K. (1988). The contraction wave in the cortex of dividing neuroblasts of the grasshopper. *Zoological Science*, **5**, 677–684.

Kawamura, K. & Carlson, J. G. (1962). Studies on cytokinesis in neuroblasts of the grasshopper, *Chortophaga viridifasciata* (De Geer). III. Factors determining the location of the cleavage furrow. *Experimental Cell Research*, **26**, 411–423.

Kawamura, K. & Yamashiki, N. (1990). Factors inducing unequal cytokinesis in grasshopper neuroblasts. In *Cytokinesis*, ed. G. W. Conrad and T. E. Schroeder. *Annals of the New York Academy of Sciences*, **582**, 304–306.

Kawamura, K. & Yamashiki, N. (1992). The determination of spindle polarity in early mitotic stages of the dividing grasshopper neuroblasts. *Development, Growth & Differentiation*, **34**, 427–435.

Kelly, R. B. (1990). Microtubules, membrane traffic, and cell organization. *Cell*, **61**, 5–7.

Kiehart, D. P., Ketchum, A., Young, P., Lutz, D., Alfenito, M. R., Chang, X.-j., Awobuluyi, M., Pesacreta, T. C., Inoué, S., Stewart, C. T. & Chen, T.-L. (1990). Contractile proteins in *Drosophila* development. In *Cytokinesis*, ed. G. W. Conrad and T. E. Schroeder. *Annals of the New York Academy of Sciences*, **582**, 233–251.

Kirkham, W. B. (1907). Maturation of the egg of the white mouse. *Transactions of the Connecticut Academy of Arts & Sciences*, **13**, 65–87.

Kitanishi-Yumura, T. & Fukui, Y. (1989). Actomyosin organization during cytokinesis: reversible translocation and differential redistribution in *Dictyostelium*. *Cell Motility and the Cytoskeleton*, **12**, 78–89.

Kite, G. L. (1913). Studies on the physical properties of protoplasm. I. The physical properties of the protoplasm of certain animal and plant cells. *American Journal of Physiology*, **32**, 146–164.

Knecht, D. A. & Loomis, W. F. (1987). Antisense RNA inactivation of myosin heavy chain gene expression in *Dictyostelium discoideum*. *Science*, **236**, 1081–1086.

Knutton, S., Jackson, D., Graham, J. M., Micklem, K. J. & Pasternak, C. A. (1976). Microvilli and cell swelling. *Nature*, **262**, 52–53.

Knutton, S., Sumner, M. C. B. & Pasternak, C. A. (1975). Role of microvilli in surface changes of synchronized P815Y mastocytoma cells. *The Journal of Cell Biology*, **66**, 568–576.

Kobayakawa, Y. (1988). Role of mitotic asters in accumulation of pigment granules around nuclei in early amphibian embryos. *The Journal of Experimental Zoology*, **248**, 232–237.

Kobayashi, N. (1962). Cleavage of the sea urchin egg recovered from the cleavage-blocking effect of demecolcine. *Embryologia*, **7**, 68–80.

Kojima, M. K. (1959). Relation between the vitally stained granules and cleavage activity in the sea urchin egg. *Embryologia*, **4**, 191–209.

Kolega, J. (1986). Effects of mechanical tension on protrusive activity and microfilament and intermediate filament organization in an epidermal epithelium moving in culture. *The Journal of Cell Biology*, **102**, 1400–1411.

Koppel, D. E., Oliver, J. M. & Berlin, R. D. (1982). Surface functions during mitosis. III. Quantitative analysis of ligand-receptor movement into the cleavage furrow: diffusion vs. flow. *The Journal of Cell Biology*, **93**, 950–960.

Krishan, A. (1971). Fine structure of cytochalasin-induced multinucleated cells. *Journal of Ultrastructure Research*, **36**, 191–204.

Kubai, D. F. (1982). Meiosis in *Sciara coprophila*: structure of the spindle and chromosome behavior during the first meiotic division. *The Journal of Cell Biology*, **93**, 655–669.

Kubota, T. (1966). Studies of the cleavage in the frog egg. I. On the temporal relation between furrow determination and nuclear division. *Journal of Experimental Biology*, **44**, 545–552.

Kubota, T. & Sakamoto, M. (1993). Furrow formation in the vegetal hemisphere of *Xenopus* eggs. *Development, Growth & Differentiation*, **35**, 403–407.

Kumé, M. & Dan, K. (1968). *Invertebrate Embryology.* Trans. J. C. Dan. Belgrade, Yugoslavia: NOLIT Publishing House.

Kuno, M. (1954). On the nature of the egg surface during cleavage of the sea urchin egg. *Embryologia*, **2**, 33–41.

Kuno-Kojima, M. (1957). On the regional difference in the nature of the cortex of the sea urchin egg during cleavage. *Embryologia*, **3**, 279–293.

Kuriyama, R. (1986). Effect of taxol on first and second meiotic spindle formation in oocytes of the surf clam, *Spisula solidissima. Journal of Cell Science*, **84**, 153–164.

Larochelle, D. A. & Epel, D. (1993). Myosin heavy chain dephosphorylation during cytokinesis in dividing sea urchin embryos. *Cell Motility & the Cytoskeleton*, **25**, 369–380.

Leslie, R. J. (1990). Recruitment: the ins and outs of spindle pole formation. *Cell Motility and the Cytoskeleton*, **16**, 225–228.

Lewis, C. A., Chia, F.-S. & Schroeder, T. E. (1973). Peristaltic constriction in fertilized Barnacle eggs (*Pollicipes polymerus*). *Experientia*, **29**, 1533–1535.

Lewis, W. H. (1939). The role of a superficial plasmagel layer in changes of form, locomotion and division of cells in tissue cultures. *Archiv für Experimentelle Zellforschung*, **23**, 1–7.

Lewis, W. H. (1942). The relation of viscosity changes of protoplasm to ameboid locomotion and cell division. In *The Structure of Protoplasm*, ed. W. Seifriz, pp. 163–197. Ames: The Iowa State College Press.

Lillie, R. S. (1903). Fusion of blastomeres and nuclear division without cell-division in solutions of non-electrolytes. *The Biological Bulletin*, **4**, 164–178.

Lillie, R. S. (1916). The physiology of cell-division. VI. Rhythmical changes in the resistance of the dividing sea-urchin egg to hypotonic sea water and their physiological significance. *The Journal of Experimental Zoology*, **21**, 369–402.

Liu, T., Williams, J. G. & Clarke, M. (1992). Inducible expression of calmodulin antisense RNA in *Dictyostelium* cells inhibits the completion of cytokinesis. *Molecular Biology of the Cell*, **3**, 1403–1413.

Longo, F. J. (1972). The effects of cytochalasin B on the events of fertilization in the surf clam, *Spisula solidissima.* I. Polar body formation. *The Journal of Experimental Zoology*, **182**, 321–329.

Longo, F. J. & Anderson, E. (1970). An ultrastructural analysis of fertilization in the surf clam, *Spisula solidissima. Journal of Ultrastructure Research*, **33**, 495–514.

Longo, F. J. & Chen, D.-Y. (1985). Development of cortical polarity in mouse eggs: involvement of the meiotic apparatus. *Developmental Biology*, **107**, 382–394.

Lorch, I. J. (1952). Enucleation of sea-urchin blastomeres with or without removal of asters. *Quarterly Journal of Microscopical Science*, **93**, 475–486.

Lutz, D. A., Hamaguchi, Y. & Inoué, S. (1988). Micromanipulation studies of the asymmetric positioning of the maturation spindle in *Chaetopterus* sp.

oocytes. I. Anchorage of the spindle to the cortex and migration of a displaced spindle. *Cell Motility & the Cytoskeleton*, **11**, 83–96.

Lutz, D. A. & Inoué, S. (1982). Colcemid but not cytochalasin inhibits asymmetric nuclear positioning prior to unequal cell division. *The Biological Bulletin*, **163**, 373–374.

McCaig, C. D. & Robinson, K. R. (1982). The distribution of lectin receptors on the plasma membrane of the fertilized sea urchin egg during first and second cleavage. *Developmental Biology*, **92**, 197–202.

McClendon, J. F. (1907). Experiments on the eggs of *Chaetopterus* and *Asterias* in which the chromatin was removed. *The Biological Bulletin*, **12**, 141–145.

McClendon, J. F. (1908). The segmentation of the eggs of *Asterias forbesii* deprived of chromatin. *Archiv für Entwicklungsmechanik der Organismen*, **26**, 662–668.

McClendon, J. F. (1912). A note on the dynamics of cell division: a reply to Robertson. *Archiv für Entwicklungsmechanik der Organismen*, **34**, 263–266.

McIntosh, J. R. & Pfarr, C. M. (1991). Mitotic motors. *The Journal of Cell Biology*, **115**, 577–585.

McNeil, P. L. (1991). Cell wounding and healing. *American Scientist*, **79**, 222–235.

Mabuchi, I. (1982). Effects of muscle proteins on the interaction between actin and an actin-depolymerizing protein from starfish oocytes. *Journal of Biochemistry*, **92**, 1439–1447.

Mabuchi, I. (1986). Biochemical aspects of cytokinesis. *International Review of Cytology*, **101**, 175–213.

Mabuchi, I. (1990). Cleavage furrow formation and actin-modulating proteins. In *Cytokinesis*, ed. by G. W. Conrad and T. E. Schroeder. *Annals of the New York Academy of Sciences*, **582**, 131–146.

Mabuchi, I. (1993). Regulation of cytokinesis in animal cells: possible involvement of protein phosphorylation. *Biomedical Research*, **14**, Supplement 2, 155–159.

Mabuchi, I. (1994). Cleavage furrow: timing of emergence of contractile ring actin filaments and establishment of the contractile ring by filament bundling in sea urchin eggs. *Journal of Cell Science*, **107**, 1853–1862.

Mabuchi, I., Hamaguchi, Y., Fujimoto, H., Morii, N., Mishima, M. & Narumiya, S. (1993). A rho-like protein is involved in the organisation of the contractile ring in dividing sand dollar eggs. *Zygote*, **1**, 325–331.

Mabuchi, I., Hosoya, H. & Sakai, H. (1980). Actin in the cortical layer of the sea urchin egg. Changes in its content during and after fertilization. *Biomedical Research*, **1**, 417–426.

Mabuchi, I. & Okuno, M. (1977). The effect of myosin antibody on the division of starfish blastomeres. *The Journal of Cell Biology*, **74**, 251–263.

Mabuchi, I. & Takano-Ohmuro, H. (1990). Effects of inhibitors of myosin light chain kinase and other protein kinases on the first cell division of sea urchin eggs. *Development, Growth & Differentiation*, **32**, 549–556.

Marsland, D. A. (1938). The effects of high hydrostatic pressure upon cell division in *Arbacia* eggs. *Journal of Cellular & Comparative Physiology*, **12**, 57–70.

Marsland, D. A. (1939). The mechanism of cell division: hydrostatic pressure effects upon dividing egg cells. *Journal of Cellular & Comparative Physiology*, **13**, 15–22.

Marsland, D. A. (1942). Protoplasmic streaming in relation to gel structure in the cytoplasm. In *The Structure of Protoplasm*, ed. W. Seifriz, pp. 127–161. Ames: The Iowa State College Press.

Marsland, D. A. (1950). The mechanisms of cell division: temperature–pressure experiments on the cleaving eggs of *Arbacia punctulata. Journal of Cellular & Comparative Physiology*, **36**, 205–227.

Marsland, D. A. & Landau, J. V. (1954). The mechanisms of cytokinesis: temperature–pressure studies on the cortical gel system in various marine eggs. *The Journal of Experimental Zoology*, **125**, 507–539.

Mascher, I. B. (1989). Aspectos biofisicos de la divisi6n celular. Teses, Universidad Nacional Autonoma de Mexico, pp. 1–121.

Maupin, P. & Pollard, T. D. (1986). Arrangement of actin filaments and myosin-like filaments in the contractile ring and of actin-like filaments in the mitotic spindle of dividing HeLa cells. *Journal of Ultrastructure & Molecular Structure Research*, **94**, 92–103.

Mazia, D. (1984). Centrosomes and mitotic poles. *Experimental Cell Research*, **153**, 1–15.

Mercer, E. H. & Wolpert, L. (1958). Electron microscopy of cleaving sea urchin eggs. *Experimental Cell Research*, **14**, 629–632.

Mercer, E. H. & Wolpert, L. (1962). An electron microscope study of the cortex of the sea urchin (*Psammechinus miliaris*) egg. *Experimental Cell Research*, **27**, 1–13.

Meves, F. (1897). Über den Vorgang der Zelleinschnürung. *Archiv für Entwicklungsmechanik*, **5**, 378–386.

Miller, A. L., Fluck, R. A., McLaughlin, J. A. & Jaffe, L. F. (1993). Calcium buffer injections inhibit cytokinesis in *Xenopus* eggs. *Journal of Cell Science*, **106**, 523–534.

Mitchison, J. M. (1952). Cell membranes and cell division. *Symposium of the Society of Experimental Biology*, **6**, 105–127.

Mitchison, J. M. (1953). Microdissection experiments on sea-urchin eggs at cleavage. *Journal of Experimental Biology*, **30**, 515–524.

Mitchison, J. M. & Swann, M. M. (1954a). The mechanical properties of the cell surface. I. The cell elastimeter. *Journal of Experimental Biology*, **31**, 443–460.

Mitchison, J. M. & Swann, M. M. (1954b). The mechanical properties of the cell surface. II. The unfertilized sea urchin egg. *Journal of Experimental Biology*, **31**, 461–472.

Mitchison, J. M. & Swann, M. M. (1955). The mechanical properties of the cell surface. III. The sea-urchin egg from fertilization to cleavage. *Journal of Experimental Biology*, **32**, 734–750.

Mitchison, T. J. & Kirschner, M. W. (1984). Dynamic instability of microtubule growth. *Nature*, **312**, 237–242.

Moens, P. B. (1976). Spindle and kinetochore morphology of *Dictyostelium discoideum*. *The Journal of Cell Biology*, **68**, 113–122.
Moll, E. & Paweletz, N. (1980). Membranes of the mitotic apparatus of mammalian cells. *European Journal of Cell Biology*, **21**, 280–287.
Morgan, T. H. (1897). *The Development of the Frog's Egg*. London: The Macmillan Company.
Morgan, T. H. (1899). The action of salt solutions on the unfertilized and fertilized eggs of *Arbacia* and of other animals. *Archiv für Entwicklungsmechanik*, **8**, 448–539.
Morgan, T. H. (1927). *Experimental Embryology*. New York: Columbia University Press.
Morgan, T. H. (1935). Centrifuging the eggs of *Ilyanassa* in reverse. *The Biological Bulletin*, **68**, 268–279.
Morgan, T. H. (1937). The formation of the antipolar lobe in *Ilyanassa*. *The Journal of Experimental Zoology*, **64**, 433–467.
Morgan, T. H. & Spooner, G. B. (1909). The polarity of the centrifuged egg. *Archiv für Entwicklungsmechanik*, **28**, 104–117.
Mota, M. (1959). Karyokinesis without cytokinesis in the grasshopper. *Experimental Cell Research*, **17**, 76–83.
Motomura, I. (1935). Determination of the embryonic axis in the eggs of amphibia and echinoderms. *Science Reports of the Tôhoku Imperial University*, 4th Series, **10**, 211–245.
Motomura, I. (1940). Studies of cleavage. I. Changes in surface area of different regions of eggs of a sea urchin in the course of the first cleavage. *Science Reports of the Tôhoku Imperial University*, 4th Series, Biology **15**, 121–130.
Murray, R. G., Murray, A. S. & Pizzo, A. (1965). The fine structure of mitosis in rat thymic lymphocytes. *The Journal of Cell Biology*, **26**, 601–619.
Nagano, H., Hirai, S., Okano, K. & Ikegami, S. (1981). Achromosomal cleavage of fertilized starfish eggs in the presence of aphidicolin. *Developmental Biology*, **85**, 409–415.
Nakamura, S. & Hiramoto, Y. (1978). Mechanical properties of the cell surface in starfish eggs. *Development, Growth & Differentiation*, **20**, 317–327.
Needham, J. (1950). *Biochemistry & Morphogenesis*. Cambridge: Cambridge University Press.
Neufeld, T. P. & Rubin, G. M. (1994). The *Drosophila peanut* gene is required for cytokinesis and encodes a protein similar to yeast putative bud neck filament proteins. *Cell*, **77**, 371–379.
Ohshima, H. & Kubota, T. (1985). Cell surface changes during cleavage of newt eggs: scanning electron microsope studies. *Journal of Embryology & Experimental Morphology*, **85**, 21–31.
Ohtsubo, M. & Hiramoto, Y. (1985). Regional difference in mechanical properties of the cell surface in dividing echinoderm eggs. *Development, Growth & Differentiation*, **27**, 371–383.
Oka, M. T., Arai, T. & Hamaguchi, Y. (1990). Heterogeneity of microtubules in dividing sea urchin eggs revealed by immunofluorescence microscopy: Spindle microtubules are composed of tubulin isotypes dif-

ferent from those of astral microtubules. *Cell Motility & the Cytoskeleton*, **16**, 239–250.

Opas, J. & Soltynska, M. S. (1978). Reorganization of the cortical layer during cytokinesis in mouse blastomeres. *Experimental Cell Research*, **113**, 208–211.

Osanai, K. (1975). Handling Japanese sea urchins and their embryos. In *The Sea Urchin Embryo*, ed. G. Czihak, pp. 26–40. Berlin, Heidelberg, New York: Springer-Verlag.

Otey, C. A., Pavalko, F. M. & Burridge, K. (1990). An interaction between ∂-actinin and the β_1 integrin subunit in vitro. *The Journal of Cell Biology*, **111**, 721–729.

Painter, T. S. (1918). Contributions to the study of cell mechanics. II. Monaster eggs and narcotized eggs. *The Journal of Experimental Zoology*, **24**, 445–497.

Pasteels, J. J. & de Harven, E. (1962). Étude au microscope électronique du cortex de l'oeuf de *Barnea candida* (mollusque bivalve), et son evolution au moment de la fécondation, de la maturation et de la segmentation. *Archives de Biologie*, **73**, 465–490.

Pasternak, C. A. (1987). Surface changes in relation to cytokinesis and other stages during the cell cycle. In *Biomechanics of Cell Division*, NATO ASI Series, ed. N. Akkas, pp. 187–196. New York & London: Plenum Press.

Pavalko, F. M. & Burridge, K. (1991). Disruption of the actin cytoskeleton after microinjection of proteolytic fragments of ∂-actinin. *The Journal of Cell Biology*, **114**, 481–491.

Paweletz, N. (1981). Membranes in the mitotic apparatus. *Cell Biology International Reports*, **5**, 323–336.

Paweletz, N. & Finze, E.-M. (1981). Membranes and microtubules of the mitotic apparatus of mammalian cells. *Journal of Ultrastructure Research*, **76**, 127–133.

Paweletz, N., Mazia, D. & Finze, E.-M. (1984). The centrosomal cycle and the mitotic cycle of sea urchin eggs. *Experimental Cell Research*, **152**, 47–65.

Paweletz, N., Mazia, D. & Finze, E.-M. (1987a). Fine structural studies of the bipolarization of the mitotic apparatus in the fertilized sea urchin egg. I. The structure and behavior of centrosomes before fusion of the pronuclei. *European Journal of Cell Biology*, **44**, 195–204.

Paweletz, N., Mazia, D. & Finze, E.-M. (1987b). Fine structural studies of the bipolarization of the mitotic apparatus in the fertilized sea urchin egg. II. Bipolarization before the first mitosis. *European Journal of Cell Biology*, **44**, 205–213.

Pereyaslawzewa, S. (1888). Le developpement de *Gammarus poecilurus*. *Rthk. Bulletin Society Nat. Mos., N. S.*, **2**, 185–219.

Perry, M. M., John, H. A. & Thomas, N. S. T. (1971). Actin-like filaments in the cleavage furrow of newt egg. *Experimental Cell Research*, **65**, 249–253.

Platner, G. (1886). Die Karyokinese bei den Lepidopteren als Grundlage für ein Theorie der Zellteilung. *Internationale Monatsschrift für Anatomie und Physiologie*, **3**, 341–398.

Pochapin, M. B., Sanger, J. M. & Sanger, J. W. (1983). Microinjection of Lucifer yellow CH into sea urchin eggs and embryos. *Cell Tissue Research*, **234**, 309–318.

Pollard, T. D. & Mooseker, M. S. (1981). Direct measurement of actin polymerization rate constants by electron microscopy of actin filaments nucleated by isolated microvillus cores. *The Journal of Cell Biology*, **88**, 654–659.

Pollard, T. D., Satterwhite, L., Cisek, L., Corden, J., Sato, M. & Maupin, P. (1990). Actin and myosin biochemistry in relation to cytokinesis. In *Cytokinesis*, ed. G. W. Conrad and T. E. Schroeder. *Annals of the New York Academy of Sciences*, **582**, 120–130.

Porter, K., Prescott, D. & Frye, J. (1973). Changes in surface morphology of Chinese ovary cells during the cell cycle. *The Journal of Cell Biology*, **57**, 815–836.

Portzehl, H. (1951). Muskelkontraction und Modellkontraction. *Zellforschung Naturforsch*, **6**, 355–361.

Poste, G. & Nicolson, G. L. (1981). *Cytoskeletal elements and plasma membrane organization.* In *Cytoskeletal Elements and Plasma Membrane Organization*, Cell Surface Reviews, vol. 7, ed. G. Poste and G. L. Nicolson. Amsterdam: Elsevier / North-Holland Biomedical Press.

Postner, M. A., Miller, K. G. & Wieschaus, E. F. (1992). Maternal effect mutations of the *sponge* locus affect actin cytoskeletal rearrangements in *Drosophila melanogaster* embryos. *The Journal of Cell Biology*, **119**, 1205–1218.

Raff, J. W. & Glover, D. M. (1988). Nuclear and cytoplasmic mitotic cycles continue in *Drosophila* embryos in which DNA synthesis is inhibited with aphidicolin. *The Journal of Cell Biology*, **107**, 2009–2019.

Rappaport, R. (1960). Cleavage of sand dollar eggs under constant tensile stress. *The Journal of Experimental Zoology*, **144**, 225–231.

Rappaport, R. (1961). Experiments concerning the cleavage stimulus in sand dollar eggs. *The Journal of Experimental Zoology*, **148**, 81–89.

Rappaport, R. (1964). Geometrical relations of the cleavage stimulus in constricted sand dollar eggs. *The Journal of Experimental Zoology*, **155**, 225–230.

Rappaport, R. (1965). Geometrical relations of the cleavage stimulus in invertebrate eggs. *Journal of Theoretical Biology*, **9**, 51–66.

Rappaport, R. (1966). Experiments concerning the cleavage furrow in invertebrate eggs. *The Journal of Experimental Zoology*, **161**, 1–8.

Rappaport, R. (1967). Cell division: direct measurement of maximum tension exerted by furrow of echinoderm eggs. *Science*, **156**, 1241–1243.

Rappaport, R. (1968). Geometrical relations of the cleavage stimulus in flattened, perforated sea urchin eggs. *Embryologia*, **10**, 115–130.

Rappaport, R. (1969a). Division of isolated furrows and furrow fragments in invertebrate eggs. *Experimental Cell Research*, **56**, 87–91.

Rappaport, R. (1969b). Aster–equatorial surface relations and furrow establishment. *The Journal of Experimental Zoology*, **171**, 59–67.

Rappaport, R. (1970). An experimental analysis of the role of cytoplasmic fountain streaming in furrow establishment. *Development, Growth & Differentiation*, **12**, 31–40.

Rappaport, R. (1971a). Reversal of chemical cleavage inhibition in echinoderm eggs. *The Journal of Experimental Zoology*, **176**, 249–255.

Rappaport, R. (1971b). Cytokinesis in animal cells. *International Review of Cytology*, **31**, 169–213.

Rappaport, R. (1973). On the rate of movement of the cleavage stimulus in sand dollar eggs. *The Journal of Experimental Zoology*, **183**, 115–119.

Rappaport, R. (1975). Establishment and organization of the cleavage mechanism. In *Molecules and Cell Movement*, ed. S. Inoué and R. E. Stephens, pp. 287–303. New York: Raven Press.

Rappaport, R. (1976). Furrowing in altered cell surfaces. *The Journal of Experimental Zoology*, **195**, 271–277.

Rappaport, R. (1977). Tensiometric studies of cytokinesis in cleaving sand dollar eggs. *The Journal of Experimental Zoology*, **201**, 375–378.

Rappaport, R. (1978). Effects of continual mechanical agitation prior to cleavage in echinoderm eggs. *The Journal of Experimental Zoology*, **206**, 1–11.

Rappaport, R. (1981). Cytokinesis: cleavage furrow establishment in cylindrical sand dollar eggs. *The Journal of Experimental Zoology*, **217**, 365–375.

Rappaport, R. (1982). Cytokinesis: the effect of initial distance between the mitotic apparatus and surface on the rate of subsequent cleavage furrow progress. *The Journal of Experimental Zoology*, **221**, 399–403.

Rappaport, R. (1983). Cytokinesis: furrowing activity in nucleated endoplasmic fragments of fertilized sand dollar eggs. *The Journal of Experimental Zoology*, **227**, 247–253.

Rappaport, R. (1985). Repeated furrow formation from a single mitotic apparatus in cylindrical sand dollar eggs. *The Journal of Experimental Zoology*, **234**, 167–171.

Rappaport, R. (1986a). Simple methods and devices for handling echinoderm eggs. In *Echinoderm Gametes and Embryos*, ed. T. E. Schroeder. *Methods in Cell Biology*, **27**, 345–358.

Rappaport, R. (1986b). Establishment of the mechanism of cytokinesis in animal cells. *International Review of Cytology*, **105**, 245–281.

Rappaport, R. (1988). Surface behavior in artificially constricted sand dollar eggs and egg fragments. *The Journal of Experimental Zoology*, **246**, 253–257.

Rappaport, R. (1990). Role of the mitotic apparatus in furrow initiation. In *Cytokinesis*, ed. G. W. Conrad and T. E. Schroeder. *Annals of the New York Academy of Sciences*, **582**, 15–21.

Rappaport, R. (1991). Enhancement of aster-induced furrowing activity by a factor associated with the nucleus. *The Journal of Experimental Zoology*, **257**, 87–95.

Rappaport, R. (1992). Cleavage inhibition by cell shape change. *The Journal of Experimental Zoology*, **264**, 26–31.

Rappaport, R. & Conrad, G. W. (1963). An experimental analysis of unilateral cleavage in invertebrate eggs. *The Journal of Experimental Zoology*, **153**, 99–112.

Rappaport, R. & Ebstein, R. P. (1965). Duration of stimulus and latent periods preceding furrow formation in sand dollar eggs. *The Journal of Experimental Zoology*, **158**, 373–382.

Rappaport, R. & Rappaport, B. N. (1968). An analysis of cytokinesis in cultured newt cells. *The Journal of Experimental Zoology*, **168**, 187–195.

Rappaport, R. & Rappaport, B. N. (1974). Establishment of cleavage furrows by the mitotic spindle. *The Journal of Experimental Zoology*, **189**, 189–196.

Rappaport, R. & Rappaport, B. N. (1976). Prefurrow behavior of the equatorial surface in *Arbacia lixula* eggs. *Development, Growth & Differentiation*, **18**, 189–193.

Rappaport, R. & Rappaport, B. N. (1983). Cytokinesis: effects of blocks between the mitotic apparatus and the surface on furrow establishment in flattened echinoderm eggs. *The Journal of Experimental Zoology*, **227**, 213–227.

Rappaport, R. & Rappaport, B. N. (1984). Division of constricted and urethane-treated sand dollar eggs: a test of the polar stimulation hypothesis. *The Journal of Experimental Zoology*, **231**, 81–92.

Rappaport, R. & Rappaport, B. N. (1985a). Surface contractile activity associated with isolated asters in cylindrical sand dollar eggs. *The Journal of Experimental Zoology*, **235**, 217–226.

Rappaport, R. & Rappaport, B. N. (1985b). Experimental analysis of polar body formation in starfish eggs. *The Journal of Experimental Zoology*, **235**, 87–103.

Rappaport, R. & Rappaport, B. N. (1986). Experimental analysis of cytokinesis in *Amoeba proteus*. *The Journal of Experimental Zoology*, **240**, 55–63.

Rappaport, R. & Rappaport, B. N. (1988). Reversing cytoplasmic flow in nucleated, constricted sand dollar eggs. *The Journal of Experimental Zoology*, **247**, 92–98.

Rappaport, R. & Rappaport, B. N. (1993). Duration of division-related events in cleaving sand dollar eggs. *Developmental Biology*, **158**, 265–273.

Rappaport, R. & Rappaport, B. N. (1994). Cleavage in conical sand dollar eggs. *Developmental Biology*, **164**, 258–266.

Rappaport, R. & Ratner, J. H. (1967). Cleavage of sand dollar eggs with altered patterns of new surface formation. *The Journal of Experimental Zoology*, **165**, 89–100.

Rashevsky, N. (1948). Can elastic stresses in gels cause the elongation and division of a cell? *Bulletin of Mathematical Biophysics*, **10**, 85–89.

Rashevsky, N. (1952). Some suggestions for a new theory of cell division. *Bulletin of Mathematical Biophysics*, **14**, 293–305.

Rebhun, L. I. (1959). Studies of early cleavage in the surf clam, *Spisula solidissima*, using methylene blue and toluidine blue as vital stains. *The Biological Bulletin*, **117**, 518–545.

Rebhun, L. I. (1960). Aster-associated particles in the cleavage of marine invertebrate eggs. In *Second Conference on the Mechanisms of Cell*

Division, ed. P. W. Gross. *Annals of the New York Academy of Sciences*, **90**, 357–380.

Rebhun, L. I. & Sander, G. (1971). Electron microscope studies of frozen-substituted marine eggs. III. Structure of the mitotic apparatus of the first meiotic division. *American Journal of Anatomy,* **130**, 35–53.

Rebhun, L. I. & Sawada, N. (1969) Augmentation and dispersion in the *in vivo* mitotic apparatus of living marine eggs. *Protoplasma,* **68**, 1–22.

Render, J. A. (1983). The second polar lobe of the *Sabellaria cementarium* embryo plays an inhibitory role in apical tuft formation. *Roux's Archives Developmental Biology*, **192**, 120–129.

Render, J. A. & Guerrier, P. (1984). Size regulation and morphogenetic localization in the *Dentalium* polar lobe. *The Journal of Experimental Zoology,* **232**, 79–86.

Rhumbler, L. (1897). Stemmen die Strahlen der Astrosphäre oder ziehen sie? *Archiv für Entwicklungsmechanik der Organismen*, **4**, 659–730.

Rhumbler, L. (1903). Mechanische Erklärung der Ähnlichkeit zwischen magnetischen Kraftliniensystemen und Zelltheilungsfiguren. *Archiv für Entwicklungsmechanik der Organismen*, **16**, 476–535.

Rice, T. B. & Garen, A. (1975). Localized defects of blastoderm formation in maternal effect mutants of *Drosophila. Developmental Biology*, **43**, 277–286.

Rieder, C. L. (1990). Formation of the astral mitotic spindle: ultrastructural basis for the centrosome–kinetochore interaction. *Electron Microscopical Reviews*, **3**, 269–300.

Rieder, C. L., Davison, E. A., Jensen, L. C. W., Cassimeris, L. & Salmon, E. D. (1986). Oscillatory movements of monooriented chromosomes and their position relative to the spindle pole result from the ejection properties of the aster and half-spindle. *The Journal of Cell Biology*, **103**, 581–591.

Rieder, C. L. & Hard, R. (1990). Newt lung epithelial cells: cultivation, use and advantages for biomedical research. *International Review of Cytology,* **122**, 153–220.

Ris, H. (1949). The anaphase movement of chromosomes in the spermatocytes of the grasshopper. *The Biological Bulletin,* **96**, 90–106.

Roberts, H. S. (1955). The mechanism of cytokinesis in neuroblasts of *Chortophaga viridifasciata* (De Geer). *The Journal of Experimental Zoology*, **130**, 83–106.

Roberts, H. S. (1961). Mechanisms of cytokinesis: a critical review. *Quarterly Review of Biology*, **36**, 155–177.

Roberts, H. S. & Johnson, N. S. (1956). Cytokinesis of multi-spindle cells. *The Biological Bulletin*, **110**, 334–343.

Robertson, T. B. (1909). Note on the chemical mechanics of cell-division. *Archiv für Entwicklungsmechanik des Organismen*, **27**, 29–34.

Rubin, R. W. & Everhart, L. B. (1973). The effect of cell-to-cell contact on the surface morphology of Chinese hamster ovary cells. *The Journal of Cell Biology*, **57**, 837–844.

Rubino, S., Fighetti, M., Unger, E. & Cappuccinelli, P. (1984). Location of actin, myosin, and microtubular structures during directed locomotion of

Dictyostelium amebae. *The Journal of Cell Biology*, **98**, 382–390.

Rugh, R. (1948). *Experimental Embryology.* Minneapolis: Burgess Publishing Company.

Rugh, R. (1951). *The Frog.* Philadelphia & Toronto: The Blakiston Company.

Rustad, R., Yuyama, S. & Rustad, L. C. (1970). Nuclear–cytoplasmic relations in the mitosis of sea urchin eggs. II. The division times of whole eggs and haploid and diploid half-eggs. *The Biological Bulletin*, **138**, 184–193.

Saiki, T. & Hamaguchi, Y. (1993). Difference between maturation division and cleavage in starfish oocytes: dependency of induced cytokinesis on the size of the aster as revealed by transplantation of the centrosome. *Development, Growth & Differentiation*, **35**, 181–188.

Saiki, T., Kyozuka, K., Osanai, K. & Hamaguchi, Y. (1991). Chromosomal behavior in starfish (*Asterina pectinifera*) zygotes under the effect of aphidicolin, an inhibitor of DNA polymerase. *Experimental Cell Research*, **192**, 380–388.

Salmon, E. D., Leslie, R. J., Saxton, W. M., Karow, M. L. & McIntosh, J. R. (1984). Spindle microtubule dynamics in sea urchin embryos: analysis using a fluorescein-labeled tubulin and measurements of fluorescence redistribution after laser photobleaching. *The Journal of Cell Biology*, **99**, 2165–2174.

Salmon, E. D. & Wolniak, S. M. (1990). Role of microtubules in stimulating cytokinesis in animal cells. In *Cytokinesis*, ed. G. W. Conrad and T. E. Schroeder. *Annals of the New York Academy of Sciences*, **582**, 88–98.

Sanders, E. J. (1975). Aspects of furrow membrane formation in the cleaving *Drosophila* embryo. *Cell & Tissue Research*, **156**, 463–474.

Sanger, J. M., Dome, J. S., Hock, R. S., Mittal, B. & Sanger, J. W. (1994). Occurrence of fibers and their association with talin in the cleavage furrows of PtK_2 cells. *Cell Motility & the Cytoskeleton*, **27**, 26–40.

Sanger, J. M., Mittal, B., Dome, J. S. & Sanger, J. W. (1989). Analysis of cell division using fluorescently labeled actin and myosin in living PtK_2 cells. *Cell Motility & the Cytoskeleton*, **14**, 201–219.

Sanger, J. M., Mittal, B., Pochapin, M. B. & Sanger, J. W. (1987). Stress fiber and cleavage furrow formation in living cells microinjected with fluorescently labeled ∂-actinin. *Cell Motility & the Cytoskeleton*, **7**, 209–220.

Sanger, J. M., Mittal, B. & Sanger, J. W. (1990). Use of fluorescently labeled probes to analyze cell division in living cells. In *Cytokinesis*, ed. G. W. Conrad and T. E. Schroeder. *Annals of the New York Academy of Sciences*, **582**, 185–198.

Sanger, J. M., Reingold, A. M. & Sanger, J. W. (1984). Cell surface changes during mitosis and cytokinesis of epithelial cells. *Cell & Tissue Research*, **237**, 409–417.

Sanger, J. M. & Sanger, J. W. (1980). Banding and polarity of actin filaments in interphase and cleaving cells. *The Journal of Cell Biology*, **86**, 568–575.

Sato, N., Yonemura, S., Obinata, T., Tsukita, Sachiko & Tsukita, Shoichiro. (1991). Radixin, a barbed end-capping actin-modulating protein, is concentrated at the cleavage furrow during cytokinesis. *The Journal of Cell Biology*, **113**, 321–330.

Satoh, N. & Deno, T. (1984). Periodic appearance and disappearance of microvilli associated with cleavage cycles in the egg of the ascidian, *Halocynthia roretzi*. *Developmental Biology*, **102**, 488–492.

Satoh, S. K., Oka, M. T. & Hamaguchi, Y. (1994). Asymmetry in the mitotic spindle induced by the attachment to the cell surface during maturation in the starfish oocyte. *Development, Growth & Differentiation*, **36**, 557–565.

Satterwhite, L. L., Lohka, M. J., Wilson, K. L., Scherson, T. Y., Cisek, L. J., Corden, J. L. & Pollard, T. D. (1992). Phosphorylation of myosin-II regulatory light chain by cyclin-p34^{cdc2}: a mechanism for the timing of cytokinesis. *The Journal of Cell Biology*, **118**, 595–605.

Satterwhite, L. L. & Pollard, T. D. (1992). Cytokinesis. *Current Opinion in Cell Biology*, **4**, 43–52.

Sawai, T. (1987). Surface movement in the region of the cleavage furrow of amphibian eggs. *Zoological Science*, **4**, 825–832.

Sawai, T. (1992). Effect of microtubular poisons on cleavage furrow formation and induction of furrow-like dent in amphibian eggs. *Development, Growth & Differentiation*, **34**, 669–675.

Sawai, T. & Yomota, A. (1990). Cleavage plane determination in amphibian eggs. In *Cytokinesis*, ed. G. W. Conrad and T. E. Schroeder. *Annals of the New York Academy of Sciences*, **582**, 40–49.

Schantz, A. R. (1985). Cytosolic free calcium-ion concentration in cleaving embryonic cells of *Oryzias latipes* measured with calcium-selective microelectrodes. *The Journal of Cell Biology*, **100**, 947–954.

Schechtman, A. M. (1937). Localized cortical growth as the immediate cause of cell division. *Science*, **85**, 222–223.

Schejter, E. D. & Wieschaus, E. (1993). Functional elements of the cytoskeleton in the early *Drosophila* embryo. *Annual Review of Cell Biology*, **9**, 67–99.

Schmidt, B. A., Kelly, P. T., May, M. C., Davis, S. E. & Conrad, G. W. (1980). Characterization of actin from fertilized eggs of *Ilyanassa obsoleta* during polar lobe formation and cytokinesis. *Developmental Biology*, **76**, 126–140.

Schnapp, B. J., Reese, T. S. & Bechtold, R. (1992). Kinesin is bound with high affinity to squid axon organelles that move to the plus-end of microtubules. *The Journal of Cell Biology*, **119**, 389–399.

Schroeder, T. E. (1968). Cytokinesis: filaments in the cleavage furrow. *Experimental Cell Research*, **53**, 272–276.

Schroeder, T. E. (1970). The contractile ring. I. Fine structure of dividing mammalian (HeLa) cells and effects of cytochalasin B. *Zeitschrift für Zellforschung und Mikroskopische Anatomie*, **109**, 431–449.

Schroeder, T. E. (1972). The contractile ring. II. Determining its brief existence, volumetric changes and vital role in cleaving *Arbacia* eggs. *The Journal of Cell Biology*, **53**, 419–434.

Schroeder, T. E. (1973). Actin in dividing cells: Contractile ring filaments bind heavy meromyosin. *Proceedings of the National Academy of Sciences*, **70**, 1688–1692.

Schroeder, T. E. (1974). Ionophore-induced contractions in frog eggs: role of actin. *The Journal of Cell Biology*, **63**, 305a.

Schroeder, T. E. (1975). Dynamics of the contractile ring. In *Molecules and Cell Movement*, ed. S. Inoué and R. E. Stephens, pp. 305–332. New York: Raven Press.
Schroeder, T. E. (1981a). The origin of cleavage forces in dividing eggs: a mechanism in two steps. *Experimental Cell Research*, **134**, 231–240.
Schroeder, T. E. (1981b). Interrelations between the cell surface and the cytoskeleton in cleaving sea urchin eggs. In *Cytoskeletal Elements and Plasma Membrane Organization*, ed. G. Poste and G. L. Nicolson, pp. 169–216. Amsterdam: Elsevier/North-Holland Biomedical Press.
Schroeder, T. E. (1985a). Physical interactions between asters and the cortex in echinoderm eggs. In *The Cellular and Molecular Biology of Invertebrate Development*, ed. R. H. Sawyer and R. M. Showman, pp. 69–89. Columbia: University of South Carolina Press.
Schroeder, T. E. (1985b). Cortical expressions of polarity in the starfish oocyte. *Development, Growth & Differentiation*, **27**, 311–321.
Schroeder, T. E. (1986). The egg cortex in early development of sea urchins and starfish. In *Developmental Biology*, vol. 2, ed. L. W. Browder, pp. 59–100. Plenum.
Schroeder, T. E. (1987). Fourth cleavage of sea urchin blastomeres: microtubule patterns and myosin localization in equal and unequal cell divisions. *Developmental Biology*, **124**, 9–22.
Schroeder, T. E. (1988). Contact-independent polarization of the cell surface and cortex of free sea urchin blastomeres. *Developmental Biology*, **125**, 255–264.
Schroeder, T. E. (1990). The contractile ring and furrowing in dividing cells. In *Cytokinesis*, ed. G. W. Conrad and T. E. Schroeder. *Annals of the New York Academy of Sciences*, **582**, 78–87.
Schroeder, T. E. & Otto, J. J. (1984). Cyclic assembly–disassembly of cortical microtubules during maturation and early development of starfish oocytes. *Developmental Biology*, **103**, 493–503.
Schroeder, T. E. & Otto, J. J. (1988). Immunofluorescent analysis of actin and myosin in isolated contractile rings of sea urchin eggs. *Zoological Science*, **5**, 713–725.
Schroeder, T. E. & Strickland, D. L. (1974). Ionophore A23187, calcium and contractility in frog eggs. *Experimental Cell Research*, **83**, 139–142.
Schroer, T. A. & Sheetz, M. P. (1991). Functions of microtubule-based motors. *Annual Review of Physiology*, **53**, 629–652.
Scott, A. C. (1946). The effect of low temperature and of hypotonicity on the morphology of the cleavage furrow in *Arbacia* eggs. *The Biological Bulletin*, **91**, 272–287.
Scott, A. C. (1960a). Furrowing in flattened sea urchin eggs. *The Biological Bulletin*, **119**, 246–259.
Scott, A. C. (1960b). Surface changes during cell division. *The Biological Bulletin*, **119**, 260–272.
Scott, A. C. (1964). A new concept of the mechanism of cytokinesis. *The Biological Bulletin*, **127**, 389.
Selman, G. G. (1966). Cell cleavage in polar body formation. *Nature*, **210**, 750–751.

Selman, G. G. (1982). Determination of the first two cleavage furrows in developing eggs of *Triturus Alpestris* compared with other forms. *Development, Growth & Differentiation*, **24**, 1–6.

Selman, G. G. & Perry, M. M. (1970). Ultrastructural changes in the surface layers of the newt's egg in relation to the mechanism of its cleavage. *Journal of Cell Science*, **6**, 207–227.

Selman, G. G. & Waddington, C. H. (1955). The mechanism of cell division in the cleavage of the newt's egg. *Journal of Experimental Biology*, **32**, 700–733.

Semik, D. & Kilarski, W. (1986). Surface morphology of eggs and early embryos of three species of newts. *Zeitschrift für Mikroskopisch-Anatomisch Forschung*, **100**, 881–904.

Shelden, E. & Wadsworth, P. (1990). Interzonal microtubules are dynamic during spindle elongation. *Journal of Cell Science*, **97**, 273–281.

Shimizu, T. (1981a). Cortical differentiation of the animal pole during maturation division in fertilized eggs of *Tubifex* (Annelida, Oligochaeta). I. Meiotic apparatus formation. *Developmental Biology*, **85**, 65–76.

Shimizu, T. (1981b). Cortical differentiation of the animal pole during maturation division in fertilized eggs of *Tubifex* (Annelida, Oligochaeta). II. Polar body formation. *Developmental Biology*, **85**, 77–88.

Shimizu, T. (1990). Polar body formation in *Tubifex* eggs. In *Cytokinesis*, ed. G. W. Conrad & T. E. Schroeder. *Annals of the New York Academy of Sciences*, **582**, 260–272.

Shinagawa, A., Kono, S., Yoshimoto, Y. & Hiramoto, Y. (1989). Nuclear involvement in localization of the initiation site of surface contraction waves in *Xenopus* eggs. *Development, Growth & Differentiation*, **31**, 249–255.

Shirai, H. & Kanatani, H. (1980). Effect of local application of 1-methyladenine on the site of polar body formation in starfish oocyte. *Development, Growth & Differentiation*, **22**, 555–560.

Shôji, Y., Hamaguchi, Y. & Hiramoto, Y. (1981). Quantitative studies on polarization optical properties of living cells. III. Cortical birefringence of the dividing sea urchin egg. *Cell Motility*, **1**, 387–397.

Sichel, F. J. M. & Burton, A. C. (1936). A kinetic method of studying surface forces in the egg of *Arbacia*. *The Biological Bulletin*, **71**, 397–398.

Singer, S. J. & Nicholson, G. L. (1972). The fluid mosaic model of the structure of cell membranes. *Science*, **175**, 720.

Sit, K. H., Bay, B. H. & Wong, K. P. (1993). Reduced surface area in mitotic rounding of human Chang liver cells. *The Anatomical Record*, **235**, 183–190.

Sluder, G. (1990). Experimental analysis of centrosome reproduction in echinoderm eggs. *Advances in Cell Biology*, **3**, 221–250.

Sluder, G. & Begg, D. A. (1983). Control mechanisms of the cell cycle: role of the spatial arrangment of spindle components in the timing of mitotic events. *The Journal of Cell Biology*, **97**, 877–886.

Sluder, G., Miller, F. J. & Rieder, C. L. (1986). The reproduction of centrosomes: nuclear versus cytoplasmic controls. *The Journal of Cell Biology*, **103**, 1873–1881.

Snodgrass, R. E. (1935). *Principles of Insect Morphology*. New York & London: McGraw-Hill Book Company, Inc.

Soltys, B. J. & Borisy, G. G. (1985). Polymerization of tubulin in vivo: direct evidence for assembly onto microtubule ends and from centrosomes. *The Journal of Cell Biology*, **100**, 1682–1689.

Spek, J. (1918). Oberflächenspannungsdifferenzen als eine Ursache der Zellteilung. *Archiv für Entwicklungsmechanik der Organismen*, **44**, 5–113.

Spudich, A. & Spudich, J. A. (1979). Actin in triton-treated cortical preparations of unfertilized and fertilized sea urchin eggs. *The Journal of Cell Biology*, **82**, 212–226.

Stafstrom, J. P. & Staehelin, L. A. (1984). Dynamic changes of the nuclear envelope and nuclear pore complexes during mitosis in the *Drosophila* embryo. *European Journal of Cell Biology*, **34**, 179–189.

Stefanini, M., Ōura, C. & Zamboni, L. (1969). Meiotic cleavage of the mammalian ovum. *The Journal of Cell Biology*, **43**, 138a.

Stossel, T. P., Hartwig, J. H. & Yin, H. L. (1981). Actin gelatin and the structure and movement of cortical cytoplasm. In *Cytoskeletal Elements and Plasma Membrane Organization*, Cell Surface Reviews, vol. 7, ed. G. Poste and G. L. Nicolson, pp. 140–168. Amsterdam: Elsevier/North-Holland Biomedical Press.

Sullivan, W., Fogarty, P. & Theurkauf, W. (1993). Mutations affecting the cytoskeletal organization of syncytial *Drosophila* embryos. *Development*, **118**, 1245–1254.

Sullivan, W., Minden, J. S. & Alberts, B. M. (1990). *daughterless-abo-like*, a *Drosophila* maternal-effect mutation that exhibits abnormal centrosome separation during the late blastoderm divisions. *Development*, **110**, 311–323.

Swann, M. M. (1951). Protoplasmic structure and mitosis. II. The nature and cause of birefringence changes in the sea urchin egg at anaphase. *Journal of Experimental Biology*, **28**, 434–444.

Swann, M. M. & Mitchison, J. M. (1953). Cleavage of sea-urchin eggs in colchicine. *Journal of Experimental Biology*, **30**, 506–514.

Swann, M. M. & Mitchison, J. M. (1958). The mechanism of cleavage in animal cells. *Biological Reviews*, **33**, 103–135.

Szollosi, D. (1968). The contractile ring and changes of the cell surface during cleavage. *The Journal of Cell Biology*, **39**, 133a.

Szollosi, D. (1970). Cortical cytoplasmic filaments of cleaving eggs: a structural element corresponding to the contractile ring. *The Journal of Cell Biology*, **44**, 192–209.

Szollosi, D., Calarco, P. & Donahue, R. P. (1972). Absence of centrioles in the first and second meiotic spindles of mouse oocytes. *Journal of Cell Science*, **11**, 521–541.

Tadano, Y. (1962). Studies on the mechanism of cleavage with reference to the nuclear division. *Japanese Journal of Zoology*, **13**, 307–328.

Tahmisian, T. N., Devine, R. L. & Wright, B. J. (1967). The ultrastructure of the plasma membrane at the division furrow of grasshopper germ cells. *Zeitschrift für Zellforschung und Mikroskopische Anatomie*, **77**, 316–324.

Tanaka, Y. (1976). Effects of the surfactants on the cleavage and further development of the sea urchin embryos. I. The inhibition of micromere formation at the fourth cleavage. *Development, Growth & Differentiation*, **18**, 113–122.

Tanaka, Y. (1979). Effects of the surfactants on the cleavage and further development of the sea urchin embryos. II. Disturbance in the arrangement of cortical vesicles and change in cortical appearance. *Development, Growth & Differentiation*, **21**, 331–342.

Teichmann, E. (1903). Über die Beziehung zwischen Astrosphären und Furchen: Experimentelle Untersuchungen am Seeigelei. *Archiv für Entwicklungsmechanik der Organismen*, **16**, 243–327.

Terasaki, M. & Jaffe, L. A. (1991). Organization of the sea urchin egg endoplasmic reticulum and its reorganization at fertilization. *The Journal of Cell Biology*, **114**, 929–940.

Thomas, R. J. (1968). Cytokinesis during early development of a teleost embryo: *Brachydanio rerio. Journal of Ultrastructure Research*, **24**, 232–238.

Tilney, L. G. & Marsland, D. (1969). A fine structural analysis of cleavage induction and furrowing in the eggs of *Arbacia punctulata. The Journal of Cell Biology*, **42**, 170–184.

Timourian, H., Clothier, G. & Watchmaker, G. (1972). Cleavage furrow: calcium as determinant of site. *Experimental Cell Research*, **75**, 296–298.

Tsukahara, J. & Ishikawa, M. (1980). Supression of polar body formation by cytochalasin B and fusion of the resulting plural female pronuclei in eggs of the starfish, *Asterina pectinifera. European Journal of Cell Biology*, **21**, 288–295.

Turner, F. R. & Mahowald, A. P. (1976). Scanning electron microscopy of *Drosophila* embryogenesis. I. The structure of the egg envelopes and the formation of the cellular blastoderm. *Developmental Biology*, **50**, 95–108.

Usui, N. & Yoneda, M. (1982). Ultrastructural basis of the tension increase in sea-urchin eggs prior to cytokinesis. *Development, Growth & Differentiation*, **24**, 453–465.

Vacquier, V. D. (1975). The isolation of intact cortical granules from sea urchin eggs: Calcium ions trigger granule discharge. *Developmental Biology*, **43**, 62–74.

Vale, R. D. (1993). Measuring single protein motors at work. *Science*, **260**, 169–170.

Vallee, R. B., Shpetner, H. S. & Paschal, B. M. (1990). Potential roles of microtubule-associated motor molecules in cell division. In *Cytokinesis*, ed. G. W. Conrad and T. E. Schroeder. *Annals of the New York Academy of Sciences*, **582**, 99–107.

Waddington, C. H. (1952). Preliminary observations on the mechanism of cleavage in the amphibian egg. *The Journal of Experimental Biology*, **29**, 484–489.

Wadsworth, P. (1987). Microinjected carboxylated beads move predominantly poleward in sea urchin eggs. *Cell Motility & the Cytoskeleton*, **8**, 293–301.

Wang, Y.-L., Silverman, J. D. & Cao, L.-G. (1994). Single particle tracking of surface receptor movement during cell division. *The Journal of Cell Biology*, **127**, 963–971.

Warn, R. M., Flegg, L. & Warn, A. (1987). An investigation of microtubule organization and functions in living *Drosophila* embryos by injection of a fluorescently labeled antibody against tyrosinated ∂-tubulin. *The Journal of Cell Biology*, **105**, 1721–1730.

Warn, R. M., Magrath, R. & Webb, S. (1984). Distribution of F-actin during cleavage of the *Drosophila* syncytial blastoderm. *The Journal of Cell Biology*, **98**, 156–162.

Warn, R. M. & Robert-Nicoud, M. (1990). F-actin organization during the cellularization of the *Drosophila* embryo as revealed with a confocal laser scanning microscope. *Journal of Cell Science*, **96**, 35–42.

Warn, R. M., Warn, A., Planques, V. & Robert-Nicoud, M. (1990). Cytokinesis in the early *Drosophila* embryo. In *Cytokinesis*, ed. G. W. Conrad and T. E. Schroeder. *Annals of the New York Academy of Sciences*, **582**, 222–232.

Weinstein, R. S. (1965). Models of the dense layer associated with cleavage in the sea urchin. *Journal of Applied Physics*, **36**, 2621.

Weinstein, R. S. & Hebert, R. B. (1964). Electron microscopy of cleavage furrows in sea urchin blastomeres. *The Journal of Cell Biology*, **23**, 101A.

White, J. G. (1985). The astral relaxation theory of cytokinesis revisited. *BioEssays*, **2**, 267–272.

White, J. G. (1990). Laterally mobile, cortical tension elements can self-assemble into a contractile ring. In *Cytokinesis*, ed. G. W. Conrad and T. E. Schroeder. *Annals of the New York Academy of Sciences*, **582**, 50–59.

White, J. G. & Borisy, G. G. (1983). On the mechanism of cytokinesis in animal cells. *Journal of Theoretical Biology*, **101**, 289–316.

Wilson, E. B. (1883). The development of *Renilla*. *Philosophical Transactions of the Royal Society of London*, **174**, 723–815.

Wilson, E. B. (1900). *The Cell in Development and Heredity*, 2d ed. New York: The Macmillan Company.

Wilson, E. B. (1901a). Experimental studies in cytology. I. A cytological study of artificial parthenogenesis in sea urchin eggs. *Archiv für Entwicklungsmechanik der Organismen*, **12**, 529–596.

Wilson, E. B. (1901b). Experimental studies in cytology. II. Some phenomena of fertilization and cell division in etherized eggs. III. The effect on cleavage of artificial obliteration of the first cleavage furrow. *Archiv für Entwicklungsmechanik der Organismen*, **13**, 353–395.

Wilson, E. B. (1903). Notes on merogony and regeneration in *Renilla*. *The Biological Bulletin*, **4**, 215–226.

Wilson, E. B. (1904). Experimental studies on germinal localization. *The Journal of Experimental Zoology*, **1**, 1–72.

Wilson, E. B. (1928). *The Cell in Development and Heredity*, 3d ed. New York: The Macmillan Company.

Wohlfarth-Botterman, K.-E. & Fleischer, M. (1976). Cycling aggregation patterns of cytoplasmic F-actin coordinated with oscillating tension force generation. *Cell & Tissue Research*, **165**, 327–344.

Wolf, R. (1973). Kausalmechanismen der Kernbewegung und-teilung während der frühen Furchung im Ei der Gallmücke *Wachtliella persicariae* L. I. Kinematische Darstellung des "Migrationsasters" wandernder Energiden und der Steuerung seiner Aktivität durch den Initialbereich der Furchung. *Wilhelm Roux's Archives*, **172**, 28–57.

Wolf, R. (1978). The cytaster, a colchicine-sensitive migration organelle of cleavage nuclei in an insect egg. *Developmental Biology*, **62**, 464–472.

Wolf, R. (1980). Migration and division of cleavage nuclei in the gall midge, *Wachtliella persicariae*. *Wilhelm Roux's Archives*, **188**, 65–73.

Wolpert, L. (1960). The mechanics and mechanism of cleavage. *International Review of Cytology*, **10**, 163–216.

Wolpert, L. (1963). Some problems of cleavage in relation to the cell membrane. In *Cell Growth and Cell Division*, ed. R. J. C. Harris. *International Society for Cell Biology*, **2**, 277–298. New York and London: Academic Press.

Wolpert, L. (1966). The mechanical properties of the membrane of the sea urchin egg during cleavage. *Experimental Cell Research,* **41**, 385–396.

Wright, B., Terasaki, M. & Scholey, J. (1993). Roles of kinesin and kinesin-like proteins in sea urchin embryonic cell division: evaluation using antibody microinjection. *The Journal of Cell Biology*, **123**, 681–689.

Yamamoto, K. & Yoneda, M. (1983). Cytoplasmic cycle in meiotic division of starfish oocytes. *Developmental Biology*, **96**, 166–172.

Yamao, W. & Miki-Noumura, T. (1988). Effect of hexyleneglycol on meiotic division of starfish oocytes. *Zoological Science*, **5**, 563–572.

Yamashiki, N. (1981). The role of the spindle body in unequal division of the grasshopper neuroblast. *Zoological Magazine*, **90**, 93–101.

Yamashiki, N. (1990). Microtubule distribution in unequal cell division of the grasshopper neuroblasts. *Journal of Rakuno Gakuen University*, **14**, 169–181.

Yamashiki, N. & Kawamura, K. (1986). Microdissection studies on the polarity of unequal division in grasshopper neuroblasts. I. Subsequent divisions in neuroblast-type cells produced against the polarity by micromanipulation. *Experimental Cell Research,* **166**, 127–138.

Yamashiki, N. & Kawamura, K. (1990). Formation of the contractile ring in dividing neuroblasts of *Chortophaga*. In *Cytokinesis*, ed. G. W. Conrad and T. E. Schroeder. *Annals of the New York Academy of Sciences*, **582**, 314–320.

Yatsu, N. (1908). Some experiments on cell-division in the egg of *Cerebratulus lacteus. Annotationes Zoologicae Japonenses*, **6**, Part 4, 267–276.

Yatsu, N. (1909). Observations on ookinesis in *Cerebratulus lacteus,* Verrill. *Journal of Morphology*, **20**, 353–401.

Yatsu, N. (1912a). Observations and experiments on the Ctenophore egg. I. The structure of the egg and experiments on cell-division. *Journal of the College of Science, Imperial University of Tokyo*, **32**, Article 3, 1–21.

Yatsu, N. (1912b). Observations and experiments on the Ctenophore egg. III. Experiments on germinal localization of the egg of *Beroë ovata*. *Annotationes Zoologicae Japonenses*, **8**, Part I, 5–13.

Yoneda, M. (1964). Tension at the surface of sea-urchin egg: a critical examination of Cole's experiment. *Journal of Experimental Biology*, **41**, 893–906.

Yoneda, M. & Dan, K. (1972). Tension at the surface of the dividing sea urchin egg. *Journal of Experimental Biology*, **57**, 575–587.

Yoneda, M., Ikeda, M. & Washitani, S. (1978). Periodic changes in the tension at the surface of activated non-nucleate fragments of sea urchin eggs. *Development, Growth & Differentiation*, **20**, 329–336.

Yoneda, M. & Schroeder, T. E. (1984). Cell cycle timing in colchicine-treated sea urchin eggs: persistent coordination between the nuclear cycles and the rhythm of cortical stiffness. *The Journal of Experimental Zoology*, **231**, 367–378.

Yoneda, M. & Yamamoto, K. (1985). Periodicity of cytoplasmic cycle in non-nucleate fragments of sea urchin and starfish eggs. *Development, Growth & Differentiation*, **27**, 385–391.

Yonemura, S. & Kinoshita, S. (1986). Actin filament organization in the sand dollar egg cortex. *Developmental Biology*, **115**, 171–183.

Yonemura, S., Nagafuchi, A., Sato, N. & Tsukita, S. (1993). Concentration of an integral membrane protein, CD43 (Leukosialin, Sialophorin), in the cleavage furrow through the interaction of its cytoplasmic domain with actin-based cytoskeletons. *The Journal of Cell Biology*, **120**, 437–449.

Yumura, S. & Fukui, Y. (1985). Reversible cyclic AMP-dependent change in distribution of myosin thick filaments in *Dictyostelium*. *Nature*, **314**, 194–196.

Ziegler, H. E. (1894). Über Furchung unter Pressung. *Anatomische Anzeiger*, **9**, 132–146.

Ziegler, H. E. (1895). Untersuchungen über die Zelltheilung. *Deutsche Zoologische Gesellschaft Verhandlungen*, 62–83.

Ziegler, H. E. (1898a). Experimentelle Studien über die Zelltheilung. I. Die Zerschnürung der Seeigeleier. II. Furchung ohne Chromosomen. *Archiv für Entwicklungsmechanik der Organismen*, **6**, 249–293.

Ziegler, H. E. (1898b). Experimentelle Studien über die Zelltheilung. III. Die Furchungszellen von *Beroë ovata*. *Archiv für Entwicklungsmechanik der Organismen*, **7**, 34–63.

Ziegler, H. E. (1903). Experimentelle Studien über Zelltheilung. IV. Die Zelltheilung der Furchungzellen bei *Beroë* und *Echinus*. *Archiv für Entwicklungsmechanik der Organismen*, **16**, 155–175.

Zimmerman, A. M., Landau, J. V. & Marsland, D. (1957). Cell division: a pressure–temperature analysis of the effects of sulfhydryl reagents on the cortical plasmagel structure and furrowing strength of dividing eggs (*Arbacia & Chaetopterus*). *Journal of Cellular & Comparative Physiology*, **49**, 395–435.

Zotin, A. I. (1964). The mechanism of cleavage in amphibian and sturgeon eggs. *Journal of Embryology & Experimental Morphology*, **12**, 247–262.

Zotin, A. I. & Pagnaeva, R. V. (1963). The time of determination of the position of cleavage furrow in sturgeon and axolotl eggs. *Dokl. Akad. Nauk. USSR*, **152**, 765–768.

Author Index

Subject Index

For further volumes:
http://www.springer.com/series/8431

Graduate Texts in Physics

Graduate Texts in Physics publishes core learning/teaching material for graduate- and advanced-level undergraduate courses on topics of current and emerging fields within physics, both pure and applied. These textbooks serve students at the MS- or PhD-level and their instructors as comprehensive sources of principles, definitions, derivations, experiments and applications (as relevant) for their mastery and teaching, respectively. International in scope and relevance, the textbooks correspond to course syllabi sufficiently to serve as required reading. Their didactic style, comprehensiveness and coverage of fundamental material also make them suitable as introductions or references for scientists entering, or requiring timely knowledge of, a research field.

Andrey Grozin

Introduction to Mathematica® for Physicists

Andrey Grozin
Theory Division
Budker Institute of Nuclear Physics
Novosibirsk, Russia

ISSN 1868-4513 ISSN 1868-4521 (electronic)
ISBN 978-3-319-03284-9 ISBN 978-3-319-00894-3 (eBook)
DOI 10.1007/978-3-319-00894-3
Springer Cham Heidelberg New York Dordrecht London

Softcover re-print of the Hardcover 1st edition 2014

Printed on acid-free paper

Springer is part of Springer Science+Business Media (www.springer.com)

Preface

Computer algebra systems are widely used in pure and applied mathematics, physics, and other natural sciences, engineering, economics, as well as in higher and secondary education (see, e.g., [1–5]). For example, many important calculations in theoretical physics could never be done by hand, without wide use of computer algebra. Polynomial or trigonometric manipulations using paper and pen are becoming as obsolete as school long division in the era of calculators.

There are several powerful general-purpose computer algebra systems. The system *Mathematica* is most popular. It contains a huge amount of mathematical knowledge in its libraries. The fundamental book on this system [6] has more than 1,200 pages. Fortunately, the same information (more up-to-date than in a printed book) is available in the help system and hence is always at the fingertips of any user. Many books about *Mathematica* and its application in various areas have been published; see, for example, the series [7–10] of four books (each more than 1,000 pages long) or [11]. The present book does not try to replace these manuals. Its first part is a short systematic introduction to computer algebra and *Mathematica*; it can (and should) be read sequentially. The second part is a set of unrelated examples from physics and mathematics which can be studied selectively and in any order. Having understood the statement of a problem, try to solve it yourself. Have a look at the book to get a hint only when you get stuck. Explanations in this part are quite short.

This book[1] is a result of teaching at the physics department of Novosibirsk State University. Starting from 2004, the course "Symbolic and numeric computations in physics applications" is given to students preparing for M.Sc., and an introduction to *Mathematica* is the first part of this course (the second part is mainly devoted to Monte Carlo methods). Practical computer classes form a required (and most important) part of the course. Most students have no problems with mastering the basics of *Mathematica* and applying it to problems in their own areas of interest.

The book describes *Mathematica* 9. Most of the material is applicable to other versions too. The *Mathematica* Book (fifth edition) [6], as well as, e.g., the book

---

[1] Work partially supported by the Russian Ministry of Education and Science.

series [7–10], describes *Mathematica* 5. The main source of up-to-date information is the *Mathematica* Help system.

The whole book (except Lecture 1 and Problems for students) consists of *Mathematica* notebooks. They can be found at

```
http://www.inp.nsk.su/~grozin/mma/mma.zip
```

The zip file is password protected. The password is the last sentence of Lecture 7 (case-sensitive, including the trailing period). The reader is encouraged to experiment with these notebook files. In the printed version of the book, plots use different curve styles (dashed, dotted, etc.) instead of colors.

The book will be useful for students, Ph.D. students, and researchers in the area of physics (and other natural sciences) and mathematics.

Novosibirsk, Russia Andrey Grozin

Contents

Part I
Lectures

Catching a lion, the computer-algebra method: catch a cat, put it into the cage and lock it; then substitute a lion for the cat.

Chapter 1
Computer Algebra Systems

First attempts to use computers for calculations not only with numbers but also with mathematical expressions (e.g., symbolic differentiation) were made in the 1950s. In the 1960s research in this direction became rather intensive. This area was known under different names: symbolic calculations, analytic calculations, and computer algebra. Recently this last name is most widely used. Why algebra and not, say, calculus? The reason is that it is most useful to consider operations usually referred to calculus (such as differentiation) as algebraic operations in appropriate algebraic structures (differential fields).

First universal (i.e., not specialized for some particular application area) computer algebra systems appeared at the end of the 1960s. Not many such systems have been constructed; they are shown in the Table 1.1. Creating a universal computer algebra system is a huge amount of work, at the scale of hundreds of man-years. Some projects of this kind were not sufficiently developed and subsequently died; they are not shown in Table 1.1.

Table 1.1 Universal computer algebra systems

<table>
<tr><th>System</th><th>Year</th><th>Implementation language</th><th>Current name</th><th>Status</th></tr>
<tr><td>REDUCE</td><td>1968</td><td rowspan="4">Lisp</td><td></td><td>Free (BSD)</td></tr>
<tr><td>Macsyma</td><td>1969</td><td>Maxima</td><td>Free (GPL)</td></tr>
<tr><td>Scratchpad</td><td>1974</td><td>Axiom
OpenAxiom
FriCAS</td><td>Free (BSD)</td></tr>
<tr><td>muMATH</td><td>1979</td><td>Derive</td><td>Dead</td></tr>
<tr><td>Maple</td><td>1983</td><td rowspan="3">C, C++</td><td></td><td>Proprietary</td></tr>
<tr><td>Mathematica</td><td>1988</td><td></td><td>Proprietary</td></tr>
<tr><td>MuPAD</td><td>1992</td><td>MATLAB
symbolic toolbox</td><td>Proprietary</td></tr>
</table>

Theoretical physicist A. Hearn (known to specialists for the Drell–Hearn sum rule) has written a Lisp program REDUCE to automatize some actions in

A. Grozin, *Introduction to Mathematica® for Physicists*, Graduate Texts in Physics,
DOI 10.1007/978-3-319-00894-3_1,

calculating Feynman diagrams. It quickly grew into a universal system. At first, it was distributed free (it was sufficient to ask for Hearn's permission) and became widely used by physicists. Later it became commercial. At the end of 2008 it has become free, with a modified BSD license.

Macsyma was born in the MAC project at MIT (1969), the name means MAC SYmbolic MAnipulator. The project has nothing to do with Macintosh computers, which appeared much later. Its name had several official meanings (Multiple-Access Computer, Man And Computer, Machine Aided Cognition) and some unofficial ones (Man Against Computer, Moses And Company, Maniacs And Clowns, etc.). The work was done on a single PDP-6, later PDP-10 computer (about 1 MByte memory; there were no bytes back then, but 36-bit words). One of the first time-sharing operating systems, ITS, was written for this computer, and many users at once worked on it interactively. Later this computer became one of the first nodes of ARPANET, the ancestor if Internet, and users from other universities could use Macsyma.

The company Symbolics was spun off MIT. It produced Lisp machines—computers with a hardware support of Lisp, as well as software for these computers, including Macsyma—the largest Lisp program at that time. Later production of Lisp machines became unprofitable, because general-purpose workstations (Sun, etc.) became faster and cheaper. Symbolics went bankrupt; Macsyma business was continued by Macsyma Inc., who sold Macsyma for a number of platforms and operating systems. Its market share continued to shrink because of the success of Maple and *Mathematica*, and finally the company was sold in 1999 to Andrew Topping. The new owner stopped Macsyma development and marketing. Then he died, and the rights to the commercial Macsyma now belong to his inheritors. All efforts spent on improving this branch of Macsyma are irreversibly lost.

Fortunately, this was not the only branch. Macsyma development at MIT was largely funded by DOE, and MIT transferred this codebase to DOE who distributed it. This version was ported to several platforms. All these ports died except one. Professor William Schelter ported DOE Macsyma to Common Lisp, the new Lisp standard, and developed this version until he died in 2001. This version was called Maxima, to avoid trademark problems. In 1998 he obtained permission from DOE to release Maxima under GPL. He also developed GCL (GNU Common Lisp). Currently Maxima is an active free software project and works on many Common Lisp implementations.

Macsyma has played a huge role in the development of computer algebra systems. It was the first system in which modern algorithms for polynomials, integration in elementary functions, etc., were implemented (REDUCE and Macsyma influenced each other strongly and are rather similar to each other). Macsyma was designed as an interactive system. For example, if the form of an answer depends on the sign of a parameter, it will ask the user

Is a positive or negative?

Scratchpad was born in IBM research laboratories (1974). At first it did not differ from other systems (Macsyma, REDUCE) very much and borrowed chunks of code from them. It was radically redesigned in the version Scratchpad II (1985).

And this design, perhaps, still remains the most beautiful one from a mathematical point of view. It is a strongly typed system (the only one among universal computer algebra systems). Any object (formula) in it belongs to some domain (e.g., it is a single-variable polynomial with integer coefficients). Each domain belongs to some category (e.g., it is a ring, or a commutative group, or a totally ordered set). New domains can be constructed from existing ones. For example, a matrix of elements belonging to any ring can be constructed. It is sufficient to program a matrix multiplication algorithm once. This algorithm calls the operations of addition and multiplication of the elements. If matrices of rational numbers are being multiplied, then addition and multiplication of rational numbers are called; and if matrices of polynomials—then addition and multiplication of polynomials.

Scratchpad was never distributed to end users by IBM. At last, IBM decided to stop wasting money for nought (or for basic research) and sold Scratchpad II to the English company NAG (famous for its numerical libraries). It marketed this system under the name Axiom. However, the product did not bring enough profit and was withdrawn in 2001. Axiom development took about 300 man-years of work of researchers having highest qualification. All this could easily disappear without a trace. Fortunately, one of old-time Scratchpad II developers at IBM, Tim Daly, has succeeded in convincing NAG to release Axiom under the modified BSD license. Now it is a free software project and still the most beautiful system from mathematical point of view. But unfortunately, due to incompatible visions of the directions of the future development, two forks appeared—OpenAxiom and FriCAS. And it is not clear which one is better.

muMATH (Soft Warehouse, Hawaii, 1979) got to the list of universal computer algebra systems with some stretch. It was written for microprocessor systems with a very limited memory (later called personal computers); mu in its name, of course, means μ, i.e., micro. This system never implemented advanced modern algorithms. It used heuristic methods instead, as taught in university calculus courses: let's try this and that, and if you can't get it, you can't get it. But it was surprisingly powerful at its humble size. The system has been essentially rewritten in 1988 and got a menu interface, graphics, and the new name, Derive. Then Soft Warehouse was bought by Texas Instruments, who presented a calculator with a (Derive-based) computer algebra system in 1995. Derive was withdrawn from market in 2007.

All these systems can be referred to the first generation. They are all written in various dialects of Lisp. They were considered related to the area of artificial intelligence.

The first representative of the second generation is the Canadian system Maple. It has a small kernel written in C, which implements an interpreted procedural language convenient for writing computer algebra algorithms. The major part of its mathematical knowledge is contained in the library written in this language. Maple can work on many platforms. It quickly became popular. In 2009 Maplesoft (Waterloo Maple Inc.) has been acquired by the Japanese company Cybernet Systems Group; development of Maple is not affected. By the way, numerical program MathCAD used a cut-down version of Maple to provide some computer algebraic capabilities.

In the beginning of the 1980s, a young theoretical physicist Steven Wolfram, an active Macsyma user, together with a few colleagues, has written a system SMP (Symbolic Manipulation Program). The project was a failure (I still have a huge SMP manual sent to me by S. Wolfram). After that, he understood what mass users want—they want a program to look pretty. He, together with a few colleagues, has rewritten the system, paying a lot of attention to the GUI and graphics (the symbolic part was largely based on SMP). The result was *Mathematica*, version 1 (1988). And Wolfram got his first million in three months of selling it.

Mathematica heavily relies on substitutions. Even a procedure call is a substitution. Pattern matching and their replacing by right-hand sides of substitutions are highly advanced in *Mathematica*. Often a set of mathematical concepts can be easily and compactly implemented via substitutions. On the other hand, this can lead to inefficiency: pattern matching is expensive.

The latest arrival in the list of universal computer algebra systems is MuPAD (its name initially meant Multi-Processor Algebra Data tool, and indeed early versions contained experimental support of multiprocessor systems, which later disappeared). The system was designed and implemented by a research group at the University of Paderborn in Germany (this is one more meaning of PAD in the name) in 1992 and later was distributed commercially by the company SciFace. Initially, MuPAD was quite similar to Maple. Later it borrowed many ideas from Axiom (domains, categories; however, MuPAD is dynamically typed). During a long period, it was allowed to download and use MuPAD Light for free; it had no advanced GUI, but its symbolic functionality was not cut down. Funding of the University project was stopped in 2005; in 2008, SciFace was bought by Mathworks, the makers of MATLAB. After that, MuPAD is available only as a MATLAB addon.

It seems that *Mathematica* dominates the market of commercial computer algebra systems, with Maple being number two. *Mathematica* is highly respected for the huge amount of mathematical knowledge accumulated in its libraries. It is not bug-free (this is true for all systems). Often it requires more resources (memory, processor time) for solving a given problem than other systems. But it is very convenient and allows a user to do a lot in a single framework.

In addition to universal systems, there are a lot of specialized computer algebra systems. Here we'll briefly discuss just one example important for theoretical physics.

In the 1960s, a well-known Dutch theoretical physicist M. Veltman, a future Nobel prize winner, has written a system Schoonschip in the assembly language of CDC-6000 computers (in Dutch Schoonschip means "to clean a ship," in a figurative sense "to put something in order," "to throw unneeded things overboard"). This system was designed for handling very long sums (millions of terms) whose size can be much larger than the main memory and is limited only by the available disk space. All operations save one are local: they are substitutions which replace a single term by several new ones. The system gets a number of terms from the disk, applies the substitution to them, and puts the results back to the disk. The only unavoidable nonlocal operation is collecting similar terms; it is done with advanced disk sorting algorithms. Built-in mathematical knowledge of the system is very limited; the user has to program everything from scratch. Many nontrivial algorithms, such as polynomial factorization, are highly nonlocal and impossible to implement. On the other hand, this was the only system which could work with very large expressions, orders of magnitude larger than in other systems. Later Schoonschip was ported to IBM-360 (in PL/I; you can guess that this was not done by Veltman :–). Then Veltman has rewritten it from the CDC assembly language to the 680x0 assembly language. When 680x0-based personal computers (Amiga, Atari) became extinct, it became clear that something similar but more portable is needed.

In 1989 another well-known Dutch theoretical physicist, Vermaseren, has written (in C) a new system, Form. It follows the same ideology, but many details differ. It was distributed free of charge as binaries for a number of platforms; recently it became free software (GPL). Development of Form continues. A parallel version for multiprocessor computers and for clusters with fast connections now exists. Many important Feynman diagram calculations could never have been done without Schoonschip and later Form.

The percentage of theoretical physicists among authors of computer algebra systems is suspiciously high. Some of them remained physicists (and even got a Nobel prize); some completely switched to development of their systems (and even became millionaires).

In conclusion we'll discuss a couple of important computer algebra concepts. For some (sufficiently simple) classes of expressions an algorithm of reduction to a *canonical form* can be constructed. Two equal expressions reduce to the same canonical form. In particular, any expression equal to 0, in whatever form it is written, has the canonical form 0.

For example, it is easy to define a canonical form for polynomials of several variables with integer (or rational) coefficients: one has to expand all brackets and collect similar terms. What's left is to agree upon an unambiguous order of terms, and we have a canonical form (this can be done in more than one way).

It is more difficult, but possible, to define a canonical form for rational expressions (ratios of polynomials). One has to expand all brackets and to bring the whole expression to a common denominator (collecting similar terms, of course). However, this is not sufficient: one can multiply both the numerator and the denominator by the same polynomial and obtain another form of the rational expression. It is necessary to cancel the greatest common divisor (gcd) of the numerator and the denominator. Calculating polynomial gcd's is an algorithmic operation, but it can be computationally expensive. What's left is to fix some minor details—an unambiguous order of terms in both the numerator and the denominator and, say, the requirement that the coefficient of the first term in the denominator is 1, and we obtain a canonical form.

A *normal form* for a class of expressions satisfies a weaker requirement: any expression equal to 0 must reduce to the normal form 0. For example, bringing to common denominator (without canceling gcd) defines a normal form for rational expressions.

For more general classes of expressions containing elementary functions, not only canonical but even normal form does not exist. Richardson has proved that it is algorithmically undecidable if such an expression is equal to 0.

Chapter 2
Overview of *Mathematica*

2.1 Symbols

Let's assign an expression containing a symbol x to a variable a. No value is assigned to x.

In[1] := a = x^2 − 1

Out[1] $= -1 + x^2$

Now let's assign some value to x and see what happens to a.

In[2] := x = z + 1; a

Out[2] $= -1 + (1 + z)^2$

The value of a has not really changed. It is still the same expression containing x. What if we assign another value to x?

In[3] := x = z − 1; a

Out[3] $= -1 + (-1 + z)^2$

We can delete the value of the variable x, thus returning it to its initial state in which it means just the symbol x. We see that indeed the value of a has not changed.

In[4] := Clear[x]; a

Out[4] $= -1 + x^2$

Now let's try to assign an expression containing x to the variable x.

In[5] := $RecursionLimit = 32; x = x + 1

$RecursionLimit :: reclim : Recursion depth of 32 exceeded.

Out[5] = 1 + (1 + (1 + (1 + (1 + (1 + (1 + (1 + (1 + (1 + (1 + (1 + (1 + (1 + (1+ (1+(1+(1+(1+(1+(1+(1+(1+(1+(1+(1+(1+(1+(1+(1+(1+Hold[1+ x]))))))))))))))))))))))))))))))))

Mathematica complains. What has happened? *Mathematica* wants to print x, and to this end it calculates the value of x. It sees x in this value and substitutes the value of x. In this value, it again sees x and substitutes its value. And so on ad infinitum. In reality, the depth of such substitutions is limited; the default value of the limit is 1,024 (we have temporarily changed it to 32). The value of the expression $1 + x$ which failed to evaluate is returned as the function Hold; we shall discuss it in Sect. 6.9.

In[6] := $RecursionLimit = 1024; Clear[x]

A. Grozin, *Introduction to Mathematica® for Physicists*, Graduate Texts in Physics, DOI 10.1007/978-3-319-00894-3_2, © Springer International Publishing Switzerland 2014

2.2 Numbers

Mathematica can work with arbitrarily long integer numbers.

In[7] := Factorial[100]

Out[7] = 9332621544394415268169923885626670049071596826438162146859\
92963895217599993229915608941463976156518286253697920827223758\
825118521091686400000000000000000000000000

When working with a rational number, the greatest common divisors of its numerator and denominator are canceled.

In[8] := *a* = 1234567890/987654321

Out[8] $= \dfrac{137174210}{109739369}$

Calculations with rational numbers are exact.

In[9] := *a*^5

Out[9] = 4856935528628288552276518549160311010000/
159152070653457846182379862366702459078 49

How much is this numerically? Say, with 30 significant digits?

In[10] := *N*[*a*, 30]

Out[10] = 1.24999998860937500014238281250

Mathematica can work with real (floating-point) numbers having arbitrarily high precision.

In[11] := *a* = 12345678909876543 21.12345678909876543 21

Out[11] = $1.23456789098765432112345678909876543 2 \times 10^{18}$

In[12] := *a*^5

Out[12] = $2.86797187177160567275921531725363508 \times 10^{90}$

Here are π and e with 100 significant digits.

In[13] := *N*[Pi, 100]

Out[13] = 3.141592653589793238462643383279502884197169399375105820974\
944592307816406286208998628034825342117068

In[14] := *N*[*E*, 100]

Out[14] = 2.718281828459045235360287471352662497757247093699959574966\
967627724076630353547594571382178525166427

2.3 Polynomials and Rational Functions

Let's take a polynomial.

In[15] := *a* = (*x* + *y* + *z*)^6

Out[15] = $(x+y+z)^6$

Expand it.

In[16] := a = Expand[a]
Out[16] = $x^6 + 6x^5y + 15x^4y^2 + 20x^3y^3 + 15x^2y^4 + 6xy^5 + y^6 + 6x^5z + 30x^4yz + 60x^3y^2z + 60x^2y^3z + 30xy^4z + 6y^5z + 15x^4z^2 + 60x^3yz^2 + 90x^2y^2z^2 + 60xy^3z^2 + 15y^4z^2 + 20x^3z^3 + 60x^2yz^3 + 60xy^2z^3 + 20y^3z^3 + 15x^2z^4 + 30xyz^4 + 15y^2z^4 + 6xz^5 + 6yz^5 + z^6$

The degree in x.

In[17] := Exponent[a,x]
Out[17] = 6

The coefficient of x^2.

In[18] := Coefficient[$a,x,2$]
Out[18] = $15y^4 + 60y^3z + 90y^2z^2 + 60yz^3 + 15z^4$

Collect terms with the same power of x together.

In[19] := Collect[a,x]
Out[19] = $x^6 + y^6 + 6y^5z + 15y^4z^2 + 20y^3z^3 + 15y^2z^4 + 6yz^5 + z^6 + x^5(6y + 6z) + x^4\left(15y^2 + 30yz + 15z^2\right) + x^3\left(20y^3 + 60y^2z + 60yz^2 + 20z^3\right) + x^2\left(15y^4 + 60y^3z + 90y^2z^2 + 60yz^3 + 15z^4\right) + x\left(6y^5 + 30y^4z + 60y^3z^2 + 60y^2z^3 + 30yz^4 + 6z^5\right)$

Factorize it.

In[20] := a = Factor[a]
Out[20] = $(x + y + z)^6$

Suppose we want to factorize polynomials $x^n - 1$ with various n. The parameter n can be varied from 2 to 10 by dragging the marker with the mouse.

In[21] := Manipulate[Factor[x^n − 1], {n, 2, 10, 1, Appearance−>"Labeled"}]

Out[21] = n [slider] 6

$(-1+x)(1+x)(1-x+x^2)(1+x+x^2)$

There exists an algorithm which completely factorizes any polynomial with integer coefficients into factors which also have integer coefficients.

In[22] := Factor[x^4 − 1]
Out[22] = $(-1 + x)(1 + x)\left(1 + x^2\right)$

If we want to get factors whose coefficients come from an extension of the ring of integers, say, by the imaginary unit i, we should say so explicitly.

In[23] := Factor[x^4 − 1, Extension−>I]
Out[23] = $(-1 + x)(-\mathrm{i} + x)(\mathrm{i} + x)(1 + x)$

This polynomial factorizes into two factors with integer coefficients.

In[24] := a = x^4 − 4; Factor[a]
Out[24] = $\left(-2 + x^2\right)\left(2 + x^2\right)$

If coefficients from the extension of the ring of integers by $\sqrt{2}$ are allowed—into three factors.

In[25] := Factor[a, Extension−>Sqrt[2]]
Out[25] = $-\left(\sqrt{2} - x\right)\left(\sqrt{2} + x\right)\left(2 + x^2\right)$

And if the ring of coefficients is extended by both $\sqrt{2}$ and i—into four factors.

In[26] := Factor[*a*, Extension−>{Sqrt[2], *I*}]

Out[26] $= -\left(\sqrt{2}-x\right)\left(\sqrt{2}-\mathrm{i}x\right)\left(\sqrt{2}+\mathrm{i}x\right)\left(\sqrt{2}+x\right)$

And this is a rational function.

In[27] := (*x*^3 − *y*^3)/(*x*^2 − *y*^2)

Out[27] $= \dfrac{x^3-y^3}{x^2-y^2}$

It is not canceled by the greatest common divisor of its numerator and denominator; this should be done explicitly.

In[28] := Cancel[%]

Out[28] $= \dfrac{x^2+xy+y^2}{x+y}$

(% means the result of the previous calculation). A sum of rational functions.

In[29] := *a* = *x*/(*x*+*y*)+*y*/(*x*−*y*)

Out[29] $= \dfrac{y}{x-y}+\dfrac{x}{x+y}$

Let's put it over the common denominator.

In[30] := *a* = Together[*a*]

Out[30] $= \dfrac{x^2+y^2}{(x-y)(x+y)}$

Partial fraction decomposition with respect to x.

In[31] := Apart[*a*, *x*]

Out[31] $= 1+\dfrac{y}{x-y}-\dfrac{y}{x+y}$

In[32] := Clear[*a*]

2.4 Elementary Functions

Mathematica knows some simple properties of elementary functions.

In[33] := Sin[−*x*]

Out[33] = −Sin[*x*]

In[34] := Cos[Pi/4]

Out[34] $= \dfrac{1}{\sqrt{2}}$

In[35] := Sin[5 ∗ Pi/6]

Out[35] $= \dfrac{1}{2}$

In[36] := Log[1]

Out[36] = 0

In[37] := Log[*E*]

Out[37] = 1

In[38] := Exp[Log[*x*]]

Out[38] = *x*

In[39] := Log[Exp[*x*]]

Out[39] = $\mathrm{Log}[e^x]$
And why not x? Because this simplification is not always correct. Try to substitute $2\pi i$.
In[40] := Sqrt[0]
Out[40] = 0
In[41] := Sqrt[x]^2
Out[41] = x
In[42] := Sqrt[x^2]
Out[42] = $\sqrt{x^2}$
And why not x? Try to substitute -1.
In[43] := a = Sqrt[12 ∗ x^2 ∗ y]
Out[43] = $2\sqrt{3}\sqrt{x^2 y}$
This result can be improved, if we know that $x > 0$.
In[44] := Simplify[a, x > 0]
Out[44] = $2\sqrt{3}x\sqrt{y}$
And this is the case $x < 0$.
In[45] := Simplify[a, x < 0]
Out[45] = $-2\sqrt{3}x\sqrt{y}$
Expansion of trigonometric functions of multiple angles, sums, and differences:
In[46] := TrigExpand[Cos[2 ∗ x]]
Out[46] = $\mathrm{Cos}[x]^2 - \mathrm{Sin}[x]^2$
In[47] := TrigExpand[Sin[x − y]]
Out[47] = $\mathrm{Cos}[y]\,\mathrm{Sin}[x] - \mathrm{Cos}[x]\,\mathrm{Sin}[y]$
The inverse operation—transformation of products and powers of trigonometric functions into linear combinations of such functions—is used more often. Let's take a truncated Fourier series.
In[48] := a = a1 ∗ Cos[x] + a2 ∗ Cos[2 ∗ x] + b1 ∗ Sin[x] + b2 ∗ Sin[2 ∗ x]
Out[48] = $\mathrm{a1}\,\mathrm{Cos}[x] + \mathrm{a2}\,\mathrm{Cos}[2x] + \mathrm{b1}\,\mathrm{Sin}[x] + \mathrm{b2}\,\mathrm{Sin}[2x]$
Its square is again a truncated Fourier series.
In[49] := TrigReduce[a^2]
Out[49] = $\frac{1}{2}(\mathrm{a1}^2 + \mathrm{a2}^2 + \mathrm{b1}^2 + \mathrm{b2}^2 + 2\,\mathrm{a1}\,\mathrm{a2}\,\mathrm{Cos}[x] + 2\,\mathrm{b1}\,\mathrm{b2}\,\mathrm{Cos}[x] + \mathrm{a1}^2\,\mathrm{Cos}[2x] -$
$\mathrm{b1}^2\,\mathrm{Cos}[2x] + 2\,\mathrm{a1}\,\mathrm{a2}\,\mathrm{Cos}[3x] - 2\,\mathrm{b1}\,\mathrm{b2}\,\mathrm{Cos}[3x] + \mathrm{a2}^2\,\mathrm{Cos}[4x] - \mathrm{b2}^2\,\mathrm{Cos}[4x] -$
$2\,\mathrm{a2}\,\mathrm{b1}\,\mathrm{Sin}[x] + 2\,\mathrm{a1}\,\mathrm{b2}\,\mathrm{Sin}[x] + 2\,\mathrm{a1}\,\mathrm{b1}\,\mathrm{Sin}[2x] + 2\,\mathrm{a2}\,\mathrm{b1}\,\mathrm{Sin}[3x] +$
$2\,\mathrm{a1}\,\mathrm{b2}\,\mathrm{Sin}[3x] + 2\,\mathrm{a2}\,\mathrm{b2}\,\mathrm{Sin}[4x])$

2.5 Calculus

Let's take a function.
In[50] := f = Log[x^5 + x + 1] + 1/(x^5 + x + 1)
Out[50] = $\frac{1}{1+x+x^5} + \mathrm{Log}\left[1+x+x^5\right]$
Calculate its derivative.

In[51] := g = D[f,x]

$$\text{Out[51]} = -\frac{1+5x^4}{(1+x+x^5)^2}+\frac{1+5x^4}{1+x+x^5}$$

Put over the common denominator.

In[52] := g = Together[g]

$$\text{Out[52]} = \frac{(1+5x^4)(x+x^5)}{(1+x+x^5)^2}$$

A stupid integration algorithm would try to solve the fifth degree equation in the denominator, in order to decompose the integrand into partial fractions. *Mathematica* is more clever than that.

In[53] := Integrate[g,x]

$$\text{Out[53]} = \frac{1}{1+x+x^5}+\text{Log}\left[1+x+x^5\right]$$

Let's expand our function in x at 0 up to x^{10}.

In[54] := Series[f, {x, 0, 10}]

$$\text{Out[54]} = 1+\frac{x^2}{2}-\frac{2x^3}{3}+\frac{3x^4}{4}-\frac{4x^5}{5}+\frac{11x^6}{6}-\frac{20x^7}{7}+\frac{31x^8}{8}-\frac{44x^9}{9}+\frac{32x^{10}}{5}+O[x]^{11}$$

Mathematica can calculate many definite integrals even when the corresponding indefinite integral cannot be taken. Here is an integral from 0 to 1.

In[55] := Integrate[Log[x]^2/(x+1), {x, 0, 1}]

$$\text{Out[55]} = \frac{3\,\text{Zeta}[3]}{2}$$

Mathematica knows how to sum many series.

In[56] := Sum[1/n^4, {n, 1, Infinity}]

$$\text{Out[56]} = \frac{\pi^4}{90}$$

Let's clear all the garbage we have generated—a very good habit.

In[57] := Clear[f, g]

2.6 Lists

We have already encountered this construct several times:

In[58] := a = {x, y, z}

Out[58] = $\{x, y, z\}$

This is a list. And here are its elements.

In[59] := a[[1]]

Out[59] = x

In[60] := a[[2]]

Out[60] = y

In[61] := a[[3]]

Out[61] = z

In[62] := Clear[a]

2.7 Plots

A simple plot of a function.
In[63] := Plot[Sin[x]/x, {x, −10, 10}]

Out[63] =

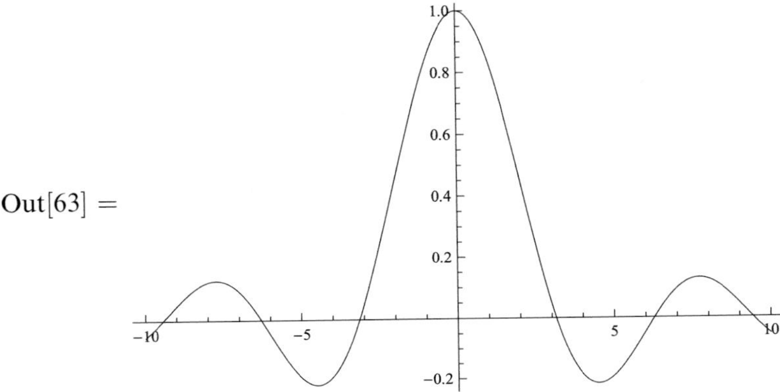

A curve given parametrically—x and y are functions of t. This particular curve contains a parameter a, which can be adjusted by the mouse. If you click the small plus sign near the marker, a control panel will open. There you can start (and stop) animation.
In[64] := Manipulate[ParametricPlot[{Exp[a ∗ t] ∗ Cos[t], Exp[a ∗ t] ∗ Sin[t]}, {t,0,20}, PlotRange−>{{−10, 10}, {−10, 10}}], {{a,0.1},0,0.2}]

Out[64] =

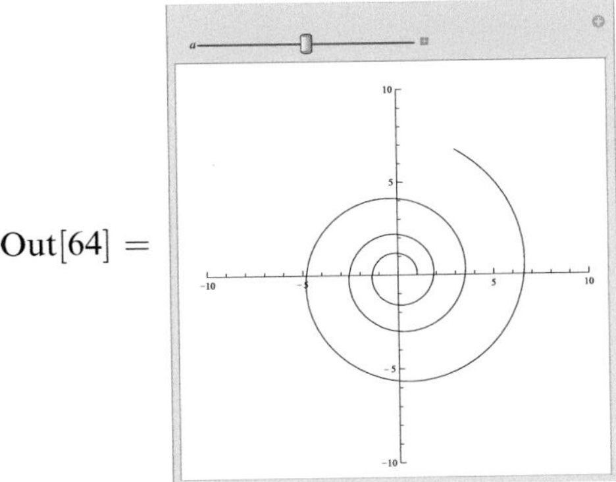

A three-dimensional plot of a function of two variables. It can be rotated by the mouse.

In[65] := Plot3D[x^2 + y^2, {x, −1, 1}, {y, −1, 1}]

Out[65] =

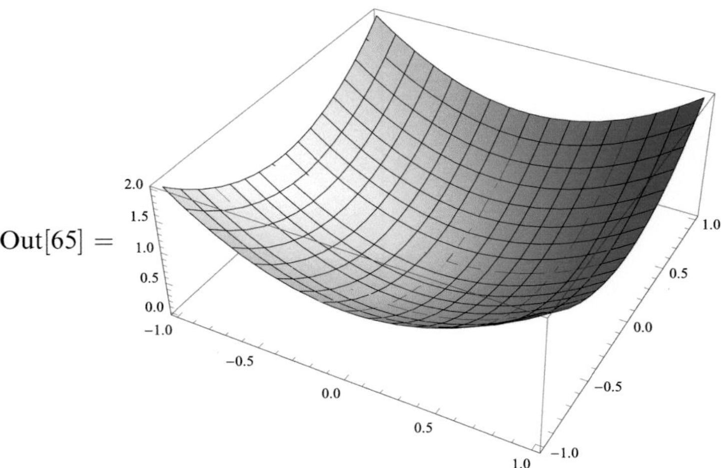

A three-dimensional curve given parametrically. The parameter a can be adjusted by the mouse.

**In[66] := Manipulate[ParametricPlot3D[{Cos[t], Sin[t], a ∗ t}, {t, 0, 20},
PlotRange−>{{−1, 1}, {−1, 1}, {0, 2}}], {{a, 0.1}, 0, 0.2}]**

Out[66] =

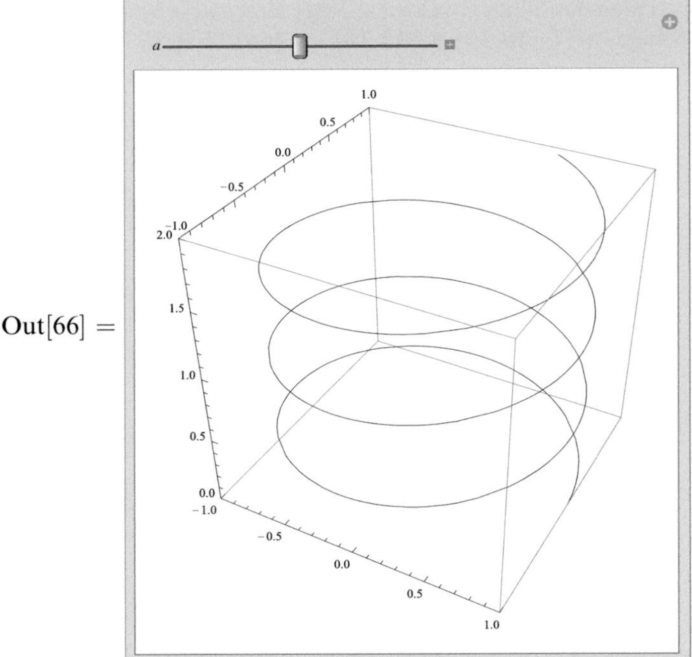

A surface given parametrically.

In[67] := ParametricPlot3D[{Sin[*t*] ∗ Cos[*u*], Sin[*t*] ∗ Sin[*u*], Cos[*t*]}, {*t*, 0, Pi}, {*u*, 0, 2 ∗ Pi}]

Out[67] =

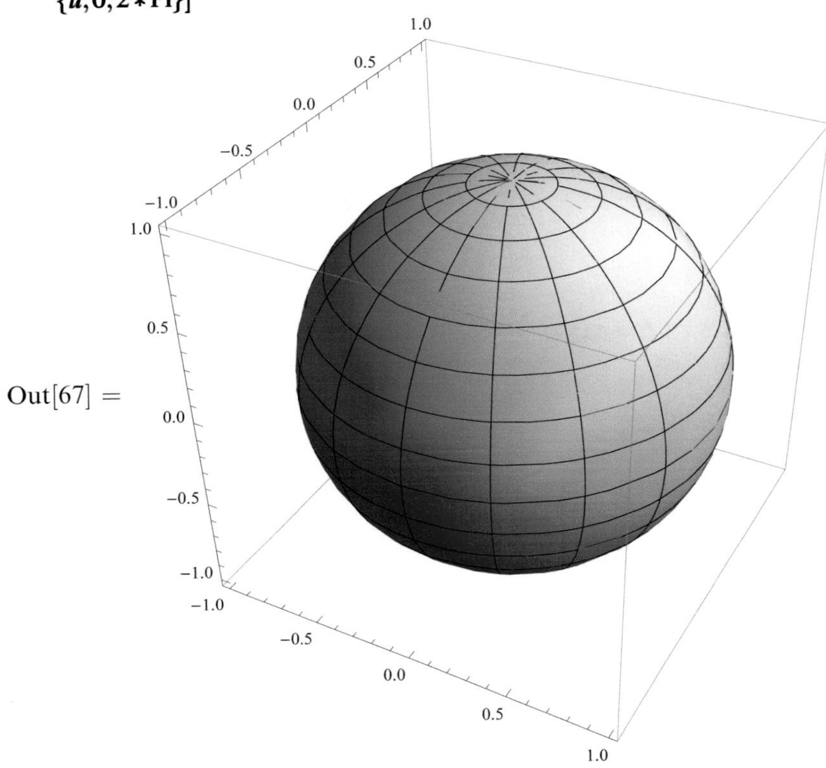

2.8 Substitutions

Substitutions are a fundamental concept in *Mathematica*, its main working instrument. This substitution replaces $f[x]$ by x^2.

In[68] := *S* = *f*[*x*]−>*x*^2

Out[68] = $f[x] \to x^2$

Let's apply it to the expression $f[x]$.

In[69] := *f*[*x*]/.*S*

Out[69] = x^2

We've got x^2, as expected. And what if we apply it to $f[y]$?

In[70] := *f*[*y*]/.*S*

Out[70] = $f[y]$

It hasn't triggered. The following substitution replaces the function f with an arbitrary argument by the square of this argument.

In[71] := $S = f[x\_]{-}{>}x$^2
Out[71] $= f[x\_] \to x^2$
Let's check.
In[72] := {f[x],f[y],f[2]}/.S
Out[72] $= \{x^2, y^2, 4\}$
In[73] := Clear[S]

2.9 Equations

Here is an equation.
In[74] := Eq = a∗x+b==0
Out[74] $= b + ax == 0$
Let's solve it for x.
In[75] := S = Solve[Eq,x]
Out[75] $= \left\{\left\{x \to -\frac{b}{a}\right\}\right\}$
We've got a list of solutions, in this particular case having a single element. Each solution is a list of substitutions, which replaces our unknowns by the corresponding expressions. And how can we extract the value of x from this result? Let's take the first (and the only) element of the list S.
In[76] := S1 = First[S]
Out[76] $= \left\{x \to -\frac{b}{a}\right\}$
And now we apply this list of substitutions (in this particular case, it's single element) to the unknown x.
In[77] := x/.S1
Out[77] $= -\frac{b}{a}$
Here is a more advanced example—a quadratic equation. It has two solutions.
In[78] := S = Solve[a∗x^2+b∗x+c==0,x]
Out[78] $= \left\{\left\{x \to \frac{-b-\sqrt{b^2-4ac}}{2a}\right\}, \left\{x \to \frac{-b+\sqrt{b^2-4ac}}{2a}\right\}\right\}$
How can we extract the value of x in the second solution? Let's apply the second element of the solutions list S (which is a single-element list of substitutions) to the unknown x.
In[79] := x/.S[[2]]
Out[79] $= \frac{-b+\sqrt{b^2-4ac}}{2a}$
And here is a system of 2 linear equations.
In[80] := Eq = {a∗x+b∗y==e,c∗x+d∗y==f}
Out[80] $= \{ax + by == e, cx + dy == f\}$
It has a single solution.

In[81] := S = Solve[Eq, {x,y}]

Out[81] = $\left\{\left\{x \to -\frac{de-bf}{bc-ad}, y \to -\frac{-ce+af}{bc-ad}\right\}\right\}$

This (first and the only) solution is a list of two substitutions.

In[82] := S1 = S[[1]]

Out[82] = $\left\{x \to -\frac{de-bf}{bc-ad}, y \to -\frac{-ce+af}{bc-ad}\right\}$

How to find the values of x and y in this solution? Apply this list of substitutions to the unknowns x and y.

In[83] := {x/.S1,y/.S1}

Out[83] = $\left\{-\frac{de-bf}{bc-ad}, -\frac{-ce+af}{bc-ad}\right\}$

In[84] := Clear[Eq,S,S1]

Chapter 3
Expressions

All objects with which *Mathematica* works are expressions. There are two classes of them—atoms and composite expressions.

3.1 Atoms

There are three kinds of atoms—numbers, symbols, and strings.

Numbers

Integer numbers (of unlimited size).
In[1] := 1234567890
Out[1] = 1234567890
A rational number consists of the numerator and the denominator.
In[2] := 1234567890/987654321
Out[2] = $\dfrac{137174210}{109739369}$
A complex number consists of the real and imaginary parts.
In[3] := 1 + 2 * *I*
Out[3] = 1 + 2i
Real numbers can have arbitrarily high precision.
In[4] := 1234567890.987654321
Out[4] = $1.23456789098765432 \times 10^{9}$

A. Grozin, *Introduction to Mathematica® for Physicists*, Graduate Texts in Physics, DOI 10.1007/978-3-319-00894-3_3, © Springer International Publishing Switzerland 2014

Symbols

A variable can be in one of two states. Initially it is free—it means itself (a symbol).
In[5] := *x*
Out[5] = x
Assigning a value to it, we make it bound.
In[6] := *x* = 123
Out[6] = 123
Now, when we use it (e.g., just by asking *Mathematica* to print it), its value is substituted.
In[7] := *x*
Out[7] = 123
How to make it free again?
In[8] := Clear[*x*]
Let's check.
In[9] := *x*
Out[9] = x

Strings

In[10] := "This is a string"
Out[10] = This is a string

3.2 Composite Expressions

A composite expression is a function of a number of arguments, each of which is an expression (i.e., an atom or a composite expression).
In[11] := *a* = *f*[*g*[*x*, 1], *h*[*y*, *z*, 2]]
Out[11] = $f[g[x,1],h[y,z,2]]$
Each composite expression has a head—the function which is applied to arguments.
In[12] := Head[*a*]
Out[12] = f
The number of arguments is given by the function Length.
In[13] := Length[*a*]
Out[13] = 2
Arguments are extracted by the function Part.
In[14] := Part[*a*, 1]
Out[14] = $g[x,1]$
In[15] := Part[*a*, 2]
Out[15] = $h[y,z,2]$
And this is the first part of the second part of the expression a.

In[16] := Part[*a*,2,1]
Out[16] = *y*
An alternative syntax.
In[17] := *a*[[2,1]]
Out[17] = *y*
Zeroth part of an expression is its head.
In[18] := Part[*a*,0]
Out[18] = *f*
By the way, a head can be any expression, not just a symbol.
In[19] := *b* = *f*[*x*][*y*, 1]
Out[19] = *f*[*x*][*y*, 1]
In[20] := Head[*b*]
Out[20] = *f*[*x*]
Expressions are trees whose leaves are atoms.
In[21] := TreeForm[*a*]

Out[21]//TreeForm =
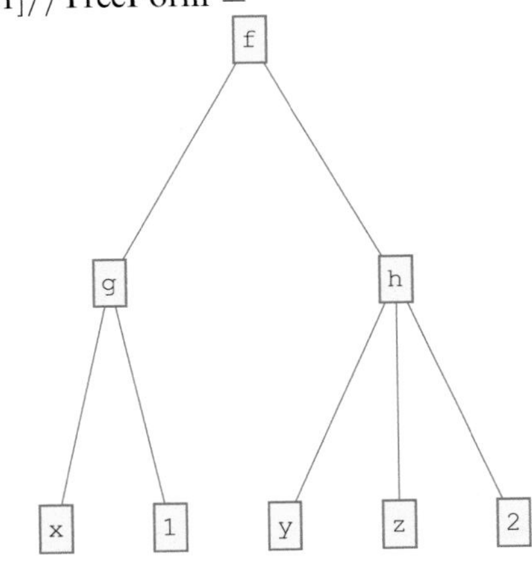

Parts of an expressions can be changed.
In[22] := *a*[[1,2]] = 0; *a*
Out[22] = *f*[*g*[*x*,0],*h*[*y*,*z*,2]]
In[23] := *a*[[0]] = *j*; *a*
Out[23] = *j*[*g*[*x*,0],*h*[*y*,*z*,2]]
A group of arguments can be selected, not just a single argument.
In[24] := *b* = *f*[x1,x2,x3,x4,x5,x6]
Out[24] = *f*[x1,x2,x3,x4,x5,x6]
In[25] := Part[*b*,Span[2,4]]
Out[25] = *f*[x2,x3,x4]
An alternative syntax.
In[26] := *b*[[2;;4]]
Out[26] = *f*[x2,x3,x4]

From the beginning to 3:
In[27] := *b*[[;;3]]
Out[27] = *f*[x1, x2, x3]
From 4 to the end:
In[28] := *b*[[4;;]]
Out[28] = *f*[x4, x5, x6]
From 1 to 5 by 2:
In[29] := *b*[[1;;5;;2]]
Out[29] = *f*[x1, x3, x5]
If such a form is used in the left-hand side of an assignment, each of the selected arguments will be replaced:
In[30] := *b*[[1;;5;;2]] = *x*; *b*
Out[30] = *f*[*x*, x2, *x*, x4, *x*, x6]
In[31] := Clear[*a*, *b*]

3.3 Queries

Let's define an integer number, a rational number, a real (floating point) number, and a complex number.
In[32] := *i* = −1234567890; *r* = −1234567890/987654321;
**   *f* = −12345678909876543 21.1234567890987654321; *c* = 1 − 2 ∗ *I*;**
The query AtomQ (Q from Query) returns the symbol True if its argument is an atom and False if it is a composite expression.
In[33] := {AtomQ[*i*], AtomQ[*r*], AtomQ[*c*], AtomQ[*f*[*x*]]}
Out[33] = {True, True, True, False}
The function Head can be applied even to atoms.
In[34] := {Head[*i*], Head[*r*], Head[*c*], Head[*f*]}
Out[34] = {Integer, Rational, Complex, Real}
The function FullForm shows the internal form of an expression with which *Mathematica* operates (to some approximation). For example, a rational number has the head Rational and two arguments—its numerator and denominator.
In[35] := FullForm[*r*]
Out[35]//FullForm =
    Rational[−137174210, 109739369]
A complex number has the head Complex and two arguments—its real and imaginary parts.
In[36] := FullForm[*c*]
Out[36]//FullForm =
    Complex[1, −2]
The internal representation of a floating point number is rather complicated. It contains the mantissa and the exponent and also the number of significant (decimal) digits. In this particular case, there are 37 significant digits.

In[37] := FullForm[*f*]
Out[37]//FullForm =
  − 1.23456789098765432112345678909876543213\`37.09151497751671 *^ 18

The query IntegerQ checks if its argument is an integer number.
In[38] := {IntegerQ[*i*], IntegerQ[*r*], IntegerQ[*c*]}
Out[38] = {True, False, False}
The functions Numerator and Denominator extract the parts of a rational number.
In[39] := {Numerator[*r*], Denominator[*r*]}
Out[39] = {−137174210, 109739369}
The functions Re and Im extract the real and imaginary parts of a complex number.
In[40] := {Re[*c*], Im[*c*]}
Out[40] = {1, −2}
In[41] := Clear[*i*, *r*, *f*, *c*]

3.4 Forms of an Expression

FullForm is a very useful function. It shows what *Mathematica* really thinks about an expression. Use it often, and you will learn a lot. For example, the following expression is a sum of 4 terms, one of which is the number −1 multiplied by the symbol z.
In[42] := FullForm[*x* + *y* − *z* − 1]
Out[42]//FullForm =
  Plus[−1, *x*, *y*, Times[−1, *z*]]
And this one is a product of 4 factors, among which are the rational number 2/3 and the negative power z^{-1}.
In[43] := *a* = 2 * *x* * *y*/(3 * *z*)
Out[43] = $\frac{2xy}{3z}$
In[44] := FullForm[*a*]
Out[44]//FullForm =
  Times[Rational[2, 3], *x*, *y*, Power[*z*, −1]]
Nevertheless, the functions Numerator and Denominator work as expected.
In[45] := {Numerator[*a*], Denominator[*a*]}
Out[45] = {2*xy*, 3*z*}
In[46] := Clear[*a*]
We have already handled lists many times. A list appears to be just the function List with arguments—elements of the list.
In[47] := FullForm[{*x*, *y*, *z*}]
Out[47]//FullForm =
  List[*x*, *y*, *z*]

Any *Mathematica* command can be written as a function with arguments (sometimes, it can also be written in some other way). For example, assignment is the function Set. In order to see this, we'll have to put an assignment inside the function Hold. Otherwise it would be executed immediately, and the function FullForm would receive only the result returned by the assignment—the symbol x.

In[48] := FullForm[Hold[$a = x$]]

Out[48]//FullForm =
Hold[Set[a,x]]

Here is a rational expression.

In[49] := a = Together[$x/(x+y)+y/(x-y)$]

$$\text{Out[49]} = \frac{x^2+y^2}{(x-y)(x+y)}$$

Its full form:

In[50] := FullForm[a]

Out[50]//FullForm =
Times[Power[Plus[x, Times[-1, y]], -1], Power[Plus[x, y], -1],
Plus[Power[x, 2], Power[y, 2]]]

And this is the same expression as a tree.

In[51] := TreeForm[a]

Out[51]//TreeForm =

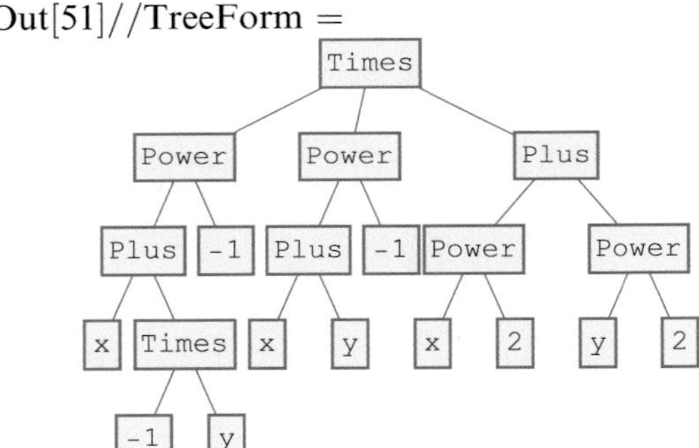

In[52] := Clear[a]

Chapter 4
Patterns and Substitutions

Substitution is the most fundamental operation in *Mathematica*. Its left-hand side is a pattern. In a given expression, all subexpressions matching the pattern are found and replaced by the right-hand side of the substitution.

4.1 Simple Patterns

$f[x]$ with a specific argument x.

In[1] := {f[x],f[y]}/.f[x]->x^2

Out[1] = $\{x^2, f[y]\}$

f with an arbitrary argument.

In[2] := {f[x],f[y]}/.f[x_]->x^2

Out[2] = $\{x^2, y^2\}$

f with two identical (arbitrary) arguments.

In[3] := {f[x,x],f[x,y]}/.f[x_,x_]->g[x]

Out[3] = $\{g[x], f[x,y]\}$

An example of a more complicated pattern.

In[4] := f[g[f[x],y],h[f[x]]]/.f[g[x_,y],h[x_]]->F[x,y]

Out[4] = $F[f[x],y]$

f with an argument being an arbitrary integer number.

In[5] := {f[x],f[2]}/.f[x_Integer]->x^2

Out[5] = $\{f[x],4\}$

In fact, such a form of an arbitrary argument checks its head. This substitution applies when the argument's head is g.

In[6] := {f[g[x,y]],f[h[x,y]]}/.f[x_g]->x^2

Out[6] = $\{g[x,y]^2, f[h[x,y]]\}$

And this one—when the argument is a sum.

In[7] := {f[{x,y}],f[x+y]}/.f[x_Plus]->x^2

Out[7] = $\{f[\{x,y\}], (x+y)^2\}$

A. Grozin, *Introduction to Mathematica® for Physicists*, Graduate Texts in Physics, DOI 10.1007/978-3-319-00894-3_4, © Springer International Publishing Switzerland 2014

And this one—when the argument is a list. By the way, note what happens when a list is being squared.

In[8] := {f[{x,y}],f[x+y]}/.f[x_List]−>x^2

$\mathrm{Out}[8] = \{\{x^2, y^2\}, f[x+y]\}$

One more example.

In[9] := a = Sqrt[x]/Sqrt[y]

$\mathrm{Out}[9] = \dfrac{\sqrt{x}}{\sqrt{y}}$

In[10] := a/.{Sqrt[x]−>u,Sqrt[y]−>v}

$\mathrm{Out}[10] = \dfrac{u}{\sqrt{y}}$

Why hasn't the second substitution triggered?

In[11] := FullForm[a]

Out[11]//FullForm =

Times[Power[x,Rational[1,2]],Power[y,Rational[−1,2]]]

a does not contain $y^{1/2}$, only $y^{-1/2}$; therefore, the substitution $y^{1/2} \to v$ does not work.

Out[11] = **Clear**[a]

4.2 One-Shot and Repeated Substitutions

Here is an expression.

In[12] := a = x^2 + y^2

$\mathrm{Out}[12] = x^2 + y^2$

Let's increase x by 1 in it. This example demonstrates that a substitution is not applied repeatedly. *Mathematica* searches for subexpressions matching the pattern (in this case x) in the expression a. After finding such a subexpression, *Mathematica* replaces it by the right-hand side. The result of such a replacement is not searched again for subexpressions matching the pattern.

In[13] := a = a/.x−>x+1

$\mathrm{Out}[13] = (1+x)^2 + y^2$

A list of substitutions can be applied to an expression. They are all applied in parallel—*Mathematica* searches for subexpressions matching some pattern from the list and replaces these subexpressions by the corresponding right-hand side. Therefore two symbols can be interchanged in an expression in this simple way.

In[14] := a = a/.{x−>y,y−>x}

$\mathrm{Out}[14] = x^2 + (1+y)^2$

In[15] := Clear[a]

The operator //. (in contrast to /.) applies a substitution repeatedly, while it is applicable. If several substitutions are applicable to some subexpression, *Mathematica* first applies the most specific one (it is not always easy to determine which substitution is more specific and which is more general; in simple cases, this is clear).

In[16] := fac[10]//.{fac[0]−>1,fac[n_]−>n∗fac[n−1]}
Out[16] = 3628800
By the way, what are the real names of /. and //. ?
In[17] := FullForm[Hold[a/.x−>y]]
Out[17]//FullForm =
Hold[ReplaceAll[a,Rule[x,y]]]
In[18] := FullForm[Hold[a//.x−>y]]
Out[18]//FullForm =
Hold[ReplaceRepeated[a,Rule[x,y]]]

4.3 Products

Let's take a product.
In[19] := FullForm[a = 2∗x∗y∗z]
Out[19]//FullForm =
Times[2,x,y,z]
The pattern xy is considered contained in this product, though the internal representation of a does not contain Times[x,y] explicitly.
In[20] := a/.x∗y−>z
Out[20] = $2z^2$
And this product does not contain xy.
In[21] := FullForm[a = 2∗x^2∗y∗z]
Out[21]//FullForm =
Times[2,Power[x,2],y,z]
In[22] := a/.x∗y−>z
Out[22] = $2x^2yz$
This product contains powers of x and y.
In[23] := FullForm[a = 2∗x^2∗y^3∗z]
Out[23]//FullForm =
Times[2,Power[x,2],Power[y,3],z]
We want to replace each product of powers of x and y by the function f of these powers. Such a problem occurs very often. For example, we want to integrate some class of expressions, and we know the result of integration as a function of powers of some variables (or subexpressions).
In[24] := a/.x^n_∗y^m_−>f[n,m]
Out[24] = $2zf[2,3]$
This works OK. And here?
In[25] := FullForm[a = 2∗x^2∗y∗z]
Out[25]//FullForm =
Times[2,Power[x,2],y,z]
In[26] := a/.x^n_∗y^m_−>f[n,m]
Out[26] = $2x^2yz$

This doesn't work. The product a does not contain a product of *powers* of x and y: the symbol y is not in the argument of the function Power. In the next example the substitution works again—as we have seen, *Mathematica* considers dividing by y as multiplying by y^{-1}.

In[27] := FullForm[*a* = 2 ∗ *x*^2 ∗ *z*/*y*]
Out[27]//FullForm =
  Times[2, Power[x, 2], Power[y, −1], z]
In[28] := *a*/.*x*^*n*_ ∗ *y*^*m*_−>*f*[*n*, *m*]
Out[28] = $2zf[2,-1]$

Let's return to the previous expression. How can we instruct *Mathematica* to consider y as a particular case of the pattern "y to an arbitrary power"? This is what an optional arbitrary argument m_. is for. When it is used in an exponent, its default value (which is used when there is no power at all) is 1.

In[29] := FullForm[*a* = 2 ∗ *x*^2 ∗ *y* ∗ *z*]
Out[29]//FullForm =
  Times[2, Power[x, 2], y, z]
In[30] := *a*/.*x*^*n*_. ∗ *y*^*m*_.−>*f*[*n*, *m*]
Out[30] = $2zf[2,1]$
In[31] := FullForm[*a* = 2 ∗ *x* ∗ *y* ∗ *z*]
Out[31]//FullForm =
  Times[2, x, y, z]
In[32] := *a*/.*x*^*n*_. ∗ *y*^*m*_.−>*f*[*n*, *m*]
Out[32] = $2zf[1,1]$

So far so good. But what if the symbol y is absent? Will *Mathematica* consider this as a particular case of the pattern "y to an arbitrary power" with the power equal 0? It will not.

In[33] := FullForm[*a* = 2 ∗ *x*^2 ∗ *z*]
Out[33]//FullForm =
  Times[2, Power[x, 2], z]
In[34] := *a*/.*x*^*n*_. ∗ *y*^*m*_.−>*f*[*n*, *m*]
Out[34] = $2x^2z$

The following method will work always. Let's collect several test expressions to a list.

In[35] := *a* = {2 ∗ *x* ∗ *y* ∗ *z*, 2 ∗ *x*^2 ∗ *y* ∗ *z*, 2 ∗ *x*^2 ∗ *y*^3 ∗ *z*, 2 ∗ *x*^2 ∗ *z*/*y*, 2 ∗ *x*^2 ∗ *z*, 2 ∗ *z*}

Out[35] = $\left\{2xyz, 2x^2yz, 2x^2y^3z, \frac{2x^2z}{y}, 2x^2z, 2z\right\}$

The method is as follows. Multiply our expression by $f[0,0]$, and apply a list of substitutions. If f with some arguments is multiplied by an arbitrary power of x, then the first argument of f is increased by this power. Of course, if x is not raised to any power, we want to use the default power equal to 1. Powers of y are treated in the same way. We need to use the repeated substitution //.—after one substitution from the list has been applied (e.g., the one about x), we want the other one to be applied to the result (to take y into account).

In[36] := s = {x^l_. * f[n_,m_]->f[n+l,m],y^l_. * f[n_,m_]->f[n,m+l]}
Out[36] = $\left\{x^{l_{-}\cdot}f[n\_,m\_] \to f[l+n,m], y^{l_{-}\cdot}f[n\_,m\_] \to f[n,l+m]\right\}$
In[37] := a * f[0,0]//.s
Out[37] = $\{2zf[1,1], 2zf[2,1], 2zf[2,3], 2zf[2,-1], 2zf[2,0], 2zf[0,0]\}$
In[38] := Clear[a,s]

4.4 Sums

Substitutions for sums are similar to those for products. They are used much more rarely. Don't use them if you can avoid this.
In[39] := FullForm[a = x + y + z + 2]
Out[39]//FullForm =
Plus[2,x,y,z]
In[40] := a/.x + y->z
Out[40] = $2+2z$
In[41] := FullForm[a = 2 * x + y + z + 2]
Out[41]//FullForm =
Plus[2,Times[2,x],y,z]
In[42] := a/.x + y->z
Out[42] = $2+2x+y+z$
This substitution replaces a sum of x and y with arbitrary coefficients by the function f of these coefficients.
In[43] := FullForm[a = 2 * x + 3 * y + z + 2]
Out[43]//FullForm =
Plus[2,Times[2,x],Times[3,y],z]
In[44] := a/.n_ * x + m_ * y->f[n,m]
Out[44] = $2+z+f[2,3]$
In[45] := FullForm[a = 2 * x + y + z + 2]
Out[45]//FullForm =
Plus[2,Times[2,x],y,z]
In[46] := a/.n_ * x + m_ * y->f[n,m]
Out[46] = $2+2x+y+z$
In[47] := FullForm[a = 2 * x − y + z + 2]
Out[47]//FullForm =
Plus[2,Times[2,x],Times[−1,y],z]
In[48] := a/.n_ * x + m_ * y->f[n,m]
Out[48] = $2+z+f[2,-1]$
Here again an optional arbitrary argument can be used. When it is used as a factor, a subexpression is considered matching this pattern even if there is no such a factor, and its value in this case is taken to be 1.

In[49] := FullForm[*a* = 2 ∗ *x* + *y* + *z* + 2]
Out[49]//FullForm =
    Plus[2, Times[2, *x*], *y*, *z*]
In[50] := *a*/.*n*_. ∗ *x* + *m*_. ∗ *y*−>*f*[*n*, *m*]
Out[50] = 2 + *z* + *f*[2, 1]
In[51] := FullForm[*a* = *x* + *y* + *z* + 2]
Out[51]//FullForm =
    Plus[2, *x*, *y*, *z*]
In[52] := *a*/.*n*_. ∗ *x* + *m*_. ∗ *y*−>*f*[*n*, *m*]
Out[52] = 2 + *z* + *f*[1, 1]
In[53] := FullForm[*a* = *x* + *z* + 2]
Out[53]//FullForm =
    Plus[2, *x*, *z*]
In[54] := *a*/.*n*_. ∗ *x* + *m*_. ∗ *y*−>*f*[*n*, *m*]
Out[54] = 2 + *x* + *z*
And here is our method which always works.
**In[55] := *a* = {*x* + *y* + *z* + 2, 2 ∗ *x* + *y* + *z* + 2, 2 ∗ *x* + 3 ∗ *y* + *z* + 2, 2 ∗ *x* − *y* + *z* + 2,
    x + *z* + 2, *z* + 2}**
Out[55] = $\{2+x+y+z, 2+2x+y+z, 2+2x+3y+z, 2+2x-y+z, 2+x+z, 2+z\}$
In[56] := *s* = {*l*_. ∗ *x* + *f*[*n*_, *m*_]−>*f*[*n* + *l*, *m*], *l*_. ∗ *y* + *f*[*n*_, *m*_]−>*f*[*n*, *m* + *l*]}
Out[56] = {*f*[*n*_, *m*_] + *xl*_. → *f*[*l* + *n*, *m*], *f*[*n*_, *m*_] + *yl*_. → *f*[*n*, *l* + *m*]}
In[57] := *a* + *f*[0, 0]//.*s*
Out[57] = $\{2+z+f[1,1], 2+z+f[2,1], 2+z+f[2,3], 2+z+f[2,-1],$
    $2+z+f[1,0], 2+z+f[0,0]\}$
In[58] := Clear[*a*, *s*]

4.5 Conditions

Substitutions which apply only when an arbitrary variable satisfies some condition are often needed.
In[59] := {*f*[1.5], *f*[3/2], *f*[*x*/2]}/.*f*[*x*_?NumberQ]−>*x*^2
Out[59] = $\left\{2.25, \frac{9}{4}, f\left[\frac{x}{2}\right]\right\}$
But this method is not very general. It checks a condition depending on a single variable. The operator /; can be applied to a pattern (or its part). It can be read as "such that." The condition in it can depend on several arbitrary variables.
In[60] := *s* = {fac[0]−>1, fac[*n*_Integer/; *n* > 0]−>*n* ∗ fac[*n* − 1]}
Out[60] = {fac[0] → 1, fac[*n*_Integer/; *n* > 0] → *n* fac[−1 + *n*]}
In[61] := {fac[10], fac[−10]}//.*s*
Out[61] = {3628800, fac[−10]}
Internally, this operator is the function Condition.
In[62] := FullForm[*s*]
Out[62]//FullForm =
    List[Rule[fac[0], 1],

Rule[fac[Condition[Pattern[n, Blank[Integer]], Greater[n, 0]]],
Times[n, fac[Plus[−1, n]]]]]

In[63] := Clear[s]

One common case is when you want to replace $f[x]$ by $g[x]$ only for some values of x.

In[64] := f[a] + f[b] + f[c]/.f[x_/;x == a||x == b]−>g[x]

Out[64] = f[c] + g[a] + g[b]

4.6 Variable Number of Arguments

A pattern can involve a construct which matches not a single subexpression but an arbitrary-length subsequence of arguments of a function. This is very convenient for working with functions having an arbitrary number of arguments. Let's consider an example. The function f has any number of arguments. We want to shuffle them in the opposite order. First, let's put a fence at the end of the argument list.

In[65] := a = f[x, y, z]

Out[65] = f[x, y, z]

In[66] := a = a/.f[x___]−>f[x, Fence]

Out[66] = f[x, y, z, Fence]

Now we take the arguments from the left one by one and throw them over the fence (placing them immediately after the fence).

In[67] := a = a//.f[x_, y___, Fence, z___]−>f[y, Fence, x, z]

Out[67] = f[Fence, z, y, x]

Now the fence is at the left, and the arguments are after it in the opposite order. What's left is to remove the fence.

In[68] := a = a/.f[Fence, x___]−>f[x]

Out[68] = f[z, y, x]

Of course, this method only works when the symbol Fence is not present among the arguments. Here is the method which always works. Let the part of the arguments which has not yet been processed be in the first list and the processed part—in the second one. We take the arguments one by one from the beginning of the first list and move them to the beginning of the second one.

In[69] := a = a/.f[x___]−>f[{x}, {}]

Out[69] = f[{z, y, x}, {}]

In[70] := a = a//.f[{x_, y___}, {z___}]−>f[{y}, {x, z}]

Out[70] = f[{}, {x, y, z}]

In[71] := a = a/.f[{}, {x___}]−>f[x]

Out[71] = f[x, y, z]

In addition to x___ (with three underscores), which means an arbitrary subsequence of arguments of a function (maybe an empty one), there is also x__ (with two underscores)—an arbitrary nonempty subsequence of arguments. I find the first construct more useful.

Chapter 5
Functions

5.1 Immediate and Delayed Assignment

This is an ordinary (immediate) assignment. The result of calculation of the right-hand side (in this case, a quadratic polynomial) is assigned to the variable a.

In[1] := a = Expand[(x + 1)^2]

Out[1] $= 1 + 2x + x^2$

And this is a delayed assignment. The unevaluated right-hand side (in this case, an expression with the function Expand) is assigned to the variable b.

In[2] := b := Expand[(x + 1)^2]

Note that a delayed assignment returns no value (an ordinary assignment returns the result of calculation of its right-hand side). The difference between a and b can be seen if we assign something to the variable x. In the first case, the value of x is substituted into the quadratic polynomial.

In[3] := $x = z + 1$; a

Out[3] $= 1 + 2(1 + z) + (1 + z)^2$

In the second case, the value of x is substituted into the expression with the function Expand, and then this expression is calculated.

In[4] := b

Out[4] $= 4 + 4z + z^2$

In[5] := Clear[a, b, x]

The real name of the operator := is SetDelayed.

In[6] := FullForm[Hold[a := x]]

Out[6]//FullForm =

Hold[SetDelayed[a,x]]

Similarly, in addition to ordinary substitutions $a->b$ (where the right-hand side is calculated at the moment the substitution is defined), there are delayed substitutions $a :> b$ (where the substitution keeps the right-hand side unevaluated, and it is calculated each time the substitution is applied).

A. Grozin, *Introduction to Mathematica® for Physicists*, Graduate Texts in Physics, DOI 10.1007/978-3-319-00894-3_5, © Springer International Publishing Switzerland 2014

In[7] := $f[z+1]/.f[x\_]->\text{Expand}[(x+1)\wedge 2]$
Out[7] $= 1+2(1+z)+(1+z)^2$
In[8] := $f[z+1]/.f[x\_]:>\text{Expand}[(x+1)\wedge 2]$
Out[8] $= 4+4z+z^2$

5.2 Functions

Left-hand side of an assignment can be a pattern, not just a variable. In this case, in all subsequent calculations, any subexpression matching the pattern will be replaced by the right-hand side of the assignment. This can be canceled by the command Clear. A pattern can contain arbitrary variables. This is how functions are defined. Here is an example—a function f. In all subsequent calculations, all subexpressions of the form $f[x]$ with arbitrary arguments x will be replaced by the right-hand side (in this case, a quadratic polynomial), in which the value of the actual argument is substituted for x.
In[9] := $f[x\_] = \text{Expand}[(x+1)\wedge 2]$
Out[9] $= 1+2x+x^2$
And this is another function. Its body is an unevaluated expression with Expand. Expanding brackets will take place each time the function g with some argument is calculated.
In[10] := $g[x\_] := \text{Expand}[(x+1)\wedge 2]$
Note the difference between them.
In[11] := $\{f[z+1], g[z+1]\}$
Out[11] $= \{1+2(1+z)+(1+z)^2, 4+4z+z^2\}$
In[12] := $\text{Clear}[f, g]$

5.3 Functions Remembering Their Values

Let's consider a useful trick—a function remembering its calculated values. If it is called again with the same argument, it will not perform calculations, but just return the remembered result. For example, take the factorial. We know fac[0].
In[13] := fac[0] = 1
Out[13] = 1
And now attention—the main trick. A delayed assignment to the pattern fac[$n\_$] (with an arbitrary n). And what do we have in the right-hand side? An immediate assignment to the pattern fac[n] (for a specific n, namely, the value of the actual argument with which the function fac was called). What happens when we call fac[10]? If the function was never calculated with this argument, then this definition for an arbitrary argument will be used. The right-hand side with 10 substituted for n will be calculated, namely, an immediate assignment fac[10] $= \cdots$. Its right-hand side is calculated (it is the factorial of 10), and it is remembered as the value of fac[10].

The immediate assignment returns the calculated value of its right-hand side, and this value becomes the result of the function call. If we ask *Mathematica* to calculate fac[10] again, then this specific definition for fac[10] (generated during the first calculation) will be used, and not the general definition for fac[n_].

In[14] := fac[*n*_] := fac[*n*] = *n* ∗ fac[*n* − 1]

What does *Mathematica* know about the symbol fac?

In[15] :=?fac

Global\`fac
fac[0] = 1
fac[n_] := fac[n] = n fac[$n-1$]

Only two definitions—for the argument 0 and for an arbitrary argument. Now let's calculate the factorial of 10.

In[16] := fac[10]

Out[16] = 3628800

And what does *Mathematica* know about this symbol now?

In[17] :=?fac

Global\`fac
fac[0] = 1
fac[1] = 1
fac[2] = 2
fac[3] = 6
fac[4] = 24
fac[5] = 120
fac[6] = 720
fac[7] = 5040
fac[8] = 40320
fac[9] = 362880
fac[10] = 3628800
fac[n_] := fac[n] = n fac[$n-1$]

In addition to the general definition, we see also specific ones for all integer values of the argument from 0 to 10. If we ask for the value of fac for one of these arguments, then the corresponding specific definition will be used, and the calculation will not be performed again.

In[18] := Clear[fac]

5.4 Fibonacci Numbers

This method is useful but not vital for the factorial, because the time of calculation of fac[n] grows linearly with n. For Fibonacci numbers the difference is crucial. For a naive definition, the calculation time grows exponentially. This means you will never get a Fibonacci number with a large n. When results are remembered, the calculation time grows linearly—the result for each value of the argument from 2 to n is calculated once.

In[19] := fib[0] = fib[1] = 1
Out[19] = 1
In[20] := fib[*n*_] := fib[*n*] = fib[*n* − 1] + fib[*n* − 2]
In[21] :=?fib
Global`fib
fib[0] = 1
fib[1] = 1
fib[*n*_] := fib[*n*] = fib[*n* − 1] + fib[*n* − 2]
In[22] := fib[10]
Out[22] = 89
In[23] :=?fib
Global`fib
fib[0] = 1
fib[1] = 1
fib[2] = 2
fib[3] = 3
fib[4] = 5
fib[5] = 8
fib[6] = 13
fib[7] = 21
fib[8] = 34
fib[9] = 55
fib[10] = 89
fib[*n*_] := fib[*n*] = fib[*n* − 1] + fib[*n* − 2]
In[24] := Clear[fib]

5.5 Functions from Expressions

In most cases, a delayed assignment is used when defining a function. But there are situations when an immediate assignment is needed. Here is one of them. Suppose you have derived an expression a containing a symbol x as a result of some calculation.
In[25] := *a* = *D*[Expand[(*x* + 1)^3], *x*]
Out[25] = $3 + 6x + 3x^2$
Now you want to calculate it many times with different values of x. This can be done by substitutions.
In[26] := *a*/.*x*−>*z* + 1
Out[26] = $3 + 6(1 + z) + 3(1 + z)^2$
But this is not very convenient. It would be nice to have a function f with the argument x which is given by the calculated expression a. Such a function can be defined by an immediate assignment. The calculated value is substituted for a in the right-hand side.

In[27] := $f[x\_] = a$
Out[27] = $3+6x+3x^2$
In[28] := $f[z+1]$
Out[28] = $3+6(1+z)+3(1+z)^2$
In[29] := Clear[a,f**]**

5.6 Antisymmetric Functions

It is often useful to give a partial definition of a function. To this end we write not just a function with all arguments being arbitrary but a more restrictive pattern in the left-hand side of an assignment. Then, if the values of the actual arguments match the pattern, the function is calculated (i.e., replaced by the right-hand side of the assignment). Otherwise the function remains unevaluated. Here is a simple example. Let's define an antisymmetric function of two arguments. If the arguments are equal, it vanishes.
In[30] := $f[x\_,x\_] := 0$
If they are not equal, we have to decide if they need to be interchanged. The expressions $f[x,y]$ and $f[y,x]$ should reduce to either the first form or the second one, for any x and y. It does not matter to which form, as long as the result is always the same. To this end the function OrderedQ is useful. Its argument is a list. It returns True if the list is ordered, i.e., each element is "greater than or equal to" the previous one in the sense of some internal ordering *Mathematica* uses for expressions. Details of this ordering are not important.
In[31] := {OrderedQ[{x,y**}], OrderedQ[{**y,x**}], OrderedQ[{**x,x**}]}**
Out[31] = {True, False, True}
Now it is easy to write a substitution which interchanges the arguments if they are not properly ordered.
In[32] := $f[x\_,y\_]$**/;Not[OrderedQ[{**x,y**}]] :=** $-f[y,x]$
In[33] := $\{f[a,a],f[a,b],f[b,a]\}$
Out[33] = $\{0,f[a,b],-f[a,b]\}$
In[34] := $\{f[a+b,a-b],f[a-b,a+b]\}$
Out[34] = $\{-f[a-b,a+b],f[a-b,a+b]\}$
In[35] := Clear[f**]**

Of course, a symmetric function can be defined similarly. An odd function of a single argument can be defined in the same way.
In[36] := $f[0] = 0$
Out[36] = 0
In[37] := $f[x\_]$**/;Not[OrderedQ[{**$-x,x$**}]] :=** $-f[-x]$
In[38] := $\{f[0],f[a],f[-a]\}$
Out[38] = $\{0,f[a],-f[a]\}$
In[39] := $\{f[a-b],f[b-a]\}$
Out[39] = $\{-f[-a+b],f[-a+b]\}$
In[40] := Clear[f**]**
Of course, an even function can be defined similarly.

5.7 Functions with Options

You have undoubtedly noted that many *Mathematica* functions (e.g., Plot) have options. They can be specified in any order; each option is given by a substitution with its name in the left-hand side and its value in the right-hand side. If they are not given, their default values are used. Suppose you want your own function f to have options. This can be done in the following way. Let's assign a list of substitutions giving default values of all options to Options[f]. Define the function f with some mandatory arguments and an arbitrary sequence of arguments opts___ (it may be empty). At the point in the function body where you need the value of the option opt1 use opt1/.{opts}/.Options[f]. The operations /. are executed left to right. Therefore, if the user has included a substitution opt1 $\to \cdots$ among the arguments, the left /. will trigger, and the result will be some value which contains no option names; the right /. will not change it. If the user has not given such a substitution, the left /. will do nothing, and the right one will replace opt1 by the default value of this option.

In[41] := Options[*f*] = {opt1−>1,opt2−>2}
Out[41] = {opt1 → 1,opt2 → 2}
In[42] := *f*[*x*_,opts___] := *g*[*x*,opt1/.{opts}/.Options[*f*],
opt2/.{opts}/.Options[*f*]]
In[43] := {*f*[*a*],*f*[*a*,opt2−>0],*f*[*a*,opt2−>*b*,opt1−>*c*]}
Out[43] = {*g*[*a*,1,2],*g*[*a*,1,0],*g*[*a*,*c*,*b*]}
In[44] := Clear[*f*]
In recent versions of *Mathematica* this can also be written as follows:
In[45] := *f*[*x*_,OptionsPattern[*f*]] := *g*[*x*,OptionValue[opt1],OptionValue[opt2]]
In[46] := Options[*f*] = {opt1−>1,opt2−>2};
In[47] := {*f*[*a*],*f*[*a*,opt2−>0],*f*[*a*,opt2−>*b*,opt1−>*c*]}
Out[47] = {*g*[*a*,1,2],*g*[*a*,1,0],*g*[*a*,*c*,*b*]}
In[48] := Clear[*f*]

5.8 Attributes

A function can have attributes which affect simplification of expressions with this function. The attribute Flat removes nested function calls (e.g., Plus and Times have this attribute).

In[49] := Attributes[*f*] = {Flat}
Out[49] = {Flat}
In[50] := *f*[*x*,*f*[*y*,*z*],*u*]
Out[50] = *f*[*x*,*y*,*z*,*u*]
The attribute Orderless means that the function is symmetric in all arguments, and *Mathematica* may interchange them at will (Plus and Times have also this attribute).

In[51] := Attributes[*f*] = {Orderless}
Out[51] = {Orderless}
In[52] := {*f*[*x*,*y*,*z*],*f*[*z*,*x*,*y*],*f*[*y*,*z*,*x*]}
Out[52] = {*f*[*x*,*y*,*z*],*f*[*x*,*y*,*z*],*f*[*x*,*y*,*z*]}
The attribute Listable means that if the first argument is a list, then the function is applied to each element, and the list of results is returned (Plus and Times have this attribute, too).
In[53] := Attributes[*f*] = {Listable}
Out[53] = {Listable}
In[54] := *f*[{*x*,*y*,*z*}]
Out[54] = {*f*[*x*],*f*[*y*],*f*[*z*]}
In[55] := *f*[{*x*,*y*,*z*},*a*]
Out[55] = {*f*[*x*,*a*],*f*[*y*,*a*],*f*[*z*,*a*]}
There exist a few attributes more. Of course, a function can have several attributes at once. The command Clear[*f*] removes only substitutions for *f* (with any arguments), but not its attributes. In order to remove attributes too, use ClearAll.
In[56] := ClearAll[*f*]; Attributes[*f*]
Out[56] = {}

5.9 Upvalues

Suppose we want to define a function f such that $f[x]*f[y]$ is replaced by $f[x+y]$ for arbitrary x and y. This can be done by the assignment $f[x\_]*f[y\_] := f[x+y]$. This definition will be associated with the function Times; *Mathematica* will have to check it each time it multiplies something, i.e., very often, and performance will degrade. It is possible to associate this definition with the function f instead. Then it will be used only when processing a product containing at least one function f.
In[57] := *f*[*x*_] * *f*[*y*_]^ := *f*[Expand[*x* + *y*]]
In[58] := *f*[(*x*+*y*)^2] * *f*[(*x*−*y*)^2] * *f*[*x*^2] * *g*[*y*^2]
Out[58] = $f\left[3x^2+2y^2\right] g\left[y^2\right]$
Here the left-hand side is Times[*f*[*x*_],*f*[*y*_]], and the definition is associated with f. If we want a definition for Times[*f*[*x*_],*g*[*y*_]], we can associate it with either f or g.
In[59] := *f* / : *f*[*x*_] * *g*[*y*_] := *f*[Expand[*x* − *y*]]
In[60] := *f*[(*x*+*y*)^2] * *f*[(*x*−*y*)^2] * *f*[*x*^2] * *g*[*y*^2]
Out[60] = $f\left[3x^2+y^2\right]$
In[61] :=?*f*
Global\`*f*
f[*x*_]*f*[*y*_]^ := *f*[Expand[*x* + *y*]]
f / : *f*[*x*_]*g*[*y*_] := *f*[Expand[*x* − *y*]]

When processing an expression *f*[g1[...],g2[...],g3[...]], *Mathematica* uses definitions associated with f and also definitions associated with g1, g2, g3, and having the form $f[\ldots] := \cdots$ (*upvalues* of g1, g2, g3). It does not look deeper, into arguments of g1, g2, and g3—this would be too inefficient. In addition to the delayed assignments ^ := and $f/$: lhs := rhs there are also immediate assignments ^ = and $f/$: lhs = rhs.

In[62] := Clear[*f*]

Chapter 6
Mathematica as a Programming Language

6.1 Compound Expressions

A compound expression consists of several expressions separated by the operator ;. They are calculated left to right. The value of a compound expression is the value of the last (rightmost) expression. The values of all the other expressions are thrown away; they are calculated only for side effects. The operator ; has a low priority, so that it is often necessary to put a compound expression inside brackets. The last expression may be empty. Its value (and hence the value of the compound expression) is the symbol Null which is not printed. Therefore, if you want to suppress printing of the result of some calculation (e.g., because it is lengthy), put ; after it.

```
In[1] := fac[0] = 1;
In[2] := fac[n_] := (Print["n=", n]; n * fac[n - 1])
In[3] := fac[4]
n=4
n=3
n=2
n=1
Out[3] = 24
In[4] := Clear[fac]
In[5] := Null
In[6] := FullForm[x;]
Out[6]//FullForm =
    Null
In[7] := FullForm[Hold[a; b]]
Out[7]//FullForm =
    Hold[CompoundExpression[a, b]]
```

A. Grozin, *Introduction to Mathematica® for Physicists*, Graduate Texts in Physics,
DOI 10.1007/978-3-319-00894-3_6, © Springer International Publishing Switzerland 2014

6.2 Conditional Expressions

If

In[8] := del[x_, y_] := If[x == y, 1, 0]
In[9] := del[a, a]
Out[9] = 1
In[10] := del[1, 2]
Out[10] = 0
When *Mathematica* cannot determine if the condition is true, a conditional expression is returned unevaluated. If such a possibility will appear later, an unevaluated If will be simplified.
In[11] := u = del[a, b]
Out[11] = If[a == b, 1, 0]
In[12] := a = b = x; u
Out[12] = 1
And what to do if several actions should be performed in the branches of If? Use compound expressions, of course! The priority of the operator; is higher than that of, (which separates function arguments, in particular, those of If).
In[13] := f[x_] := If[x > 0, Print["x>0"]; 1, Print["x<=0"]; 0]
In[14] := f[1]
x>0
Out[14] = 1
In[15] := Clear[a, b, u, del, f]

Which

This is a choice with many branches. Arguments of the function Which form pairs: a condition and a result. The conditions are evaluated left to right. As soon as a true one is found, the corresponding result is evaluated and returned. Often (but not always) the last condition is True; the corresponding result is returned when none of the previous conditions is satisfied. When *Mathematica* cannot decide if the conditions are true, the function Which returns unevaluated.
In[16] := sign[x_] := Which[x > 0, 1, x < 0, −1]
In[17] := sign[0.1]
Out[17] = 1
In[18] := sign[0]
In[19] := sign[a]
Out[19] = Which[a > 0, 1, a < 0, −1]
In[20] := Clear[sign]

Conditions

What can be used as conditions in If and Which? The operator == returns True if its left-hand side and right-hand side are the same expression. Let's stress: mathematically equivalent expressions written in different forms don't qualify. If the left-hand side and the right-hand one are not identical, this operator returns unevaluated. It is used for writing equations, for example, for the function Solve.
In[21] := {a == a,a == b}
Out[21] = {True, $a == b$}
In contrast to this, the operator === returns False if its arguments are not identical (even if they are mathematically equivalent).
In[22] := {a === a,a === b}
Out[22] = {True, False}
Not[$a == b$] is written as a!=b, and Not[$a === b$] as a=!=b.

The function NumberQ checks if its argument is a number (integer, rational, real, complex).
In[23] := {NumberQ[3.14],NumberQ[Pi]}
Out[23] = {True, False}
The function NumericQ returns True also for symbolic mathematical constants.
In[24] := {NumericQ[3.14],NumericQ[Pi]}
Out[24] = {True, True}

The function FreeQ returns True if its first argument contains no subexpressions given by the second argument. It is often used to check if an expression contains a given symbol.
In[25] := {FreeQ[Sin[a+b],a],FreeQ[Sin[b+c],a]}
Out[25] = {False, True}
The function MatchQ checks if the expression—its first argument— matches the pattern, its second argument. When designing a system of substitutions, use this function often in order to check if your ideas about the structure of expressions agree with those of *Mathematica*.
In[26] := MatchQ[x^2,a_^b_]
Out[26] = True
In[27] := MatchQ[1/x,a_^b_]
Out[27] = True
In[28] := MatchQ[x,a_^b_]
Out[28] = False

Switch

The function Switch starts from evaluating its first argument. All the remaining arguments form couples: a pattern and a result. The first argument is matched against the patterns from left to right. As soon as a match is found, the corresponding result is evaluated and returned. Often (but not always) the last pattern is x_ (or just _ because we don't need the value of x).

In[29] := f[x_] := Switch[x, _Plus, "A sum", _Times, "A product", _,
"Neither a sum nor a product"]
In[30] := f[$a+b$]
Out[30] = A sum
In[31] := f[$a*b$]
Out[31] = A product
In[32] := f[a^b]
Out[32] = Neither a sum nor a product
In[33] := Clear[f]

6.3 Loops

Do

This loop is very convenient to those pupils who were ordered by a teacher to write "I shall behave well" 100 times.
In[34] := Do[Print["OK"], {4}]
OK
OK
OK
OK
In this loop the parameter varies from 1 to an upper limit.
In[35] := Do[Print[x^i], {i, 4}]
x
x^2
x^3
x^4
And here—from a lower limit to an upper one.
In[36] := Do[Print[x^i], {i, 0, 4}]
1
x
x^2
x^3
x^4
And now with a given step.
In[37] := Do[Print[x^i], {i, 0, 4, 2}]
1
x^2
x^4
This loop takes the elements of a list.
In[38] := Do[Print[x^i], {i, {0, 1, 4}}]
1
x
x^4

While

While the list is not empty, we print and remove its first element.
In[39] := $l = \{a,b,c\}$;
In[40] := While[l!={},Print[First[l]]; l = Rest[l]]
a
b
c
In[41] := l
Out[41] = {}
In[42] := Clear[l]

For

This is a C style loop. First the initialization (the first argument) is executed. If the condition (the second argument) is satisfied, then the loop body (the fourth argument) is executed. Then the increment (the third argument) is performed. The condition is checked again, and so on.
In[43] := For[$i = 0, i < 5, i{+}{+}$,Print[x^i]]
1
x
x^2
x^3
x^4
In[44] := i
Out[44] = 5
A loop running through several parameters (or data structures) in parallel can be easily written.
In[45] := For[$i = 0; j = 1, i + j < 20, i{+}{+}; j{*} = 2$,Print[$x$^$i + y$^$j$]]
$1 + y$
$x + y^2$
$x^2 + y^4$
$x^3 + y^8$
In[46] := Clear[i, j]

6.4 Functions

The function Function returns an anonymous function. Its first argument is a formal parameter (or a list of formal parameters), and the second one is an expression containing these formal parameters. When the function is called, the actual parameters are substituted for the formal ones in this expression, and the result of its evaluation is returned as the value of the function. A note for experts: the function Function is a λ-expression. Of course, an anonymous function can be assigned to a variable. This is similar to a function defined by $f[x\_] := \cdots$, but more efficient. The usual method of assigning a function body to a pattern is more general, because it is possible to construct a function which is defined only for arguments which satisfy some condition (such a function returns unevaluated if the conditions are not satisfied). This is not possible in the case of Function.

In[47] := f = Function[x,x^2]
Out[47] = Function $[x, x^2]$

An anonymous function can be just applied to some arguments.

In[48] := Function[{x,y},x^2 + y^3][a,b]
Out[48] = $a^2 + b^3$

The function Map applies the function given by its first argument to each element of the list given by the second argument and returns the list of results.

In[49] := Map[f,{a,b,c}]
Out[49] = $\{a^2, b^2, c^2\}$
In[50] := Clear[f]

An anonymous function can be the first argument of Map. Any function with arguments (e.g., Plus) can be the second argument, not just a list. The function given by the first argument is applied to each argument of the expression—the second argument. An expression having the same Head (as the second argument) and the calculated results is constructed and returned.

In[51] := Map[Function[x,x^2],a + b + c]
Out[51] = $a^2 + b^2 + c^2$

The function Apply$[f,l]$ applies f to the list of arguments l; this simply means that the Head of l is replaced by f.

In[52] := Apply[f,{a,b,c}]
Out[52] = $f[a,b,c]$
In[53] := Apply[Times,a + b + c]
Out[53] = abc

The first argument of the function Select is a list. It returns the list of those elements which satisfy the condition given by the second argument. In order to avoid inventing names for such disposable things, anonymous functions are often used as the second argument.

In[54] := Select[{1,5,3,6},Function[x,x > 4]]
Out[54] = $\{5,6\}$

Function Generator

Here's an interesting example. The function Adder has a formal parameter n and returns a function which adds n to its argument, that is, Adder is a function generator.

In[55] := Adder = Function[n,Function[x,x + n]]
Out[55] = Function$[n, \text{Function}[x, x+n]]$

Specific functions can be obtained from it. This one, for example, adds 2 to its argument.

In[56] := Add2 = Adder[2]
Out[56] = Function[$x\$,x\$+2$]
In[57] := Map[Add2, {3,*x*}]
Out[57] = $\{5,2+x\}$
In[58] := Clear[Add2, Adder]

6.5 Local Variables

When writing a function which can be used as a black box by a user, it is crucial to use local variables. Assigning a value to a local variable does not change the global one with the same name (which can store some value precious for the user). To this end the function Module is used. Its first argument is a list of local variables.

In[59] := *x* = 1
Out[59] = 1
In[60] := Module[{*x*},*x* = 2;*x*]
Out[60] = 2
In[61] := *x*
Out[61] = 1
In[62] := Clear[*x*]

Coding functions like this means inviting big troubles.

In[63] := *f* = Function[{*a*,*b*},*x* = *a*;*x* ∗ *b*]
Out[63] = Function[$\{a,b\},x=a;xb$]
In[64] := *f*[*c*,*x*]
Out[64] = c^2
In[65] := *x*
Out[65] = c
In[66] := Clear[*f*,*x*]

This is much better.

In[67] := *f* = Function[{*a*,*b*},Module[{*x* = *a*},*x* ∗ *b*]]
Out[67] = Function[$\{a,b\}$,Module[$\{x=a\},xb$]]
In[68] := *f*[*c*,*x*]
Out[68] = cx
In[69] := *x*
Out[69] = x
In[70] := Clear[*f*]

What happens if a local variable escapes from its scope? We can see that *Mathematica* implements local variables in the most trivial way—by renaming.

In[71] := Module[{*x*},*x*]
Out[71] = x$103

The function Block introduces another kind of local variables. As you value your life or your reason keep away from the function Block. Especially in those dark hours when the powers of evil are exalted.

Local Constants

Local variables which cannot be changed after initialization can be introduced. This is done by the function With. Such local constants can be considered temporary notations introduced to make writing a single expression easier.
In[72] := $x = 1$
Out[72] = 1
In[73] := With[{$x = a+1$}, Print[x^2]]
Out[73] = $(1+a)^2$
In[74] := x
Out[74] = 1
In[75] := Clear[x]

6.6 Table

The function Table constructs a list of values of an expression where a parameter varies in a given way (like in the Do loop).
In[76] := Table[0, {4}]
Out[76] = {0,0,0,0}
In[77] := Table[x^i, {i,4}]
Out[77] = $\{x, x^2, x^3, x^4\}$
In[78] := Table[x^i, {i,0,4}]
Out[78] = $\{1, x, x^2, x^3, x^4\}$
In[79] := Table[x^i, {i,0,4,2}]
Out[79] = $\{1, x^2, x^4\}$
In[80] := Table[x^i, {i, {0,1,4}}]
Out[80] = $\{1, x, x^4\}$
Let's turn the list into a product.
In[81] := Table[x + i, {i,0,4}]/.List->Times
Out[81] = $x(1+x)(2+x)(3+x)(4+x)$

6.7 Parallelization

Now most computers have multi-core processors. The function Parallelize tries to calculate its argument faster by starting several *Mathematica* kernels and ordering them to calculate parts of the expression and then collecting these parts together.
In[82] := Parallelize[Table[$KernelID, {n,0,7}]]
Out[82] = {4,4,3,3,2,2,1,1}
($KernelID is the number of the kernel in which a particular list element has been evaluated). In addition to Table, it can handle Map[f, {...}] (or f[{...}] where f is a Listable function) and some other cases. Note that the slave *Mathematica*

kernels started by Parallelize don't know definitions made in the master process; only built-in functions can be used. If you need a user-defined function f to be available in slave processes, you should explicitly distribute it.
In[83] := f[n_] := Integrate[x^n * Sin[x], {x,0,Pi}]
In[84] := DistributeDefinitions[f]
Out[84] = $\{f\}$
In[85] := Parallelize[Table[{f[n], $KernelID}, {n,0,7}]]
Out[85] = $\{\{2,4\},\{\pi,4\},\{-4+\pi^2,3\},\{\pi\left(-6+\pi^2\right),3\},\{48-12\pi^2+\pi^4,2\},$
$\{\pi\left(120-20\pi^2+\pi^4\right),2\},\{-1440+360\pi^2-30\pi^4+\pi^6,1\},$
$\{\pi(-5040+840\pi^2-42\pi^4+\pi^6),1\}\}$
In[86] := Clear[f]

6.8 Functions with an Index

Something like an array of functions can be constructed. Let $f[1]$ be the function adding 1 to its argument and $f[2]$—the function adding 2 to its argument; $f[n]$ is undefined for other values of n.
In[87] := f[1] = Function[x,x+1]
Out[87] = Function$[x,x+1]$
In[88] := f[2] = Function[x,x+2]
Out[88] = Function$[x,x+2]$
In[89] := {f[1][a], f[2][a], f[n][a]}
Out[89] = $\{1+a,2+a,f[n][a]\}$
In[90] := Clear[f]

6.9 Hold and Evaluate

Assignment

Assignment does not evaluate its left-hand side. This is natural: the value of the right-hand side is assigned to the variable given by the left-hand side, not to its value.
In[91] := a = x
Out[91] = x
In[92] := a = y
Out[92] = y
In[93] := a
Out[93] = y
In[94] := x
Out[94] = x

The attribute HoldFirst is responsible for this.
In[95] := Attributes[Set]
Out[95] = {HoldFirst, Protected, SequenceHold}

Evaluate

It is possible to assign a value to *the value* of the left-hand side. To this end the function Evaluate is used.
In[96] := *a* = *x*
Out[96] = x
In[97] := Evaluate[*a*] = *y*
Out[97] = y
In[98] := *x*
Out[98] = y
In[99] := *a*
Out[99] = y
In[100] := Clear[*x*]; *a*
Out[100] = x
In[101] := Clear[*a*]

Delayed Assignment

Delayed assignment does not evaluate also its right-hand side. The attribute HoldAll is responsible for this.
In[102] := Attributes[SetDelayed]
Out[102] = {HoldAll, Protected, SequenceHold}
You can use these attributes for your functions, too.
In[103] := *x* = 1; *y* = 2; *z* = 3;
In[104] := Attributes[*f*] = {HoldAll}; *f*[*x*, *y*, *z*]
Out[104] = $f[x, y, z]$
In[105] := Attributes[*f*] = {HoldFirst}; *f*[*x*, *y*, *z*]
Out[105] = $f[x, 2, 3]$
In[106] := Clear[*x*, *y*, *z*]
In[107] := ClearAll[*f*]

The First Argument of the Function Plot

The function Plot does not evaluate its arguments.
In[108] := Attributes[Plot]
Out[108] = {HoldAll, Protected}
The first argument $f[x]$ can be meaningful only for numerical values of x, not for symbolic x. Therefore it is better not to evaluate $f[x]$ before the function Plot will

call it for numerical values of x. But if the first argument is a command which generates the list of expressions to draw, it will not work. We want the command to be executed; to this end Evaluate is used.

In[109] := Plot[Evaluate[Table[Sin[*n* ∗ *x*], {*n*, 1, 3}]], {*x*, 0, 2 ∗ Pi}]

Out[109] =

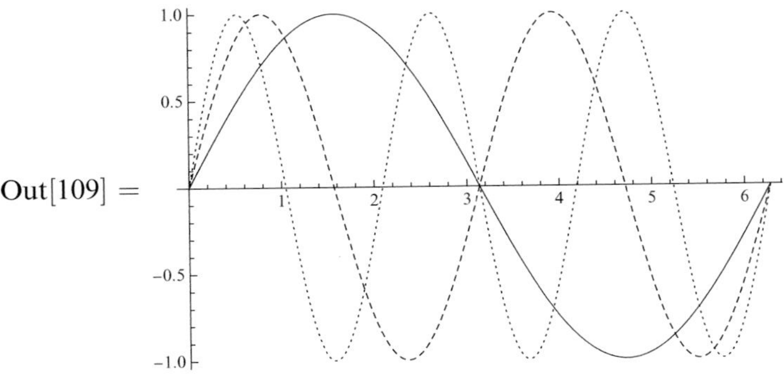

Hold

The function Hold suppresses evaluation of its argument.

In[110] := *a* = *x*

Out[110] = x

In[111] := *b* = Hold[*a*]

Out[111] = Hold[a]

This suppression can be removed by the function ReleaseHold.

In[112] := ReleaseHold[*b*]

Out[112] = x

The function Hold is simple—it has the attribute HoldAll, i.e., it does not evaluate its arguments.

In[113] := Attributes[Hold]

Out[113] = {HoldAll, Protected}

In[114] := Clear[*a*, *b*]

Chapter 7
Gröbner Bases

7.1 Statement of the Problem

In this lecture we shall consider (in a slightly vulgarized form, without rigorous mathematical terms) an important mathematical achievement of the second half of the last century—Gröbner bases, the Buchberger algorithm (which constructs them), and their applications (see [12, 13] for an introduction).

Suppose we have n variables $x_1, \ldots, x_n$. They are not independent, but satisfy some polynomial equations $p_1 = 0, \ldots, p_m = 0$ (p_j are polynomials of x_i). Let's consider some polynomial q of the same variables. It is natural to ask if this polynomial is equal to 0 due to the constraints on our variables or not. If there is another polynomial q_2, there is the question of their equality.

These questions would become very easy if we had an algorithm reducing polynomials of dependent variables to a canonical form. Two equal polynomials reduce to the same canonical form; a polynomial equal to 0 reduces to the canonical form 0.

We can try to use the equations $p_j = 0$ for simplifying the polynomial q, i.e., for replacing its more complicated terms by combinations of simpler ones. But to do so we first have to accept some convention in which terms are more complicated and which are more simple.

7.2 Monomial Orders

We need a total order of monomials (i.e., products of powers of the variables $x_1^{n_1} \cdots x_n^{n_n}$). An order is total if for any monomials s and t either $s < t$ or $s > t$ or $s = t$ is true. An order is admissible if two properties are satisfied:

- $1 \leqslant s$ for any monomial s.
- If $s < t$ then $s u < t u$ for any monomial u.

Three admissible orders are most popular.

A. Grozin, *Introduction to Mathematica® for Physicists*, Graduate Texts in Physics,
DOI 10.1007/978-3-319-00894-3_7, © Springer International Publishing Switzerland 2014

Lexicographic

Anybody who has ever seen a dictionary knows what is lexicographic order. We are comparing two monomials: $s = x_1^{n_1} x_2^{n_2} \cdots x_n^{n_n}$ and $t = x_1^{m_1} x_2^{m_2} \cdots x_n^{m_n}$. If the degree of the main variable x_1 in s is larger than in t ($n_1 > m_1$), then $s > t$. If it is smaller ($n_1 < m_1$), then $s < t$. If $n_1 = m_1$, we compare the degrees of the next variable x_2: if $n_2 > m_2$, then $s > t$; if $n_2 < m_2$, then $s < t$; if $n_2 = m_2$, we compare the degrees of x_3; and so on.

By Total Degree than Lexicographic

First we compare the total degree $n = n_1 + n_2 + \cdots + n_n$ of the monomial s and the total degree $m = m_1 + m_2 + \cdots + m_n$ of the monomial t. If $n > m$ then $s > t$; if $n < m$ then $s < t$; if the total degrees are equal, we compare s and t lexicographically.

By Total Degree than Reverse Lexicographic

First we compare the total degrees. If they are equal, then we begin from the junior variable x_n: if its degree in s is larger than in t ($n_n > m_n$), then $s < t$; if it is smaller ($n_n < m_n$), then $s > t$; if $n_n = m_n$, we compare the degrees of the previous variable x_{n-1}; and so on, that is, this is (within some total degree) the reverse lexicographic order with respect to the reverse list of variables.

7.3 Reduction of Polynomials

Let's fix some admissible monomial order. We'll write polynomials in descending order: the leading term first, followed by the rest ones. We'll normalize all polynomials p_i in such a way that the coefficient of the leading term is 1. Now they can be used as substitutions which replace the leading term by minus sum of the remaining ones, that is, if some term of a polynomial q is divisible by the leading term of some polynomial p_i, we remove this leading term and insert minus sum of the remainder terms of p_i instead. This is called reduction of the polynomial q with respect to the set of polynomials p_i; if none of the substitutions is applicable, the polynomial q is called reduced. For example, let's consider a set of polynomials

```
In[1] := p1 = x^2 + y^2 - 1; p2 = x * y - 1/4;
```

Let's try to reduce the polynomial

```
In[2] := q = x^2 * y;
```

(we use the lexicographic order with $x > y$). This can be done in different ways.

Let's first reduce q with respect to p_1:
In[3] := PolynomialReduce[q, {p1}, {x,y}]
Out[3] $= \{\{y\}, y - y^3\}$
The result means that if we subtract the polynomial p_1 multiplied by y from q, then the reduced polynomial $y - y^3$ is obtained. This is what we are interested in.
In[4] := q1 = %[[2]]
Out[4] $= y - y^3$
Now let's reduce q with respect to p_2:
In[5] := PolynomialReduce[q, {p2}, {x,y}]
Out[5] $= \left\{\{x\}, \frac{x}{4}\right\}$
In[6] := q2 = %[[2]]
Out[6] $= \frac{x}{4}$
So, we have obtained two different results, q_1 and q_2. In fact they are equal due to $p_1 = 0$ and $p_2 = 0$, but this is not evident. Every time when more than one substitution can be applied to a term of a polynomial q (in this particular case, we can replace either x^2 or xy in x^2y), a fork appears; maybe, its branches join later, but maybe, they don't (as in this case).

A set of polynomials $p_1, \ldots, p_n$ is called a **Gröbner basis** (for a given monomial order) if reduction of any polynomial q with respect to this set is unique.

This definition is not constructive: it does not say how to check if a given set of polynomials forms a Gröbner basis. Presently we shall formulate Buchberger algorithm which transforms a set of polynomials (constraints on variables) into an equivalent system of constraints which is a Gröbner basis.

7.4 S-Polynomials

In our example, the constraints $p_1 = 0$ and $p_2 = 0$ allow us to simplify the monomials x^2 and xy. Do these constraints contain an extra information usable for simplification but not obvious? Yes, they do! Let's multiply p_1 and p_2 by monomials (i.e., products of powers of variables) in such a way that their leading terms become identical (equal to the least common multiple of the leading terms of p_1 and p_2). Then we subtract the second polynomial from the first one. The leading terms cancel, and we get a new polynomial with a new leading term which can be used for simplifying terms in q (because this new polynomial also vanishes). This polynomial is called the S-polynomial $S[p_1, p_2]$ (from the word subtraction). In our example
In[7] := S = Expand[y ∗ p1 − x ∗ p2]
Out[7] $= \frac{x}{4} - y + y^3$
This polynomial can be added to the system of constraints $p_1 = 0$, $p_2 = 0$. Let's normalize its leading coefficient to 1:
In[8] := p3 = Expand[4 ∗ S]
Out[8] $= x - 4y + 4y^3$
In[9] := Clear[S]

Now we have a new possibility for reduction:
In[10] := PolynomialReduce[q2, {p3}, {x, y}]

$\mathrm{Out}[10] = \left\{\left\{\frac{1}{4}\right\}, y - y^3\right\}$

Now we've got the same result q_1. The polynomials $\{p_1, p_2, p_3\}$ form a Gröbner basis. This set can be simplified by reducing them with respect to each other:
In[11] := PolynomialReduce[p1, {p3}, {x, y}]

$\mathrm{Out}[11] = \left\{\left\{x + 4y - 4y^3\right\}, -1 + 17y^2 - 32y^4 + 16y^6\right\}$

In[12] := p1a = Expand[%[[2]]/16]

$\mathrm{Out}[12] = -\frac{1}{16} + \frac{17y^2}{16} - 2y^4 + y^6$

In[13] := PolynomialReduce[p2, {p3}, {x, y}]

$\mathrm{Out}[13] = \left\{\{y\}, \frac{1}{4}\left(-1 + 16y^2 - 16y^4\right)\right\}$

In[14] := p2a = Expand[−%[[2]]/4]

$\mathrm{Out}[14] = \frac{1}{16} - y^2 + y^4$

In[15] := PolynomialReduce[p1a, p2a, {x, y}]

$\mathrm{Out}[15] = \{\{-1 + y^2\}, 0\}$

The polynomial p_{1a} reduces to 0, and hence it can be excluded from the system of constraints on our variables x, y. The polynomials p_{2a} and p_3 form a reduced Gröbner basis (with respect to the lexicographic order with $x > y$). Reduced Gröbner basis is unique (for a given monomial order), if we accept the convention that the coefficients of the leading terms are 1.

7.5 Buchberger Algorithm

Generalizing this example, we can formulate an algorithm for construction of the Gröbner basis of a set of n polynomials $P = \{p_i\}$:

1. $S = \{$the set of pairs (p_i, p_j) of these polynomials with $i < j \leqslant n\}$
2. **while** S is not empty
3. choose and remove some pair (p_i, p_j) from S;
4. calculate the S-polynomial $S[p_i, p_j]$;
5. reduce it with respect to P;
6. if the result is not 0, add this polynomial to P, and the corresponding pairs to S.

The set of pairs S alternatingly shrinks and grows. But it can be proved that this process terminates after a finite number of steps and produces a Gröbner basis P. Reducing these polynomials with respect to each other and throwing zeros away, one can get the reduced Gröbner basis. Some variations can improve the efficiency of the algorithm. For example, when adding a new polynomial to the set P, we can reduce all polynomials already in P with respect to the new one; if some of them changes, reduce other ones with respect to them, and so on (throwing zeros away while doing so). The order in which pairs are selected from the set S is very important—a good choice can reduce the amount of computations drastically.

Let's ask *Mathematica* to construct the Gröbner basis for the system $\{p_1, p_2\}$ with respect to the lexicographic order with $x > y$:

In[16] := *B* = GroebnerBasis[{p1,p2},{*x*,*y*}]
Out[16] = $\{1-16y^2+16y^4, x-4y+4y^3\}$

Let's reduce the polynomial q to the canonical form, i.e., reduce it with respect to the Gröbner basis (the result is unique).

In[17] := PolynomialReduce[*q*,*B*,{*x*,*y*}]
Out[17] = $\left\{\left\{-\frac{x}{4}, \frac{1}{4}+xy\right\}, y-y^3\right\}$

It is difficult to predict the complexity of the Buchberger algorithm. In worst cases it can be very high, i.e., constructing the Gröbner basis of a moderately large system can require a huge amount of calculations. The complexity strongly depends on the monomial order being used. In the case of ordering by the total degree (and then something) reduction tries to lower the total degree of a polynomial. The number of possible terms in a polynomial of a low total degree is small. In the case of the lexicographic order, a polynomial of y of an arbitrarily large degree is considered simpler than x to the first power. Therefore reduction does not lower the number of terms in a polynomial as strongly as in the case of total-degree orders, and the complexity of Gröbner basis calculations is higher. On the other hand, a reduced Gröbner basis with respect to a lexicographic order provides more information useful for solving the system, as we shall see soon. *Mathematica* knows how to construct Gröbner bases with respect to monomial orders we discussed.

**In[18] := *B* = GroebnerBasis[{p1,p2},{*x*,*y*},
MonomialOrder → DegreeLexicographic]**
Out[18] = $\{-1+4xy, -1+x^2+y^2, x-4y+4y^3\}$
In[19] := PolynomialReduce[*q*,*B*,MonomialOrder → DegreeLexicographic]
Out[19] = $\left\{\left\{\frac{x}{4}, 0, 0\right\}, \frac{x}{4}\right\}$
In[20] := Clear[p1,p2,p3,p1a,p2a,*q*,q1,q2,*B*]

7.6 Is the System Compatible?

Consider the system

**In[21] := p1 = *x*^2 ∗ *y* + 4 ∗ *y*^2 − 17; p2 = 2 ∗ *x* ∗ *y* − 3 ∗ *y*^3 + 8;
p3 = *x* ∗ *y*^2 − 5 ∗ *x* ∗ *y* + 1;**

Let's construct its Gröbner basis—an equivalent system of equations.

In[22] := GroebnerBasis[{p1,p2,p3},{*x*,*y*,*z*}]
Out[22] = $\{1\}$

This system contains the equation $1 = 0$. This means that it has no solutions. If the Gröbner basis contains 1, the system is incompatible. The inverse statement can be

also proved—the Gröbner basis of an incompatible system always contains 1 (if we normalize all leading coefficients to 1; otherwise, just some nonzero constant).
In[23] := Clear[p1, p2, p3]

7.7 Gröbner Bases with Respect to Lexicographic Order

Reduction with respect to the lexicographic order first of all tries to lower the degree of the main variable (x in our examples), and if possible, down to 0. Therefore usually there is a subset of polynomials in a reduced Gröbner bases which don't contain x. When x is absent, reduction tries to lower the degree of y, and if possible, down to 0. Therefore usually among these polynomials there are those which don't contain y, and so on. In other words, a lexicographic Gröbner bases has a triangular structure. For example,
In[24] := *B* = GroebnerBasis[{*x*^2 + *y*^2 + *z*^2, *x* + *y* − *z*, *y* + *z*^2}, {*x*, *y*, *z*}]
Out[24] $= \{z^2+z^3+z^4, y+z^2, x-z-z^2\}$
The polynomial
In[25] := p1 = *B*[[1]]
Out[25] $= z^2+z^3+z^4$
depends only on the most junior variable z. This means that projections of all solutions of our system on the z axis form a finite set of points—roots of this equation. In our example, they are $z = 0$ and
In[26] := p1 = Expand[p1/*z*^2]; *s* = Solve[p1 == 0, *z*]
Out[26] $= \left\{\left\{z \to -(-1)^{1/3}\right\}, \left\{z \to (-1)^{2/3}\right\}\right\}$
In[27] := z1 = ComplexExpand[*z*/.*s*[[1]]]
Out[27] $= -\frac{1}{2} - \frac{i\sqrt{3}}{2}$
In[28] := z2 = ComplexExpand[*z*/.*s*[[2]]]
Out[28] $= -\frac{1}{2} + \frac{i\sqrt{3}}{2}$
Substituting any of these z values to
In[29] := p2 = *B*[[2]]
Out[29] $= y+z^2$
we find the corresponding y value. Substituting these z and y into
In[30] := p3 = *B*[[3]]
Out[30] $= x-z-z^2$
we find the corresponding x value. Thus solving any system of polynomial equations with several unknowns reduces to solving single-variable polynomial equations sequentially, thanks to lexicographic Gröbner bases. Even when some of them cannot be solved in radicals, it is easy to solve them numerically to any desired precision.
In[31] := Clear[*B*, p1, p2, p3, z1, z2]

And here is another example.

In[32] := $\mathbf{B = GroebnerBasis[\{x\hat{}2 - 2*x*y + 2*y\hat{}2 - 1, x*y - y*z + z\hat{}2 - 1, x*z + y\hat{}2 - y*z - 1\}, \{x, y, z\}]}$
Out[32] $= \{1 - y^2 - 2z^2 + y^2z^2 + z^4, -y + y^3 + z - y^2z + yz^2 - z^3, x - y - 2z + y^2z + z^3\}$
Now we have no equations with a single variable z; there are 2 equations containing z and y:
In[33] := p1 = Factor[*B*[[1]]]
Out[33] $= (-1+z)(1+z)\left(-1+y^2+z^2\right)$
In[34] := p2 = Factor[*B*[[2]]]
Out[34] $= (y-z)\left(-1+y^2+z^2\right)$
The common set of their solutions is the circle $y^2 + z^2 = 1$. Substituting a point on this circle into
In[35] := p3 = *B*[[3]]
Out[35] $= x - y - 2z + y^2z + z^3$
we find the corresponding x value, that is, the set of solutions of this system is one-dimensional.
In[36] := Clear[*B*, p1, p2, p3]

For solving a system of polynomial equations it is useful to construct its Gröbner basis and then to factorize its elements.

7.8 Is the Number of Solutions Finite?

Gröbner bases with respect to other monomial orders don't have such simple triangular structure. But any Gröbner basis can tell us not only if the system is compatible but also if the number of its solutions is finite. Let's consider the same examples.
In[37] := $\mathbf{GroebnerBasis[\{x\hat{}2 + y\hat{}2 + z\hat{}2, x + y - z, y + z\hat{}2\}, \{x, y, z\},}$ **MonomialOrder → DegreeLexicographic]**
Out[37] $= \{x + y - z, y + z^2, -y + y^2 - yz\}$
The leading terms of the polynomials forming this basis are x, z^2, and y^2. What is the dimensionality of the space of polynomials which cannot be reduced with respect to this basis? Only monomials which are not divisible by these leading terms cannot be reduced, namely, 1, y, and z. So the space of polynomials reduced to the canonical form is three-dimensional for our system of constraints on the variables. Therefore our system has 3 solutions (there explicit form can be obtained more easily from the lexicographic Gröbner basis, as we have seen).

If each variable raised to some power is the leading term of some element of a Gröbner basis, then any monomials with this (or higher) degree of this variable are reducible. Irreducible monomials are inside the parallelepiped bounded by these powers, and their number is finite. Therefore the space of polynomials reduced to the canonical form is finite-dimensional, and the system has a finite number of solutions.

And here is our second example:

```
In[38] := GroebnerBasis[{x^2 - 2*x*y + 2*y^2 - 1, x*y - y*z + z^2 - 1,
    x*z + y^2 - y*z - 1}, {x, y, z}, MonomialOrder → DegreeLexicographic]
```

Out[38] $= \{-1+y^2+xz-yz, -1+xy-yz+z^2, -3+x^2+2y^2-2yz+2z^2,$
$x-y-2z+y^2z+z^3, x-2y+y^3-z+yz^2\}$

The leading terms are xz, xy, x^2, y^2z, and y^3. Among them there are powers of x and of y, but not of z. Therefore the space of polynomials in the canonical form (i.e., reduced with respect to this basis) is infinite-dimensional. This space contains, e.g., the directions $1, z, z^2, z^3 \ldots$ (and not only them). This means that the set of solutions of our system is infinite.

So, the criterion works in the opposite direction, too. If there exists a variable no power of which appears as the leading term of some element of the Gröbner basis (not being multiplied by some other variable), then all powers of this variable are irreducible, and the space of polynomials in the canonical form is infinite-dimensional. And hence the set of solutions of the equation system is infinite.

Knowing the reduced Gröbner basis (for any monomial order) one can also find the dimensionality of the set of solutions [14]. Consider sets of variables satisfying the following condition: none of the leading terms of the elements of the basis is a product of powers of these variables. The number of variables in the longest set gives the dimensionality of the set of solutions. In our example there is just one such set—$\{z\}$. Therefore the set of solutions of this system is one-dimensional.

Chapter 8
Calculus

8.1 Series

Let's expand a function in x at the point $x = 0$ up to the fifth order.

In[1] := s = Series[Exp[x], {x, 0, 5}]

$$\text{Out}[1] = 1 + x + \frac{x^2}{2} + \frac{x^3}{6} + \frac{x^4}{24} + \frac{x^5}{120} + O[x]^6$$

How are series represented in *Mathematica*? By the function SeriesData. Its first argument is the expansion variable; the second one—the expansion point; the third one—the list of coefficients; the fourth one—the minimum degree (here 0); the fifth one—the power of $O[x]$; the sixth one is 1 for series with integer degrees (all degrees are divided by it if it's not 1). Thus a series is not a sum (Plus) in spite of its appearance.

In[2] := FullForm[s]

Out[2]//FullForm =
SeriesData[x, 0, List[1, 1, Rational[1, 2], Rational[1, 6], Rational[1, 24], Rational[1, 120]], 0, 6, 1]

Coefficients are extracted by the function SeriesCoefficient.

In[3] := Do[Print[SeriesCoefficient[s, n]], {n, 0, 5}]

1

1

$\frac{1}{2}$

$\frac{1}{6}$

$\frac{1}{24}$

$\frac{1}{120}$

This series begins with degree -1.

In[4] := s = Series[Cot[x], {x, 0, 5}]

$$\text{Out}[4] = \frac{1}{x} - \frac{x}{3} - \frac{x^3}{45} - \frac{2x^5}{945} + O[x]^6$$

A. Grozin, *Introduction to Mathematica® for Physicists*, Graduate Texts in Physics,
DOI 10.1007/978-3-319-00894-3_8, © Springer International Publishing Switzerland 2014

In[5] := FullForm[s]
Out[5]//FullForm =
SeriesData[x, 0, List[1, 0, Rational[−1, 3], 0, Rational[−1, 45], 0, Rational[−2, 945]], −1, 6, 1]

This is a series with half-integer degrees.

In[6] := s = Series[Sqrt[x ∗ (1 − x)], {x, 0, 5}]

$$\text{Out[6]} = \sqrt{x} - \frac{x^{3/2}}{2} - \frac{x^{5/2}}{8} - \frac{x^{7/2}}{16} - \frac{5x^{9/2}}{128} + O[x]^{11/2}$$

In[7] := FullForm[s]
Out[7]//FullForm =
SeriesData[x, 0, List[1, 0, Rational[−1, 2], 0, Rational[−1, 8], 0, Rational[−1, 16], 0, Rational[−5, 128]], 1, 11, 2]

This is an expansion at infinity.

In[8] := s = Series[Log[x + 1], {x, Infinity, 4}]

$$\text{Out[8]} = -\text{Log}\left[\frac{1}{x}\right] + \frac{1}{x} - \frac{1}{2x^2} + \frac{1}{3x^3} - \frac{1}{4x^4} + O\left[\frac{1}{x}\right]^5$$

In[9] := FullForm[s]
Out[9]//FullForm =
SeriesData[x, DirectedInfinity[1], List[Times[−1, Log[Power[x, −1]]], 1, Rational[−1, 2], Rational[1, 3], Rational[−1, 4]], 0, 5, 1]

Coefficients of a series in x may depend on x, but only weakly, weaker than any degree.

In[10] := s = Series[x^x, {x, 0, 3}]

$$\text{Out[10]} = 1 + \text{Log}[x]x + \frac{1}{2}\text{Log}[x]^2x^2 + \frac{1}{6}\text{Log}[x]^3x^3 + O[x]^4$$

In[11] := FullForm[s]
Out[11]//FullForm =
SeriesData[x, 0, List[1, Log[x], Times[Rational[1, 2], Power[Log[x], 2]], Times[Rational[1, 6], Power[Log[x], 3]]], 0, 4, 1]

In[12] := Clear[s]

Operations with Series

Let's take three series.

In[13] := sinx = Series[Sin[x], {x, 0, 7}]

$$\text{Out[13]} = x - \frac{x^3}{6} + \frac{x^5}{120} - \frac{x^7}{5040} + O[x]^8$$

In[14] := cosx = Series[Cos[x], {x, 0, 7}]

$$\text{Out[14]} = 1 - \frac{x^2}{2} + \frac{x^4}{24} - \frac{x^6}{720} + O[x]^8$$

In[15] := tanx = Series[Tan[x], {x, 0, 7}]

$$\text{Out[15]} = x + \frac{x^3}{3} + \frac{2x^5}{15} + \frac{17x^7}{315} + O[x]^8$$

Series can be added, multiplied, divided, etc.

In[16] := tanx ∗ cosx

$\text{Out}[16] = x - \frac{x^3}{6} + \frac{x^5}{120} - \frac{x^7}{5040} + O[x]^8$

In[17] := sinx/cosx

$\text{Out}[17] = x + \frac{x^3}{3} + \frac{2x^5}{15} + \frac{17x^7}{315} + O[x]^8$

In[18] := sinx^2 + cosx^2

$\text{Out}[18] = 1 + O[x]^8$

If a series occurs as an argument of a function, the function is expanded automatically.

In[19] := Exp[sinx]

$\text{Out}[19] = 1 + x + \frac{x^2}{2} - \frac{x^4}{8} - \frac{x^5}{15} - \frac{x^6}{240} + \frac{x^7}{90} + O[x]^8$

In[20] := (1 − cosx)/x^2

$\text{Out}[20] = \frac{1}{2} - \frac{x^2}{24} + \frac{x^4}{720} + O[x]^6$

Here is an interesting method to expand a function in x.

In[21] := X = Series[x, {x, 0, 7}]

$\text{Out}[21] = x + O[x]^8$

In[22] := Sin[X]

$\text{Out}[22] = x - \frac{x^3}{6} + \frac{x^5}{120} - \frac{x^7}{5040} + O[x]^8$

In[23] := Clear[X]

Series can be differentiated and integrated.

In[24] := D[cosx, x]

$\text{Out}[24] = -x + \frac{x^3}{6} - \frac{x^5}{120} + O[x]^7$

In[25] := Integrate[tanx, x]

$\text{Out}[25] = \frac{x^2}{2} + \frac{x^4}{12} + \frac{x^6}{45} + \frac{17x^8}{2520} + O[x]^9$

A series (beginning from a small term) can be substituted for the expansion variable of another series. This is Sin[Tan[x]].

In[26] := st = sinx/.x−>tanx

$\text{Out}[26] = x + \frac{x^3}{6} - \frac{x^5}{40} - \frac{55x^7}{1008} + O[x]^8$

An alternative syntax.

In[27] := ComposeSeries[sinx, tanx]

$\text{Out}[27] = x + \frac{x^3}{6} - \frac{x^5}{40} - \frac{55x^7}{1008} + O[x]^8$

Let's subtract Tan[Sin[x]]; this expression is expanded automatically, i.e., series are contagious.

In[28] := st − Tan[Sin[x]]

$\text{Out}[28] = -\frac{x^7}{30} + O[x]^8$

In[29] := Clear[st]

Series inversion—solving the equation tanx $= y$ for x as a series in y.

In[30] := atany = InverseSeries[tanx, y]

Out[30] $= y - \frac{y^3}{3} + \frac{y^5}{5} - \frac{y^7}{7} + O[y]^8$

The result should be the arctangent.

In[31] := Series[ArcTan[y], {y, 0, 7}]

Out[31] $= y - \frac{y^3}{3} + \frac{y^5}{5} - \frac{y^7}{7} + O[y]^8$

In[32] := ComposeSeries[tanx, atany]

Out[32] $= y + O[y]^8$

It is not allowed to substitute a numerical value for the expansion variable into a series. The function Normal converts a series into a normal expression by dropping $+O[x]^n$. Here is a plot of sine and a few truncations of its series.

In[33] := Plot[Evaluate[Prepend[Table[Normal[Series[Sin[x], {x, 0, n}]], {n, 1, 5, 2}], Sin[x]]], {x, −Pi, Pi}]

Out[33] =

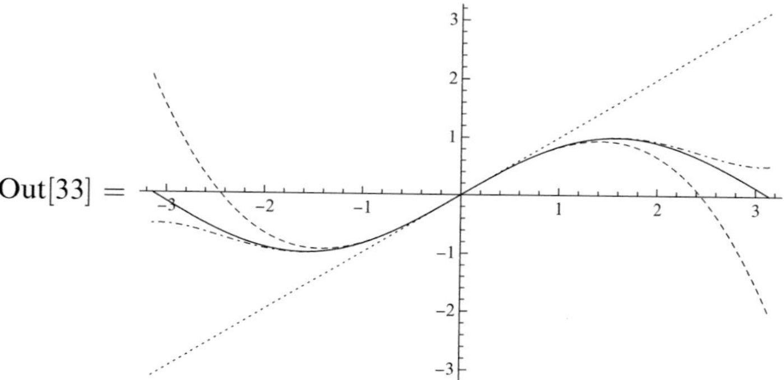

In[34] := Clear[sinx, cosx, tanx, atany]

You should work with series as long as possible, converting them into normal polynomials only at the very end. Then terms of too high orders of smallness are dropped automatically. At any moment you know exactly what is the order of the neglected term $O[x]^n$.

Arbitrary-Order Term

The function SeriesCoefficient can also be used in this way.

In[35] := SeriesCoefficient[Exp[x], {x, 0, 4}]

Out[35] $= \frac{1}{24}$

This is the 4th coefficient in the expansion of Exp[x] in x at 0. The number of the series term can be given symbolically.

In[36] := cn = SeriesCoefficient[Exp[x], {x,0,n}]

$$\text{Out}[36] = \begin{cases} \frac{1}{n!} & n \geq 0 \\ 0 & \text{True} \end{cases}$$

In[37] := Sum[cn ∗ x^n, {n,0,Infinity}]

Out[37] = e^x

In[38] := cn = SeriesCoefficient[Cos[x], {x,0,n}]

$$\text{Out}[38] = \begin{cases} \frac{i^n(1+(-1)^n)}{2n!} & n \geq 0 \\ 0 & \text{True} \end{cases}$$

In[39] := Sum[cn ∗ x^n, {n,0,Infinity}]

Out[39] = Cos[x]

In[40] := Clear[cn]

8.2 Differentiation

In[41] := f = x ∗ Sin[x + y]

Out[41] = x Sin[x + y]

The derivative in x; in y; in x and y; the second derivative in x; the second derivative in x and the first one in y:

In[42] := {D[f,x], D[f,y], D[f,x,y], D[f,{x,2}], D[f,{x,2},y]}

Out[42] = {x Cos[x + y] + Sin[x + y], x Cos[x + y], Cos[x + y] − x Sin[x + y],
2 Cos[x + y] − x Sin[x + y], −x Cos[x + y] − 2 Sin[x + y]}

In[43] := Clear[f]

Unknown Functions

Expressions with unknown functions can be differentiated.

In[44] := D[x ∗ f[x^2], x]

Out[44] = $f[x^2] + 2x^2 f'[x^2]$

Mathematica represents the first derivative of an unknown function f as the operator Derivative[1] applied to f.

In[45] := FullForm[%]

Out[45]//FullForm =
Plus[f[Power[x,2]], Times[2, Power[x,2], Derivative[1][f][Power[x,2]]]]

And this is the second derivative.

In[46] := Expand[D[x ∗ f[x^2], {x,2}]]

Out[46] = $6xf'[x^2] + 4x^3 f''[x^2]$

In[47] := FullForm[%]

Out[47]//FullForm =
Plus[Times[6, x, Derivative[1][f][Power[x,2]]],
Times[4, Power[x,3], Derivative[2][f][Power[x,2]]]]

Derivative[2,3] means the second derivative in the first argument and the third one in the second.

In[48] := D[f[x,y],{x,2},{y,3}]

Out[48] = $f^{(2,3)}[x,y]$

In[49] := FullForm[%]

Out[49]//FullForm =

Derivative[2,3][f][x,y]

Defining Derivatives

Let's tell *Mathematica* that the derivative of the function f is g.

In[50] := f'[x_] := g[x]

In[51] := D[x * f[x^2],x]

Out[51] = $f\left[x^2\right]+2x^2 g\left[x^2\right]$

The second derivative is not substituted automatically.

In[52] := Expand[D[x * f[x^2],{x,2}]]

Out[52] = $6xg\left[x^2\right]+4x^3 f''\left[x^2\right]$

we can tell Mathematica that $\frac{\partial^3 f[x,y]}{\partial x \partial^2 y}$ is a function g.

In[53] := Derivative[1,2][f][x_,y_] := g[x,y]

In[54] := D[x * f[x,y],{x,2},{y,2}]

Out[54] = $2g[x,y]+xf^{(2,2)}[x,y]$

8.3 Integration

Indefinite Integrals

In[55] := Integrate[1/(x * (x^2 − 2)^2),x]

Out[55] = $-\frac{1}{4(-2+x^2)}+\frac{\text{Log}[x]}{4}-\frac{1}{8}\text{Log}\left[2-x^2\right]$

In[56] := Integrate[1/(Exp[x] + 1),x]

Out[56] = $x-\text{Log}\left[1+e^x\right]$

In[57] := Integrate[x/(Exp[x] + 1),x]

Out[57] = $\frac{x^2}{2}-x\text{Log}\left[1+e^x\right]-\text{PolyLog}\left[2,-e^x\right]$

In[58] := Integrate[Log[x],x]

Out[58] = $-x+x\text{Log}[x]$

In[59] := Integrate[1/Log[x],x]

Out[59] = LogIntegral[x]

In[60] := Integrate[Exp[x^2],x]

$\text{Out}[60] = \frac{1}{2}\sqrt{\pi}\,\text{Erfi}[x]$

In[61] := Integrate[x * Exp[x^2],x]

$\text{Out}[61] = \frac{e^{x^2}}{2}$

In[62] := Integrate[1/Sqrt[(1 − x^2) * (1 − k^2 * x^2)],x]

$\text{Out}[62] = \frac{\sqrt{1-x^2}\sqrt{1-k^2x^2}\,\text{EllipticF}\left[\text{ArcSin}[x],k^2\right]}{\sqrt{(-1+x^2)(-1+k^2x^2)}}$

In[63] := Simplify[Integrate[x/Sqrt[(1 − x^2) * (1 − k^2 * x^2)],x],x > 1]

$\text{Out}[63] = \frac{\text{Log}\left[k\left(k\sqrt{-1+x^2}+\sqrt{-1+k^2x^2}\right)\right]}{k}$

Definite Integrals

Here *Mathematica* produces a result with some assumptions about the parameter n.

In[64] := Integrate[x^n, {x,0,1}]

$\text{Out}[64] = \text{ConditionalExpression}\left[\frac{1}{1+n},\text{Re}[n] > -1\right]$

Let's tell it that $n > -1$.

In[65] := Integrate[x^n, {x,0,1},Assumptions−>{n > −1}]

$\text{Out}[65] = \frac{1}{1+n}$

In[66] := Integrate[Exp[a * Sin[x]], {x,0,2 * Pi}]

$\text{Out}[66] = 2\pi\,\text{BesselI}[0,a]$

In[67] := Integrate[Log[x]/(1 − x), {x,0,1}]

$\text{Out}[67] = -\frac{\pi^2}{6}$

The default value of the option Assumptions for Simplify, Integrate, etc. can be given in the variable $Assumptions.

In[68] := $Assumptions = {t > 0,t < 1,a > −1,b > −1};

In[69] := Integrate[x^a * (1 − x)^b * (1 − t * x)^c, {x,0,1}]

Out[69] = −(π Csc[aπ] Gamma[1 + b] Hypergeometric2F1Regularized[1 + a, −c, 2 + a + b, t])/Gamma[−a]

Now we can clear $Assumptions.

In[70] := $Assumptions = True;

Multiple integral

In[71] := Integrate[1/(1 + x * y), {x,0,1}, {y,0,1}]

$\text{Out}[71] = \frac{\pi^2}{12}$

8.4 Summation

Finite Sums

In[72] := Sum[n, {n, 0, k}]
Out[72] $= \frac{1}{2}k(1+k)$
In[73] := Sum[n^2, {n, 0, k}]
Out[73] $= \frac{1}{6}k(1+k)(1+2k)$
In[74] := Sum[x^n, {n, 0, k}]
Out[74] $= \frac{-1+x^{1+k}}{-1+x}$
In[75] := Sum[Binomial[k, n], {n, 0, k}]
Out[75] $= 2^k$
In[76] := Sum[(−1)^n ∗ Binomial[k, n], {n, 0, k}]
Out[76] = KroneckerDelta[k]
In[77] := Sum[Binomial[k, n]^2, {n, 0, k}]
Out[77] = Binomial[2k, k]

Series

In[78] := Sum[1/n^2, {n, 1, Infinity}]
Out[78] $= \frac{\pi^2}{6}$
In[79] := Sum[1/n^4, {n, 1, Infinity}]
Out[79] $= \frac{\pi^4}{90}$
In[80] := Sum[(−1)^n/n^2, {n, 1, Infinity}]
Out[80] $= -\frac{\pi^2}{12}$
In[81] := Sum[x^n/n!, {n, 0, Infinity}]
Out[81] $= e^x$

8.5 Differentiol Equations

A first-order differential equation.
In[82] := DSolve[D[x[t], t] + x[t] == 0, x[t], t]
Out[82] $= \{\{x[t] \to e^{-t}C[1]\}\}$
The solution contains an arbitrary constant $C[1]$. Let's add an initial condition:

In[83] := DSolve[{D[x[t], t] + x[t] == 0, x[0] == 1}, x[t], t]
Out[83] = $\{\{x[t] \to e^{-t}\}\}$

A second-order differential equation.

In[84] := DSolve[D[x[t], {t, 2}] + x[t] == 0, x[t], t]
Out[84] = $\{\{x[t] \to C[1]\,\mathrm{Cos}[t] + C[2]\,\mathrm{Sin}[t]\}\}$

Initial conditions.

In[85] := DSolve[{D[x[t], {t, 2}] + x[t] == 0, x[0] == 0, x'[0] == 1}, x[t], t]
Out[85] = $\{\{x[t] \to \mathrm{Sin}[t]\}\}$

Boundary conditions.

In[86] := DSolve[{D[x[t], {t, 2}] + x[t] == 0, x[0] == 0, x[1] == 1}, x[t], t]
Out[86] = $\{\{x[t] \to \mathrm{Csc}[1]\,\mathrm{Sin}[t]\}\}$

A system of differential equations.

In[87] := DSolve[{D[x[t], t] == p[t], D[p[t], t] == −x[t]}, {x[t], p[t]}, t]
Out[87] = $\{\{p[t] \to C[1]\,\mathrm{Cos}[t] - C[2]\,\mathrm{Sin}[t], x[t] \to C[2]\,\mathrm{Cos}[t] + C[1]\,\mathrm{Sin}[t]\}\}$

Chapter 9
Numerical Calculations

9.1 Approximate Numbers in *Mathematica*

Mathematica usually works with exact numbers, either symbolic (π, e) or rational, and derives exact analytical results. However, it can also perform approximate numerical calculations. Many problems cannot be solved analytically, but numerical solution is possible. On the other side, an analytical result can depend on symbolic parameters; in order to do a numerical calculation, you have to substitute some numerical values for all parameters. It is often useful to check the correctness of a complicated analytical derivation by a direct numerical calculation for a few sets of values of the parameters.

There are two kinds of approximate real numbers in *Mathematica*. The first one is *machine numbers*.

In[1] := *p* = *N*[Pi]
Out[1] = 3.14159
In[2] := FullForm[*p*]
Out[2]//FullForm =
3.141592653589793`

Precision is the number of significant decimal digits. For machine numbers it is a symbolic constant:

In[3] := Precision[*p*]
Out[3] = MachinePrecision
In[4] := *N*[MachinePrecision]
Out[4] = 15.9546

It is 53 bits (or about 16 decimal digits) of precision. In other languages (C, Fortran) such numbers are usually called double precision. Operations with such numbers in *Mathematica* are performed by hardware, as in C or Fortran. They are less efficient than in these languages, but nevertheless rather efficient.

In[5] := MachineNumberQ[*p*]
Out[5] = True

A. Grozin, *Introduction to Mathematica® for Physicists*, Graduate Texts in Physics,
DOI 10.1007/978-3-319-00894-3_9, © Springer International Publishing Switzerland 2014

There are also *arbitrary-precision* numbers. Operations with them are implemented in software and are far less efficient.

In[6] := p = N[Pi, 25]
Out[6] = 3.141592653589793238462643
In[7] := FullForm[p]
Out[7]//FullForm =
3.141592653589793238462643`25.
In[8] := Precision[p]
Out[8] = 25.

Precision is a part of a value, not of a variable, as in some other languages. If it is equal to n, then the estimated relative error of the value is 10^{-n}. There is also *accuracy*—the number of significant decimal digits after the point. If it is equal to m, then the estimated absolute error of the value is 10^{-m}.

In[9] := Accuracy[p]
Out[9] = 24.5029

When approximate numbers are added or subtracted, the absolute errors are added. The difference of two approximately equal numbers has a lower precision than the operands.

In[10] := q = N[355/113, 20]
Out[10] = 3.1415929203539823009
In[11] := Accuracy[q]
Out[11] = 19.5029
In[12] := d = p − q
Out[12] = $-2.667641890624 \times 10^{-7}$
In[13] := {Precision[d], Accuracy[d]}
Out[13] = {12.929, 19.5028}

When approximate numbers are multiplied or divided, the relative errors are added.

In[14] := r = p/q
Out[14] = 0.99999991508632855520
In[15] := {Precision[r], Accuracy[r]}
Out[15] = {20., 20.}
In[16] := Clear[p, q, d, r]

This error handling sometimes may be too pessimistic. Let's consider an example [15]. The sequence $x_n = f(x_{n-1})$, $x_1 = 1$,

In[17] := f[x_] := (x^2 + 4)/(2 ∗ x)

converges to 2. With machine numbers

**In[18] := f[x_, n_] := Module[{t = Table[x, {n}]},
Do[t[[i]] = f[t[[i − 1]]], {i, 2, n}]; t]**
In[19] := x = f[1.0, 60];

**In[20] := ListPlot[x, PlotRange−>{0.95, 2.55},
PlotMarkers−>{Automatic, Medium}]**

Out[20] =

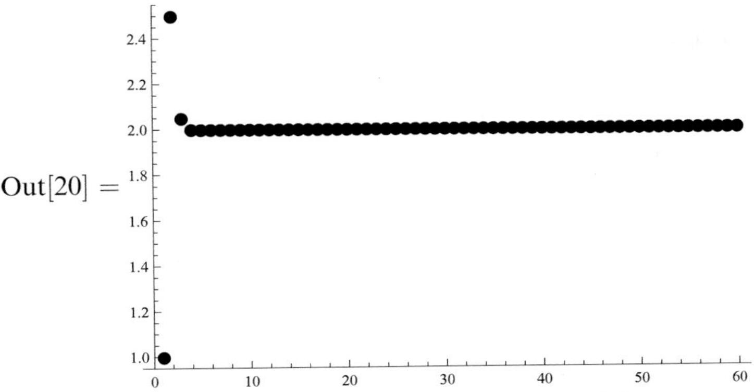

this is indeed so. Now let x_1 be 1.0 with 17-digits precision:

In[21] := x = f[1.0`17, 60];

**In[22] := ListPlot[x, PlotRange−>{−0.05, 2.55},
PlotMarkers−>{Automatic, Medium}]**

Out[22] =

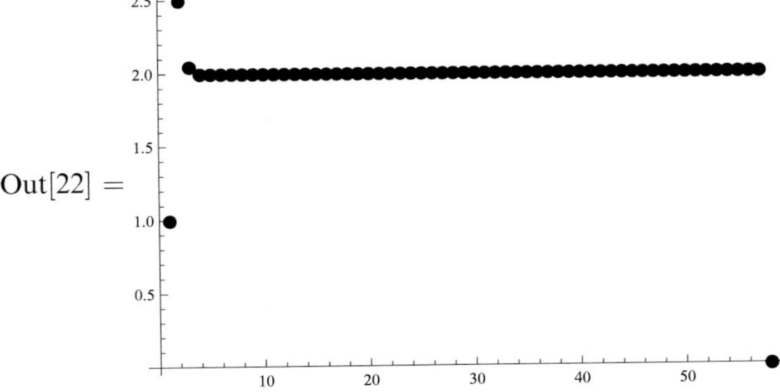

**In[23] := ListPlot[Map[Precision, x],
PlotMarkers−>{Automatic, Medium}]**

Out[23] =

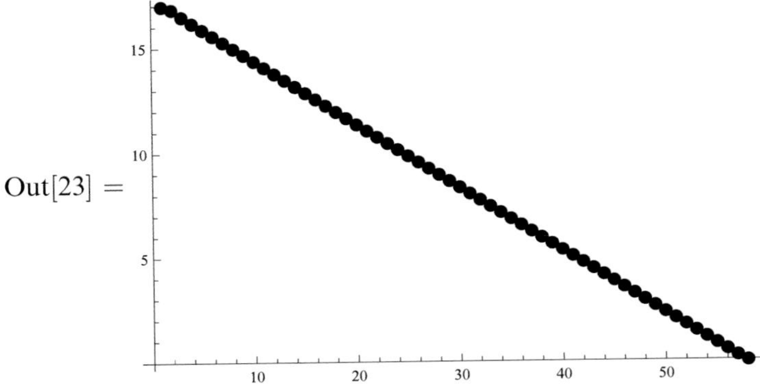

The tail of this list is
In[24] := x[[55;;60]]
Out[24] = {2.,2.,0.,0.,ComplexInfinity, Indeterminate}
In[25] := Map[Precision,x[[55;;60]]]
Out[25] = {0.84866, 0.54763, 0.2466, 0., ∞, ∞}
The initial precision is completely lost in 58 iterations.
In[26] := Clear[f,x]

9.2 Solving Equations

NSolve tries to solve equations numerically. They must not contain symbolic parameters, only numbers and unknowns. Only very limited classes of equations can be solved analytically; numerical solution is possible nearly always. The option Reals says to find only real roots.
In[27] := NSolve[x^5 + x + 1 == 0,x]
Out[27] = {{x → −0.754878}, {x → −0.5 − 0.866025i}, {x → −0.5 + 0.866025i}, {x → 0.877439 − 0.744862i}, {x → 0.877439 + 0.744862i}}
In[28] := NSolve[x^5 + x + 1 == 0,x,Reals]
Out[28] = {{x → −0.754878}}
We can add an interval in which we want to find solutions.
In[29] := NSolve[{Exp[−x] == Sin[x],0 < x < Pi},x]
Out[29] = {{x → 0.588533}, {x → 3.09636}}
In[30] := Plot[{Exp[−x],Sin[x]}, {x,0,Pi}]

Out[30] =

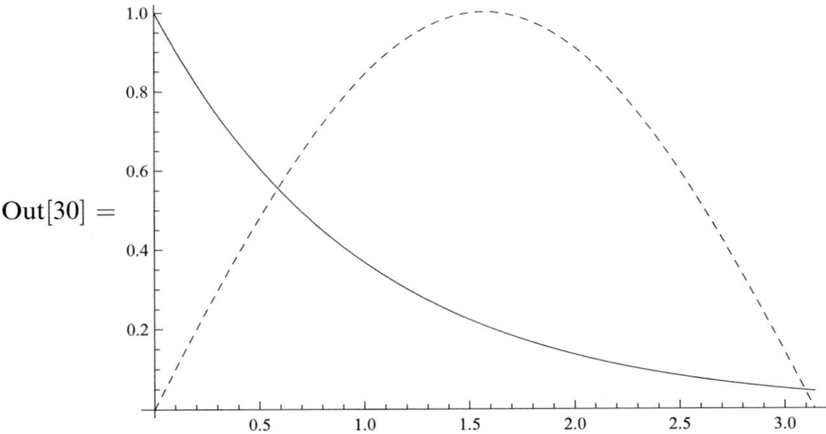

9.3 Numerical Integration and Summation

The function NIntegrate integrates numerically, without trying to do it analytically first—it uses an appropriate numerical method right away. Of course, integration limits must be numbers, and there must be no symbolic parameters.

In[31] := NIntegrate[Sin[*x*]/*x*, {*x*, 0, Infinity}]

Out[31] = 1.5708

The option PrecisionGoal states the desired precision of the result. If the result is close to 0 due to strong cancellations, it may be difficult to attain a high precision (i.e., a small relative error). Then it is better to specify AccuracyGoal, i.e., the desired absolute error. The option WorkingPrecision specifies the precision level at which internal calculations are done; it must be ⩾ PrecisionGoal.

In[32] := NIntegrate[Exp[−*x*^2], {*x*, 0, Infinity}, WorkingPrecision−>30]

Out[32] = 0.886226925452758013649083741785

The integration method is selected automatically; however, we can specify it:

In[33] := NIntegrate[1/(1 + *x* ∗ *y*), {*x*, 0, 1}, {*y*, 0, 1},
Method−>"AdaptiveMonteCarlo"]

Out[33] = 0.822237

In[34] := *i* = NIntegrate[Log[*x*]^2/(*x* + 1), {*x*, 0, 1}, PrecisionGoal−>30,
WorkingPrecision−>35]

Out[34] = 1.8030853547393914280996072422671750

Suppose we suspect that the integral i is a linear combination of $\zeta(3)$ and 1 with rational coefficients. FindIntegerNullVector tries to find integer coefficients such that the linear combination vanishes:

In[35] := FindIntegerNullVector[{*i*, Zeta[3], 1}]

Out[35] = {2, −3, 0}

This means that $2i - 3\zeta(3) = 0$, i.e., our integral is $\frac{3}{2}\zeta(3)$. Of course, this is not a mathematical proof. However, if we increase precision, and the linear combination stays the same, we can be practically sure that the result is correct (this is called *experimental mathematics*).

NSum is similar.

In[36] := *s* = NSum[1/*n*^4, {*n*, 1, Infinity}, PrecisionGoal−>30,
WorkingPrecision−>35, NSumTerms−>30]

Out[36] = 1.0823232337111381915160036965412

If we suspect that the sum s is a linear combination of π^4 and 1 with rational coefficients, we can do

In[37] := FindIntegerNullVector[{*s*, Pi^4, 1}]

Out[37] = {90, −1, 0}

This means that our sum is, probably, $\frac{\pi^4}{90}$.

In[38] := Clear[*i*, *s*]

9.4 Differential Equations

NDSolve solves differential equations numerically, for a finite interval of the independent variable. It returns results in terms of InterpolatingFunction; this result can be numerically evaluated for any value of the independent variable in the given interval.

In[39] := $a = 1/2$;

In[40] := ns = NDSolve[{$y''[t] + a * y'[t] + y[t] == 0, y'[0] == 0, y[0] == 1$}, $y[t], \{t, 0, 10\}$]

Out[40] = {{$y[t] \rightarrow$ InterpolatingFunction[{{0., 10.}}, <>][t]}}

In[41] := Plot[$y[t]$/.ns[[1]], {t, 0, 10}]

Out[41] =

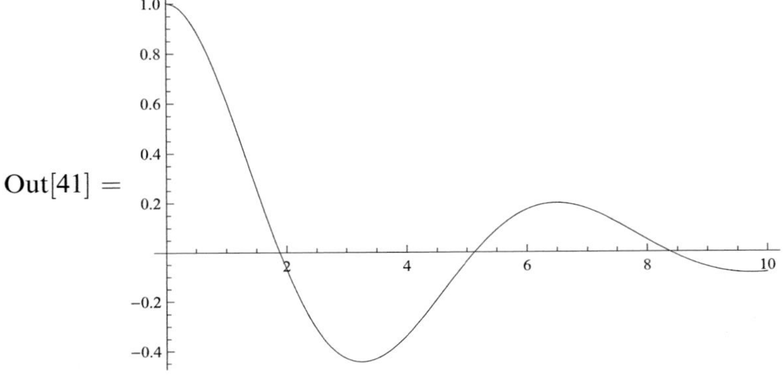

Let's compare with the analytical solution.

In[42] := s = DSolve[{y"[t] + $a * y'[t] + y[t] == 0, y'[0] == 0, y[0] == 1$}, $y[t], t$]

$$\text{Out[42]} = \left\{\left\{y[t] \rightarrow \frac{1}{15}\mathrm{e}^{-t/4}\left(15\mathrm{Cos}\left[\frac{\sqrt{15}t}{4}\right] + \sqrt{15}\mathrm{Sin}\left[\frac{\sqrt{15}t}{4}\right]\right)\right\}\right\}$$

In[43] := Plot[$y[t]$/.s[[1]], {t, 0, 10}]

Out[43] =

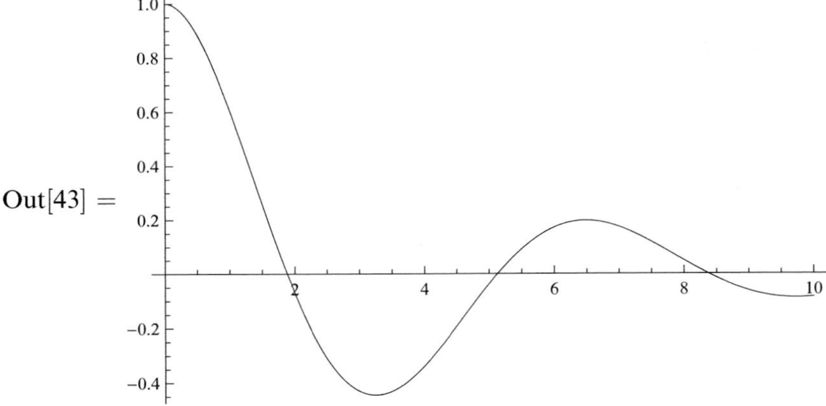

In[44] := Clear[a, s, ns]

Chapter 10
Risch Algorithm

We were taught at calculus classes that integration is an art, not a science (in contrast to differentiation—even a monkey can be trained to take derivatives). And we were taught wrong. The Risch algorithm (which is known for decades) allows one to find, in a finite number of steps, if a given indefinite integral can be taken in elementary functions, and if so, to calculate it. This algorithm has been constructed in works by an American mathematician Risch near 1970; many cases were not analyzed completely in these works and were later considered by other mathematicians. The algorithm is very complicated, and no computer algebra system implements it fully. Its implementation in *Mathematica* is rather complete, even with extensions to some classes of special functions, but details are not publicly known. Strictly speaking, it is not quite an algorithm, because it contains algorithmically unsolvable subproblems, such as finding out if a given combination of elementary functions vanishes. But in practice computer algebra systems are quite good in solving such problems. Here we shall consider, at a very elementary level, the main ideas of the Risch algorithm; see [16] for more details.

10.1 Rational Functions

We begin with a very simple case—integration of rational functions. Better methods than the partial fraction decomposition exist for this problem. And these methods can be generalized to much wider classes of integrands. Let's consider an integral

$$\int \frac{N(x)}{D(x)} \mathrm{d}x,$$

where $N(x)$ and $D(x)$ are polynomials. If $\deg N \geq \deg D$, we can divide with remainder; integration of a polynomial is trivial. Therefore we'll assume $\deg N < \deg D$. The integration result consists of a rational part and a logarithmic one:

A. Grozin, *Introduction to Mathematica® for Physicists*, Graduate Texts in Physics,
DOI 10.1007/978-3-319-00894-3_10,

$$\int \frac{N(x)}{D(x)}\,\mathrm{d}x = \frac{P(x)}{\hat{D}(x)} + \sum c_i \log(x-a_i)\,,$$

where a_i are the roots of the denominator $D(x)$, and c_i are constants. If

$$D(x) = \prod (x-a_i)^{d_i}\,,$$

then

$$\hat{D}(x) = \prod (x-a_i)^{d_i-1}\,.$$

Indeed, at $x \to a_i$ the rational part has a pole of the order $d_i - 1$; when differentiated, it becomes a pole of the order d_i, as needed. The numerator $P(x)$ is a polynomial of degree $\deg P < \deg \hat{D}$:

$$P(x) = \sum p_n x^n\,.$$

Substituting all these parts and differentiating, we can find the unknown coefficients c_i and p_n by solving a linear system.

For example, let's calculate

$$\int \frac{\mathrm{d}x}{x^2(x-1)} = \frac{p_0}{x} + c_1 \log(x) + c_2 \log(x-1)\,.$$

In[1] := Res = p[0]/x + c[1] ∗ Log[x] + c[2] ∗ Log[x − 1]
Out[1] = $c[2]\mathrm{Log}[-1+x] + c[1]\mathrm{Log}[x] + \frac{p[0]}{x}$
In[2] := Eq = Together[x^2 ∗ (x − 1) ∗ D[Res, x] − 1]
Out[2] = $-1 - xc[1] + x^2c[1] + x^2c[2] + p[0] - xp[0]$
In[3] := Eqs = Table[Coefficient[Eq, x, n] == 0, {n, 0, 2}]
Out[3] = $\{-1+p[0]==0, -c[1]-p[0]==0, c[1]+c[2]==0\}$
In[4] := Sol = Solve[Eqs, {p[0], c[1], c[2]}][[1]]
Out[4] = $\{p[0]\to 1, c[1]\to -1, c[2]\to 1\}$
In[5] := Res/. Sol
Out[5] = $\frac{1}{x} + \mathrm{Log}[-1+x] - \mathrm{Log}[x]$
In[6] := Clear[Res, Eq, Eqs, Sol]

10.2 Logarithmic Extension

Now we begin to extend the class of integrands and consider

$$\int \frac{N(x,y)}{D(x,y)}\,\mathrm{d}x\,,$$

where y depends on x (in a nonrational way). The extension is called algebraic if y is a root of a polynomial equation $p(x,y)=0$. For example, $y=\sqrt[n]{p(x)/q(x)}$ is a root of the equation $q(x)y^n-p(x)=0$. An algorithm for integration of expressions belonging to algebraic extensions has been constructed, but it requires an advanced mathematical apparatus [17], and we shall not discuss it here.

Extensions which are not algebraic are called transcendental. There are two important classes of such extensions. A logarithmic extension $y=\log r(x)$ (where $r(x)$ is a rational function) is characterized by the property $y'=r'/r$. An exponential extension $y=\exp r(x)$—by $y'=r'y$.

If an integral of an expression from a logarithmic extension with some $y=\log r(x)$ can be taken in elementary functions, it has the form

$$\int \frac{N(x,y)}{D(x,y)}\,\mathrm{d}x = \frac{P(x,y)}{\hat{D}(x,y)} + \sum c_i \log q_i\,,$$

where

$$D=\prod D_i^{d_i} \quad \Rightarrow \quad \hat{D}=\prod D_i^{d_i-1}\,,$$

q_i are the irreducible factors of all D_i, c_i are constants,

$$P(x,y)=\sum p_n(x)y^n=\sum p_{mn}x^m y^n$$

is a polynomial with unknown (so far) coefficients. Differentiating this general form of the result, putting everything over a common denominator, and equating coefficients of $x^m y^n$, we obtain a linear system for finding all unknown coefficients. If this system is incompatible, this means that the integral cannot be taken in elementary functions.

Example 1

We shall consider several examples. Let $y=\log x$ so that $y'=1/x$. Let's calculate the integral

$$\int y\,\mathrm{d}x = p_2(x)y^2+p_1(x)y+p_0(x)\,.$$

When differentiated, the degree in y reduces by 1, so that the result is quadratic in y (there is no denominator, and hence no q_i).

In[7] := y′[x_] := 1/x
In[8] := Res = Sum[p[n][x] * y[x]^n, {n, 0, 2}]
Out[8] $= p[0][x]+y[x]p[1][x]+y[x]^2p[2][x]$
In[9] := Eq = D[Res, x] − y[x]
Out[9] $= -y[x]+\dfrac{p[1][x]}{x}+\dfrac{2y[x]p[2][x]}{x}+p[0]'[x]+y[x]p[1]'[x]+y[x]^2p[2]'[x]$

In[10] := Eqs = Table[Coefficient[Eq, y[x], n] == 0, {n, 0, 2}]

$$\text{Out}[10] = \left\{\frac{p[1][x]}{x} + p[0]'[x] == 0, -1 + \frac{2p[2][x]}{x} + p[1]'[x] == 0, p[2]'[x] == 0\right\}$$

We see that $p_2(x)$ is a constant:

In[11] := p[2][x_] := p[2,0]; Eqs

$$\text{Out}[11] = \left\{\frac{p[1][x]}{x} + p[0]'[x] == 0, -1 + \frac{2p[2,0]}{x} + p[1]'[x] == 0, \text{True}\right\}$$

Since $p_1(x)$ is a polynomial, the second equation can be satisfied only if $p_{20} = 0$:

In[12] := p[2,0] = 0; Eqs

$$\text{Out}[12] = \left\{\frac{p[1][x]}{x} + p[0]'[x] == 0, -1 + p[1]'[x] == 0, \text{True}\right\}$$

The second equation gives

In[13] := p[1][x_] := x + p[1,0]

Now the first equation

In[14] := Eq1 = ExpandAll[Eqs[[1]]]

$$\text{Out}[14] = 1 + \frac{p[1,0]}{x} + p[0]'[x] == 0$$

gives $p_{10} = 0$:

In[15] := p[1,0] = 0; Eq1

$\text{Out}[15] = 1 + p[0]'[x] == 0$

Therefore $p_0(x) = -x$ (omitting the integration constant):

In[16] := p[0][x_] := −x; Eq1

Out[16] = True

In[17] := Res

$\text{Out}[17] = -x + xy[x]$

In[18] := Clear[Res, Eq, Eqs, Eq1, p]

We have derived the well-known result

$$\int \log(x)\mathrm{d}x = x\log(x) - x\,.$$

Consider the integrals

$$\int x\log(x)\mathrm{d}x\,, \qquad \int \log^2(x)\mathrm{d}x$$

in a similar way.

Example 2

Let's calculate the integral

$$\int \frac{y}{x}\mathrm{d}x = p_2(x)y^2 + p_1(x)y + p_0(x)\,.$$

Here $\hat{D} = 1$; it seems that there is a single q, namely x, but $\log(x) = y$, so that the logarithmic part contributes nothing new.

In[19] := Res = Sum[p[n][x] * y[x]^n, {n,0,2}];

In[20] := Eq = D[Res,x] − y[x]/x

$$\text{Out[20]} = -\frac{y[x]}{x} + \frac{p[1][x]}{x} + \frac{2y[x]p[2][x]}{x} + p[0]'[x] + y[x]p[1]'[x] + y[x]^2 p[2]'[x]$$

In[21] := Eqs = Table[Coefficient[Eq, y[x], n] == 0, {n,0,2}]

$$\text{Out[21]} = \left\{\frac{p[1][x]}{x} + p[0]'[x] == 0, -\frac{1}{x} + \frac{2p[2][x]}{x} + p[1]'[x] == 0, p[2]'[x] == 0\right\}$$

In[22] := p[2][x_] := p[2,0]; Eqs

$$\text{Out[22]} = \left\{\frac{p[1][x]}{x} + p[0]'[x] == 0, -\frac{1}{x} + \frac{2p[2,0]}{x} + p[1]'[x] == 0, \text{True}\right\}$$

From the second equation, $p_{20} = 1/2$; then p_1 is a constant:

In[23] := p[2,0] = 1/2; Eqs[[2]]

$\text{Out[23]} = p[1]'[x] == 0$

In[24] := p[1][x_] = p[1,0]; Eqs

$$\text{Out[24]} = \left\{\frac{p[1,0]}{x} + p[0]'[x] == 0, \text{True}, \text{True}\right\}$$

From the first equation, $p_{10} = 0$; then p_0 is a constant (which may be omitted):

In[25] := p[1,0] = 0; Eqs[[1]]

$\text{Out[25]} = p[0]'[x] == 0$

In[26] := p[0][x_] := 0; Res

$$\text{Out[26]} = \frac{y[x]^2}{2}$$

In[27] := Clear[Res, Eq, Eqs, p]

We have derived the well-known result

$$\int \frac{\log(x)}{x}\,\mathrm{d}x = \frac{1}{2}\log^2(x)\,.$$

Example 3

Let's change the previous integral a little:

$$\int \frac{y}{x+1}\,\mathrm{d}x = p_2(x)y^2 + p_1(x)y + p_0(x) + c\log(x+1)\,.$$

In[28] := Res = Sum[p[n][x] * y[x]^n, {n,0,2}] + c * Log[x + 1]

$\text{Out[28]} = c, \text{Log}[1+x] + p[0][x] + y[x]p[1][x] + y[x]^2 p[2][x]$

In[29] := Eq = D[Res,x] − y[x]/(x + 1)

$$\text{Out[29]} = \frac{c}{1+x} - \frac{y[x]}{1+x} + \frac{p[1][x]}{x} + \frac{2y[x]p[2][x]}{x} + p[0]'[x] + y[x]p[1]'[x] + y[x]^2 p[2]'[x]$$

In[30] := Eqs = Table[Coefficient[Eq, y[x], n] == 0, {n, 0, 2}]

$$\text{Out[30]} = \left\{ \frac{c}{1+x} + \frac{p[1][x]}{x} + p[0]'[x] == 0, -\frac{1}{1+x} + \frac{2p[2][x]}{x} + p[1]'[x] == 0, p[2]'[x] == 0 \right\}$$

As in the previous examples, $p_2(x)$ is a constant:

In[31] := p[2][x_] := p[2,0]; Eqs

$$\text{Out[31]} = \left\{ \frac{c}{1+x} + \frac{p[1][x]}{x} + p[0]'[x] == 0, -\frac{1}{1+x} + \frac{2p[2,0]}{x} + p[1]'[x] == 0, \text{True} \right\}$$

A polynomial $p_1(x)$ satisfying the second equation does not exist. Therefore, this integral cannot be taken in elementary functions.

In[32] := Clear[Res, Eq, Eqs, p]

Example 4

Let's consider

$$\int \frac{\mathrm{d}x}{y} = p_1(x)y + p_0(x) + c\log(y)$$

(it is not quite clear what the degree of the right-hand side in y should be; we shall see in a moment that this is irrelevant).

In[33] := Res = Sum[p[n][x] * y[x]^n, {n, 0, 1}] + c * Log[y[x]]

Out[33] = c, Log[y[x]] + p[0][x] + y[x]p[1][x]

In[34] := Eq = Expand[y[x] * D[Res, x] − 1]

$$\text{Out[34]} = -1 + \frac{c}{x} + \frac{y[x]p[1][x]}{x} + y[x]p[0]'[x] + y[x]^2 p[1]'[x]$$

In[35] := Eqs = Table[Coefficient[Eq, y[x], n] == 0, {n, 0, 2}]

$$\text{Out[35]} = \left\{ -1 + \frac{c}{x} == 0, \frac{p[1][x]}{x} + p[0]'[x] == 0, p[1]'[x] == 0 \right\}$$

From the last equation, $p_1(x)$ is a constant. The previous equation shows that it is 0, and $p_0(x)$ is a constant (it may be set to 0). It is clear that if we started from some other degree in y, we would all the same find that all $p_n(x) = 0$. And the first equation cannot be solved for c.

In[36] := Clear[Res, Eq, Eqs]

Consider the integrals

$$\int \frac{\mathrm{d}x}{xy}, \qquad \int \frac{\mathrm{d}x}{y+1}$$

in a similar way.

10.3 Exponential Extension

Now we shall consider an exponential extension with some $y = \exp r(x)$. If an integral of an expression from this extension can be taken in elementary functions, it has the form

$$\int \frac{N(x,y)}{D(x,y)}\,\mathrm{d}x = \frac{P(x,y)}{\hat{D}(x,y)} + \sum c_i \log q_i\,,$$

$$D = \prod D_i^{d_i} \quad \Rightarrow \quad \hat{D} = \prod D_i^{\hat{d}_i}\,,$$

where $\hat{d}_i = d_i - 1$ always except the case $D_i = y$ in which $\hat{d}_i = d_i$. This is because the derivative of $1/y$ is proportional to $1/y$, the degree of y in the denominator does not increase. We exclude y from the list of the factors q_i (if it was present, of course) because $\log y = r(x)$ is a rational function, and such a contribution is already accounted for. As usual, we differentiate the result and equate to the integrand to obtain a linear system. If it cannot be solved, then the integral does not exist in elementary functions.

Example 1

Let $y = \mathrm{e}^x$:

In[37] := y'[x_] := y[x]

Let's calculate

$$\int y\,\mathrm{d}x = p_1(x)y + p_0(x)$$

(the degree in y does not change when differentiating; therefore the polynomial in y in the right-hand side should have the same degree as the integrand).

In[38] := Res = Sum[p[n][x] * y[x]^n, {n, 0, 1}]

Out[38] = p[0][x] + y[x]p[1][x]

In[39] := Eq = D[Res, x] − y[x]

Out[39] = −y[x] + y[x]p[1][x] + p[0]'[x] + y[x]p[1]'[x]

In[40] := Eqs = Table[Coefficient[Eq, y[x], n] == 0, {n, 0, 1}]

Out[40] = {p[0]'[x] == 0, −1 + p[1][x] + p[1]'[x] == 0}

In[41] := p[1][x_] := Sum[p[1, m] * x^m, {m, 0, 1}]; Eqs[[2]]

Out[41] = −1 + p[1, 0] + p[1, 1] + xp[1, 1] == 0

From this equation:

In[42] := p[1, 1] = 0; p[1, 0] = 1; Eqs

Out[42] = {p[0]'[x] == 0, True}

Therefore $p_0(x)$ is a constant (which may be omitted):

In[43] := p[0][x_] := 0; Res

Out[43] = y[x]

We've got the expected result.
In[44] := Clear[Res, Eq, Eqs, *p*]

Example 2

Now let's calculate

$$\int xy\,\mathrm{d}x = p_1(x)y + p_0(x)\,.$$

In[45] := Res = Sum[*p*[*n*][*x*] ∗ *y*[*x*]^*n*, {*n*, 0, 1}]
Out[45] = $p[0][x] + y[x]p[1][x]$
In[46] := Eq = *D*[Res, *x*] − *x* ∗ *y*[*x*]
Out[46] = $-xy[x] + y[x]p[1][x] + p[0]'[x] + y[x]p[1]'[x]$
In[47] := Eqs = Table[Coefficient[Eq, *y*[*x*], *n*] == 0, {*n*, 0, 1}]
Out[47] = $\{p[0]'[x] == 0, -x + p[1][x] + p[1]'[x] == 0\}$
In[48] := *p*[1][*x*_] := Sum[*p*[1, *m*] ∗ *x*^*m*, {*m*, 0, 1}]; Eqs[[2]]
Out[48] = $-x + p[1,0] + p[1,1] + xp[1,1] == 0$
From this equation:
In[49] := *p*[1, 1] = 1; *p*[1, 0] = −1; Eqs
Out[49] = $\{p[0]'[x] == 0, \mathrm{True}\}$
In[50] := *p*[0][*x*_] := 0; Res
Out[50] = $(-1 + x)y[x]$
And without integration by parts!
In[51] := Clear[Res, Eq, Eqs, *p*]

Consider

$$\int x^2 y\,\mathrm{d}x$$

in a similar way.

Example 3

Now let's try to calculate

$$\int \frac{y}{x}\,\mathrm{d}x = p_1(x)y + p_0(x) + c\log x\,.$$

In[52] := Res = Sum[*p*[*n*][*x*] ∗ *y*[*x*]^*n*, {*n*, 0, 1}] + *c* ∗ Log[*x*]
Out[52] = $c, \mathrm{Log}[x] + p[0][x] + y[x]p[1][x]$
In[53] := Eq = *D*[Res, *x*] − *y*[*x*]/*x*
Out[53] = $\frac{c}{x} - \frac{y[x]}{x} + y[x]p[1][x] + p[0]'[x] + y[x]p[1]'[x]$

In[54] := Eqs = Table[Coefficient[Eq, y[x], n] == 0, {n, 0, 1}]

$\text{Out}[54] = \left\{\frac{c}{x} + p[0]'[x] == 0, -\frac{1}{x} + p[1][x] + p[1]'[x] == 0\right\}$

A polynomial $p_1(x)$ satisfying the second equation does not exist. Therefore this integral cannot be taken in elementary functions.

In[55] := Clear[Res, Eq, Eqs]

Example 4

Let's calculate

$$\int \frac{\mathrm{d}x}{y} = \frac{p_1(x)y + p_0(x)}{y}$$

(here $\hat{D} = y$, and there are no q_i). Of course, we could denote e^{-x} as y, and the problem would reduce to the Example 1 with trivial modifications; but we want to observe how the algorithm works in this new case.

In[56] := Res = Sum[p[n][x] * y[x]^n, {n, 0, 1}]/y[x]

$\text{Out}[56] = \frac{p[0][x] + y[x]p[1][x]}{y[x]}$

In[57] := Eq = Expand[y[x] * D[Res, x] − 1]

$\text{Out}[57] = -1 - p[0][x] + p[0]'[x] + y[x]p[1]'[x]$

In[58] := Eqs = Table[Coefficient[Eq, y[x], n] == 0, {n, 0, 1}]

$\text{Out}[58] = \{-1 - p[0][x] + p[0]'[x] == 0, p[1]'[x] == 0\}$

From the second equation, $p_1(x)$ is a constant, which may be omitted—this is the integration constant.

In[59] := p[1][x_] := 0

In[60] := p[0][x_] := Sum[p[0, m] * x^m, {m, 0, 1}]; Eqs[[1]]

$\text{Out}[60] = -1 - p[0,0] + p[0,1] - xp[0,1] == 0$

Therefore

In[61] := p[0, 1] = 0; p[0, 0] = −1; Res

$\text{Out}[61] = -\frac{1}{y[x]}$

In[62] := Clear[Res, Eq, Eqs, p]

Example 5

Let's consider

$$\int \frac{\mathrm{d}x}{y-1} = p_1(x)y + p_0(x) + c\log(y-1)$$

(now $\hat{D} = 1$, and there is a single q, namely $y - 1$).

In[63] := Res = Sum[p[n][x] ∗ y[x]^n, {n,0,1}] + c ∗ Log[y[x] − 1]
Out[63] = c, Log[−1 + y[x]] + p[0][x] + y[x]p[1][x]
In[64] := Eq = Cancel[(y[x] − 1) ∗ D[Res,x]] − 1
Out[64] = −1 + cy[x] − y[x]p[1][x] + y[x]^2p[1][x] − p[0]′[x] + y[x]p[0]′[x] − y[x]p[1]′[x] + y[x]^2p[1]′[x]
In[65] := Eqs = Table[Coefficient[Eq,y[x],n] == 0, {n,0,2}]
Out[65] = {−1 − p[0]′[x] == 0, c − p[1][x] + p[0]′[x] − p[1]′[x] == 0, p[1][x] + p[1]′[x] == 0}
In[66] := p[1][x_] := Sum[p[1,m] ∗ x^m, {m,0,1}]; Eqs[[3]]
Out[66] = p[1,0] + p[1,1] + xp[1,1] == 0
Therefore
In[67] := p[1,1] = 0; p[1,0] = 0; Eqs
Out[67] = {−1 − p[0]′[x] == 0, c + p[0]′[x] == 0, True}
From the first equation, $p_0(x) = -x$ (omitting the integration constant).
In[68] := p[0][x_] := −x; Eqs
Out[68] = {True, −1 + c == 0, True}
In[69] := c = 1; Res
Out[69] = −x + Log[−1 + y[x]]
In[70] := Clear[Res, Eq, Eqs, p, c]

Consider

$$\int \frac{x}{y-1}\,\mathrm{d}x$$

in a similar way and demonstrate that this integral does not exist in elementary functions.

Example 6

Of course, the method can be also used for other exponential extensions. For example, let $y = \exp x^2$:
In[71] := y′[x_] := 2 ∗ x ∗ y[x]
Let's consider

$$\int y\,\mathrm{d}x = p_1(x)y + p_0(x)\,.$$

In[72] := Res = Sum[p[n][x] ∗ y[x]^n, {n,0,1}];
In[73] := Eq = D[Res,x] − y[x]
Out[73] = −y[x] + 2xy[x]p[1][x] + p[0]′[x] + y[x]p[1]′[x]
In[74] := Eqs = Table[Coefficient[Eq,y[x],n] == 0, {n,0,1}]
Out[74] = {p[0]′[x] == 0, −1 + 2xp[1][x] + p[1]′[x] == 0}
In[75] := p[1][x_] := Sum[p[1,m] ∗ x^m, {m,0,1}]; ExpandAll[Eqs[[2]]]
Out[75] = −1 + 2xp[1,0] + p[1,1] + 2x^2p[1,1] == 0
This equation cannot be solved. A larger degree of $p_1(x)$ does not help. Therefore this integral cannot be taken in elementary functions.
In[76] := Clear[Res, Eq, Eqs, p]

Example 7

And what about

$$\int xy\,\mathrm{d}x = p_1(x)y + p_0(x) \ ?$$

In[77] := Res = Sum[*p*[*n*][*x*] ∗ *y*[*x*]^*n*, {*n*, 0, 1}];
In[78] := Eq = *D*[Res, *x*] − *x* ∗ *y*[*x*]
Out[78] $= -xy[x] + 2xy[x]p[1][x] + p[0]'[x] + y[x]p[1]'[x]$
In[79] := Eqs = Table[Coefficient[Eq, *y*[*x*], *n*] == 0, {*n*, 0, 1}]
Out[79] $= \{p[0]'[x] == 0, -x + 2xp[1][x] + p[1]'[x] == 0\}$
In[80] := *p*[1][x_] := Sum[*p*[1, *m*] ∗ *x*^*m*, {*m*, 0, 1}]; ExpandAll[Eqs[[2]]]
Out[80] $= -x + 2xp[1,0] + p[1,1] + 2x^2p[1,1] == 0$
From this equation, $p_{11} = 0$, $p_{10} = 1/2$. The first equation says that $p_0(x)$ is a constant (it may be omitted).
In[81] := *p*[1, 1] = 0; *p*[1, 0] = 1/2; *p*[0][*x*_] := 0; Res
Out[81] $= \frac{y[x]}{2}$
In[82] := Clear[Res, Eq, Eqs, *p*]

Consider

$$\int x^2y\,\mathrm{d}x, \qquad \int \frac{y}{x}\,\mathrm{d}x$$

in a similar way.

10.4 Elementary Functions

A tower of extensions can be constructed. We start from the set of rational functions $N(x)/D(x)$. Then we introduce y_1, which is either a root of a polynomial equation $p(x,y_1) = 0$, or logarithm or exponent of some rational function of x, and we obtain a first extension—the set of functions $N(x,y_1)/D(x,y_1)$. Then we introduce y_2, which is either a root of a polynomial equation $p(x,y_1,y_2) = 0$, or logarithm or exponent of some function from the previous extension, and we get a next extension—the set of rational functions of x, y_1, y_2. And so on. We should take care that an extension which seems transcendental is not in fact algebraic; if we neglect this, the methods designed for transcendental extensions may break down, e.g., produce divisions by 0. For example, if $y_1 = \mathrm{e}^x$, then it would not be a good idea to introduce the exponential extension with $y_2 = e^{2x}$, because y_2 is not algebraically independent: $y_2 = y_1^2$. Similarly, if $y_1 = \log x$, then it's not reasonable to introduce the logarithmic extension with $y_2 = \log 2x$, because $y_2 = y_1 + \log 2$ (in this case the field of constants needs to be extended by the transcendental number $\log 2$). In simple cases such restrictions are obvious, but in complicated ones it is necessary to decide if some function from some extension is identical 0, and this problem is algorithmically unsolvable in general.

Such towers of algebraic, logarithmic, and exponential extensions include all functions called elementary. And even some extra ones: an algebraic extension can be defined, e.g., by a root of a fifth degree polynomial unsolvable in radicals. Indeed, trigonometric functions reduce to exponentials, and inverse trigonometric ones—to logarithms. For each elementary function there exists a tower of extensions to which it belongs (it is not unique).

Suppose we want to integrate an elementary function. We construct a tower of extensions to which it belongs. If the indefinite integral exists in elementary functions, it belongs to some further extension of our tower by some extra logarithms (their number may be zero). The Risch algorithm allows one to decide in a finite number of steps if the result exists in this further extension, and if so, to find it; if it does not exist, the algorithm proves this fact. In its classical form, the algorithm is recursive in extensions—it calls itself for solving integration subproblems in previous (smaller) extensions, until rational functions. There is a simpler and more efficient version of the Risch algorithm—to write down the general form of the result with unknown coefficients, differentiate it and equate to the integrand. Then the problem reduces to solving a linear system. This approach is guaranteed to be correct if we know upper bounds on the degrees of the polynomial $P(x, y_1, y_2, \ldots)$ in its variables. But such upper bounds are not always known (as we have seen in the examples, they are known if there is no denominator). Therefore some heuristic rules to bound the degrees of P are used. This can give a situation when no result is found, though it really exists (but has a larger degree in some variables).

The Risch algorithm is an outstanding achievement of mathematics in the twentieth century. But it does not solve all problems with indefinite integration. The answer that no result exists in elementary functions is not very useful. It would be much better to get the result with some special functions. There were attempts to generalize the Risch algorithm to some special functions (the error function, polylogarithms). Some of them are implemented in *Mathematica*.

Chapter 11
Linear Algebra

11.1 Constructing Matrices

A matrix in *Mathematica* is a list of lists; all the lists—rows of the matrix—must have the same length.

In[1] := $\boldsymbol{M = \{\{a,b\},\{c,d\}\}}$
Out[1] $= \{\{a,b\},\{c,d\}\}$
MatrixForm is used to print a matrix nicely.
In[2] := MatrixForm[*M*]
Out[2]//MatrixForm =

$$\begin{pmatrix} a & b \\ c & d \end{pmatrix}$$

If the (i,j)th element of a matrix is given by an expression depending on i and j, this matrix can be constructed by the function Table.
In[3] := $\boldsymbol{A = }$ **Table[*a*[*i*, *j*], {*i*, 1, 2}, {*j*, 1, 2}]**
Out[3] $= \{\{a[1,1],a[1,2]\},\{a[2,1],a[2,2]\}\}$
In[4] := $\boldsymbol{B = }$ **Table[1/(*i* + *j* + 1), {*i*, 1, 2}, {*j*, 1, 2}]**

$$\text{Out[4]} = \left\{\left\{\frac{1}{3},\frac{1}{4}\right\},\left\{\frac{1}{4},\frac{1}{5}\right\}\right\}$$

In[5] := MatrixForm[*A*]
Out[5]//MatrixForm =

$$\begin{pmatrix} a[1,1] & a[1,2] \\ a[2,1] & a[2,2] \end{pmatrix}$$

In[6] := MatrixForm[*B*]
Out[6]//MatrixForm =

$$\begin{pmatrix} \frac{1}{3} & \frac{1}{4} \\ \frac{1}{4} & \frac{1}{5} \end{pmatrix}$$

The function Array is similar, but its first argument is a function of two parameters, not an expression. It may be just a symbol or an anonymous function (Function).
In[7] := $\boldsymbol{A = }$ **Array[*a*, {2, 2}]**
Out[7] $= \{\{a[1,1],a[1,2]\},\{a[2,1],a[2,2]\}\}$

A. Grozin, *Introduction to Mathematica® for Physicists*, Graduate Texts in Physics,
DOI 10.1007/978-3-319-00894-3_11,

In[8] := B = Array[Function[{i, j}, 1/($i+j+1$)], {2,2}]
$\text{Out}[8] = \left\{\left\{\frac{1}{3}, \frac{1}{4}\right\}, \left\{\frac{1}{4}, \frac{1}{5}\right\}\right\}$
Mathematica does not distinguish column vectors and row vectors; both are just lists.
In[9] := V = Table[v[i], {i, 1, 2}]
Out[9] = {v[1], v[2]}
In[10] := V = Array[v, 2]
Out[10] = {v[1], v[2]}
In[11] := U = Array[u, 2]
Out[11] = {u[1], u[2]}

11.2 Parts of a Matrix

A matrix element.
In[12] := A[[1,2]]
Out[12] = a[1,2]
A row.
In[13] := A[[1]]
Out[13] = {a[1,1], a[1,2]}
A column.
In[14] := A[[All,2]]
Out[14] = {a[1,2], a[2,2]}
A new value can be assigned to a matrix element.
In[15] := M[[1,2]] = 0
Out[15] = 0
In[16] := MatrixForm[M]
Out[16]//MatrixForm =
$$\begin{pmatrix} a & 0 \\ c & d \end{pmatrix}$$
Add 1 to the first row.
In[17] := M[[1]] + +
Out[17] = {a, 0}
In[18] := MatrixForm[M]
Out[18]//MatrixForm =
$$\begin{pmatrix} 1+a & 1 \\ c & d \end{pmatrix}$$
Add the second column to the first one.
In[19] := M[[All, 1]]+ = M[[All, 2]]
Out[19] = {$2+a$, $c+d$}
In[20] := MatrixForm[M]
Out[20]//MatrixForm =
$$\begin{pmatrix} 2+a & 1 \\ c+d & d \end{pmatrix}$$

11.3 Queries

The function VectorQ checks if its argument is a vector, i.e., a list whose elements are not lists.

In[21] := {VectorQ[V], VectorQ[M]}

Out[21] = {True, False}

The function MatrixQ checks if its argument is a matrix, i.e., a list of same-length lists.

In[22] := {MatrixQ[V], MatrixQ[M], MatrixQ[{{a, b}, {x, y, z}}]}

Out[22] = {False, True, False}

The argument of the function Dimensions must be a matrix. This function returns a two-element list—the numbers of rows and columns of the matrix.

In[23] := Dimensions[M]

Out[23] = {2, 2}

It can be conveniently used for simultaneous assignment to two variables.

In[24] := {n1, n2} = Dimensions[{{a, b, c}, {x, y, z}}]

Out[24] = {2, 3}

In[25] := n1

Out[25] = 2

In[26] := n2

Out[26] = 3

In[27] := Clear[M, n1, n2]

11.4 Operations with Matrices and Vectors

Vectors can be added and multiplied by scalar expressions.

In[28] := 2 ∗ V + U

Out[28] = $\{u[1]+2v[1], u[2]+2v[2]\}$

Matrices can be added and multiplied by scalar expressions.

In[29] := MatrixForm[A + 2 ∗ B]

Out[29]//MatrixForm =

$$\begin{pmatrix} \frac{2}{3}+a[1,1] & \frac{1}{2}+a[1,2] \\ \frac{1}{2}+a[2,1] & \frac{2}{5}+a[2,2] \end{pmatrix}$$

The scalar product of two vectors.

In[30] := $V.U$

Out[30] = $u[1]v[1]+u[2]v[2]$

The product of a matrix and a column vector.

In[31] := $A.V$

Out[31] = $\{a[1,1]v[1]+a[1,2]v[2], a[2,1]v[1]+a[2,2]v[2]\}$

The product of a row vector and a matrix.

In[32] := $V.A$

Out[32] = $\{a[1,1]v[1]+a[2,1]v[2], a[1,2]v[1]+a[2,2]v[2]\}$

The product of two matrices.

In[33] := MatrixForm[$A.B$]

Out[33]//MatrixForm =

$$\begin{pmatrix} \frac{1}{3}a[1,1]+\frac{1}{4}a[1,2] & \frac{1}{4}a[1,1]+\frac{1}{5}a[1,2] \\ \frac{1}{3}a[2,1]+\frac{1}{4}a[2,2] & \frac{1}{4}a[2,1]+\frac{1}{5}a[2,2] \end{pmatrix}$$

It is not commutative.

In[34] := MatrixForm[$A.B - B.A$]

Out[34]//MatrixForm =

$$\begin{pmatrix} \frac{1}{4}a[1,2]-\frac{1}{4}a[2,1] & \frac{1}{4}a[1,1]-\frac{2}{15}a[1,2]-\frac{1}{4}a[2,2] \\ -\frac{1}{4}a[1,1]+\frac{2}{15}a[2,1]+\frac{1}{4}a[2,2] & -\frac{1}{4}a[1,2]+\frac{1}{4}a[2,1] \end{pmatrix}$$

Determinant (the matrix must be square).

In[35] := Det[A]

Out[35] $= -a[1,2]a[2,1]+a[1,1]a[2,2]$

Trace (the matrix must be square).

In[36] := Tr[A]

Out[36] $= a[1,1]+a[2,2]$

Transposing.

In[37] := MatrixForm[Transpose[A]]

Out[37]//MatrixForm =

$$\begin{pmatrix} a[1,1] & a[2,1] \\ a[1,2] & a[2,2] \end{pmatrix}$$

The inverse matrix.

In[38] := MatrixForm[Inverse[A]]

Out[38]//MatrixForm =

$$\begin{pmatrix} \frac{a[2,2]}{-a[1,2]a[2,1]+a[1,1]a[2,2]} & -\frac{a[1,2]}{-a[1,2]a[2,1]+a[1,1]a[2,2]} \\ -\frac{a[2,1]}{-a[1,2]a[2,1]+a[1,1]a[2,2]} & \frac{a[1,1]}{-a[1,2]a[2,1]+a[1,1]a[2,2]} \end{pmatrix}$$

A square matrix can be raised to an integer power.

In[39] := MatrixForm[MatrixPower[B, 3]]

Out[39]//MatrixForm =

$$\begin{pmatrix} \frac{197}{2160} & \frac{1009}{14400} \\ \frac{1009}{14400} & \frac{323}{6000} \end{pmatrix}$$

The power -1 is the inverse matrix.

In[40] := MatrixForm[MatrixPower[B, -1]]

Out[40]//MatrixForm =

$$\begin{pmatrix} 48 & -60 \\ -60 & 80 \end{pmatrix}$$

This is the solution of the linear system $A.X = V$.

In[41] := Together[Inverse[A].V]

$$\text{Out[41]} = \left\{\frac{a[2,2]v[1]-a[1,2]v[2]}{-a[1,2]a[2,1]+a[1,1]a[2,2]}, \frac{a[2,1]v[1]-a[1,1]v[2]}{a[1,2]a[2,1]-a[1,1]a[2,2]}\right\}$$

The same can be done using LinearSolve.

In[42] := LinearSolve[A, V]

$$\text{Out[42]} = \left\{\frac{a[2,2]v[1]-a[1,2]v[2]}{-a[1,2]a[2,1]+a[1,1]a[2,2]}, \frac{a[2,1]v[1]-a[1,1]v[2]}{a[1,2]a[2,1]-a[1,1]a[2,2]}\right\}$$

In[43] := Clear[A, B, U, V]

11.5 Eigenvalues and Eigenvectors

Here is some symbolic matrix.

In[44] := MatrixForm[*M* =
{{(1 − *x*)^3 ∗ (3 + *x*), 4 ∗ *x* ∗ (1 − *x*^2), −2 ∗ (1 − *x*^2) ∗ (3 − *x*)},
{4 ∗ *x* ∗ (1 − *x*^2), −(1 + *x*)^3 ∗ (3 − *x*), 2 ∗ (1 − *x*^2) ∗ (3 + *x*)},
{−2 ∗ (1 − *x*^2) ∗ (3 − *x*), 2 ∗ (1 − *x*^2) ∗ (3 + *x*), 16 ∗ *x*}}]

Out[44]//MatrixForm =

$$\begin{pmatrix} (1-x)^3(3+x) & 4x\left(1-x^2\right) & -2(3-x)\left(1-x^2\right) \\ 4x\left(1-x^2\right) & -(3-x)(1+x)^3 & 2(3+x)\left(1-x^2\right) \\ -2(3-x)\left(1-x^2\right) & 2(3+x)\left(1-x^2\right) & 16x \end{pmatrix}$$

It is singular.

In[45] := Det[*M*]

Out[45] = 0

Its rank.

In[46] := MatrixRank[*M*]

Out[46] = 2

The function NullSpace returns a list of vectors forming a basis of the null space of the matrix, i.e., the subspace of vectors nullified by the matrix.

In[47] := *s* = NullSpace[*M*]

$$\text{Out[47]} = \left\{\left\{-\frac{2}{-1+x}, \frac{2}{1+x}, 1\right\}\right\}$$

In this case, the null space is one-dimensional—it has a single basis vector. Let's check it.

In[48] := Together[*M*.*s*[[1]]]

Out[48] = {0,0,0}

The function Eigenvalues returns a list of eigenvalues of a matrix.

In[49] := Simplify[Eigenvalues[*M*], Element[*x*, Reals]]

$$\text{Out[49]} = \left\{0, \left(3+x^2\right)^2, -\left(3+x^2\right)^2\right\}$$

We have added the second argument to Simplify which informs *Mathematica* that the variable x is real. The function Eigenvectors returns the list of the corresponding eigenvectors (in the same order).

In[50] := Simplify[Eigenvectors[*M*], Element[*x*, Reals]]

$$\text{Out[50]} = \left\{\left\{-\frac{2}{-1+x}, \frac{2}{1+x}, 1\right\}, \left\{\frac{-1+x}{1+x}, \frac{1-x}{2}, 1\right\}, \left\{\frac{1+x}{2}, \frac{1+x}{-1+x}, 1\right\}\right\}$$

The function Eigensystem returns both. It is convenient for simultaneous assignment to two variables.

In[51] := {val, vec} = Simplify[Eigensystem[*M*], Element[*x*, Reals]];

Let's check.

In[52] := Do[Print[Simplify[*M*.vec[[*i*]] − val[[*i*]] ∗ vec[[*i*]]]], {*i*, 1, 3}]

{0,0,0}
{0,0,0}
{0,0,0}

In[53] := Clear[*M*, *s*, val, vec]

11.6 Jordan Form

Here is a matrix of rational numbers.

In[54] := MatrixForm[*M* =
{{13/9, −2/9, 1/3, 4/9, 2/3},
{−2/9, 10/9, 2/15, −2/9, −11/15},
{1/5, −2/5, 41/25, −2/5, 12/25},
{4/9, −2/9, 14/15, 13/9, −2/15},
{−4/15, 8/15, 12/25, 8/15, 34/25}}]

Out[54]//MatrixForm =

$$\begin{pmatrix} \frac{13}{9} & -\frac{2}{9} & \frac{1}{3} & \frac{4}{9} & \frac{2}{3} \\ -\frac{2}{9} & \frac{10}{9} & \frac{2}{15} & -\frac{2}{9} & -\frac{11}{15} \\ \frac{1}{5} & -\frac{2}{5} & \frac{41}{25} & -\frac{2}{5} & \frac{12}{25} \\ \frac{4}{9} & -\frac{2}{9} & \frac{14}{15} & \frac{13}{9} & -\frac{2}{15} \\ -\frac{4}{15} & \frac{8}{15} & \frac{12}{25} & \frac{8}{15} & \frac{34}{25} \end{pmatrix}$$

The function JordanDecomposition returns a pair of matrices—the Jordan form J and the transformation matrix P which reduces our matrix to its Jordan form.

In[55] := {*P*, *J*} = JordanDecomposition[*M*];

In[56] := MatrixForm[*J*]

Out[56]//MatrixForm =

$$\begin{pmatrix} 1 & 0 & 0 & 0 & 0 \\ 0 & 2 & 1 & 0 & 0 \\ 0 & 0 & 2 & 0 & 0 \\ 0 & 0 & 0 & 1-\mathrm{i} & 0 \\ 0 & 0 & 0 & 0 & 1+\mathrm{i} \end{pmatrix}$$

In[57] := MatrixForm[*P*]

Out[57]//MatrixForm =

$$\begin{pmatrix} -2 & 1 & 0 & \frac{5\mathrm{i}}{12} & -\frac{5\mathrm{i}}{12} \\ -2 & -\frac{1}{2} & 0 & -\frac{5\mathrm{i}}{6} & \frac{5\mathrm{i}}{6} \\ 0 & 0 & \frac{6}{5} & -\frac{3}{4} & -\frac{3}{4} \\ 1 & 1 & 0 & -\frac{5\mathrm{i}}{6} & \frac{5\mathrm{i}}{6} \\ 0 & 0 & \frac{9}{10} & 1 & 1 \end{pmatrix}$$

Let's check.

In[58] := MatrixForm[*P*.*J*.Inverse[*P*] − *M*]

Out[58]//MatrixForm =

$$\begin{pmatrix} 0 & 0 & 0 & 0 & 0 \\ 0 & 0 & 0 & 0 & 0 \\ 0 & 0 & 0 & 0 & 0 \\ 0 & 0 & 0 & 0 & 0 \\ 0 & 0 & 0 & 0 & 0 \end{pmatrix}$$

Here are the eigenvalues and the eigenvectors of our matrix. Note that only one eigenvector corresponds to the eigenvalue 2, because the corresponding Jordan block has the size 2×2. The eigenvectors are the columns of the transformation matrix P.

In[59] := {val, vec} = Eigensystem[M];
In[60] := val
Out[60] = $\{2, 2, 1+\mathrm{i}, 1-\mathrm{i}, 1\}$
In[61] := vec
Out[61] = $\left\{\left\{1, -\frac{1}{2}, 0, 1, 0\right\}, \{0, 0, 0, 0, 0\}, \left\{-\frac{5\mathrm{i}}{12}, \frac{5\mathrm{i}}{6}, -\frac{3}{4}, \frac{5\mathrm{i}}{6}, 1\right\},\right.$
$\left.\left\{\frac{5\mathrm{i}}{12}, -\frac{5\mathrm{i}}{6}, -\frac{3}{4}, -\frac{5\mathrm{i}}{6}, 1\right\}, \{-2, -2, 0, 1, 0\}\right\}$
In[62] := Clear[M, J, P, val, vec]

11.7 Symbolic Vectors, Matrices, and Tensors

Let's inform *Mathematica* that u, v, and w are symbolic three-dimensional vectors with real components. The scalar product is $u.v$, and the vector product is Cross$[u, v]$. The function TensorReduce simplifies expressions with vectors.
In[63] := \$Assumptions = Element[$u|v|w$, Vectors[3, Reals]]
Out[63] = $(u|v|w) \in$ Vectors[3, Reals]
In[64] := TensorReduce[$u.v - v.u$]
Out[64] = 0
In[65] := TensorReduce[Cross[u, v] + Cross[v, u]]
Out[65] = 0
In[66] := TensorReduce[u.Cross[v, w] + v.Cross[w, u] + w.Cross[u, v]]
Out[66] = $3u \times v.w$
In[67] := TensorReduce[Cross[u, Cross[v, w]]]
Out[67] = $-wu.v + vu.w$
In[68] := TensorReduce[$u.(2 * v + 3 * w)$]
Out[68] = $2u.v + 3u.w$

Now let's say that u and v are d-dimensional vectors, and S and A are $d \times d$ matrices, S symmetric and A antisymmetric.
In[69] := \$Assumptions = {Element[$u|v$, Vectors[d, Reals]],
Element[S, Matrices[{d, d}, Reals, Symmetric[{1, 2}]]],
Element[A, Matrices[{d, d}, Reals, Antisymmetric[{1, 2}]]]};
In[70] := TensorReduce[$v.A.(u + v)$]
Out[70] = $-u.A.v$
In[71] := TensorReduce[$u.S.v + v.S.u$]
Out[71] = $2u.S.v$
The tensor product $S_{ij}A_{kl}$ is contracted in j and k and in i and l.
In[72] := TensorReduce[TensorContract[TensorProduct[S, A], {{2, 3}, {1, 4}}]]
Out[72] = 0

The Riemann curvature tensor has the properties $R_{ijkl} = -R_{jikl}$ and $R_{ijkl} = R_{klij}$.

In[73] := $Assumptions = Element[*R*, Arrays[{4,4,4,4}, Reals, {{{2,1,3,4}, −1}, {{3,4,1,2}, 1}}]]

Out[73] = $R \in$ Arrays[{4,4,4,4}, Reals, {{Cycles[{{1,2}}], −1}, {Cycles[{{1,3},{2,4}}], 1}, {Cycles[{{3,4}}], −1}}]

In[74] := TensorReduce[TensorContract[*R*, {{1,2}}]]

Out[74] = 0

The Ricci tensor.

In[75] := R2 = TensorContract[*R*, {{1,3}}]

Out[75] = TensorContract[*R*, {{1,3}}]

In[76] := {TensorRank[R2], TensorDimensions[R2], TensorSymmetry[R2]}

Out[76] = {2, {4,4}, Symmetric[{1,2}]}

$R_{ijkl}R_{ijkl} + R_{ijkl}R_{klji} + R_{ijkl}R_{ikjl}$.

**In[77] := TensorReduce[
TensorContract[TensorProduct[*R*, *R*], {{1,5},{2,6},{3,7},{4,8}}]+
TensorContract[TensorProduct[*R*, *R*], {{1,7},{2,8},{3,6},{4,5}}]+
TensorContract[TensorProduct[*R*, *R*], {{1,5},{2,7},{3,6},{4,8}}]]**

Out[77] = TensorContract[$R \otimes R$, {{1,5},{2,7},{3,6},{4,8}}]

Unfortunately, *Mathematica* cannot take $R_{ijkl} + R_{iklj} + R_{iljk} = 0$ into account.

In[78] := $Assumptions = True;

Chapter 12
Input–Output and Strings

12.1 Reading and Writing .m Files

When developing any nontrivial *Mathematica* program, it is better to write it in a text file, using a text editor, and then to read it into a fresh *Mathematica* session. In this way, your actions will be reproducible. You can fix a bug and see what has changed. Suppose we have a text file called wrong.m. It contains the text

In[1] := FilePrint["wrong.m"]

```
a=x^2+2*x*y+y^2;
b=x^3+3*x^2*y
+3*x*y^2+y^3;
```

We can read it. Now the variables a and b have values:

In[2] := <<wrong.m

In[3] := {*a*,*b*}

Out[3] $= \{x^2+2xy+y^2, x^3+3x^2y\}$

The value of b is not what we expected. Why? When *Mathematica* sees an end of a line, it checks if the text read so far forms a syntactically correct expression. If so, the expression is considered complete; the next line starts a new expression. Otherwise, the next line is considered a continuation of the current expression. Thus our file wrong.m contains not two but three expressions: two assignments (to a and b) and a separate polynomial consisting of two terms. One way to prevent such unintended splitting of multiline expressions is to place a binary operator (e.g., $+$ or $-$) at the end of each line. *Mathematica* always writes its result in such a way. But it is easy to forget about it when writing a long expression in a text editor. Also, if we paste results from some other system into a *Mathematica* program, it would be tedious and error-prone to bring long expressions to such a form by hand. It is better to enclose each multiline expression in extra parentheses, then any incomplete subset of lines is not syntactically correct. Therefore we edit our wrong.m and obtain a file right.m:

A. Grozin, *Introduction to Mathematica® for Physicists*, Graduate Texts in Physics,
DOI 10.1007/978-3-319-00894-3_12,

In[4] := FilePrint["right.m"]
```
a=x^2+2*x*y+y^2;
b=(x^3+3*x^2*y
+3*x*y^2+y^3);
```
Now everything's all right:
In[5] := <<right.m
In[6] := {*a*,*b*}
Out[6] $= \{x^2+2xy+y^2, x^3+3x^2y+3xy^2+y^3\}$
In[7] := Clear[*a*,*b*]

Let us modify the file:
In[8] := FilePrint["right2.m"]
```
a=x^2+2*x*y+y^2;
(x^3+3*x^2*y
+3*x*y^2+y^3)
```
Now the last expression is not an assignment, but just a polynomial. What happens if we read it?
In[9] := <<right2.m
Out[9] $= x^3+3x^2y+3xy^2+y^3$
In[10] := *a*
Out[10] $= x^2+2xy+y^2$

The operator << (its full name is Get) returns the value of the last expression. Therefore, we can use it in an assignment (or as an argument of any other function):
In[11] := *b* = <<right2.m
Out[11] $= x^3+3x^2y+3xy^2+y^3$
In[12] := {*a*,*b*}
Out[12] $= \{x^2+2xy+y^2, x^3+3x^2y+3xy^2+y^3\}$
In[13] := Clear[*b*]

The function Get tries to find the file in all directories in the list $Path. Its default value contains, in particular, the current directory ".". You can add more directories to it using list operations, such as Append, Prepend, or Join.

And this operator (its full name is Put) writes the value of an expression into a file. The value is written in the input form, and hence can be later read by *Mathematica*.
In[14] := *a*>>result.m
In[15] := FilePrint["result.m"]
```
x^2 + 2*x*y + y^2
```
In[16] := Clear[*a*]
In[17] := *a* = <<result.m
Out[17] $= x^2+2xy+y^2$
In[18] := *a*
Out[18] $= x^2+2xy+y^2$

The function >> writes just an expression, not an assignment to some variable. Often it is more useful to write an assignment (or several of them) which defines some variable (or function). Suppose we have
In[19] := *f*[0] = 1; *f*[*n*_] := *n* ∗ *f*[*n* − 1]

We can save the definition of the function f into a file:

In[20] := Save["f.m", *f*]
In[21] := FilePrint["f.m"]
f[0] = 1
$f[n\_] := n * f[n-1]$
In[22] := Clear[*f*]
In[23] := <<f.m
In[24] := *f*[10]
Out[24] = 3628800
In[25] := Clear[*f*]

12.2 Output

The function Print prints expressions (including strings, plots, etc.). It does not separate them by spaces, so that it is usually a good idea to insert " " between expressions. It adds a newline at the end.

In[26] := *s* = "A strings\nwith a newline";
In[27] := *p* = Plot3D[*a*, {*x*, −1, 1}, {*y*, −1, 1}];
In[28] := Print[*s*, " ", *a*, " ", *p*]
Out[28] = A strings

with a newline $x^2 + 2xy + y^2$

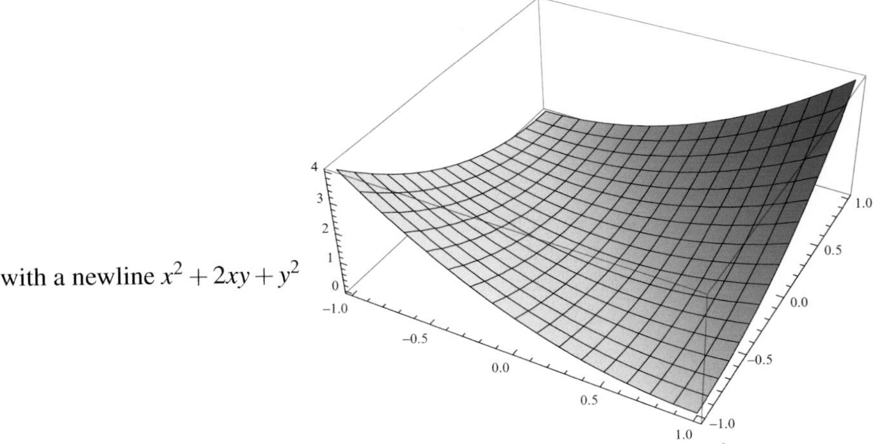

In[29] := Clear[*p*]

The function Print is most useful when you want to know what happens at some point deep inside your own function when it is called. If you want to write expressions to a file instead, use Write:

In[30] := *f* = OpenWrite["res.m"]
Out[30] = OutputStream[res.m, 19]
In[31] := Write[*f*, *s*]; Write[*f*, *a*]
In[32] := Close[*f*]
Out[32] = res.m
In[33] := FilePrint["res.m"]

A strings\nwith a newline
x^2 + 2*x*y + y^2
Expressions are written in the input form and can be read later.
In[34] := Clear[*s*, *a*]

12.3 C, Fortran, and TEX Forms

Suppose you have derived analytically a valuable expression a:
In[35] := *a* = Sin[*x*]^2/*x*^2 − 1
Out[35] $= -1 + \frac{\mathrm{Sin}[x]^2}{x^2}$
Now you want to do some large-scale numerical calculations with it. In order to avoid possible errors when translating this expression into a form suitable to inclusion into a numerical program, it is a good idea to do this step automatically. The function CForm converts an expression into a form which can be inserted into a C (or C++) program.
In[36] := CForm[*a*]
Out[36] = -1 + Power(Sin(x),2)/Power(x,2)
The macros Power, Sin, and friends are defined in the file mdefs.h which is supplied with *Mathematica*. If the expression contains special functions, you will need some extra C libraries which can calculate them numerically. Of course, it is better to write the C form of an expression into a file and then insert it into your program. For those old-fashioned enough to do numerical computations in Fortran, there is also FortranForm.
In[37] := FortranForm[*a*]
Out[37] = -1 + Sin(x)**2/x**2

At last, all computations are done, and you are writing an article to be submitted to a scientific journal. You want to include your valuable expression a and to avoid errors in process of doing so. Scientific articles are written in LATEX (in most cases). The function TeXForm converts an expression into a form which can be inserted into a LATEX file.
In[38] := TeXForm[*a*]
Out[38] = \frac{\sin ^2(x)}{x^2}-1
In[39] := Clear[*a*]

12.4 Strings

The simplest string operation is concatenation:
In[40] := s1 = "This"; s2 = "is" ; s3 = "a string";
In[41] := *s* = s1 <> " " <> s2 <> " " <> s3
Out[41] = This is a string

In[42] := FullForm[*s*]
Out[42]//FullForm =
"This is a string"
In[43] := StringLength[*s*]
Out[43] = 16
In[44] := Clear[*s*, s1, s2, s3]
Mathematica has string patterns and substitutions.
In[45] := StringReplace["abcabd", "ab" → "AB"]
Out[45] = ABcABd
*x*_ means an arbitrary (single) character.
In[46] := StringReplace["a1b a2b a3c", a ~~ *x*_ ~~ b → A ~~ *x* ~~ B]
Out[46] = A1B A2B a3c
In[47] := StringReplace["a1b2 a12b",
a ~~ *x*_ ~~ b ~~ *y*_ → A ~~ *y* ~~ B ~~ *x*]
Out[47] = A2B1 a12b
Internally, these patterns are the function StringExpression.
In[48] := FullForm[a ~~ *x*_]
Out[48]//FullForm =
StringExpression[a, Pattern[*x*, Blank[]]]
A pattern with two identical arbitrary characters.
In[49] := StringReplace["a1b1 a1b2 a2b2",
a ~~ *x*_ ~~ b ~~ *x*_ → A ~~ *x* ~~ *x* ~~ B]
Out[49] = A11B a1b2 A22B
b|c means b or c.
In[50] := StringReplace["abcd abcd", b|c → X]
Out[50] = aXXd aXXd
*x*__ means any nonempty sequence of characters.
In[51] := StringReplace["ab", a ~~ *x*__ ~~ b ~~ *x*__ → A ~~ *x* ~~ B ~~ *x*]
Out[51] = ab
In[52] := StringReplace["a12b12",
a ~~ *x*__ ~~ b ~~ *x*__ → A ~~ *x* ~~ B ~~ *x*]
Out[52] = A12B12
*x*___ means any sequence of characters (including empty).
In[53] := StringReplace["ab", a ~~ *x*___ ~~ b ~~ *x*___ → A ~~ *x* ~~ B ~~ *x*]
Out[53] = AB
In[54] := StringReplace["a12b12",
a ~~ *x*___ ~~ b ~~ *x*___ → A ~~ *x* ~~ B ~~ *x*]
Out[54] = A12B12
As in the case of general patterns, /; means a condition (such that).
In[55] := StringReplace["a1b1 a1b2",
a ~~ *x*_ ~~ b ~~ *y*_/; *x*!=*y* → A ~~ *x* ~~ *y* ~~ B]
Out[55] = a1b1 A12B
The function StringMatchQ tests if a string matches a pattern.

In[56] := {StringMatchQ["a1b1", a ~~ x_ ~~ b ~~ x_],
StringMatchQ["a1b2", a ~~ x_ ~~ b ~~ x_],
StringMatchQ["a1b1c", a ~~ x_ ~~ b ~~ x_]}

Out[56] = {True, False, False}

The function StringFreeQ tests if there are no substrings matching a pattern.

In[57] := {StringFreeQ["a1b1", a ~~ x_ ~~ b ~~ x_],
StringFreeQ["a1b2", a ~~ x_ ~~ b ~~ x_],
StringFreeQ["a1b1c", a ~~ x_ ~~ b ~~ x_]}

Out[57] = {False, True, False}

The function StringSplit splits a string at each occurrence of a pattern; if the second argument is not given, then at each white space.

In[58] := StringSplit["xxa1byya2bzz", a ~~ x_ ~~ b]

Out[58] = {xx, yy, zz}

In[59] := StringSplit["xx yy zz\nuu\tvv\t\t ww"]

Out[59] = {xx, yy, zz, uu, vv, ww}

Mathematica contains many more string manipulation functions and additional powerful features of string patterns and can be used for text processing instead of Perl. For more details, see the online help.

Chapter 13
Packages

13.1 Contexts

When writing a large program, it is easy to accidentally use one symbol for two different quantities in different parts of the program. This leads to difficult-to-find bugs. This is especially true if parts of the program are written by different persons (in particular, when some packages from the standard library, or third-party packages, are used). To avoid such problems, contexts are used.

In *Mathematica*, a full symbol name consists of two parts: the *context* and the *short name*. Two symbols in different contexts may have the same short name. For example, the global symbol x and the symbol x in the contexts a and b are unrelated.

In[1] := x = 1; a\`x = 2; {x, a\`x, b\`x}

Out[1] = {1, 2, b\`x}

Contexts may be nested. Here the variable x lives in the context b which lives in the context a (and thus is unrelated to the global context b used above).

In[2] := a\`b\`x

Out[2] = a\`b\`x

The current default context is held in the variable $Context. It is used when a new symbol is created without specifying its context.

In[3] := $Context

Out[3] = Global\`

When a symbol without an explicit context is used, it is being searched in contexts specified in $ContextPath.

In[4] := $ContextPath

Out[4] = {PacletManager\`, QuantityUnits\`, WebServices\`, System\`, Global\`}

Built-in symbols live in the context System\`.

In[5] := {Context[x], Context[a\`x], Context[b\`x], Context[a\`b\`x], Context[Pi]}

Out[5] = {Global\`, a\`, b\`, a\`b\`, System\`}

You can change $ContextPath using standard list functions. The function Remove removes symbols from the system.

In[6] := Clear[x, a\`x]; Remove[a\`x, b\`x, a\`b\`x]

A. Grozin, *Introduction to Mathematica® for Physicists*, Graduate Texts in Physics,
DOI 10.1007/978-3-319-00894-3_13, © Springer International Publishing Switzerland 2014

13.2 Packages

Mathematica comes with a library of packages extending its built-in functionality. A package can be loaded by
In[7] := <<Quaternions`
Now this context is prepended to $ContextPath.
In[8] := $ContextPath
Out[8] = {Quaternions`, PacletManager`, QuantityUnits`, WebServices`,
System`, Global`}
Short names will be searched in this context first, possibly shadowing variables and functions from Global`.
In[9] := Context[Quaternion]
Out[9] = Quaternions`
The package Quaternions lives in the directory Quaternions, which lives in the standard library directory.

Many additional packages are available at http://library.wolfram.com/infocenter/MathSource/. You can download them and install somewhere in your $Path.

You can instruct *Mathematica* to load a package whenever any function defined in it is used.
**In[10] := DeclarePackage["NumericalCalculus`",
{"EulerSum", "NLimit", "ND", "NSeries", "NResidue"}]**
Out[10] = NumericalCalculus`
In[11] := ND[*x*^2, *x*, 1.0]
Out[11] = 2.

13.3 Writing Your Own Package

Begin, End

Begin["*a*`"] changes the default $Context; all symbols defined after this will live in this context.
In[12] := Begin["*a*`"]
Out[12] = *a*`
In[13] := $Context
Out[13] = *a*`
In[14] := *z* = 0; Context[*z*]
Out[14] = *a*`
End[] restores the previous $Context.
In[15] := End[]
Out[15] = *a*`

In[16] := $Context
Out[16] = Global`
In[17] := *a*`*z*
Out[17] = 0

BeginPackage, EndPackage

BeginPackage["*a*`"] sets $Context and changes $ContextPath in such a way that only the contexts *a*` and System` are available.
In[18] := BeginPackage["*a*`"]
Out[18] = *a*`
In[19] := $Context
Out[19] = *a*`
In[20] := $ContextPath
Out[20] = {*a*`, System`}
In[21] := *u* = 1;

EndPackage[] restores the previous $Context; ContextPath gets its old value prepended by *a*`, so that symbols defined after BeginPackage[*a*`] remain available.
In[22] := EndPackage[]
In[23] := $Context
Out[23] = Global`
In[24] := $ContextPath
Out[24] = {*a*`, NumericalCalculus`, Quaternions`, PacletManager`,
QuantityUnits`, WebServices`, System`, Global`}
In[25] := *u*
Out[25] = 1
In[26] := Clear[*z*, *u*]

A Typical Package

A simple package looks like this.
In[27] := FilePrint["APackage.m"]

```
BeginPackage["APackage`"]
f::usage = "f squares its argument"
Begin["`Private`"]
g[x_] := x^2
f[x_] := Expand[g[x]]
End[]
EndPackage[]
```

After BeginPackage["APackage`"], publicly available functions and variables of the package are introduced, usually by assignments to their usage messages.

Implementation of the package is done in the context APackage`Private`, which may contain additional functions and variables (not seen by users of the package).

In[28] := <<APackage`

In[29] :=?*f*

Out[29] = f squares its argument

In[30] := *f*[*a*+*b*]

Out[30] = $a^2 + 2ab + b^2$

Part II
Computer Classes

Before spending your time and effort on catching a lion, check: maybe, somebody has already caught it, and it is available for download at http://library.wolfram.com/infocenter/MathSource/

Chapter 14
Plots

After typing a *Mathematica* command in its notebook interface, you can send it to the kernel (which performs calculations) by pressing Shift-Enter. The result appears in the output cell which follows your input cell. Both are nested in an outer cell representing a calculation step. Later you can return to this input cell, edit the command, and execute it again. The old output will be replaced by the new result. It is allowed to type several commands in a single input cell, but this is not convenient—don't do so unless you have good reasons.

Mathematica Help contains all the necessary information. The Help menu contains Documentation Center, Function Navigator, and Virtual Book (among other things). You can quickly get help for a specific function if you select it with the mouse and press F1.

14.1 2D Plots

Function Plot

See Help → Virtual Book → Visualization and Graphics → Graphics and Sound → Basic Plotting, and Help → Function Navigator → Visualization and Graphics → Function Visualization → Plot for more details.

A. Grozin, *Introduction to Mathematica® for Physicists*, Graduate Texts in Physics,
DOI 10.1007/978-3-319-00894-3_14, © Springer International Publishing Switzerland 2014

In[1] := Plot[Sin[x], {x, −10, 10}]

Out[1] =

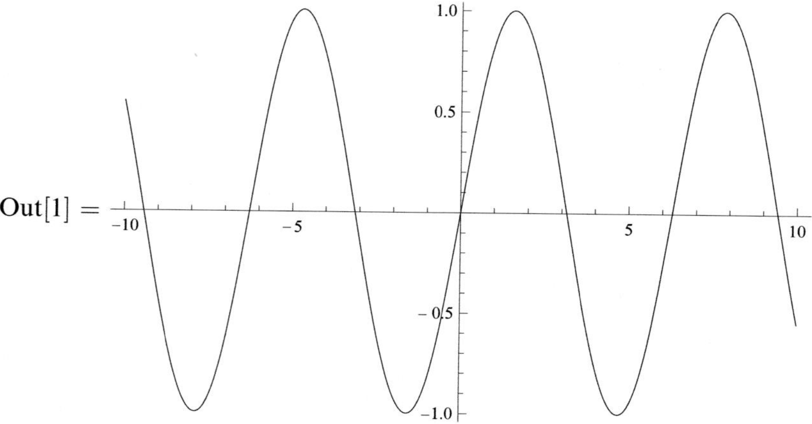

Several Functions

In[2] := Plot[{Sin[x], Cos[x]}, {x, −10, 10}]

Out[2] =

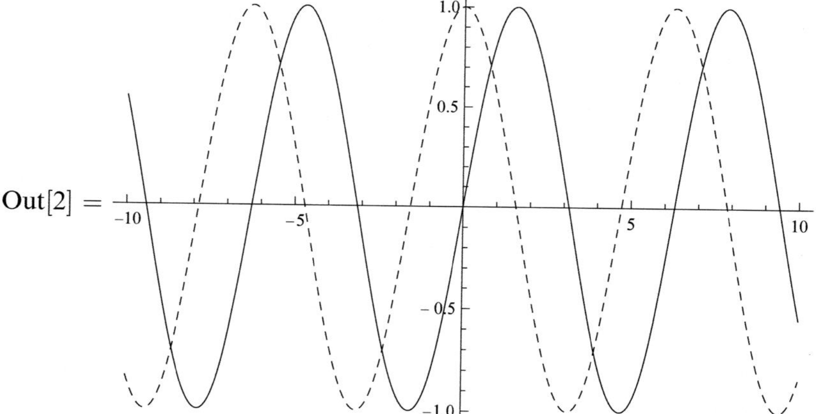

You can set colors and styles of the curves (Virtual Book → Visualization and Graphics → Graphics and Sound → Options for Graphics; Function Navigator → Visualization and Graphics → Options and Styling → Plotting Options → PlotStyle).

In[3] := Plot[{Sin[x], Cos[x]}, {x, −10, 10}, PlotStyle−>{Red, {Blue, Dashed}}]

Out[3] =

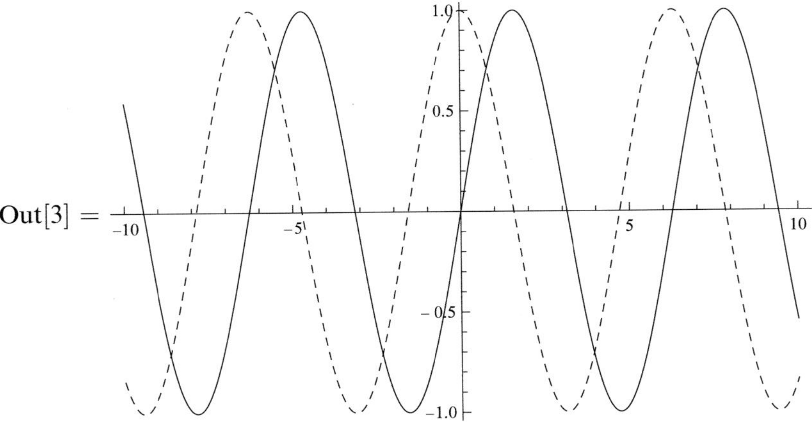

Unbounded Function

In[4] := Plot[Tan[x], {x, −10, 10}]

Out[4] =

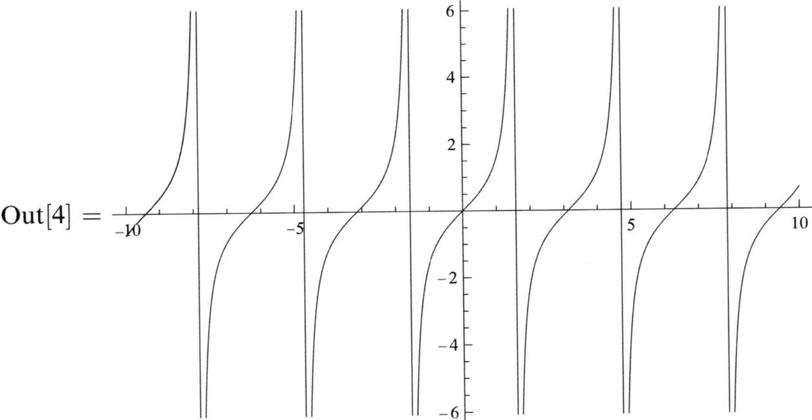

Mathematica has chosen some y scale. How to set it? Find in the Help.

In[5] := Plot[Tan[x], {x, −10, 10}, PlotRange−>{−3, 3}]

Out[5] =

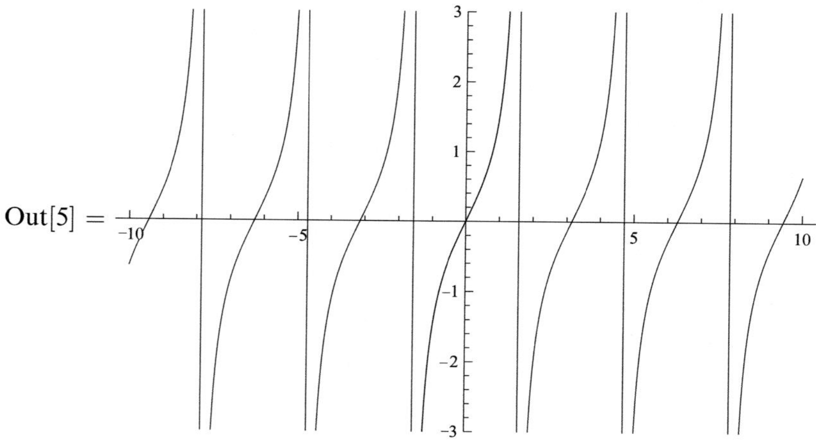

Logarithmic Scale

If our function is positive and varies by orders of magnitude in our region, it is convenient to use logarithmic scale in y. If the independent variable also varies by orders of magnitude, the x-axis scale also should be logarithmic (Function Navigator → Visualization and Graphics → Function Visualization → LogPlot, LogLogPlot).
In[6] := LogPlot[Exp[x] + 1, {x, −10, 10}]

Out[6] =

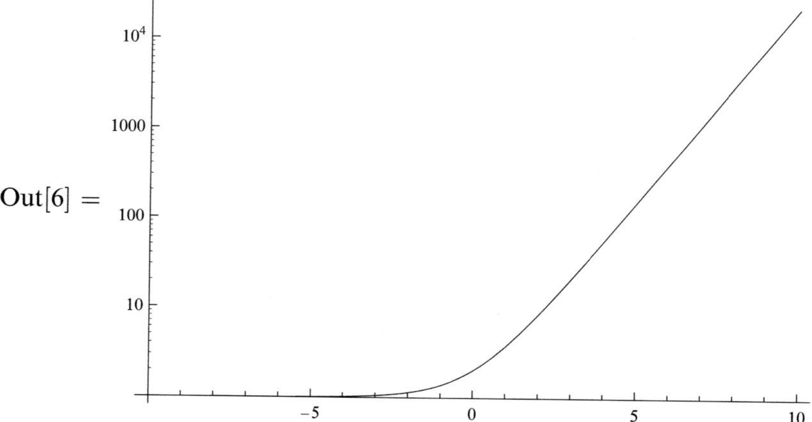

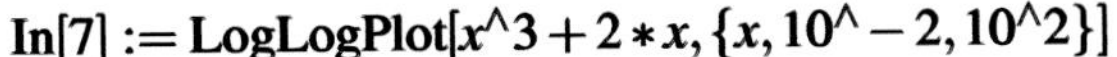

In[7] := LogLogPlot[x^3 + 2 * x, {x, 10^ − 2, 10^2}]

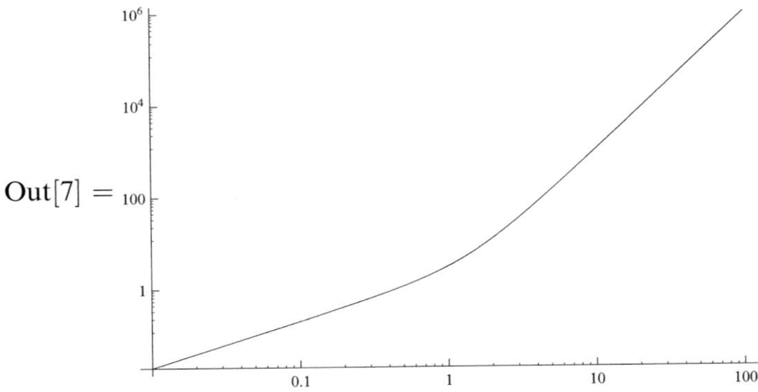

Parametric Curve

Lissajous figures.

In[8] := Manipulate[ParametricPlot[{Sin[a * t + c], Sin[b * t]}, {t, 0, 2 * Pi}],
{a, 1, 10, 1, Appearance−>"Labeled"},
{b, 1, 10, 1, Appearance−>"Labeled"}, {{c, Pi/2}, 0, Pi/2}]

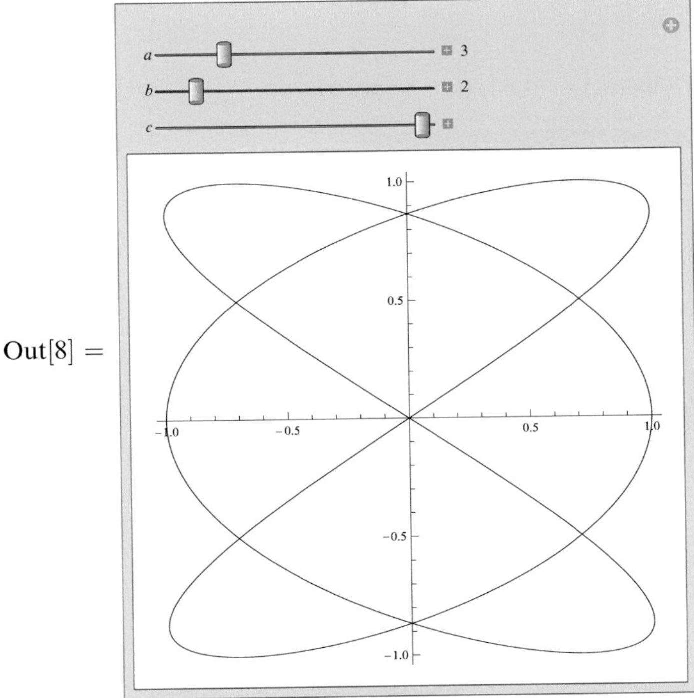

Implicit Plots

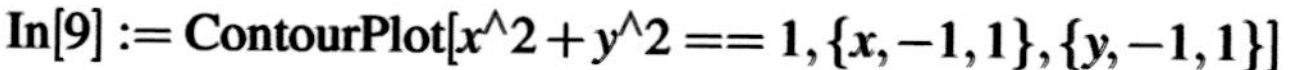

In[9] := ContourPlot[x^2 + y^2 == 1, {x, −1, 1}, {y, −1, 1}]

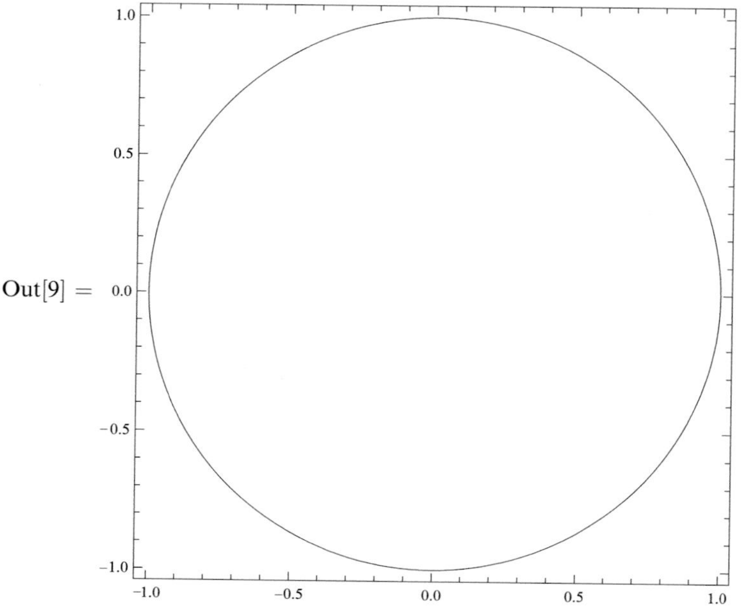

In[10] := ContourPlot[x ∗ y, {x, −4, 4}, {y, −4, 4}, Contours−>{1, 2, 3, 4}]

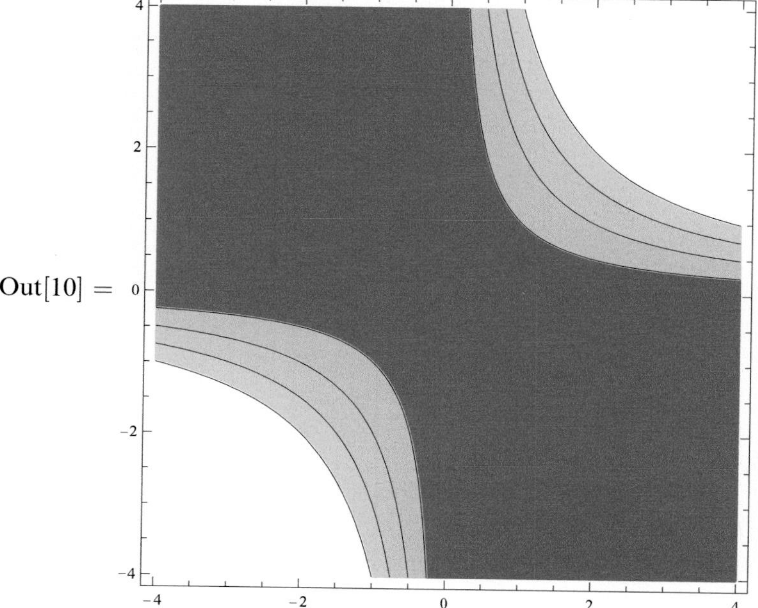

Experimental Points

In real life they are being read from a file. We don't have such a file at hand, and therefore we'll generate a list of "experimental" points according to some formulas.

In[11] := l = Table[N[Sin[Pi $*$ n/20]], {n,0,20}]

Out[11] = {0.,0.156434,0.309017,0.45399,0.587785,0.707107,0.809017, 0.891007,0.951057,0.987688,1.,0.987688,0.951057,0.891007,0.809017, 0.707107,0.587785,0.45399,0.309017,0.156434,0.}

Function Navigator → Visualization and Graphics → Data Visualization → ListPlot.

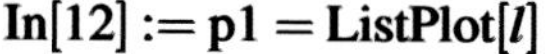

In[12] := p1 = ListPlot[l]

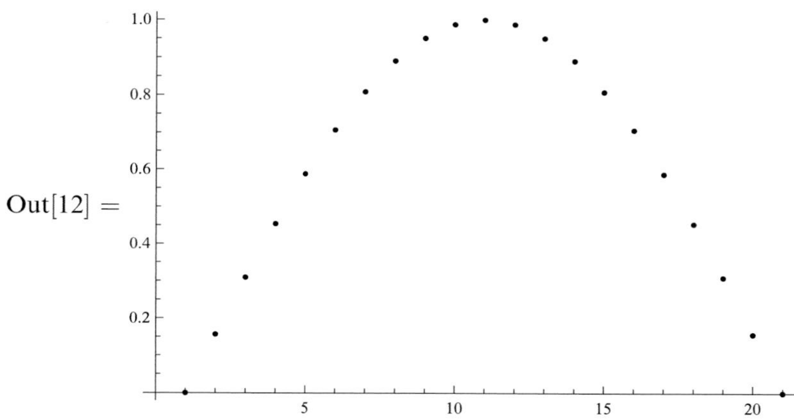

Let's try to fit these points by a quadratic polynomial.

In[13] := f = Fit[l, {1,x,x^2},x]

Out[13] = $-0.24953+0.222936x-0.0101334x^2$

In[14] := p2 = Plot[f, {x, 1, 21}]

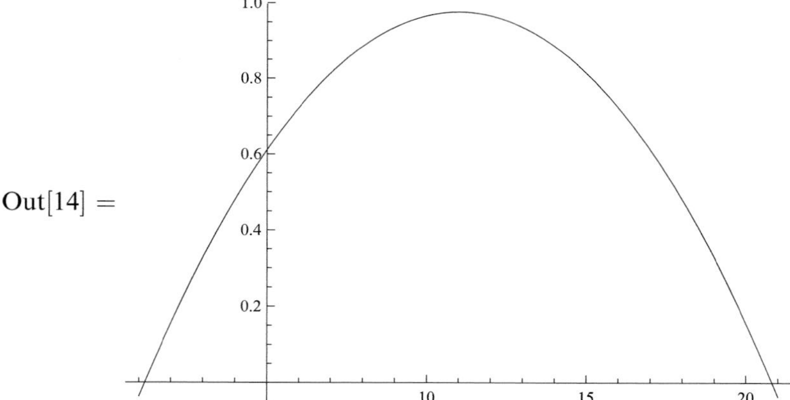

And now the curve and the points on a single plot (Function Navigator → Visualization and Graphics → Data Visualization → Annotation & Combination → Show).

```
In[15] := Show[p1,p2]
```

Out[15] =

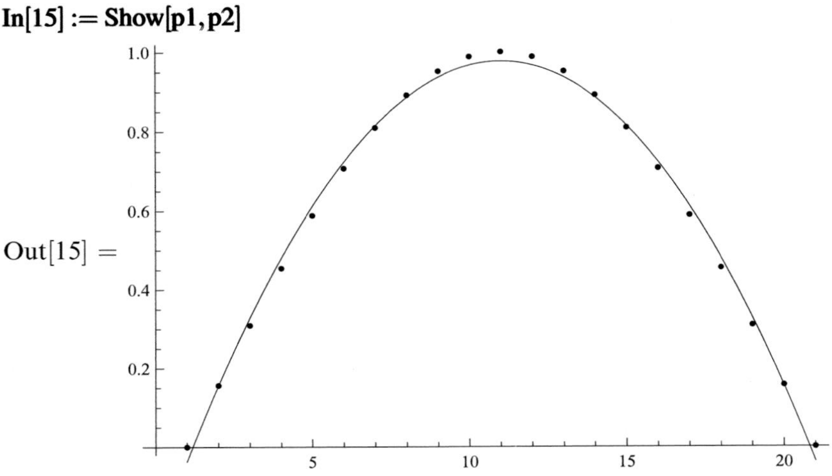


```
In[16] := Clear[l,f,p1,p2]
```

Inequalities

Function Navigator → Visualization and Graphics → Function Visualization → RegionPlot.

```
In[17] := RegionPlot[x^2+y^2<1&&x+y>0,{x,−1,1},{y,−1,1}]
```

Out[17] =

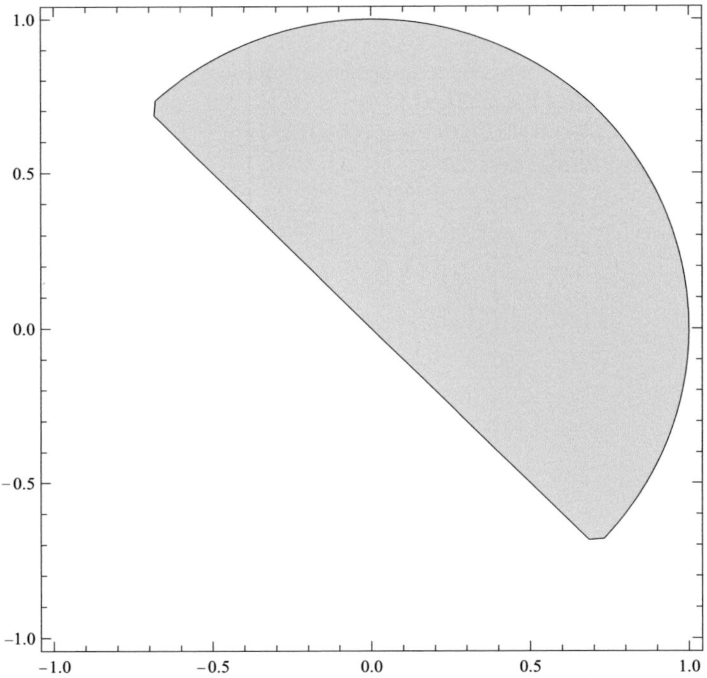

Vector Fields

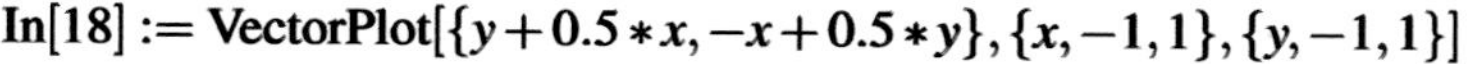

In[18] := VectorPlot[{y + 0.5 * x, −x + 0.5 * y}, {x, −1, 1}, {y, −1, 1}]

Out[18] =

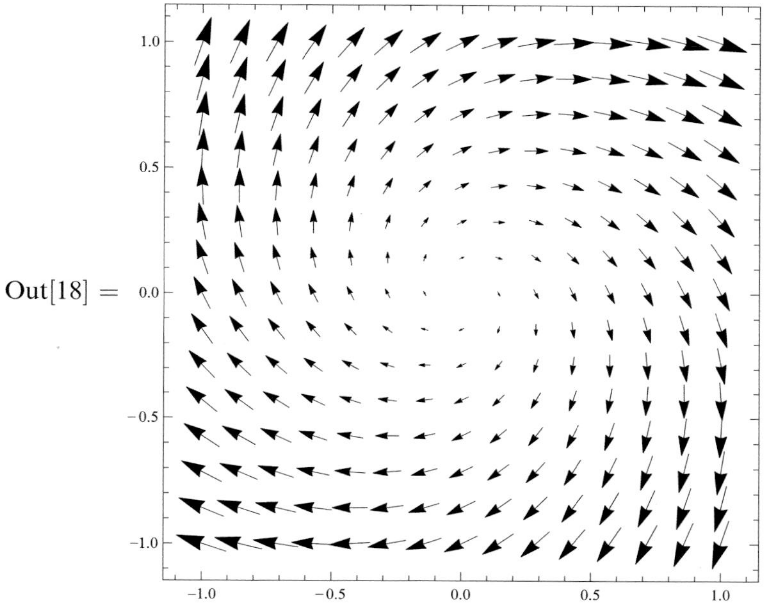

In[19] := StreamPlot[{y + 0.5 * x, −x + 0.5 * y}, {x, −1, 1}, {y, −1, 1}]

Out[19] =

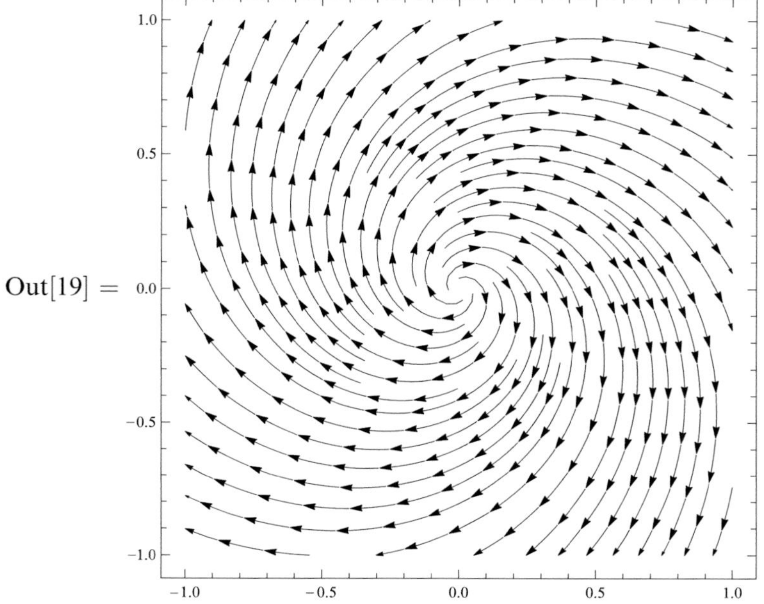

14.2 3D Plots

Hat

Let's draw a hat with a wavy pent. First define a function.
In[20] := f[x_,y_] := With[{r = Sqrt[x^2 + y^2]}, Sin[r]/r]
Virtual Book → Visualization and Graphics → Three-Dimensional Surface Plots; Function Navigator → Visualization and Graphics → Function Visualization → Plot3D.
In[21] := Plot3D[f[x,y], {x, −10, 10}, {y, −10, 10}]

Out[21] =

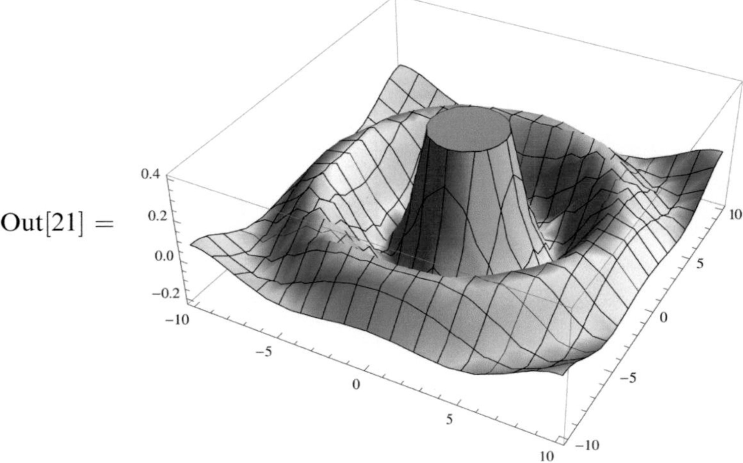

And why is the top of the hat cut off?
In[22] := Plot3D[f[x,y], {x, −10, 10}, {y, −10, 10}, PlotRange−>All]

Out[22] =

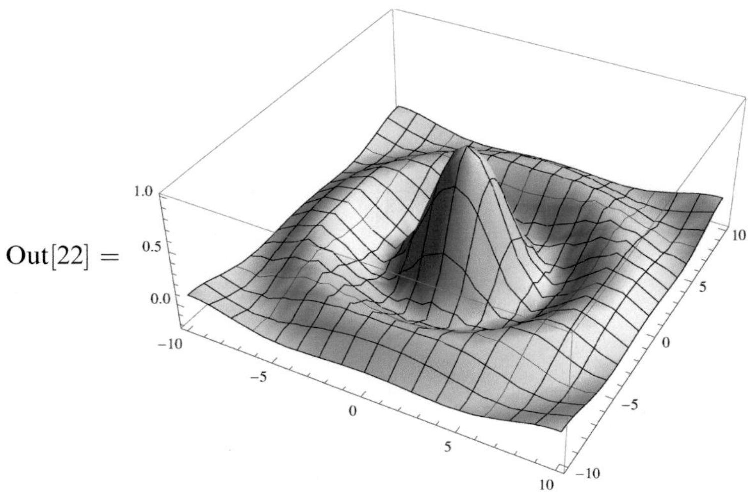

All 3D plots can be rotated by the mouse. If you press Shift, the mouse will move the plot; and if you press Ctrl, then it will resize it.

```
In[23] := Clear[f]
```

Sphere

```
In[24] := ParametricPlot3D[{Sin[θ] * Cos[ϕ], Sin[θ] * Sin[ϕ], Cos[θ]},
    {θ, 0, Pi}, {ϕ, 0, 2 * Pi}]
```

Out[24] =

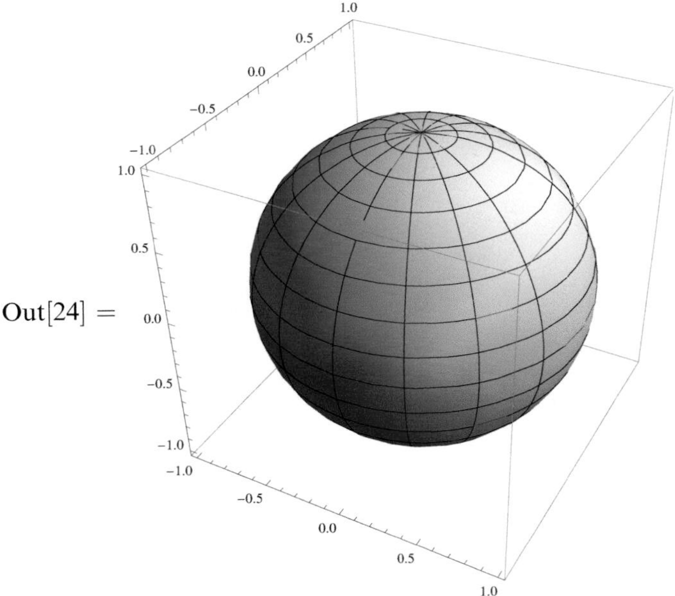

Donut

Take a point on the x-axis at a distance R from the origin; draw a circle of a radius r around it in the x, z plane; and rotate it around the z axis. You will get a donut (torus). Let R be 1; r can be tuned by the mouse from 0 to 1, with the initial value 0.3.

```
In[25] := R = 1;
In[26] := Manipulate[
    ParametricPlot3D[
        {(R + r * Cos[θ]) * Cos[ϕ], (R + r * Cos[θ]) * Sin[ϕ], r * Sin[θ]},
        {θ, 0, 2 * Pi}, {ϕ, 0, 2 * Pi}],
    {{r, 0.3}, 0, 1}]
```

Out[26] =

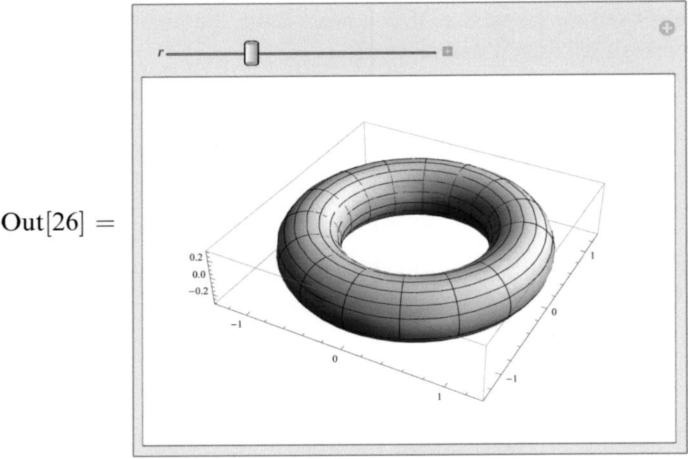

In[27] := Clear[*R*]

Spiral

ParametricPlot3D can draw curves, too.

**In[28] := Manipulate[ParametricPlot3D[{Cos[*t*], Sin[*t*], *a* ∗ *t*}, {*t*, 0, 20},
PlotRange−>{{−1, 1}, {−1, 1}, {0, 2}}],
{{*a*, 0.1}, 0, 0.2}]**

Out[28] =

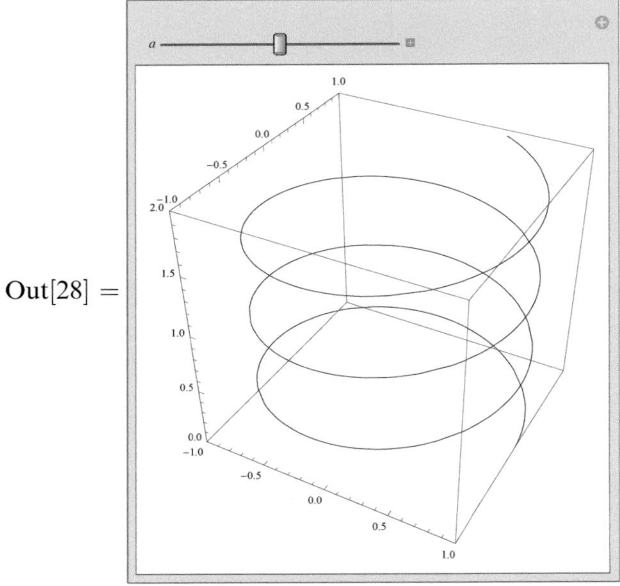

Implicit Surface

In[29] := ContourPlot3D[x^2 + y^2 + z^2 == 1, {x, −1, 1}, {y, −1, 1}, {z, −1, 1}]

Out[29] =

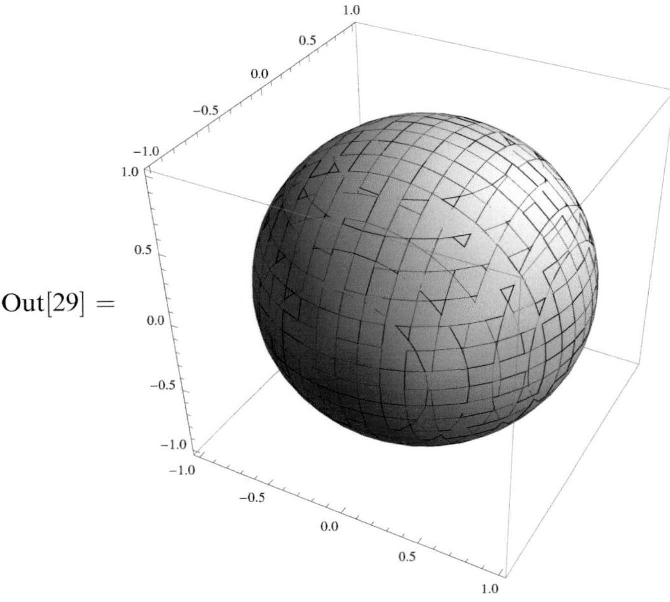

Inequalities

**In[30] := RegionPlot3D[x^2 + y^2 + z^2 < 1&&x + y + z > 0,
{x, −1, 1}, {y, −1, 1}, {z, −1, 1},
ViewPoint−> {−2, −10, 2}, PlotPoints−> 100]**

Out[30] =

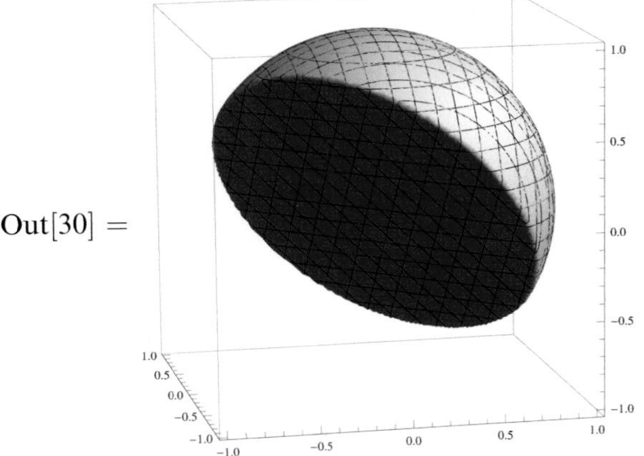

Chapter 15
Trigonometric Functions

Statement of the Problem

Mathematica can calculate Cos and Sin for many arguments equal to π times rational numbers. For example,

In[1] := Table[Cos[Pi/*n*], {*n*, 1, 12}]

$$\text{Out}[1] = \left\{-1, 0, \frac{1}{2}, \frac{1}{\sqrt{2}}, \frac{1}{4}\left(1+\sqrt{5}\right), \frac{\sqrt{3}}{2}, \text{Cos}\left[\frac{\pi}{7}\right], \text{Cos}\left[\frac{\pi}{8}\right], \text{Cos}\left[\frac{\pi}{9}\right], \sqrt{\frac{5}{8}+\frac{\sqrt{5}}{8}}, \text{Cos}\left[\frac{\pi}{11}\right], \frac{1+\sqrt{3}}{2\sqrt{2}}\right\}$$

But it does not apply the half-angle formulas in all possible cases. We'll write our own cos function which does this; then

In[2] := sin[*x*_] := cos[*x* − Pi/2]

Simple Cases

These cases should be considered separately because *Mathematica* does not treat them as rational numbers times π.

In[3] := cos[0] = 1; cos[Pi] = −1;
In[4] := cos[*n*_Integer ∗ Pi] := (−1)^*n*

General Case

In[5] := cos[*r*_Rational ∗ Pi] := Which[*r* < 0, cos[−*r* ∗ Pi],
** *r* > 2, cos[FractionalPart[*r*/2] ∗ 2 ∗ Pi], *r* > 1, cos[(2 − *r*) ∗ Pi],**
** *r* > 1/2, −cos[(1 − *r*) ∗ Pi],**

A. Grozin, *Introduction to Mathematica® for Physicists*, Graduate Texts in Physics,
DOI 10.1007/978-3-319-00894-3_15,

EvenQ[Denominator[r]], Simplify[Sqrt[(1 + cos[2 * r * Pi])/2]],
True, Cos[r * Pi]]

For example,

In[6] := {cos[Pi/32], cos[−65 * Pi/32], cos[3 * Pi/32], cos[33 * Pi/32]}

$$\text{Out[6]} = \left\{\frac{1}{2}\sqrt{2+\sqrt{2+\sqrt{2+\sqrt{2}}}}, \frac{1}{2}\sqrt{2+\sqrt{2+\sqrt{2+\sqrt{2}}}}, \frac{1}{2}\sqrt{2+\sqrt{2+\sqrt{2-\sqrt{2}}}}, -\frac{1}{2}\sqrt{2+\sqrt{2+\sqrt{2+\sqrt{2}}}}\right\}$$

In[7] := {cos[Pi/48], cos[5 * Pi/48], cos[7 * Pi/48]}

$$\text{Out[7]} = \left\{\frac{1}{2}\sqrt{2+\sqrt{2+\sqrt{2+\sqrt{3}}}}, \frac{1}{2}\sqrt{2+\sqrt{2+\sqrt{2-\sqrt{3}}}}, \frac{1}{2}\sqrt{2+\sqrt{2-\sqrt{2-\sqrt{3}}}}\right\}$$

In[8] := {cos[Pi/40], cos[3 * Pi/40]}

$$\text{Out[8]} = \left\{\frac{1}{2}\sqrt{2+\sqrt{2+\sqrt{\frac{1}{2}\left(5+\sqrt{5}\right)}}}, \frac{1}{2}\sqrt{2+\sqrt{2+\sqrt{\frac{1}{2}\left(5-\sqrt{5}\right)}}}\right\}$$

Check

```
In[9] := check[d_, n_] := Module[{e = 0, x, ex},
    Do[x = i/d * Pi; ex = Abs[N[cos[x] − Cos[x]]]; If[ex > e, e = ex],
        {i, −d * n, d * n}];
    e]
In[10] := {check[128, 5], check[192, 5], check[320, 5]}
```

$$\text{Out[10]} = \{2.220446049250313 \times 10^{-16}, 6.106226635438361 \times 10^{-16}, 1.861358289723114 \times 10^{-15}\}$$

Chapter 16
Quantum Oscillator

Catching a lion, the Schrödinger's method: At any moment, there is a nonzero probability that a lion is inside the cage. Sit and wait.

16.1 Lowering and Raising Operators

The Hamiltonian of the harmonic oscillator is [18]

$$\hat{H} = \frac{\hat{p}^2}{2m} + \frac{m\,\omega^2\,\hat{x}^2}{2}.$$

There is a dimensionful constant $\hbar$ in quantum mechanics; therefore, two quantities m and ω define a scale of length $\sqrt{\hbar/(m\,\omega)}$, momentum $\sqrt{\hbar\,m\,\omega}$, energy $\hbar\,\omega$, and any other quantity of any dimensionality. They have the meaning of the characteristic amplitude, momentum, and energy of zero oscillations. We shall put $\hbar = 1$, $m = 1$, and $\omega = 1$, thus choosing these characteristic scales as units for measurement of corresponding quantities. Then

$$\hat{H} = \frac{\hat{p}^2 + \hat{x}^2}{2}.$$

Let's introduce the operator

$$\hat{a} = \frac{\hat{x} + \mathrm{i}\hat{p}}{\sqrt{2}}$$

and its Hermitian conjugate

$$\hat{a}^+ = \frac{\hat{x} - \mathrm{i}\hat{p}}{\sqrt{2}}.$$

The commutation relation $[\hat{p},\hat{x}] = -\mathrm{i}$ implies for them

$$\left[\hat{a},\hat{a}^+\right] = 1\,.$$

A. Grozin, *Introduction to Mathematica® for Physicists*, Graduate Texts in Physics,
DOI 10.1007/978-3-319-00894-3_16,

The Hamiltonian is expressed via these operators as

$$\hat{H} = \hat{a}^+ \hat{a} + \frac{1}{2},$$

from where we obtain $\left[\hat{H},\hat{a}\right] = -\hat{a}$, $\left[\hat{H},\hat{a}^+\right] = \hat{a}^+$. Therefore, if $|\psi>$ is an eigenstate of $\hat{H}$ with the energy E: $\hat{H}\,|\psi> = E\,|\psi>$, then $\hat{a}\,|\psi>$ and $\hat{a}^+\,|\psi>$ are also eigenstates of $\hat{H}$ with the energies $E-1$ and $E+1$:

$$\hat{H}\,\hat{a}\,|\psi> = \left(\hat{a}\hat{H} - \hat{a}\right)\,|\psi> = (E-1)\hat{a}\,|\psi>, \qquad \hat{H}\,\hat{a}^+\,|\psi> = (E+1)\hat{a}^+\,|\psi>$$

(if only these states don't vanish). Hence the eigenvalues of $\hat{H}$ form an arithmetic progression with step equal to 1. It is bounded from below because $\hat{H}$ is a positive definite operator. Therefore, there exists a state $|0>$ with the lowest energy that cannot be lowered any more:

$$\hat{a}\,|0> = 0.$$

Its energy is equal to $\frac{1}{2}$:

$$\hat{H}\,|0> = \left(\hat{a}^+\hat{a} + \frac{1}{2}\right)\,|0> = \frac{1}{2}\,|0>$$

(this is the zero oscillations energy). Acting on $|0>$ by the raising operator $\hat{a}^+$ n times, we obtain a state $|n>$ with the energy

$$E_n = n + \frac{1}{2}.$$

Hence, $\hat{H}\,|n> = \left(\hat{a}^+ a + \frac{1}{2}\right)\,|n> = \left(n + \frac{1}{2}\right)\,|n>$ or

$$\hat{a}^+\,\hat{a}\,|n> = n\,|n>,$$

i.e., $\hat{a}^+\hat{a}$ acts as an operator of the level number.

We have $\hat{a}\,|n> = c_n\,|n-1>$; it is possible to make c_n real and positive by tuning the phases of the states $|n>$. These coefficients can be found from the normalization condition: $|c_n|^2 = <n|\hat{a}^+\hat{a}|n> = n$. The action of the operator $\hat{a}^+$ follows from Hermitian conjugation:

$$\hat{a}\,|n> = \sqrt{n}\,|n-1>, \qquad \hat{a}^+\,|n> = \sqrt{n+1}\,|n+1>.$$

From this we again have $\hat{a}^+\hat{a}\,|n> = n\,|n>$.

In the coordinate representation

$$\hat{x} = x, \qquad \hat{p} = -\mathrm{i}\frac{\mathrm{d}}{\mathrm{d}x}.$$

Let's implement the operators $\hat{a}$ and $\hat{a}^+$ in *Mathematica*.

```
In[1] := a[f_] := Together[(x*f+D[f,x])/Sqrt[2]]
In[2] := ac[f_] := Together[(x*f-D[f,x])/Sqrt[2]]
```

16.2 Ground State

This is the state which cannot be lowered by $\hat{a}$.

In[3] := Eq = a[ψ0[x]] == 0

$\text{Out}[3] = \frac{x\psi_0[x] + \psi_0'[x]}{\sqrt{2}} == 0$

In[4] := s = DSolve[Eq, ψ0[x], x]

$\text{Out}[4] = \left\{\left\{\psi_0[x] \to \mathrm{e}^{-\frac{x^2}{2}} C[1]\right\}\right\}$

In[5] := ψ0 = ψ0[x]/.s[[1]]

$\text{Out}[5] = \mathrm{e}^{-\frac{x^2}{2}} C[1]$

In[6] := Clear[Eq]

The normalization integral.

In[7] := NI = Integrate[ψ0^2, {x, −Infinity, Infinity}]

$\text{Out}[7] = \sqrt{\pi} C[1]^2$

In[8] := s = Solve[NI == 1, C[1]]

$\text{Out}[8] = \left\{\left\{C[1] \to -\frac{1}{\pi^{1/4}}\right\}, \left\{C[1] \to \frac{1}{\pi^{1/4}}\right\}\right\}$

In[9] := ψ0 = ψ0/.s[[2]]

$\text{Out}[9] = \frac{\mathrm{e}^{-\frac{x^2}{2}}}{\pi^{1/4}}$

In[10] := Clear[NI, s]

16.3 Excited States

In[11] := ψ[0] = ψ0;

In[12] := ψ[n_] := ψ[n] = ac[ψ[n − 1]]/Sqrt[n]

The wave functions of a few first states.

In[13] := Table[ψ[n], {n, 0, 10}]

$$\text{Out}[13] = \left\{\frac{\mathrm{e}^{-\frac{x^2}{2}}}{\pi^{1/4}}, \frac{\sqrt{2}\mathrm{e}^{-\frac{x^2}{2}} x}{\pi^{1/4}}, \frac{\mathrm{e}^{-\frac{x^2}{2}}\left(-1+2x^2\right)}{\sqrt{2}\pi^{1/4}}, \frac{\mathrm{e}^{-\frac{x^2}{2}} x\left(-3+2x^2\right)}{\sqrt{3}\pi^{1/4}},\right.$$

$$\frac{\mathrm{e}^{-\frac{x^2}{2}}\left(3-12x^2+4x^4\right)}{2\sqrt{6}\pi^{1/4}}, \frac{\mathrm{e}^{-\frac{x^2}{2}} x\left(15-20x^2+4x^4\right)}{2\sqrt{15}\pi^{1/4}},$$

$$\frac{\mathrm{e}^{-\frac{x^2}{2}}\left(-15+90x^2-60x^4+8x^6\right)}{12\sqrt{5}\pi^{1/4}}, \frac{\mathrm{e}^{-\frac{x^2}{2}} x\left(-105+210x^2-84x^4+8x^6\right)}{6\sqrt{70}\pi^{1/4}},$$

$$\frac{\mathrm{e}^{-\frac{x^2}{2}}\left(105-840x^2+840x^4-224x^6+16x^8\right)}{24\sqrt{70}\pi^{1/4}},$$

$$\frac{e^{-\frac{x^2}{2}} x \left(945 - 2520x^2 + 1512x^4 - 288x^6 + 16x^8\right)}{72\sqrt{35}\pi^{1/4}},$$
$$\left.\frac{e^{-\frac{x^2}{2}} \left(-945 + 9450x^2 - 12600x^4 + 5040x^6 - 720x^8 + 32x^{10}\right)}{720\sqrt{7}\pi^{1/4}}\right\}$$

And here the level number can be set by the mouse.

In[14] := Manipulate[ψ[n], {n, 0, 10, 1, Appearance−>"Labeled"}]

Out[14] =

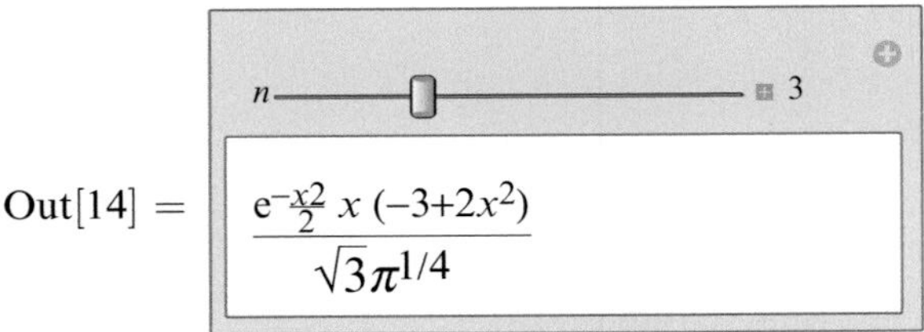

The wave functions of a few first states.

In[15] := Plot[Evaluate[Table[ψ[n], {n, 0, 3}]], {x, −5, 5}]

Out[15] =

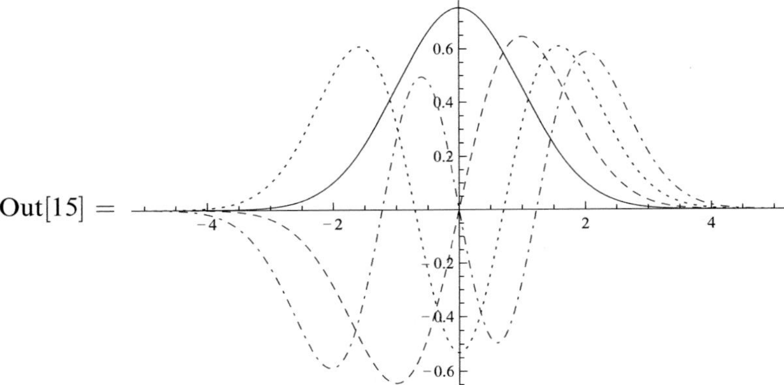

And this is a live plot: the level number can be set by the mouse.

In[16] := Manipulate[Plot[ψ[n], {x, −5, 5}, PlotRange → {−0.8, 0.8}], {n, 0, 10, 1, Appearance−>"Labeled"}]

Out[16] = 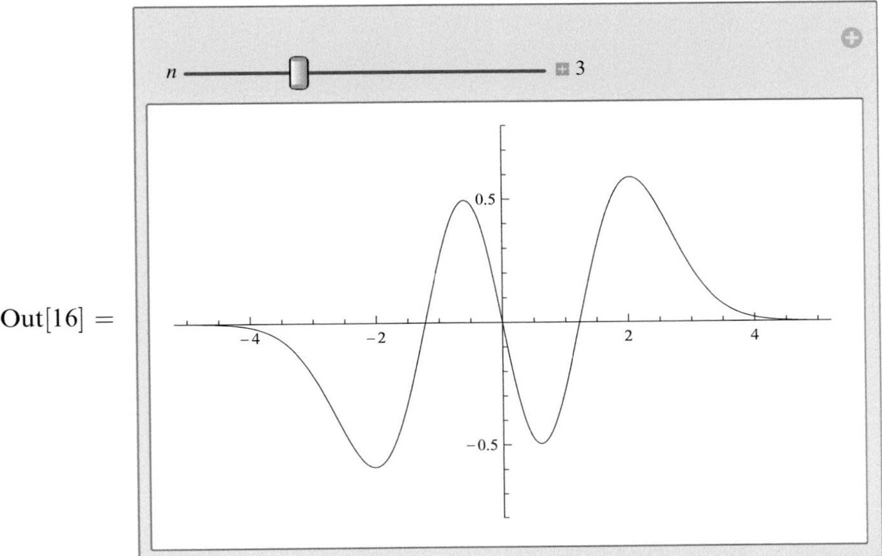

16.4 Some Properties

Orthogonality and normalization.

In[17] := Distribute[ψ]

Out[17] = ψ

In[18] := Parallelize[Table[Table[Integrate[ψ[n] ∗ ψ[m], {x, −Infinity, Infinity}], {n, 0, 10}], {m, 0, 10}]]

Out[18] = {{1,0,0,0,0,0,0,0,0,0,0}, {0,1,0,0,0,0,0,0,0,0,0}, {0,0,1,0,0,0,0,0,0,0,0}, {0,0,0,1,0,0,0,0,0,0,0}, {0,0,0,0,1,0,0,0,0,0,0}, {0,0,0,0,0,1,0,0,0,0,0}, {0,0,0,0,0,0,1,0,0,0,0}, {0,0,0,0,0,0,0,1,0,0,0}, {0,0,0,0,0,0,0,0,1,0,0}, {0,0,0,0,0,0,0,0,0,1,0}, {0,0,0,0,0,0,0,0,0,0,1}}

Wave functions in the momentum representation (Fourier transforms) are the same as in the coordinate one, up to phase factors.

In[19] := Parallelize[Table[
Cancel[Integrate[ψ[n] ∗ Exp[−I ∗ p ∗ x], {x, −Infinity, Infinity}]/
Sqrt[2 ∗ Pi]/(ψ[n]/.x → p)],
{n, 0, 10}]]

Out[19] = {1, −i, −1, i, 1, −i, −1, i, 1, −i, −1}

The probability density.

In[20] := Manipulate[Plot[ψ[n]^2, {x, −6, 6}, PlotRange → {0, 0.6}], {{n, 10}, 0, 10, 1, Appearance−>"Labeled"}]

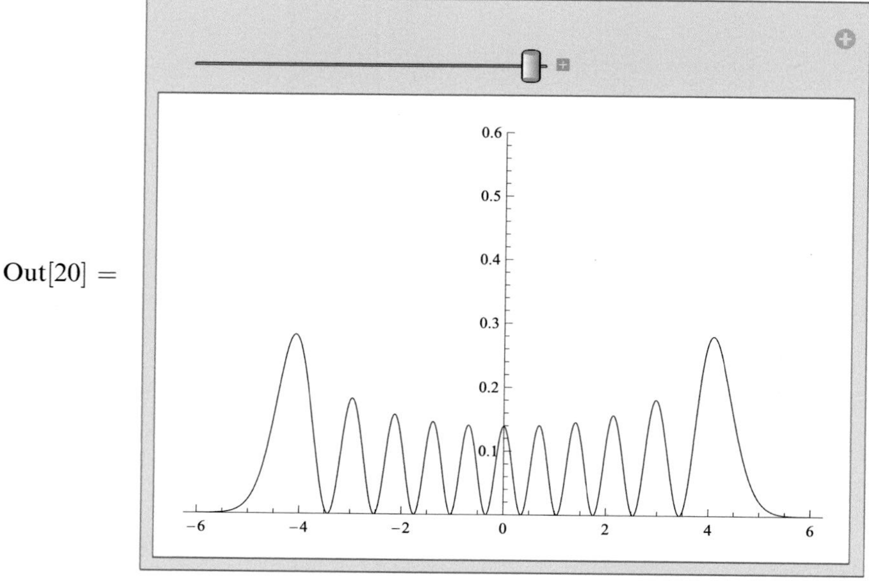

Why is it larger near the boundaries?

Chapter 17
Spherical Harmonics

17.1 Angular Momentum in Quantum Mechanics

The angular momentum operator $\hat{\boldsymbol{J}}$ is defined [18] in such a way that $\hat{U} = \exp\left(-\mathrm{i}\hat{\boldsymbol{J}}\cdot\delta\boldsymbol{\varphi}\right)$ is the operator of an infinitesimal rotation with the angle $\delta\boldsymbol{\varphi}$: if $|\psi>$ is a state, then $\hat{U}\,|\psi>$ is the same state rotated by $\delta\boldsymbol{\varphi}$. Therefore, the average value $\boldsymbol{V}'$ of a vector operator $\hat{\boldsymbol{V}}$ over $\hat{U}\,|\psi>$ is related to its average value $\boldsymbol{V}$ over $|\psi>$ by the formula $\boldsymbol{V}' = \boldsymbol{V} + \delta\boldsymbol{\varphi}\times\boldsymbol{V}$ and hence $\hat{U}^+\hat{\boldsymbol{V}}\hat{U} = \hat{\boldsymbol{V}} + \mathrm{i}\left[\hat{\boldsymbol{J}}\cdot\delta\boldsymbol{\varphi},\hat{\boldsymbol{V}}\right] = \hat{\boldsymbol{V}} + \delta\boldsymbol{\varphi}\times\hat{\boldsymbol{V}}$. Therefore, for any vector operator $\hat{\boldsymbol{V}}$ the commutation relation $\left[\hat{V}_i,\hat{J}_j\right] = \mathrm{i}\varepsilon_{ijk}\hat{V}_k$ holds. The average value of a scalar operator $\hat{S}$ does not change at rotations; hence $\left[\hat{S},\hat{J}_i\right] = 0$. In particular, the angular momentum $\hat{\boldsymbol{J}}$ is a vector operator, and its square $\hat{\boldsymbol{J}}^2 = \hat{J}_x^2 + \hat{J}_y^2 + \hat{J}_z^2$ is a scalar one:

$$\left[\hat{J}_i,\hat{J}_j\right] = \mathrm{i}\varepsilon_{ijk}\hat{J}_k\,, \qquad \left[\hat{\boldsymbol{J}}^2,\hat{J}_i\right] = 0\,.$$

Therefore, a system of common eigenstates of $\hat{\boldsymbol{J}}^2$ and $\hat{J}_z$ exists. Let's introduce the operators $\hat{J}_\pm = \hat{J}_x \pm \mathrm{i}\hat{J}_y$, $\hat{J}_\pm^+ = \hat{J}_\mp$; we have $\left[\hat{J}_z,\hat{J}_\pm\right] = \pm\hat{J}_\pm$. This means that if $|\psi>$ is an eigenstate of $\hat{J}_z$ ($\hat{J}_z\,|\psi> = m\,|\psi>$), then $\hat{J}_\pm\,|\psi>$ are also eigenstates of $\hat{J}_z$: $\hat{J}_z\hat{J}_\pm\,|\psi> = (m\pm1)\hat{J}_\pm\,|\psi>$ (if they don't vanish). Therefore, eigenvalues m of $\hat{J}_z$ form a progression with unit step, and the ladder operators $\hat{J}_\pm$ increase and decrease m.

Let's consider states with a given eigenvalue of $\hat{\boldsymbol{J}}^2$. For these states, eigenvalues of $\hat{J}_z$ are bounded from above and from below because the operator $\hat{\boldsymbol{J}}^2 - \hat{J}_z^2 = \hat{J}_x^2 + \hat{J}_y^2$ is positive definite. Let $|m_\pm>$ be the eigenstates with the maximum and the minimum eigenvalues of $\hat{J}_z$ equal to $m_\pm$. Then these eigenvalues cannot be further increased and decreased by the operators $\hat{J}_\pm$ correspondingly: $\hat{J}_\pm\,|m_\pm> = 0$. We have

$$\hat{J}_\pm\hat{J}_\mp = \hat{\boldsymbol{J}}^2 - \hat{J}_z^2 \pm \hat{J}_z\,.$$

A. Grozin, *Introduction to Mathematica® for Physicists*, Graduate Texts in Physics,
DOI 10.1007/978-3-319-00894-3_17, © Springer International Publishing Switzerland 2014

Therefore, $\hat{J}_\mp \hat{J}_\pm |m_\pm> = 0 = \left[\hat{\boldsymbol{J}}^2 - m_\pm (m_\pm \pm 1)\right] |m_\pm>$, i.e., the eigenvalue of the operator $\hat{\boldsymbol{J}}^2$ for these states (as well as for all the other states being considered) is $m_+ (m_+ + 1) = m_- (m_- - 1)$. Hence $(m_+ + m_-)(m_+ - m_- + 1) = 0$; taking into account $m_+ \geqslant m_-$ we obtain $m_- = -m_+$ or $m_\pm = \pm j$. The number j must be integer or half-integer because m_+ and m_- differ by an integer.

Finally, we have a system of common eigenstates $|j,m>$ of the operators $\hat{\boldsymbol{J}}^2$ and $\hat{J}_z$:

$$\hat{\boldsymbol{J}}^2 |j,m> = j(j+1)|j,m>\,, \qquad \hat{J}_z |j,m> = m|j,m>\,,$$

where j is integer or half-integer, and m varies from $-j$ to j by 1. The operators $\hat{J}_\pm$ increase and decrease m correspondingly: $\hat{J}_\pm |j,m> = a_\pm(j,m)|j,m\pm 1>$. Tuning the phases of $|j,m>$ we can make $a_\pm(j,m)$ real and positive. They can be found from the normalization: $|a_\pm(j,m)|^2 = <j,m|\hat{J}_\mp \hat{J}_\pm|j,m> = j(j+1) - m(m\pm 1)$. Finally we arrive at

$$\hat{J}_\pm |j,m> = \sqrt{j(j+1) - m(m\pm 1)}\,|j,m\pm 1> = \sqrt{(j\pm m+1)(j\mp m)}\,|j,m\pm 1>\,.$$

The orbital angular momentum of a particle $\hat{\boldsymbol{l}} = \boldsymbol{r} \times \hat{\boldsymbol{p}}$ (where $\hat{\boldsymbol{p}} = -\mathrm{i}\boldsymbol{\nabla}$ in the coordinate representation) is an example of angular momentum. In spherical coordinates

$$\hat{l}_z = -\mathrm{i}\frac{\partial}{\partial\varphi}\,, \qquad \hat{l}_\pm = \mathrm{e}^{\pm\mathrm{i}\varphi}\left(\pm\frac{\partial}{\partial\theta} + \mathrm{i}\cot\theta\frac{\partial}{\partial\varphi}\right)\,.$$

The eigenfunctions of $\hat{l}_z$ are $\mathrm{e}^{\mathrm{i}m\varphi}$; they must not change at $\varphi \to \varphi + 2\pi$; hence m must be integer. The common eigenfunctions of $\hat{\boldsymbol{l}}^2$ and $\hat{l}_z$ are called spherical harmonics:

$$\hat{\boldsymbol{l}}^2 Y_{lm}(\theta,\varphi) = l(l+1)Y_{lm}(\theta,\varphi)\,, \qquad \hat{l}_z Y_{lm}(\theta,\varphi) = mY_{lm}(\theta,\varphi)\,,$$

where l is integer, and m varies from $-l$ to l by 1; $Y_{lm}(\theta,\varphi) = P_{lm}(\theta)\mathrm{e}^{\mathrm{i}m\varphi}$. They are orthonormalized:

$$\int Y^*_{l'm'}(\theta,\varphi)Y_{lm}(\theta,\varphi)\mathrm{d}\Omega = \delta_{l'l}\delta_{m'm}\,.$$

Here are the operators $\hat{l}_\pm$ in *Mathematica*:

In[1] := lp[f_] := Together[Exp[*I* ∗ φ] ∗ (D[f, θ] + *I* ∗ Cot[θ] ∗ D[f, φ])]
In[2] := lm[f_] := Together[Exp[−*I* ∗ φ] ∗ (−D[f, θ] + *I* ∗ Cot[θ] ∗ D[f, φ])]

17.2 $Y_{ll}(\theta,\varphi)$

The angular momentum projection of this state cannot be raised by $\hat{l}_+$.

In[3] := Eq = lp[Exp[*I* ∗ *l* ∗ φ] ∗ *P*[θ]] == 0
Out[3] = $-\mathrm{e}^{\mathrm{i}\varphi+\mathrm{i}l\varphi}\left(l\,\mathrm{Cot}[\theta]P[\theta] - P'[\theta]\right) == 0$

In[4] := s = DSolve[Eq, P[θ], θ]
$\text{Out[4]} = \left\{\left\{P[\theta] \to C[1]\text{Sin}[\theta]^l\right\}\right\}$
In[5] := P = P[θ]/.s[[1]]
$\text{Out[5]} = C[1]\text{Sin}[\theta]^l$
The normalization integral.
In[6] := NI = 2 ∗ Pi ∗ Integrate[P^2 ∗ Sin[θ], {θ, 0, Pi}, Assumptions−>{l ≥ 0}]/. Gamma[−l]−> − Pi/(Sin[Pi ∗ l] ∗ Gamma[l + 1])
$$\text{Out[6]} = \frac{2\pi^{3/2}C[1]^2\text{Gamma}[1+l]}{\text{Gamma}\left[\frac{3}{2}+l\right]}$$
In[7] := s = Solve[NI == 1, C[1]]
$$\text{Out[7]} = \left\{\left\{C[1] \to -\frac{\sqrt{\text{Gamma}\left[\frac{3}{2}+l\right]}}{\sqrt{2}\pi^{3/4}\sqrt{\text{Gamma}[1+l]}}\right\},\left\{C[1] \to \frac{\sqrt{\text{Gamma}\left[\frac{3}{2}+l\right]}}{\sqrt{2}\pi^{3/4}\sqrt{\text{Gamma}[1+l]}}\right\}\right\}$$
In[8] := P = P/.s[[2]]
$$\text{Out[8]} = \frac{\sqrt{\text{Gamma}\left[\frac{3}{2}+l\right]}\text{Sin}[\theta]^l}{\sqrt{2}\pi^{3/4}\sqrt{\text{Gamma}[1+l]}}$$
The phase can be chosen arbitrarily. According to Landau–Lifshitz [18]:
In[9] := Y[l_, l_] = (−I)^l ∗ P ∗ Exp[I ∗ l ∗ φ]
$$\text{Out[9]} = \frac{(-\text{i})^l e^{\text{i}l\varphi}\sqrt{\text{Gamma}\left[\frac{3}{2}+l\right]}\text{Sin}[\theta]^l}{\sqrt{2}\pi^{3/4}\sqrt{\text{Gamma}[1+l]}}$$
In[10] := Clear[Eq, s, P]

17.3 $Y_{lm}(\theta,\varphi)$

These states can be obtained from $Y_{ll}(\theta,\varphi)$ by the lowering operator $\hat{l}_-$.
In[11] := S = Cos[x_]^n_−>(1 − Sin[x]^2)^Quotient[n, 2] ∗ Cos[x]^Mod[n, 2];
In[12] := Y[l_, m_]/; m < l := Y[l, m] = Factor[Expand[lm[Y[l, m + 1]]/Sqrt[(l − m) ∗ (l + m + 1)]/.S]]
In[13] := Table[Table[Y[l, m], {m, l, −l, −1}], {l, 0, 4}]
$$\text{Out[13]} = \left\{\left\{\frac{1}{2\sqrt{\pi}}\right\},\left\{-\frac{1}{2}\text{i}e^{\text{i}\varphi}\sqrt{\frac{3}{2\pi}}\text{Sin}[\theta],\frac{1}{2}\text{i}\sqrt{\frac{3}{\pi}}\text{Cos}[\theta],\frac{1}{2}\text{i}e^{-\text{i}\varphi}\sqrt{\frac{3}{2\pi}}\text{Sin}[\theta]\right\},\right.$$
$$\left\{-\frac{1}{4}e^{2\text{i}\varphi}\sqrt{\frac{15}{2\pi}}\text{Sin}[\theta]^2,\frac{1}{2}e^{\text{i}\varphi}\sqrt{\frac{15}{2\pi}}\text{Cos}[\theta]\text{Sin}[\theta],\frac{1}{4}\sqrt{\frac{5}{\pi}}\left(-2+3\text{Sin}[\theta]^2\right),\right.$$
$$\left.-\frac{1}{2}e^{-\text{i}\varphi}\sqrt{\frac{15}{2\pi}}\text{Cos}[\theta]\text{Sin}[\theta],-\frac{1}{4}e^{-2\text{i}\varphi}\sqrt{\frac{15}{2\pi}}\text{Sin}[\theta]^2\right\},$$

$$\left\{\frac{1}{8}\mathrm{i}e^{3\mathrm{i}\varphi}\sqrt{\frac{35}{\pi}}\mathrm{Sin}[\theta]^3, -\frac{1}{4}\mathrm{i}e^{2\mathrm{i}\varphi}\sqrt{\frac{105}{2\pi}}\mathrm{Cos}[\theta]\mathrm{Sin}[\theta]^2,\right.$$
$$-\frac{1}{8}\mathrm{i}e^{\mathrm{i}\varphi}\sqrt{\frac{21}{\pi}}\mathrm{Sin}[\theta]\left(-4+5\mathrm{Sin}[\theta]^2\right), \frac{1}{4}\mathrm{i}\sqrt{\frac{7}{\pi}}\mathrm{Cos}[\theta]\left(-2+5\mathrm{Sin}[\theta]^2\right),$$
$$\frac{1}{8}\mathrm{i}e^{-\mathrm{i}\varphi}\sqrt{\frac{21}{\pi}}\mathrm{Sin}[\theta]\left(-4+5\mathrm{Sin}[\theta]^2\right), -\frac{1}{4}\mathrm{i}e^{-2\mathrm{i}\varphi}\sqrt{\frac{105}{2\pi}}\mathrm{Cos}[\theta]\mathrm{Sin}[\theta]^2,$$
$$\left.-\frac{1}{8}\mathrm{i}e^{-3\mathrm{i}\varphi}\sqrt{\frac{35}{\pi}}\mathrm{Sin}[\theta]^3\right\},$$
$$\left\{\frac{3}{16}e^{4\mathrm{i}\varphi}\sqrt{\frac{35}{2\pi}}\mathrm{Sin}[\theta]^4, -\frac{3}{8}e^{3\mathrm{i}\varphi}\sqrt{\frac{35}{\pi}}\mathrm{Cos}[\theta]\mathrm{Sin}[\theta]^3,\right.$$
$$-\frac{3}{8}e^{2\mathrm{i}\varphi}\sqrt{\frac{5}{2\pi}}\mathrm{Sin}[\theta]^2\left(-6+7\mathrm{Sin}[\theta]^2\right),$$
$$\frac{3}{8}e^{\mathrm{i}\varphi}\sqrt{\frac{5}{\pi}}\mathrm{Cos}[\theta]\mathrm{Sin}[\theta]\left(-4+7\mathrm{Sin}[\theta]^2\right), \frac{3\left(8-40\mathrm{Sin}[\theta]^2+35\mathrm{Sin}[\theta]^4\right)}{16\sqrt{\pi}},$$
$$-\frac{3}{8}e^{-\mathrm{i}\varphi}\sqrt{\frac{5}{\pi}}\mathrm{Cos}[\theta]\mathrm{Sin}[\theta]\left(-4+7\mathrm{Sin}[\theta]^2\right),$$
$$-\frac{3}{8}e^{-2\mathrm{i}\varphi}\sqrt{\frac{5}{2\pi}}\mathrm{Sin}[\theta]^2\left(-6+7\mathrm{Sin}[\theta]^2\right), \frac{3}{8}e^{-3\mathrm{i}\varphi}\sqrt{\frac{35}{\pi}}\mathrm{Cos}[\theta]\mathrm{Sin}[\theta]^3,$$
$$\left.\left.\frac{3}{16}e^{-4\mathrm{i}\varphi}\sqrt{\frac{35}{2\pi}}\mathrm{Sin}[\theta]^4\right\}\right\}$$

In[14] := Manipulate[Manipulate[*Y*[*l*, *m*], {*m*, −*l*, *l*, 1, Appearance−>"Labeled"}], {*l*, 0, 4, 1, Appearance−>"Labeled"}]

Out[14] =

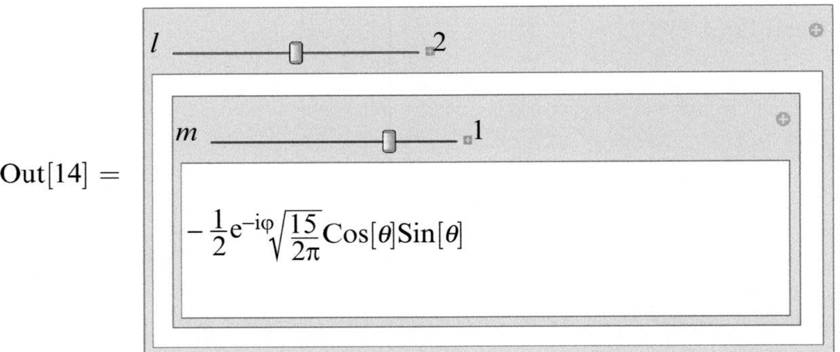

Orthogonality of $Y_{l_1 m_1}$ and $Y_{l_2 m_2}$ with $m_1 \neq m_2$ is evident; let's check all the rest.

```
In[15] := Table[Table[Table[
    Integrate[Y[l1,m] * Conjugate[Y[l2,m]] * Sin[θ], {φ,0,2*Pi}, {θ,0,Pi}],
    {m,l2,-l2,-1}], {l2,0,l1}], {l1,0,4}]
Out[15] = {{{1}}, {{0}, {1,1,1}}, {{0}, {0,0,0}, {1,1,1,1,1}},
    {{0}, {0,0,0}, {0,0,0,0,0}, {1,1,1,1,1,1,1}},
    {{0}, {0,0,0}, {0,0,0,0,0}, {0,0,0,0,0,0,0}, {1,1,1,1,1,1,1,1,1}}}
```

Chapter 18
Adding Angular Momenta in Quantum Mechanics

Let $\hat{\boldsymbol{J}}_1$ and $\hat{\boldsymbol{J}}_2$ be two angular momentum operators commuting with each other. Then the basis $|j_1,m_1;j_2,m_2>$ of common eigenstates of the operators $\hat{\boldsymbol{J}}_1^2$, $\hat{J}_{1z}$, $\hat{\boldsymbol{J}}_2^2$, $\hat{J}_{2z}$ exists. On the other hand, the total angular momentum $\hat{\boldsymbol{J}} = \hat{\boldsymbol{J}}_1 + \hat{\boldsymbol{J}}_2$ is also an angular momentum operator. Therefore, linear combinations $|j,m>$ of the states $|j_1,m_1;j_2,m_2>$ at given j_1, j_2 can be constructed in such a way that they are eigenstates of $\hat{\boldsymbol{J}}^2$ and $\hat{J}_z$. This problem [18] is called addition of the angular momenta j_1 and j_2.

We always have $m = m_1 + m_2$ because $\hat{J}_z = \hat{J}_{1z} + \hat{J}_{2z}$. The following figure illustrates addition of j_1 and j_2 (it assumes $j_1 \geqslant j_2$).

```
In[1] := Fig[j1_,j2_] := If[j1 < j2, Fig[j2,j1],
  With[{d = 0.75*j2, d2 = 0.1*j2, d3 = 0.15*j2, r = 0.05*j2},
    Graphics[Join[{Line[{{j2,j1+j2}, {j2,j2-j1}, {-j2,-j1-j2},
          {-j2,j1-j2}, {j2,j1+j2}}],
        Line[{{0,-j1-j2-d}, {0,j1+j2+d}}],
        Line[{{-j2-d,0}, {j2+d,0}}],
        Line[{{-d2,j1+j2}, {d2,j1+j2}}],
        Text[j1+j2, {-d3,j1+j2}, {1,0}],
        Line[{{-d2,-j1-j2}, {d2,-j1-j2}}],
        Text[-(j1+j2), {d3,-j1-j2}, {-1,0}],
        Line[{{-d2,j1-j2}, {d2,j1-j2}}],
        Text[j1-j2, {d3,j1-j2}, {-1,1}],
        Line[{{-d2,j2-j1}, {d2,j2-j1}}],
        Text[-(j1-j2), {-d3,j2-j1}, {1,-1}],
        Text[-j2, {-j2-d2,-d2}, {1,1}], Text[j2, {j2+d2,-d2}, {-1,1}],
        Text[m, {-d2,j1+j2+d}, {1,0}], Text[m2, {j2+d,-d2}, {0,1}]},
      Join[Table[{Disk[{m2,m},r]}, {m,j1-j2+1,j1+j2}, {m2,m-j1,j2}]],
      Join[Table[Disk[{m2,m},r], {m,-j1-j2,j2-j1-1}, {m2,-j2,m+j1}]],
      Join[Table[Disk[{m2,m},r], {m,j2-j1,j1-j2}, {m2,-j2,j2}]]]]]]
```

A. Grozin, *Introduction to Mathematica® for Physicists*, Graduate Texts in Physics, 139
DOI 10.1007/978-3-319-00894-3_18, © Springer International Publishing Switzerland 2014

In[2] := Show[Fig[4, 2]]

Out[2] =

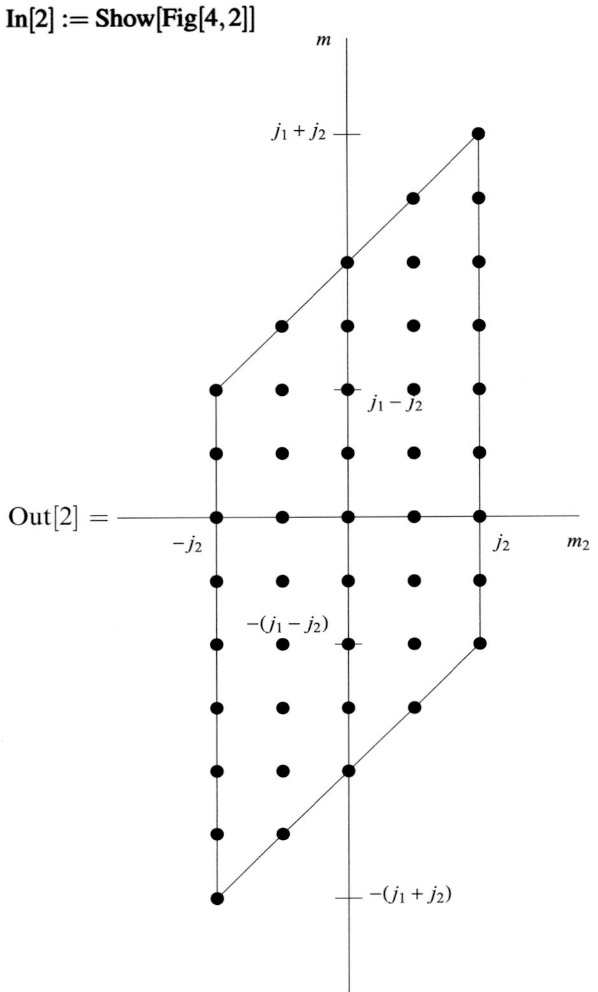

In[3] := Clear[Fig]

There is one state with $m = j_1 + j_2$, two states with $m = j_1 + j_2 - 1$, etc. Such an increase of the number of states occurs up to $m = j_1 - j_2$; further on it is constant up to $m = -(j_1 - j_2)$ and then decreases to one at $m = -(j_1 + j_2)$. Therefore, the maximum angular momentum resulting from adding j_1 and j_2 is $j = j_1 + j_2$. One state in the two-dimensional space of states with $m = j_1 + j_2 - 1$ refers to the same angular momentum, and the other one is the state with the maximum projection for the angular momentum $j = j_1 + j_2 - 1$. Continuing such reasoning, we see that all angular momenta up to $j_1 - j_2$ appear. In general, adding angular momenta j_1 and j_2 results in the angular momenta j from $|j_1 - j_2|$ to $j_1 + j_2$ in steps of 1.

This description naturally leads to the algorithm implemented below. We start from the only state with $m = j_1 + j_2$, namely the state $|j_1, j_1; j_2, j_2>$. It has $j = j_1 + j_2$, i.e., it is $|j_1 + j_2, j_1 + j_2>$. Repeatedly acting by the ladder operator $\hat{J}_- = \hat{J}_{1-} + \hat{J}_{2-}$ (and dividing by the appropriate normalization factor) we construct all the states with the total angular momentum $j = j_1 + j_2$: $|j_1 + j_2, j_1 + j_2 - 1>, \ldots,$ $|j_1 + j_2, -(j_1 + j_2)>$. Then we turn to the projection $m = j_1 + j_2 - 1$ and choose the state orthogonal to the already constructed one $|j_1 + j_2, j_1 + j_2 - 1>$. It has $j = j_1 + j_2 - 1$, i.e., it is $|j_1 + j_2 - 1, j_1 + j_2 - 1>$. Using the ladder operator we construct all the states with $j = j_1 + j_2 - 1$.

Then we proceed in a similar way. At the beginning of each step, when considering a new value of the projection m, we need to construct the state orthogonal to all the states with the same m already constructed. This is achieved as follows: we start from an arbitrary state, say, $|j_1, j_1; j_2, m - j_1>$, subtract its components along the already constructed states, and finally normalize the result. Then we construct all the states with the same total angular momentum from this state repeatedly acting by $\hat{J}_-$.

The function AddJ constructs the states $|j,m>$ (denoted Ket$[j,m]$) as linear combinations of the states $|j_1, m_1; j_2, m_2>$ (denoted ket[m1, m2]). It uses two local functions: Jm is the lowering operator $\hat{J}_-$ and ScaP is the scalar product. The procedure returns its local Ket, so that later the user will be able to inquire about Ket$[j,m]$ for specific values of j, m; in addition to this, the procedure prints all its results.

```
In[4] := AddJ = Function[{j1,j2},If[j2 > j1,AddJ[j2,j1],
  Module[{Ket, j,J,m,
      Jm = Function[{k},
        Expand[k/.ket[m1_,m2_]->
          Sqrt[(j1 - m1 + 1) * (j1+m1)] * ket[m1 - 1,m2]+
            Sqrt[(j2 - m2+ 1) * (j2+m2)] * ket[m1,m2 - 1]]],
      ScaP = Function[{k1,k2},
        Expand[k1 * k2]/.
          {ket[m1_,m2_]^2->1,ket[m1_,m2_] * ket[M1_,M2_]->0}]},
    Do[Ket[j, j] = ket[j - j2,j2];
        Do[Ket[j, j]- = Expand[ScaP[Ket[j, j],Ket[J, j]] * Ket[J, j]],
          {J,j1+j2, j+1, -1}];
        Print["Ket[", j,",", j,"] = ",
          Ket[j, j] = Expand[Ket[j, j]/Sqrt[ScaP[Ket[j, j],Ket[j, j]]]]];
        Do[Print["Ket[", j,",",m,"] = ",
            Ket[j,m] = Expand[Jm[Ket[j,m+1]]/Sqrt[(j - m) * (j+m+1)]]],
          {m, j - 1, -j, -1}],
      {j,j1+j2,j1 - j2, -1}];
    Ket]]];
In[5] := AddJ[1/2,1/2]
```

$$\text{Ket}[1,1] = \text{ket}\left[\frac{1}{2},\frac{1}{2}\right]$$

$$\text{Ket}[1,0] = \frac{\text{ket}\left[-\frac{1}{2},\frac{1}{2}\right]}{\sqrt{2}} + \frac{\text{ket}\left[\frac{1}{2},-\frac{1}{2}\right]}{\sqrt{2}}$$

$\text{Ket}[1,-1] = \text{ket}\left[-\frac{1}{2},-\frac{1}{2}\right]$

$\text{Ket}[0,0] = \frac{\text{ket}\left[-\frac{1}{2},\frac{1}{2}\right]}{\sqrt{2}} - \frac{\text{ket}\left[\frac{1}{2},-\frac{1}{2}\right]}{\sqrt{2}}$

Out[5] = Ket\$668

In[6] := AddJ[1, 1/2]

$\text{Ket}\left[\frac{3}{2},\frac{3}{2}\right] = \text{ket}\left[1,\frac{1}{2}\right]$

$\text{Ket}\left[\frac{3}{2},\frac{1}{2}\right] = \sqrt{\frac{2}{3}}\text{ket}\left[0,\frac{1}{2}\right] + \frac{\text{ket}\left[1,-\frac{1}{2}\right]}{\sqrt{3}}$

$\text{Ket}\left[\frac{3}{2},-\frac{1}{2}\right] = \frac{\text{ket}\left[-1,\frac{1}{2}\right]}{\sqrt{3}} + \sqrt{\frac{2}{3}}\text{ket}\left[0,-\frac{1}{2}\right]$

$\text{Ket}\left[\frac{3}{2},-\frac{3}{2}\right] = \text{ket}\left[-1,-\frac{1}{2}\right]$

$\text{Ket}\left[\frac{1}{2},\frac{1}{2}\right] = \frac{\text{ket}\left[0,\frac{1}{2}\right]}{\sqrt{3}} - \sqrt{\frac{2}{3}}\text{ket}\left[1,-\frac{1}{2}\right]$

$\text{Ket}\left[\frac{1}{2},-\frac{1}{2}\right] = \sqrt{\frac{2}{3}}\text{ket}\left[-1,\frac{1}{2}\right] - \frac{\text{ket}\left[0,-\frac{1}{2}\right]}{\sqrt{3}}$

Out[6] = Ket\$669

In[7] := AddJ[1, 1]

$\text{Ket}[2,2] = \text{ket}[1,1]$

$\text{Ket}[2,1] = \frac{\text{ket}[0,1]}{\sqrt{2}} + \frac{\text{ket}[1,0]}{\sqrt{2}}$

$\text{Ket}[2,0] = \frac{\text{ket}[-1,1]}{\sqrt{6}} + \sqrt{\frac{2}{3}}\text{ket}[0,0] + \frac{\text{ket}[1,-1]}{\sqrt{6}}$

$\text{Ket}[2,-1] = \frac{\text{ket}[-1,0]}{\sqrt{2}} + \frac{\text{ket}[0,-1]}{\sqrt{2}}$

$\text{Ket}[2,-2] = \text{ket}[-1,-1]$

$\text{Ket}[1,1] = \frac{\text{ket}[0,1]}{\sqrt{2}} - \frac{\text{ket}[1,0]}{\sqrt{2}}$

$\text{Ket}[1,0] = \frac{\text{ket}[-1,1]}{\sqrt{2}} - \frac{\text{ket}[1,-1]}{\sqrt{2}}$

$\text{Ket}[1,-1] = \frac{\text{ket}[-1,0]}{\sqrt{2}} - \frac{\text{ket}[0,-1]}{\sqrt{2}}$

$\text{Ket}[0,0] = \frac{\text{ket}[-1,1]}{\sqrt{3}} - \frac{\text{ket}[0,0]}{\sqrt{3}} + \frac{\text{ket}[1,-1]}{\sqrt{3}}$

Out[7] = Ket\$670

In[8] := AddJ[2, 1]

$\text{Ket}[3,3] = \text{ket}[2,1]$

$\text{Ket}[3,2] = \sqrt{\frac{2}{3}}\text{ket}[1,1] + \frac{\text{ket}[2,0]}{\sqrt{3}}$

$\text{Ket}[3,1] = \sqrt{\frac{2}{5}}\text{ket}[0,1] + 2\sqrt{\frac{2}{15}}\text{ket}[1,0] + \frac{\text{ket}[2,-1]}{\sqrt{15}}$

$$\text{Ket}[3,0] = \frac{\text{ket}[-1,1]}{\sqrt{5}} + \sqrt{\frac{3}{5}}\text{ket}[0,0] + \frac{\text{ket}[1,-1]}{\sqrt{5}}$$
$$\text{Ket}[3,-1] = \frac{\text{ket}[-2,1]}{\sqrt{15}} + 2\sqrt{\frac{2}{15}}\text{ket}[-1,0] + \sqrt{\frac{2}{5}}\text{ket}[0,-1]$$
$$\text{Ket}[3,-2] = \frac{\text{ket}[-2,0]}{\sqrt{3}} + \sqrt{\frac{2}{3}}\text{ket}[-1,-1]$$
$$\text{Ket}[3,-3] = \text{ket}[-2,-1]$$
$$\text{Ket}[2,2] = \frac{\text{ket}[1,1]}{\sqrt{3}} - \sqrt{\frac{2}{3}}\text{ket}[2,0]$$
$$\text{Ket}[2,1] = \frac{\text{ket}[0,1]}{\sqrt{2}} - \frac{\text{ket}[1,0]}{\sqrt{6}} - \frac{\text{ket}[2,-1]}{\sqrt{3}}$$
$$\text{Ket}[2,0] = \frac{\text{ket}[-1,1]}{\sqrt{2}} - \frac{\text{ket}[1,-1]}{\sqrt{2}}$$
$$\text{Ket}[2,-1] = \frac{\text{ket}[-2,1]}{\sqrt{3}} + \frac{\text{ket}[-1,0]}{\sqrt{6}} - \frac{\text{ket}[0,-1]}{\sqrt{2}}$$
$$\text{Ket}[2,-2] = \sqrt{\frac{2}{3}}\text{ket}[-2,0] - \frac{\text{ket}[-1,-1]}{\sqrt{3}}$$
$$\text{Ket}[1,1] = \frac{\text{ket}[0,1]}{\sqrt{10}} - \sqrt{\frac{3}{10}}\text{ket}[1,0] + \sqrt{\frac{3}{5}}\text{ket}[2,-1]$$
$$\text{Ket}[1,0] = \sqrt{\frac{3}{10}}\text{ket}[-1,1] - \sqrt{\frac{2}{5}}\text{ket}[0,0] + \sqrt{\frac{3}{10}}\text{ket}[1,-1]$$
$$\text{Ket}[1,-1] = \sqrt{\frac{3}{5}}\text{ket}[-2,1] - \sqrt{\frac{3}{10}}\text{ket}[-1,0] + \frac{\text{ket}[0,-1]}{\sqrt{10}}$$
$$\text{Out}[8] = \text{Ket\$671}$$

Chapter 19
Classical Nonlinear Oscillator

19.1 Statement of the Problem

Let's consider one-dimensional motion of a particle with mass m near a minimum of an arbitrary smooth potential [19]

$$U(x) = \frac{kx^2}{2} + V(x), \qquad V(x) = O\left(x^3\right)$$

(we have chosen the origin of x and the zero energy level to be at the minimum). If we neglect $V(x)$, then the equation of motion

$$m\frac{\mathrm{d}^2x}{\mathrm{d}t^2} = -\frac{\mathrm{d}U}{\mathrm{d}x}$$

becomes

$$\frac{\mathrm{d}^2x}{\mathrm{d}t^2} + \omega_0^2 x = 0, \qquad \omega_0^2 = \frac{k}{m},$$

and has the solution

$$x(t) = a\cos\omega_0 t + b\sin\omega_0 t.$$

Now we consider the effect of

$$V(x) = \sum_{n=1}^{\infty} c_n x^{n+2}.$$

Choosing units of measurement in such a way that $m = 1$ and $k = 1$, we have the equation of motion

$$\frac{\mathrm{d}^2x}{\mathrm{d}t^2} + x = R(x) \equiv -\frac{\mathrm{d}V}{\mathrm{d}x}.$$

Its solution $x(t)$ is a periodic function of t. If we choose the time origin at a maximum of $x(t)$, then $x(t)$ is an even function, due to reversibility. In the zeroth approximation $x(t) = a\cos t$.

A. Grozin, *Introduction to Mathematica® for Physicists*, Graduate Texts in Physics, DOI 10.1007/978-3-319-00894-3_19, © Springer International Publishing Switzerland 2014

In[1] := V = Series[c[1] * x^3, {x,0,3}]
Out[1] = $c[1]x^3 + O[x]^4$
In[2] := R = −D[V,x]
Out[2] = $-3c[1]x^2 + O[x]^3$
In[3] := x[t] = Series[a * Cos[t], {a,0,1}]
Out[3] = $\mathrm{Cos}[t]a + O[a]^2$
The equation of motion is satisfied at $O(a)$.
In[4] := D[x[t], {t,2}] + x[t]
Out[4] = $O[a]^2$

19.2 The First Correction

Now we want to take terms of order a^2 into account. The right-hand side is
In[5] := R1 = R/.x−>x[t]
Out[5] = $-3\left(c[1]\mathrm{Cos}[t]^2\right)a^2 + O[a]^3$
Let's expand it in harmonics.
In[6] := R1 = Map[TrigReduce, R1]
Out[6] = $-\frac{3}{2}(c[1] + c[1]\mathrm{Cos}[2t])a^2 + O[a]^3$
That is, the "driving force" contains zeroth and second harmonics. This means that we should add such harmonics to $x(t)$. We'll not add the solution of the homogeneous equation—the first harmonic: by definition of the amplitude a, it is completely given by the leading term $a\cos t$.
In[7] := x[t] = Series[a * Cos[t]+
a^2 * Sum[b[2, j] * Cos[j * t], {j,0,2,2}], {a,0,2}]
Out[7] = $\mathrm{Cos}[t]a + (b[2,0] + b[2,2]\mathrm{Cos}[2t])a^2 + O[a]^3$
Now we substitute this form of the solution into the equation of motion.
In[8] := Eq = D[x[t], {t,2}] + x[t] − (R/.x−>x[t])
Out[8] = $\left(b[2,0] + 3c[1]\mathrm{Cos}[t]^2 - 3b[2,2]\mathrm{Cos}[2t]\right)a^2 + O[a]^3$
In[9] := Eq = Map[TrigReduce, Eq]
Out[9] = $\frac{1}{2}(2b[2,0] + 3c[1] - 6b[2,2]\mathrm{Cos}[2t] + 3c[1]\mathrm{Cos}[2t])a^2 + O[a]^3$
In[10] := Eq2 = SeriesCoefficient[Eq, 2]
Out[10] = $\frac{1}{2}(2b[2,0] + 3c[1] - 6b[2,2]\mathrm{Cos}[2t] + 3c[1]\mathrm{Cos}[2t])$
This expression should vanish. How can we separate harmonics? Let's help *Mathematica* a little.
In[11] := Eq2 = Eq2/.Cos[j_. * t]−>z^j
Out[11] = $\frac{1}{2}\left(2b[2,0] - 6z^2b[2,2] + 3c[1] + 3z^2c[1]\right)$
The coefficients of z^0 and z^2 should vanish.
In[12] := Eq20 = Coefficient[Eq2, z, 0]
Out[12] = $\frac{1}{2}(2b[2,0] + 3c[1])$

In[13] := Eq22 = Coefficient[Eq2, z, 2]

$\text{Out}[13] = \frac{1}{2}(-6b[2,2] + 3c[1])$

We can find $b[2,0]$ from the first equation and $b[2,2]$ from the second one.

In[14] := b[2,0] = b[2,0]/. Solve[Eq20 == 0, b[2,0]][[1]]

$\text{Out}[14] = -\frac{3c[1]}{2}$

In[15] := b[2,2] = b[2,2]/. Solve[Eq22 == 0, b[2,2]][[1]]

$\text{Out}[15] = \frac{c[1]}{2}$

Now we know the solution.

In[16] := x[t] = x[t]

$\text{Out}[16] = \text{Cos}[t]a + \left(-\frac{3c[1]}{2} + \frac{1}{2}c[1]\text{Cos}[2t]\right)a^2 + O[a]^3$

Let's check energy conservation.

In[17] := Et = D[x[t], t]^2/2 + x[t]^2/2 + (V/.x−>x[t]);

In[18] := Map[TrigReduce, Et]

$\text{Out}[18] = \frac{a^2}{2} + O[a]^4$

In[19] := Clear[b]

19.3 The Second Correction

Now we want to find two corrections.

In[20] := n = 2;

In[21] := V = Series[Sum[c[i] * x^(i+2), {i, 1, n}], {x, 0, n+2}]

$\text{Out}[21] = c[1]x^3 + c[2]x^4 + O[x]^5$

In[22] := R = −D[V, x]

$\text{Out}[22] = -3c[1]x^2 - 4c[2]x^3 + O[x]^4$

This is $x[t]$ up to a^2.

In[23] := x[t] = Series[a * Cos[t]+
    a^2 * Sum[b[2, j] * Cos[j * t], {j, 0, 2, 2}], {a, 0, n}]

$\text{Out}[23] = \text{Cos}[t]a + (b[2,0] + b[2,2]\text{Cos}[2t])a^2 + O[a]^3$

The right-hand side of the equation of motion.

In[24] := R1 = Map[TrigReduce, ExpandAll[R/.x−>x[t]]]

$$\text{Out}[24] = -\frac{3}{2}(c[1] + c[1]\text{Cos}[2t])a^2 + (-6b[2,0]c[1]\text{Cos}[t] - 3b[2,2]c[1]\text{Cos}[t] - 3c[2]\text{Cos}[t] - 3b[2,2]c[1]\text{Cos}[3t] - c[2]\text{Cos}[3t])a^3 + O[a]^4$$

There are the first and the third harmonics at the order a^3, that is, there is a resonant term in the "driving force" which would lead to an unbounded growth of the solution. This means we have forgotten something. Namely, we have forgotten that the oscillation period depends on the amplitude (unless the potential is strictly parabolic). And our solution should contain $\cos(j\omega t)$. If we denote $\tau = \omega t$, then the equation of motion is

$$\omega^2 \frac{d^2x}{d\tau^2} + x = R.$$

Let's suppose that the variable t in the program really means τ and denote $\omega^2 = w$. It is a series in a^2 beginning with 1.

In[25] := w = Series[1 + Sum[u[i] * a^i, {i,2,n+1,2}], {a,0,n+1}]

Out[25] $= 1 + u[2]a^2 + O[a]^4$

Now we are able to cancel the first harmonic in the a^3 term of the equation of motion. And the third one should be added to the general form of the solution.

In[26] := x[t] = Series[a * Cos[t] + a^2 * Sum[b[2, j] * Cos[j * t], {j,0,2,2}]+ a^3 * Sum[b[3, j] * Cos[j * t], {j,3,3,2}], {a,0,n+1}]

Out[26] $= \mathrm{Cos}[t]a + (b[2,0] + b[2,2]\mathrm{Cos}[2t])a^2 + b[3,3]\mathrm{Cos}[3t]a^3 + O[a]^4$

The equation of motion is

In[27] := Eq = Map[TrigReduce, ExpandAll[w * D[x[t], {t,2}] + x[t] − (R/.x−>x[t])]]

Out[27] $= \frac{1}{2}(2b[2,0] + 3c[1] - 6b[2,2]\mathrm{Cos}[2t] + 3c[1]\mathrm{Cos}[2t])a^2 + (6b[2,0]c[1]\mathrm{Cos}[t] + 3b[2,2]c[1]\mathrm{Cos}[t] + 3c[2]\mathrm{Cos}[t] - 8b[3,3]\mathrm{Cos}[3t] + 3b[2,2]c[1]\mathrm{Cos}[3t] + c[2]\mathrm{Cos}[3t] - \mathrm{Cos}[t]u[2])a^3 + O[a]^4$

We already know how to solve it at the order a^2.

In[28] := Eq2 = SeriesCoefficient[Eq, 2]/.Cos[j_. * t]−>z^j

Out[28] $= \frac{1}{2}\left(2b[2,0] - 6z^2b[2,2] + 3c[1] + 3z^2c[1]\right)$

In[29] := Do[Print[b[2, j] = b[2, j]/.Solve[Coefficient[Eq2, z, j] == 0, b[2, j]][[1]]], {j,0,2,2}]

$-\frac{3c[1]}{2}$

$\frac{c[1]}{2}$

At the order a^3:

In[30] := Eq3 = SeriesCoefficient[Eq, 3]/.Cos[j_. * t]−>z^j

Out[30] $= -8z^3b[3,3] - \frac{15}{2}zc[1]^2 + \frac{3}{2}z^3c[1]^2 + 3zc[2] + z^3c[2] - zu[2]$

This is the coefficient of the first harmonic, i.e., of z^1:

In[31] := Eq31 = Coefficient[Eq3, z, 1]

Out[31] $= -\frac{15}{2}c[1]^2 + 3c[2] - u[2]$

It can be nullified by choosing $u[2]$.

In[32] := u[2] = u[2]/.Solve[Eq31 == 0, u[2]][[1]]

Out[32] $= -\frac{3}{2}\left(5c[1]^2 - 2c[2]\right)$

And this is the coefficient of the third harmonic, i.e., of z^3:

In[33] := Eq33 = Coefficient[Eq3, z, 3]

$$\text{Out}[33] = -8b[3,3] + \frac{3c[1]^2}{2} + c[2]$$

It can be nullified by choosing $b[3,3]$.

In[34] := b[3,3] = b[3,3]/. Solve[Eq33 == 0, b[3,3]][[1]]

$$\text{Out}[34] = \frac{1}{16}\left(3c[1]^2 + 2c[2]\right)$$

Now we know the oscillation frequency

In[35] := w = w

$$\text{Out}[35] = 1 - \frac{3}{2}\left(5c[1]^2 - 2c[2]\right)a^2 + O[a]^4$$

and $x[t]$:

In[36] := x[t] = x[t]

$$\text{Out}[36] = \text{Cos}[t]a + \left(-\frac{3c[1]}{2} + \frac{1}{2}c[1]\text{Cos}[2t]\right)a^2 + \frac{1}{16}\left(3c[1]^2 + 2c[2]\right)\text{Cos}[3t]a^3 + O[a]^4$$

Let's check energy conservation.

In[37] := Et = Map[TrigReduce,
ExpandAll[w * D[x[t], t]^2/2 + x[t]^2/2 + (V/.x->x[t])]]

$$\text{Out}[37] = \frac{a^2}{2} + \frac{1}{16}\left(-37c[1]^2 + 18c[2]\right)a^4 + O[a]^5$$

In[38] := Clear[b, u]

19.4 The nth Correction

Now we'll write a program which can find n corrections in a to the particle motion for any n. Just a single line should be changed for the calculation with a new value of n.

In[39] := n = 4;

The correction to the potential and the "driving force."

In[40] := V = Series[Sum[c[i] * x^(i + 2), {i, 1, n}], {x, 0, n + 2}]

$$\text{Out}[40] = c[1]x^3 + c[2]x^4 + c[3]x^5 + c[4]x^6 + O[x]^7$$

In[41] := R = −D[V, x]

$$\text{Out}[41] = -3c[1]x^2 - 4c[2]x^3 - 5c[3]x^4 - 6c[4]x^5 + O[x]^6$$

The frequency squared is a series in a^2.

In[42] := w = Series[1 + Sum[u[i] * a^i, {i, 2, n + 1, 2}], {a, 0, n + 1}]

$$\text{Out}[42] = 1 + u[2]a^2 + u[4]a^4 + O[a]^6$$

The general form of the solution. The order a^i contains harmonics up to the ith one. They are all even at even values of i and odd at odd values. The first harmonic never appears—by definition, it is entirely contained in the leading term $a\cos t$.

```
In[43] := x[t] = Series[a * Cos[t]+
        Sum[a^i * Sum[b[i, j] * Cos[j * t], {j, If[EvenQ[i], 0, 3], i, 2}], {i, 2, n + 1}],
    {a, 0, n + 1}]
```

Out[43] = $\mathrm{Cos}[t]a + (b[2,0] + b[2,2]\mathrm{Cos}[2t])a^2 + b[3,3]\mathrm{Cos}[3t]a^3 + (b[4,0] + b[4,2]\mathrm{Cos}[2t] + b[4,4]\mathrm{Cos}[4t])a^4 + (b[5,3]\mathrm{Cos}[3t] + b[5,5]\mathrm{Cos}[5t])a^5 + O[a]^6$

The equation of motion.

```
In[44] := Eq = Map[TrigReduce,
    ExpandAll[w * D[x[t], {t, 2}] + x[t] - (R/.x->x[t])]]
```

Out[44] = $\frac{1}{2}(2b[2,0] + 3c[1] - 6b[2,2]\mathrm{Cos}[2t] + 3c[1]\mathrm{Cos}[2t])a^2 +$
$(6b[2,0]c[1]\mathrm{Cos}[t] + 3b[2,2]c[1]\mathrm{Cos}[t] + 3c[2]\mathrm{Cos}[t] -$
$8b[3,3]\mathrm{Cos}[3t] + 3b[2,2]c[1]\mathrm{Cos}[3t] + c[2]\mathrm{Cos}[3t] - \mathrm{Cos}[t]u[2])a^3 +$
$\frac{1}{8}(8b[4,0] + 24b[2,0]^2c[1] + 12b[2,2]^2c[1] + 48b[2,0]c[2] + 24b[2,2]c[2] +$
$15c[3] - 24b[4,2]\mathrm{Cos}[2t] + 48b[2,0]b[2,2]c[1]\mathrm{Cos}[2t] +$
$24b[3,3]c[1]\mathrm{Cos}[2t] + 48b[2,0]c[2]\mathrm{Cos}[2t] + 48b[2,2]c[2]\mathrm{Cos}[2t] +$
$20c[3]\mathrm{Cos}[2t] - 120b[4,4]\mathrm{Cos}[4t] + 12b[2,2]^2c[1]\mathrm{Cos}[4t] +$
$24b[3,3]c[1]\mathrm{Cos}[4t] + 24b[2,2]c[2]\mathrm{Cos}[4t] + 5c[3]\mathrm{Cos}[4t] -$
$32b[2,2]\mathrm{Cos}[2t]u[2])a^4 +$
$\Big(3b[2,2]b[3,3]c[1]\mathrm{Cos}[t] + 6b[4,0]c[1]\mathrm{Cos}[t] + 3b[4,2]c[1]\mathrm{Cos}[t] +$
$12b[2,0]^2c[2]\mathrm{Cos}[t] + 12b[2,0]b[2,2]c[2]\mathrm{Cos}[t] + 6b[2,2]^2c[2]\mathrm{Cos}[t] +$
$3b[3,3]c[2]\mathrm{Cos}[t] + 15b[2,0]c[3]\mathrm{Cos}[t] + 10b[2,2]c[3]\mathrm{Cos}[t] +$
$\frac{15}{4}c[4]\mathrm{Cos}[t] - 8b[5,3]\mathrm{Cos}[3t] + 6b[2,0]b[3,3]c[1]\mathrm{Cos}[3t] +$
$3b[4,2]c[1]\mathrm{Cos}[3t] + 3b[4,4]c[1]\mathrm{Cos}[3t] + 12b[2,0]b[2,2]c[2]\mathrm{Cos}[3t] +$
$3b[2,2]^2c[2]\mathrm{Cos}[3t] + 6b[3,3]c[2]\mathrm{Cos}[3t] + 5b[2,0]c[3]\mathrm{Cos}[3t] +$
$\frac{15}{2}b[2,2]c[3]\mathrm{Cos}[3t] + \frac{15}{8}c[4]\mathrm{Cos}[3t] - 24b[5,5]\mathrm{Cos}[5t] +$
$3b[2,2]b[3,3]c[1]\mathrm{Cos}[5t] + 3b[4,4]c[1]\mathrm{Cos}[5t] + 3b[2,2]^2c[2]\mathrm{Cos}[5t] +$
$3b[3,3]c[2]\mathrm{Cos}[5t] + \frac{5}{2}b[2,2]c[3]\mathrm{Cos}[5t] + \frac{3}{8}c[4]\mathrm{Cos}[5t] -$
$9b[3,3]\mathrm{Cos}[3t]u[2] - \mathrm{Cos}[t]u[4]\Big)a^5 +$
$O[a]^6$

All terms of the orders a^i for i from 2 to $n+1$ must vanish. If i is odd, the first harmonic is present; a correction to the frequency squared is found from the condition that this harmonic vanishes. All other harmonics give us coefficients in $x(t)$.

```
In[45] := Do[Eqi = SeriesCoefficient[Eq, i]/.Cos[j_. * t]->z^j;
        If[OddQ[i],
            u[i - 1] = u[i - 1]/.Solve[Coefficient[Eqi, z, 1] == 0, u[i - 1]][[1]]];
        Do[b[i, j] = b[i, j]/.Solve[Coefficient[Eqi, z, j] == 0, b[i, j]][[1]],
            {j, If[EvenQ[i], 0, 3], i, 2}],
    {i, 2, n + 1}]
```

Now we know the frequency squared.

In[46] := w = w

$$\text{Out}[46] = 1 - \frac{3}{2}\left(5c[1]^2 - 2c[2]\right)a^2 - \frac{3}{32}\left(335c[1]^4 - 572c[1]^2c[2] - 4c[2]^2 + 280c[1]c[3] - 40c[4]\right)a^4 + O[a]^6$$

and $x(t)$

In[47] := x[t] = x[t]

$$\text{Out}[47] = \text{Cos}[t]a + \left(-\frac{3c[1]}{2} + \frac{1}{2}c[1]\text{Cos}[2t]\right)a^2 + \frac{1}{16}\left(3c[1]^2 + 2c[2]\right)\text{Cos}[3t]a^3 + \left(-\frac{3}{8}\left(19c[1]^3 - 20c[1]c[2] + 5c[3]\right) + \frac{1}{48}\left(177c[1]^3 - 186c[1]c[2] + 40c[3]\right)\text{Cos}[2t] + \frac{1}{48}\left(3c[1]^3 + 6c[1]c[2] + 2c[3]\right)\text{Cos}[4t]\right)a^4 + \left(\frac{3}{256}\left(237c[1]^4 - 172c[1]^2c[2] - 28c[2]^2 - 12c[1]c[3] + 20c[4]\right)\text{Cos}[3t] + \frac{1}{768}\left(15c[1]^4 + 60c[1]^2c[2] + 12c[2]^2 + 44c[1]c[3] + 12c[4]\right)\text{Cos}[5t]\right)a^5 + O[a]^6$$

Let's check energy conservation.

**In[48] := Et = Map[TrigReduce,
ExpandAll[w * D[x[t],t]^2/2 + x[t]^2/2 + (V/.x->x[t])]]**

$$\text{Out}[48] = \frac{a^2}{2} + \frac{1}{16}\left(-37c[1]^2 + 18c[2]\right)a^4 + \frac{1}{1536}\left(-9309c[1]^4 + 17796c[1]^2c[2] + 300c[2]^2 - 10880c[1]c[3] + 1920c[4]\right)a^6 + O[a]^7$$

It is easy to write a function with the parameter n which can be used as a black box. It should use only local variables.

Now we save the results for the energy Et and the frequency squared w to a file; later we'll compare them to the similar results in quantum mechanics.

```
In[49] := Ec = Normal[Et]/.a->Sqrt[2*A];
    Wc = Normal[Simplify[Sqrt[w]]]/.a->Sqrt[2*A];
    Save["class.m", {Ec, Wc}]
```

Chapter 20
Quantum Nonlinear Oscillator

20.1 Perturbation Theory

Suppose we know eigenvalues and eigenstates of a Hamiltonian $\hat{H}_0$ and want to find them for a Hamiltonian $\hat{H} = \hat{H}_0 + \hat{V}$ in the form of series in $\hat{V}$ [18]. Let's concentrate on a non-degenerate eigenstate of the unperturbed Hamiltonian

$$\hat{H}_0 \,|\psi_0\rangle = E_0 \,|\psi_0\rangle .$$

After switching the perturbation on it transforms to a similar eigenstate of the full Hamiltonian

$$\hat{H} \,|\psi\rangle = E \,|\psi\rangle , \qquad E = E_0 + \delta E .$$

Let's normalize $|\psi\rangle$ in such a way that $\langle\psi_0|\psi\rangle = 1$, then $|\psi\rangle = |\psi_0\rangle + |\delta\psi\rangle$, $\langle\psi_0|\delta\psi\rangle = 0$. We need to solve the equation

$$\hat{H}_0 \,|\psi\rangle + \hat{V} \,|\psi\rangle = E \,|\psi\rangle .$$

Let's separate its components parallel and orthogonal to $|\psi_0\rangle$. The parallel part is singled out by multiplying by $\langle\psi_0|$:

$$\delta E = \langle\psi_0|\hat{V}|\psi\rangle .$$

The orthogonal part is singled out by the projector $\hat{P} = 1 - |\psi_0\rangle\langle\psi_0|$:

$$\hat{H}_0 \,|\delta\psi\rangle + \hat{P}\hat{V} \,|\psi\rangle = E \,|\delta\psi\rangle ,$$

or

$$|\delta\psi\rangle = \hat{D}\hat{V} \,|\psi\rangle , \qquad \hat{D} = \frac{\hat{P}}{E - \hat{H}_0} .$$

Solving this equation by iterations, we obtain

$$|\delta\psi\rangle = \hat{D}\hat{V} \,|\psi_0\rangle + \hat{D}\hat{V}\hat{D}\hat{V} \,|\psi_0\rangle + \cdots ,$$

A. Grozin, *Introduction to Mathematica® for Physicists*, Graduate Texts in Physics,
DOI 10.1007/978-3-319-00894-3_20,

and

$$\delta E = <\psi_0|\hat{V}|\psi_0> + <\psi_0|\hat{V}\hat{D}\hat{V}|\psi_0> + <\psi_0|\hat{V}\hat{D}\hat{V}\hat{D}\hat{V}|\psi_0> + \cdots .$$

Note that $\hat{D}$ contains $E = E_0 + \delta E$ and should be expanded in δE:

$$\hat{D} = \hat{G} - \hat{G}\delta E\hat{G} + \hat{G}\delta E\hat{G}\delta E\hat{G} - \cdots , \qquad \hat{G} = \frac{\hat{P}}{E_0 - \hat{H}_0} .$$

20.2 Nonlinear Oscillator

We are going to apply the perturbation theory to the nonlinear oscillator

$$\hat{H}_0 = \frac{\hat{p}^2 + \hat{x}^2}{2} , \qquad \hat{V} = \sum_{k=1}^{\infty} c_k \hat{x}^{k+2} .$$

The oscillation amplitude is ~ 1 if $n \sim 1$; therefore, $c_k \sim a^k$, where $a \sim 1/L \ll 1$, L is the characteristic length of the potential in the oscillator units. If $n \gg 1$, the amplitude is $\sqrt{n}$ times larger, and the real expansion parameter is $a\sqrt{n}$.

We concentrate on the eigenstate $|n>$ of $\hat{H}_0$ having the energy $E_0 = n + \frac{1}{2}$. In order to calculate δE up to the Mth order of the perturbation theory, we need to sum all expressions of the form

$$\left(\hat{V}_{k_N}\right)_{n,n+j_{N-1}} \hat{G}_{n+j_{N-1}} \left(\hat{\Delta}_{k_{N-1}}\right)_{n+j_{N-1},n+j_{N-2}} \hat{G}_{n+j_{N-2}} \cdots$$
$$\hat{G}_{n+j_2} \left(\hat{\Delta}_{k_2}\right)_{n+j_2,n+j_1} \hat{G}_{n+j_1} \left(\hat{V}_{k_1}\right)_{n+j_1,n} ,$$

where $\hat{\Delta}$ is $\hat{V}$ or $-\delta E$ and the sum of the orders of smallness $k_1 + k_2 + \cdots + k_N \leqslant M$; the sum over all nonzero $j_1, j_2, \ldots\ j_{N-1}$ is assumed. The following procedure prepares the values $V[k, j]$ of $\left(\hat{V}_k\right)_{n+j,n}$ ($x[k, j]$ means $\left(\hat{x}^k\right)_{n+j,n}$):

```
In[1] := Prepare[m_] := (M = m; x[1,1] = Sqrt[(n+1)/2]; x[1,−1] = Sqrt[n/2];
    x[1,0] = 0;
    Do[x[k, j] = If[j < k − 1, (x[1,1]/.n → n + j) ∗ x[k − 1, j + 1], 0]+
             If[j > 1 − k, (x[1,−1]/.n → n + j) ∗ x[k − 1, j − 1], 0],
        {k, 2, m + 2}, {j, −k, k}];
    Do[V[k, j] = Simplify[c[k] ∗ x[k + 2, j]], {k, 1, m}, {j, −k − 2, k + 2, 2}])
```

20.3 Energy Levels

The expressions we want to generate and calculate can be visualized as paths in the following graph:

```
In[2] := ParametricPlot[{{t,3 − 3∗t},{t,−3+3∗t},
    {t/2,3/2∗t},{t/2,−3/2∗t}},{t,0,1},
    PlotRange → {{0,1},{−3,3}},AxesLabel → {l, j},
    Ticks → {{0,{1/2,M/2},{1,M}},
        {{−3,−3∗M},{−3/2,−3/2∗M},0,{3/2,3/2∗M},{3,3∗M}}}]
```

Out[2] =

j

$3\,M$

$\frac{3\,M}{2}$

0 $\frac{M}{2}$ M l

$-\frac{3\,M}{2}$

$-3\,M$

Here l is the number of orders of smallness we need to distribute. We start at the point $l = M$, $j = 0$. At each step we consider all possible $\hat{\Delta}_k$. If we choose $(\hat{V}_k)_{n+j+i,n+j}$, we move k steps to the left and i steps vertically. If we choose $-\delta E_k$, we move k steps to the left horizontally (this choice is not allowed as the first or the last step). Whenever we hit the $j = 0$ axis, we have a complete expression for a contribution to δE. The fastest movement along j at varying l occurs when $\hat{V}_1$ is used, and its velocity is 3. Hence, in order to have enough time to return to $j = 0$, we should not leave the rhombus in the figure.

Suppose we have already generated a right-hand part a of the expression up to some $\hat{G}_{n+j}$ inclusively, and there remain l orders of smallness to distribute. The procedure considers all possible $(\hat{\Delta}_k)_{n+j+i,n+j}$ which can be inserted to the left of a. It may be $-\delta E_k$ with all possible values of k (if we are not at the very first step $l = M$), the maximum k is obtained from the rhombus. Or it may be $(\hat{V}_k)_{n+j+i,n+j}$ with all possible values of k and i. $\hat{V}_k = c_k \hat{x}^{k+2}$ has nonzero matrix elements for i from $-k-2$ to $k+2$ in steps of 2. The limits of the i loop follow from the intersection of this range with the rhombus, and the k loop terminates when this intersection disappears. If we happen to return to the initial state ($j+i = 0$), this means that the generation of an expression is complete, and it should be added to the element of the list d which accumulates contributions to δE (this contribution may contain lower-order $\delta E_k = \mathrm{de}[k]$). In all other cases, the procedure is called recursively, with l replaced by $l-k$, j by $j+i$, and a multiplied by $(\hat{\Delta}_k)_{n+j+i,n+j}$ and $\hat{G}_{n+j+i}$.

```
In[3] := v[l_, j_, a_] := (If[l == M, d = Table[0, {M}],
            Do[v[l − k, j, a ∗ de[k]/j], {k, 2, l − Abs[j]/3, 2}]];
        Do[If[j + i == 0, d[[M − l + k]]+ = a ∗ (V[k, i]/.n → n + j),
                v[l − k, j + i, −a ∗ (V[k, i]/.n → n + j)/(j + i)]],
            {k, Min[l, 1 + (3 ∗ l − Abs[j])/2]},
            {i, Max[−k − 2, −3 ∗ (l − k) − j], Min[k + 2, 3 ∗ (l − k) − j], 2}])
In[4] := Prepare[6]; v[M, 0, 1];
```

Now we substitute lower-order δE into expressions for higher-order ones, to get explicit formulas.

```
In[5] := Do[de[k] = Simplify[d[[k]]]; Print[Collect[de[k], c[_], Factor]], {k, 2, M, 2}]
```

$$\frac{1}{8}\left(-11-30n-30n^2\right)c[1]^2+\frac{3}{4}\left(1+2n+2n^2\right)c[2]$$

$$-\frac{15}{32}(1+2n)\left(31+47n+47n^2\right)c[1]^4+\frac{9}{8}(1+2n)\left(19+25n+25n^2\right)c[1]^2c[2]-$$
$$\frac{1}{8}(1+2n)\left(21+17n+17n^2\right)c[2]^2-\frac{5}{8}(1+2n)\left(13+14n+14n^2\right)c[1]c[3]+$$
$$\frac{5}{8}(1+2n)\left(3+2n+2n^2\right)c[4]$$

$$\frac{1}{128}\left(-39709-162405n-278160n^2-231510n^3-115755n^4\right)c[1]^6+$$
$$\frac{3}{64}\left(15169+59385n+98160n^2+77550n^3+38775n^4\right)c[1]^4c[2]+$$
$$\frac{3}{16}\left(111+347n+472n^2+250n^3+125n^4\right)c[2]^3+$$

$$\frac{1}{16}\left(-4517-16815n-26580n^2-19530n^3-9765n^4\right)c[1]^3c[3]+$$
$$\frac{1}{32}\left(-449-1400n-2030n^2-1260n^3-630n^4\right)c[3]^2-$$
$$\frac{15}{8}\left(12+35n+46n^2+22n^3+11n^4\right)c[2]c[4]+$$
$$c[1]^2\left(\frac{1}{32}\left(-11827-43479n-68424n^2-49890n^3-24945n^4\right)c[2]^2+\right.$$
$$\left.\frac{5}{16}\left(323+1125n+1668n^2+1086n^3+543n^4\right)c[4]\right)+$$
$$c[1]\left(\frac{3}{8}\left(474+1625n+2430n^2+1610n^3+805n^4\right)c[2]c[3]-\right.$$
$$\left.\frac{105}{16}\left(5+16n+22n^2+12n^3+6n^4\right)c[5]\right)+$$
$$\frac{35}{16}\left(3+8n+10n^2+4n^3+2n^4\right)c[6]$$

20.4 Correspondence Principle

At $n \gg 1$ the expansion parameter of the perturbation series is $a\sqrt{n}$ where $c_k \sim a^k$. Keeping only the highest powers of n in each order, we have

```
In[6] := Eq = Series[n+
         Sum[(Expand[(de[2*(j-1)]/.n->1/a)*a^j]/.a->0)*n^j,
             {j,2,M/2+1}],
    {n,0,M/2+1}]
```

$$\mathrm{Out}[6] = n+\left(-\frac{15}{4}c[1]^2+\frac{3c[2]}{2}\right)n^2+$$
$$\frac{1}{16}(-705c[1]^4+900c[1]^2c[2]-68c[2]^2-280c[1]c[3]+40c[4])n^3-$$
$$\frac{5}{128}(23151c[1]^6-46530c[1]^4c[2]+19956c[1]^2c[2]^2-600c[2]^3+$$
$$15624c[1]^3c[3]-7728c[1]c[2]c[3]+504c[3]^2-4344c[1]^2c[4]+$$
$$528c[2]c[4]+1008c[1]c[5]-112c[6])n^4+$$
$$O[n]^5$$

Bohr's correspondence principle must hold. From the quantum point of view, the particle at the nth energy level can radiate a photon, jumping to the $(n-1)$th one, or more generally to the $(n-k)$th one. The frequency of this photon is $E_n - E_{n-k}$, or approximately $\frac{\mathrm{d}E_n}{\mathrm{d}n}k$. From the classical point of view, the frequencies of the emitted light are equal to the oscillation frequency ω and its harmonics. Therefore, the oscillation frequency is

$$\omega = \frac{\mathrm{d}E_n}{\mathrm{d}n}.$$

In[7] := Wq = D[Eq, n]

$$\mathrm{Out}[7] = 1+2\left(-\frac{15}{4}c[1]^2+\frac{3c[2]}{2}\right)n+$$
$$\frac{3}{16}(-705c[1]^4+900c[1]^2c[2]-68c[2]^2-280c[1]c[3]+40c[4])n^2-$$
$$\frac{5}{32}(23151c[1]^6-46530c[1]^4c[2]+19956c[1]^2c[2]^2-600c[2]^3+$$
$$15624c[1]^3c[3]-7728c[1]c[2]c[3]+504c[3]^2-4344c[1]^2c[4]+$$
$$528c[2]c[4]+1008c[1]c[5]-112c[6])n^3+$$
$$O[n]^4$$

We want to compare it with the result of the calculation in classical mechanics. But the quantum expression for ω is in terms of n and the classical one in terms of the oscillation amplitude a. We need to re-express both of them via the same quantity, the energy E.

In[8] := ne = InverseSeries[Eq, e]

$$\mathrm{Out}[8] = e+\frac{3}{4}\left(5c[1]^2-2c[2]\right)e^2+$$
$$\frac{5}{16}(231c[1]^4-252c[1]^2c[2]+28c[2]^2+56c[1]c[3]-8c[4])e^3+$$
$$\frac{35}{128}(7293c[1]^6-12870c[1]^4c[2]+5148c[1]^2c[2]^2-264c[2]^3+$$
$$3432c[1]^3c[3]-1584c[1]c[2]c[3]+72c[3]^2-792c[1]^2c[4]+$$
$$144c[2]c[4]+144c[1]c[5]-16c[6])e^4+$$
$$O[e]^5$$

In[9] := Wqe = Simplify[Wq/. n → ne]

$$\mathrm{Out}[9] = 1+\left(-\frac{15}{2}c[1]^2+3c[2]\right)e-$$
$$\frac{3}{16}(855c[1]^4-1020c[1]^2c[2]+92c[2]^2+280c[1]c[3]-40c[4])e^2+$$
$$\frac{1}{32}(-164805c[1]^6+311670c[1]^4c[2]-94920c[1]^3c[3]-$$
$$180c[1]^2\left(715c[2]^2-134c[4]\right)+5040c[1](9c[2]c[3]-c[5])+$$
$$8\left(633c[2]^3-315c[3]^2-450c[2]c[4]+70c[6]\right))e^3+$$
$$O[e]^4$$

Now we read the classical results for the energy Ec and the frequency Wc (written in terms of $A=a^2/2$, where the amplitude a is defined as the coefficient of the first harmonic $\cos\omega t$).

In[10] := <<class.m;

In[11] := Ec = Series[Ec, {A, 0, 3}]

$$\mathrm{Out}[11] = A+\left(-\frac{37}{4}c[1]^2+\frac{9c[2]}{2}\right)A^2+$$
$$\frac{1}{192}\left(-9309c[1]^4+17796c[1]^2c[2]+300c[2]^2-10880c[1]c[3]+1920c[4]\right)A^3+$$
$$O[A]^4$$

In[12] := Wc = Series[Wc, {A,0,2}]

$$\text{Out}[12] = 1 + \left(-\frac{15}{2}c[1]^2 + 3c[2]\right)A - \frac{3}{16}\left(485c[1]^4 - 692c[1]^2c[2] + 20c[2]^2 + 280c[1]c[3] - 40c[4]\right)A^2 + O[A]^3$$

In[13] := Ae = Simplify[InverseSeries[Ec, e]]

$$\text{Out}[13] = e + \left(\frac{37c[1]^2}{4} - \frac{9c[2]}{2}\right)e^2 + \left(\frac{14055c[1]^4}{64} - \frac{4147}{16}c[1]^2c[2] + \frac{623c[2]^2}{16} + \frac{170}{3}c[1]c[3] - 10c[4]\right)e^3 + O[e]^4$$

In[14] := Wce = Simplify[Wc/.A → Ae]

$$\text{Out}[14] = 1 + \left(-\frac{15}{2}c[1]^2 + 3c[2]\right)e - \frac{3}{16}\left(855c[1]^4 - 1020c[1]^2c[2] + 92c[2]^2 + 280c[1]c[3] - 40c[4]\right)e^2 + O[e]^3$$

In[15] := Wqe − Wce

$$\text{Out}[15] = O[e]^3$$

20.5 States

The following procedure accumulates contributions to δE in elements of the list d and to $|\delta\psi>$ in the list dp. Now we have to consider the large triangle in the figure, not just the rhombus.

```
In[16] := v2[l_, j_, a_] := (dp[[M − l]]+ = a ∗ ket[n + j];
      If[l < M, Do[v2[l − k, j, a ∗ de[k]/j], {k, 2, l − 1, 2}]];
      Do[If[j + i == 0, d[[M − l + k]]+ = a ∗ (V[k, i]/.n → n + j),
            v2[l − k, j + i, −a ∗ (V[k, i]/.n → n + j)/(j + i)]],
         {k, l}, {i, −k − 2, k + 2, 2}])
In[17] := Prepare[2]; d = Table[0, {M}]; dp = Table[0, {M}];
      Clear[de]; v2[M, 0, 1];
In[18] := Do[de[k] = Simplify[d[[k]]];
      Print[Collect[dp[[k]], ket[_], Simplify]], {k, M}]
```

$$\frac{\sqrt{-2+n}\sqrt{-1+n}\sqrt{n}c[1]\text{ket}[-3+n]}{6\sqrt{2}} + \frac{3n^{3/2}c[1]\text{ket}[-1+n]}{2\sqrt{2}} - \frac{3(1+n)^{3/2}c[1]\text{ket}[1+n]}{2\sqrt{2}} - \frac{\sqrt{1+n}\sqrt{2+n}\sqrt{3+n}c[1]\text{ket}[3+n]}{6\sqrt{2}}$$

$$\frac{1}{144}\sqrt{-5+n}\sqrt{-4+n}\sqrt{-3+n}\sqrt{-2+n}\sqrt{-1+n}\sqrt{n}c[1]^2\text{ket}[-6+n] + \frac{1}{32}\sqrt{-3+n}\sqrt{-2+n}\sqrt{-1+n}\sqrt{n}\left((-3+4n)c[1]^2 + 2c[2]\right)\text{ket}[-4+n] +$$

$$\frac{1}{16}\sqrt{-1+n}\sqrt{n}\left(\left(1-19n+7n^2\right)c[1]^2+4(-1+2n)c[2]\right)\text{ket}[-2+n]+$$
$$\frac{1}{16}\sqrt{1+n}\sqrt{2+n}\left(\left(27+33n+7n^2\right)c[1]^2-4(3+2n)c[2]\right)\text{ket}[2+n]+$$
$$\frac{1}{32}\sqrt{1+n}\sqrt{2+n}\sqrt{3+n}\sqrt{4+n}\left((7+4n)c[1]^2-2c[2]\right)\text{ket}[4+n]+$$
$$\frac{1}{144}\sqrt{1+n}\sqrt{2+n}\sqrt{3+n}\sqrt{4+n}\sqrt{5+n}\sqrt{6+n}c[1]^2\text{ket}[6+n]$$

As an additional problem, calculate the average values of $\hat{x}^k$ over the states just obtained, for several k. At $n \gg 1$ compare them to the classical averages obtained from the particle's motion $x(t)$.

Chapter 21
Riemann Curvature Tensor

Catching a lion, the Einstein's method: Enter the cage and lock it from inside. Then the Universe will be subdivided into two disjoint regions in such a way that you are in one of them and the lion is in the other one. It depends on one's point of view whom to consider caught; for convenience, let's say it's the lion.

Suppose we have a coordinate system x^μ in a region of an n-dimensional Riemann (or pseudo-Riemann) manifold [20]. Components of the metric tensor $g_{\mu\nu}$ are given as functions of x^μ. We want to calculate the Riemann curvature tensor $R^\mu{}_{\nu\alpha\beta}$ and related quantities (the Ricci tensor $R_{\mu\nu}$, the scalar curvature R).

The metric tensor is symmetric; therefore, it is reasonable to ask the user to provide only the components with $\mu \geq \nu$. If the user gives an argument having a wrong shape, we print an error message and abort the calculation. We shall also need the contravariant metric tensor $g^{\mu\nu}$ defined by $g^{\mu\lambda} g_{\lambda\nu} = \delta^\mu_\nu$.

```
In[1] := Metric[g0_] := Module[{n = Length[g0], g, gu},
    Do[If[Length[g0[[μ]]]=!=μ, Message[Metric :: shape]; Abort[]], {μ, n}];
    g = Table[If[μ ≥ ν, g0[[μ, ν]], g0[[ν, μ]]], {μ, n}, {ν, n}];
    gu = Simplify[Inverse[g]]; {n, g, gu}]
In[2] := Metric :: shape = "Wrong shape of the argument";
```

Next we calculate the Christoffel symbols

$$\Gamma_{\lambda\mu\nu} = \frac{1}{2}\left(\partial_\mu g_{\lambda\nu} + \partial_\nu g_{\lambda\mu} - \partial_\lambda g_{\mu\nu}\right)$$

and $\Gamma^\lambda{}_{\mu\nu} = g^{\lambda\rho}\Gamma_{\rho\mu\nu}$. They are symmetric in μ and ν; therefore, we calculate them only at $\nu \leqslant \mu$ and reuse the calculated values at $\nu > \mu$. If the optional parameter PrintNonZero is True, nonzero components are printed (following the tradition, in the printed results all indices vary from 0 to $n-1$).

```
In[3] := Christoffel[{n_, g_, gu_}, OptionsPattern[]] := Module[{Γ, Γu},
   Γ = Γu = Table[0, {λ, n}, {μ, n}, {ν, n}];
   Do[Γ[[λ, μ, ν]] = Simplify[(D[g[[λ, ν]], x[μ]] + D[g[[λ, μ]], x[ν]]−
          D[g[[μ, ν]], x[λ]])/2];
```

A. Grozin, *Introduction to Mathematica® for Physicists*, Graduate Texts in Physics,
DOI 10.1007/978-3-319-00894-3_21, © Springer International Publishing Switzerland 2014

```
        If[μ ≠ ν,Γ[[λ,ν,μ]] = Γ[[λ,μ,ν]]],
      {λ,n},{μ,n},{ν,μ}];
   Do[Γu[[λ,μ,ν]] = Simplify[Sum[gu[[λ,ρ]] ∗ Γ[[ρ,μ,ν]],{ρ,n}]];
        If[μ ≠ ν,Γu[[λ,ν,μ]] = Γu[[λ,μ,ν]]],
      {λ,n},{μ,n},{ν,μ}];
   If[OptionValue[PrintNonZero],
      Do[If[Γ[[λ,μ,ν]]=!=0,Print["Γ",λ − 1,μ − 1,ν − 1," ",Γ[[λ,μ,ν]]]],
        {λ,n},{μ,n},{ν,μ}];
      Do[If[Γu[[λ,μ,ν]]=!=0,Print["Γu",λ − 1,μ − 1,ν − 1," ",Γu[[λ,μ,ν]]]],
        {λ,n},{μ,n},{ν,μ}]];
   {Γ,Γu}]
In[4] := Options[Christoffel] = {PrintNonZero → True};
```

Finally, we calculate the Riemann tensor

$$R_{\alpha\beta\mu\nu} = g_{\alpha\lambda}\left(\partial_\mu \Gamma^\lambda{}_{\beta\nu} - \partial_\nu \Gamma^\lambda{}_{\beta\mu}\right) + \Gamma_{\alpha\lambda\mu}\Gamma^\lambda{}_{\beta\nu} - \Gamma_{\alpha\lambda\nu}\Gamma^\lambda{}_{\beta\mu}\,,$$

the Ricci tensor $R_{\mu\nu} = g^{\alpha\beta}R_{\alpha\mu\beta\nu}$, and the scalar curvature $R = g^{\mu\nu}R_{\mu\nu}$. The Riemann tensor has the properties

$$R_{\alpha\beta\mu\nu} = -R_{\beta\alpha\mu\nu} = -R_{\alpha\beta\nu\mu} = R_{\mu\nu\alpha\beta}\,,$$

and we use them to avoid unnecessary calculations.

```
In[5] := Riemann[{n_,g_,gu_},OptionsPattern[]] := Module[{Γ,Γu,
      R = Table[0,{α,n},{β,n},{μ,n},{ν,n}],R2 = Table[0,{μ,n},{ν,n}],R0},
   {Γ,Γu} = Christoffel[{n,g,gu}];
   Do[R[[α,β,μ,ν]] = R[[β,α,ν,μ]] = Simplify[Sum[
           g[[α,λ]] ∗ (D[Γu[[λ,β,ν]],x[μ]] − D[Γu[[λ,β,μ]],x[ν]])
          +Γ[[α,λ,μ]] ∗ Γu[[λ,β,ν]] − Γ[[α,λ,ν]] ∗ Γu[[λ,β,μ]],{λ,n}]];
        R[[β,α,μ,ν]] = R[[α,β,ν,μ]] = −R[[α,β,μ,ν]];
        If[μ ≠ α,R[[μ,ν,α,β]] = R[[ν,μ,β,α]] = R[[α,β,μ,ν]];
          R[[ν,μ,α,β]] = R[[μ,ν,β,α]] = −R[[α,β,μ,ν]]],
      {α,2,n},{β,α − 1},{μ,2,α},{ν,If[μ === α,β,μ − 1]}];
   Do[R2[[μ,ν]] = Simplify[Sum[gu[[α,β]] ∗ R[[α,μ,β,ν]],{α,n},{β,n}]];
        If[μ ≠ ν,R2[[ν,μ]] = R2[[μ,ν]]],
      {μ,n},{ν,μ}];
   R0 = Simplify[Sum[gu[[μ,ν]] ∗ R2[[μ,ν]] ∗ If[μ ≠ ν,2,1],{μ,n},{ν,μ}]];
   If[OptionValue[PrintNonZero],
      Do[If[R[[α,β,μ,ν]]=!=0,
           Print[R,α − 1,β − 1,μ − 1,ν − 1," ",R[[α,β,μ,ν]]]],
        {α,2,n},{β,α − 1},{μ,2,α},{ν,If[μ === α,β,μ − 1]}];
      Do[If[R2[[μ,ν]]=!=0,Print["R",μ − 1,ν − 1," ",R2[[μ,ν]]]],{μ,n},{ν,μ}];
      If[R0=!=0,Print["R ",R0]]];
   {R,R2,R0}]
In[6] := Options[Riemann] = {PrintNonZero → True};
```

Let's consider an example: the Schwarzschild metric

$$\mathrm{d}s^2 = \left(1 - \frac{r_0}{r}\right)\mathrm{d}t^2 - \frac{\mathrm{d}r^2}{1 - \frac{r_0}{r}} - r^2\left(\mathrm{d}\theta^2 + \sin^2\theta\,\mathrm{d}\varphi^2\right).$$

First we give names to the coordinates.

In[7] := Evaluate[Table[$x[\mu]$, {μ, 4}]] = {t, r, θ, φ};

Setting the Schwarzschild radius $r_0 = 1$, we obtain

**In[8] := Riemann[Metric[{{1 − 1/r}, {0, −1/(1 − 1/r)}, {0, 0, −r^2},
{0, 0, 0, −r^2 ∗ Sin[θ]^2}}]];**

Γ010 $\frac{1}{2r^2}$

Γ100 $-\frac{1}{2r^2}$

Γ111 $\frac{1}{2(-1+r)^2}$

Γ122 r

Γ133 $r\text{Sin}[\theta]^2$

Γ221 $-r$

Γ233 $r^2\text{Cos}[\theta]\text{Sin}[\theta]$

Γ331 $-r\text{Sin}[\theta]^2$

Γ332 $-r^2\text{Cos}[\theta]\text{Sin}[\theta]$

Γu010 $\frac{1}{2(-1+r)r}$

Γu100 $\frac{-1+r}{2r^3}$

Γu111 $\frac{1}{2r-2r^2}$

Γu122 $1-r$

Γu133 $-(-1+r)\text{Sin}[\theta]^2$

Γu221 $\frac{1}{r}$

Γu233 $-\text{Cos}[\theta]\text{Sin}[\theta]$

Γu331 $\frac{1}{r}$

Γu332 $\text{Cot}[\theta]$

R1010 $\frac{1}{r^3}$

R2020 $-\frac{-1+r}{2r^2}$

R2121 $\frac{1}{2(-1+r)}$

R3030 $-\frac{(-1+r)\text{Sin}[\theta]^2}{2r^2}$

R3131 $\frac{\text{Sin}[\theta]^2}{-2+2r}$

R3232 $-r\text{Sin}[\theta]^2$

The Ricci tensor (and hence the scalar curvature) vanishes. Therefore, the Schwarzschild metric satisfies the vacuum Einstein equation.

Chapter 22
Multi-ζ Functions

22.1 Definition

The Riemann ζ-function is defined by

$$\zeta_s = \sum_{n>0} \frac{1}{n^s} .$$

Mathematica knows this function; it can be expressed via powers of π for even integer values of s.

In[1] := Table[Zeta[*s*], {*s*, 2, 6}]

$\text{Out}[1] = \left\{ \frac{\pi^2}{6}, \text{Zeta}[3], \frac{\pi^4}{90}, \text{Zeta}[5], \frac{\pi^6}{945} \right\}$

Let's define

$$\zeta_{s_1 s_2} = \sum_{n_1>n_2>0} \frac{1}{n_1^{s_1} n_2^{s_2}} , \qquad \zeta_{s_1 s_2 s_3} = \sum_{n_1>n_2>n_3>0} \frac{1}{n_1^{s_1} n_2^{s_2} n_3^{s_3}} ,$$

and so on. These series converge at $s_1 > 1$. *Mathematica* does not know these multi-ζ functions. The sum $s_1 + s_2 + \cdots + s_k$ is called the *weight*. All relations we shall discuss contain terms of the same weight (the weight of a product is the sum of the weights of its factors).

22.2 Stuffling Relations

Suppose we want to multiply $\zeta_s \zeta_{s_1 s_2}$:

$$\zeta_s \zeta_{s_1 s_2} = \sum_{\substack{n>0 \\ n_1>n_2>0}} \frac{1}{n^s n_1^{s_1} n_2^{s_2}} .$$

A. Grozin, *Introduction to Mathematica® for Physicists*, Graduate Texts in Physics,
DOI 10.1007/978-3-319-00894-3_22,

Here n can be anywhere with respect to n_1, n_2. There are five contributions:

$$\sum_{n>n_1>n_2>0} \frac{1}{n^s n_1^{s_1} n_2^{s_2}} = \zeta_{s s_1 s_2},$$

$$\sum_{n=n_1>n_2>0} \frac{1}{n^s n_1^{s_1} n_2^{s_2}} = \zeta_{s+s_1, s_2},$$

$$\sum_{n_1>n>n_2>0} \frac{1}{n^s n_1^{s_1} n_2^{s_2}} = \zeta_{s_1 s s_2},$$

$$\sum_{n_1>n=n_2>0} \frac{1}{n^s n_1^{s_1} n_2^{s_2}} = \zeta_{s_1, s+s_2},$$

$$\sum_{n_1>n_2>n>0} \frac{1}{n^s n_1^{s_1} n_2^{s_2}} = \zeta_{s_1 s_2 s}.$$

This process reminds shuffling cards. The order of cards in the upper deck, as well as in the lower one, is kept fixed. We sum over all possible shufflings. Unlike real playing cards, however, two cards may be exactly on top of each other. In this case they are stuffed together: a single card (which is their sum) appears in the resulting deck. A mathematical jargon term for such shuffling with (possible) stuffing is *stuffling*.

In[2] := Show[Import["c1.jpg"]]

Out[2] =

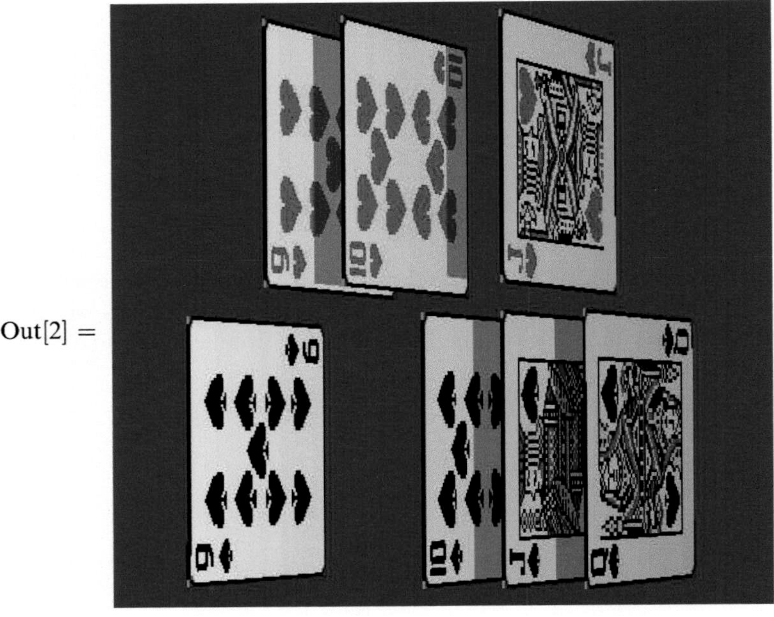

Let's implement this in *Mathematica*. The multi-ζ function will be called ζ; it can have any number of arguments. The function Stuffling first of all transforms products of ζ functions (including squares) to a local function z with three list parameters: the first two contain the arguments of the initial ζ functions, and the third one is empty. These are our two decks for shuffling and the resulting deck, initially empty. All the work is done by the following repeated substitution. Let the two unprocessed decks be nonempty: the first one contains some front "card" a_ and the remainder A___; the second one—the front "card" b_ and the remainder B___. Then there are three possibilities: either we move the front "card" from the first deck (a) to the resulting deck, or we move the front "card" from the second deck (b) to the resulting deck, or we take the front "cards" from both decks and put their sum to the resulting deck (stuffing). We need to use a delayed substitution :> here to ensure that the command Expand in its right-hand side is executed when the substitution is applied. In addition to this, we should take care of the situations when one of the source decks becomes empty. In this case we can just append the other source deck to the resulting one and yield the result.

This process can also be described as the following. The final result is a sum of many ζ functions with various argument lists. During intermediate steps, the function z[deck1, deck2, res] represents the sum of a subset of terms of the result whose argument lists begin with res. At each step we subdivide this sum into three parts, according to three possible values of the next argument.

```
In[3] := Stuffling[x_] := Module[{y,z},y = x/.
      {ζ[A___]^2->z[{A},{A},{}],ζ[A___]*ζ[B___]->z[{A},{B},{}]};
   y//.{z[{},{B___},{C___}]->ζ[C,B],z[{A___},{},{C___}]->ζ[C,A],
      z[{a_,A___},{b_,B___},{C___}] :>
         Expand[z[{A},{b,B},{C,a}]+z[{a,A},{B},{C,b}]+
            z[{A},{B},{C,a+b}]]}]
In[4] := Map[Stuffling, {ζ[x]^2, ζ[x]*ζ[y], ζ[x]*ζ[y,z]}]
Out[4] = {ζ[2x]+2ζ[x,x], ζ[x+y]+ζ[x,y]+ζ[y,x],
      ζ[y,x+z]+ζ[x+y,z]+ζ[x,y,z]+ζ[y,x,z]+ζ[y,z,x]}
```

22.3 Integral Representation

It is easy to check the integral representation of the ζ-function of an integer argument

$$\zeta_s = \int\limits_{1>x_1>\cdots>x_s>0} \frac{dx_1}{x_1} \cdots \frac{dx_{s-1}}{x_{s-1}} \frac{dx_s}{1-x_s}.$$

Let's denote

$$\omega_0 = \frac{dx}{x}, \qquad \omega_1 = \frac{dx}{1-x}.$$

All integrals will always have the integration region $1 > x_1 > \cdots > x_s > 0$. Then

$$\zeta_s = \int \omega_0^{s-1} \omega_1 \,.$$

This representation can be generalized to multi-ζ functions:

$$\zeta_{s_1 s_2} = \int \omega_0^{s_1-1} \omega_1 \omega_0^{s_2-1} \omega_1 \,, \qquad \zeta_{s_1 s_2 s_3} = \int \omega_0^{s_1-1} \omega_1 \omega_0^{s_2-1} \omega_1 \omega_0^{s_3-1} \omega_1 \,,$$

and so on. Let's write functions for transforming ζ with integer arguments to such integral representation and back. The integral representation ζi takes an arbitrary number of arguments equal to 0 or 1 corresponding to ω_0, ω_1.

```
In[5] := s2i[x_] := Module[{y,z},y=x/.ζ[A___]->z[{A},{}];
    y//.{z[{},{B___}]->ζi[B],
        z[{a_,A___},{B___}]:>
            z[{A},Append[Join[{B},Table[0,{a-1}]],1]]}]
In[6] := l = Map[s2i,{ζ[2],ζ[2,3]}]
Out[6] = {ζi[0,1],ζi[0,1,0,0,1]}
In[7] := i2s[x_] := Module[{y,z},y=x/.ζi[A___]->z[{A},{1}];
    y//.{z[{1},{B___}]->ζ[B],
        z[{a_,A___},{B___,b_}]:>
            If[a==0,z[{A},{B,b+1}],z[{A},{B,b,1}]]}]
In[8] := Map[i2s,l]
Out[8] = {ζ[2],ζ[2,3]}
In[9] := Clear[l]
```

22.4 Shuffling Relations

Suppose we want to multiply $\zeta_2 \cdot \zeta_2$:

$$\zeta_2^2 = \int\limits_{1>x_1>x_2>0} \omega_0 \omega_1 \cdot \int\limits_{1>x_1'>x_2'>0} \omega_0 \omega_1 \,.$$

The order of primed and non-primed integration variables is not fixed. There are six contributions:

$$1 > x_1 > x_2 > x_1' > x_2' > 0: \quad \int \omega_0 \omega_1 \omega_0 \omega_1 = \zeta_{22} \,;$$

$$1 > x_1 > x_1' > x_2 > x_2' > 0: \quad \int \omega_0 \omega_0 \omega_1 \omega_1 = \zeta_{31} \,;$$

$$1 > x_1 > x_1' > x_2' > x_2 > 0: \quad \int \omega_0 \omega_0 \omega_1 \omega_1 = \zeta_{31} \,;$$

$$1 > x_1' > x_1 > x_2 > x_2' > 0 : \quad \int \omega_0 \omega_0 \omega_1 \omega_1 = \zeta_{31} ;$$

$$1 > x_1' > x_1 > x_2' > x_2 > 0 : \quad \int \omega_0 \omega_0 \omega_1 \omega_1 = \zeta_{31} ;$$

$$1 > x_1' > x_2' > x_1 > x_2 > 0 : \quad \int \omega_0 \omega_1 \omega_0 \omega_1 = \zeta_{22} .$$

Now we are multiplying integrals, not sums. Therefore our "cards" are now infinitely thin and cannot be exactly on top of each other. There are just two kinds of "cards": ω_0 and ω_1, and we sum over all possible shufflings of two decks.

In[10] := Show[Import["c2.jpg"]]

Out[10] =

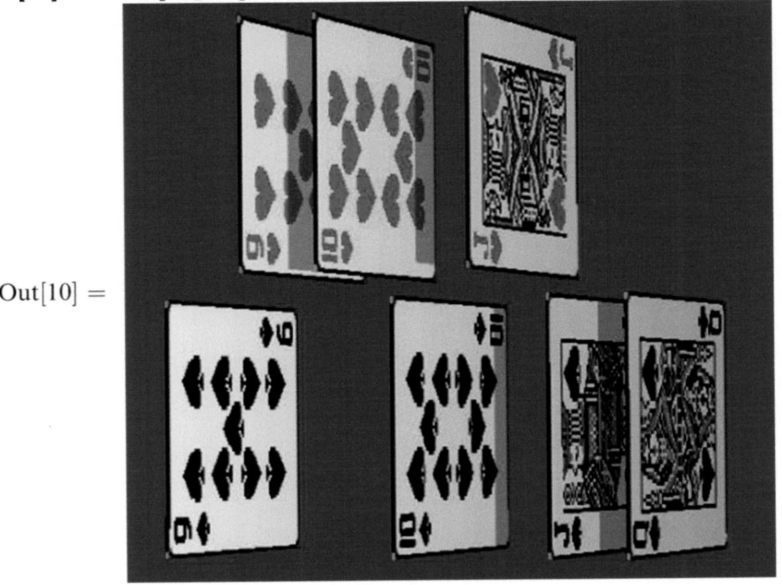

```
In[11] := shuffling[x_] := Module[{y,z},y = x/.
      {ζi[A___]^2->z[{A},{A},{}],
        ζi[A___] * ζi[B___]->z[{A},{B},{}]};
   y//.{z[{},{B___},{C___}]->ζi[C,B],z[{A___},{},{C___}]->ζi[C,A],
      z[{a_,A___},{b_,B___},{C___}] :>
        Expand[z[{A},{b,B},{C,a}] + z[{a,A},{B},{C,b}]]}]
In[12] := Shuffling[x_] := i2s[Expand[shuffling[s2i[x]]]]
In[13] := Map[Shuffling, {ζ[2]^2, ζ[2] * ζ[3], ζ[2] * ζ[2,1]}]
Out[13] = {2ζ[2,2] + 4ζ[3,1], ζ[2,3] + 3ζ[3,2] + 6ζ[4,1],
      ζ[2,1,2] + 3ζ[2,2,1] + 6ζ[3,1,1]}
```

22.5 Duality Relations

The integral representation allows us to derive another set of useful relations, even simpler than shuffling—duality relations. Let's make the substitution $x_i \to 1 - x_i$. Then $\omega_0 \longleftrightarrow \omega_1$; to preserve the order $1 > x_1 > \cdots > x_s > 0$, we have to arrange all the ω factors in the opposite order. In other words, after writing down an integral representation for a multi-ζ value, we may read it in the Arabic fashion, right to left, simultaneously replacing $\omega_0 \longleftrightarrow \omega_1$. Duality relations are the only known relations which say that two multi-ζ values with distinct arguments are just equal to each other.

```
In[14] := duality[x_] := Module[{y,z},y = x/.ζi[A___]->z[{A},{}];
      y//.{z[{},{B___}]->ζi[B],z[{a_,A___},{B___}]->z[{A},{1-a,B}]}]
In[15] := Duality[x_] := i2s[duality[s2i[x]]]
In[16] := Map[Duality,{ζ[3],ζ[4],ζ[5],ζ[4,1],ζ[3,2],ζ[2,3]}]
Out[16] = {ζ[2,1],ζ[2,1,1],ζ[2,1,1,1],ζ[3,1,1],ζ[2,2,1],ζ[2,1,2]}
```

22.6 Weight 4

There are four converging multi-ζ series of weight 4: ζ_4, ζ_{31}, ζ_{22}, and ζ_{211}. Due to duality, two of them are equal to each other.

```
In[17] := Duality[ζ[4]]
Out[17] = ζ[2,1,1]
```

We can express ζ_4 via ζ_2^2 using their explicit values:

```
In[18] := S = ζ[4]->Zeta[4]/Zeta[2]^2*ζ[2]^2
```

Out[18] = $\zeta[4] \to \frac{2\zeta[2]^2}{5}$

Two equations for ζ_2^2 follow from stuffling

```
In[19] := eq1 = ζ[2]^2 == Stuffling[ζ[2]^2]
```

Out[19] = $\zeta[2]^2 == \zeta[4] + 2\zeta[2,2]$

and shuffling

```
In[20] := eq2 = ζ[2]^2 == Shuffling[ζ[2]^2]
```

Out[20] = $\zeta[2]^2 == 2\zeta[2,2] + 4\zeta[3,1]$

They can be solved for ζ_{22} and ζ_{31} (taking the expression for ζ_4 into account).

```
In[21] := s = Solve[{eq1/.S,eq2},{ζ[2,2],ζ[3,1]}][[1]]
```

Out[21] = $\left\{\zeta[2,2] \to \frac{3\zeta[2]^2}{10}, \zeta[3,1] \to \frac{\zeta[2]^2}{10}\right\}$

Thus we have demonstrated that all multi-ζ values of weight 4 can be expressed via ζ_2^2.

```
In[22] := Clear[S]
```

22.7 Weight 5

There are four distinct multi-ζ values of weight 5,
In[23] := Map[Duality, {ζ[5], ζ[4,1], ζ[3,2], ζ[2,3]}]
Out[23] = {ζ[2,1,1,1], ζ[3,1,1], ζ[2,2,1], ζ[2,1,2]}
due to duality. The stuffling relation for $\zeta_2\zeta_3$:
In[24] := eq1 = ζ[2] * ζ[3] == Stuffling[ζ[2] * ζ[3]]
Out[24] = ζ[2]ζ[3] == ζ[5] + ζ[2,3] + ζ[3,2]
A similar equation where ζ_3 is written in the dual form.
In[25] := eq2 = Stuffling[ζ[2] * Duality[ζ[3]]]
Out[25] = ζ[2,3] + ζ[4,1] + ζ[2,1,2] + 2ζ[2,2,1]
**In[26] := eq2 = eq2/. {ζ[2,1,2] :> Duality[ζ[2,1,2]],
ζ[2,2,1] :> Duality[ζ[2,2,1]]}**
Out[26] = 2ζ[2,3] + 2ζ[3,2] + ζ[4,1]
In[27] := eq2 = ζ[2] * ζ[3] == eq2
Out[27] = ζ[2]ζ[3] == 2ζ[2,3] + 2ζ[3,2] + ζ[4,1]
The shuffling relation for $\zeta_2\zeta_3$:
In[28] := eq3 = ζ[2] * ζ[3] == Shuffling[ζ[2] * ζ[3]]
Out[28] = ζ[2]ζ[3] == ζ[2,3] + 3ζ[3,2] + 6ζ[4,1]
This system can be solved for ζ_{41}, ζ_{32}, and ζ_{23}.
In[29] := *s* = Solve[{eq1, eq2, eq3}, {ζ[4,1], ζ[3,2], ζ[2,3]}][[1]]

$$\text{Out[29]} = \left\{\zeta[4,1] \to -\zeta[2]\zeta[3] + 2\zeta[5], \zeta[3,2] \to 3\zeta[2]\zeta[3] - \frac{11\zeta[5]}{2},\right.$$
$$\left.\zeta[2,3] \to -2\zeta[2]\zeta[3] + \frac{9\zeta[5]}{2}\right\}$$

Thus we have demonstrated that all multi-ζ values of weight 5 can be expressed via $\zeta_2\zeta_3$ and ζ_5.

Chapter 23
Rainbow

23.1 Statement of the Problem

We consider scattering of light by a spherical water drop in geometrical optics. For small drops, diffraction becomes significant; our analysis is only valid for drops which are not too small. Let the drop radius be 1. The ray with the impact parameter ρ splits into the reflected ray and the refracted one. Their directions are given by the Snell law

```
In[1] := Snell = {α->ArcSin[ρ], β->ArcSin[ρ/n]};
```

where the refraction index of water is

```
In[2] := Water = n->1.333;
In[3] := col = {RGBColor[1,0,0], RGBColor[0,0,1], RGBColor[0,1,0],
    RGBColor[1,0,1], RGBColor[0,1,1], RGBColor[1,1,0]};
In[4] := With[{y = 0.6},
  With[{α = α/.Snell/.ρ->y, β = β/.Snell/.ρ->y/.Water,
    x = -Sqrt[1 - y^2]},
   With[{φ = π - α, ϑ = π - 2*α}, With[{ψ = φ + 2*β - π},
    Graphics[{Black, Circle[], Line[{{-2,0}, {1.2,0}}],
     Line[{{0,0}, {2*Cos[φ], 2*Sin[φ]}}],
     Line[{{0,0}, {Cos[ψ], Sin[ψ]}}],
     Line[{{x,y}, {x,0}}], col[[1]], Line[{{-2,y}, {x,y}}],
     col[[2]], Line[{{x,y}, {x+1.5*Cos[ϑ], y+1.5*Sin[ϑ]}}],
     col[[3]], Line[{{x,y}, {Cos[ψ], Sin[ψ]}}],
     Black, Inset[Style["α", 24], {-0.25, 0.1}],
     Inset[Style["α", 24], {x - 0.25, y + 0.1}],
     Inset[Style["α", 24], {x - 0.15, y + 0.25}],
     Inset[Style["β", 24], {x + 0.35, y - 0.15}],
     Inset[Style["ρ", 24], {x + 0.1, 0.5*y}]}]]]]
```

A. Grozin, *Introduction to Mathematica® for Physicists*, Graduate Texts in Physics, DOI 10.1007/978-3-319-00894-3_23, © Springer International Publishing Switzerland 2014

Out[4] =

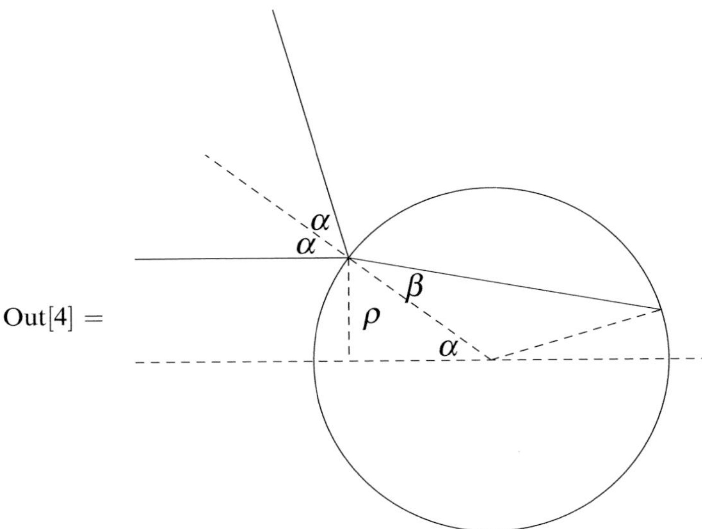

The incident ray hits the drop at the point $\{\mathrm{Cos}[\varphi], \mathrm{Sin}[\varphi]\}$, where $\varphi = \pi - \alpha$. The reflected ray has the direction $\vartheta = \pi - 2\alpha$. The refracted ray hits the drop surface again at $\{\mathrm{Cos}[\psi], \mathrm{Sin}[\psi]\}$, where $\psi = \varphi + 2\beta - \pi$.

The incident light has two polarizations, with the electric field orthogonal to the scattering plane or lying in this plane. The reflection coefficients for these polarizations are given by the Fresnel formulas [21] (they don't depend on the direction, i.e., are the same for a ray entering water and a ray leaving it).

```
In[5] := Rs = (Sin[α − β]/Sin[α + β])^2; Rp = (Tan[α − β]/Tan[α + β])^2;
```

23.2 0 Ray Segments Inside the Drop

First let's consider rays reflected by the drop immediately after they hit its surface.

```
In[6] := Ray0[y_, a_] := With[{α = α/.Snell/.ρ−>y,
           β = β/.Snell/.ρ−>y/.Water, x = −Sqrt[1 − y^2]},
      With[{ϑ = π − 2 ∗ α}, {col[[1]], Line[{{−1 − a, y}, {x, y}}],
           col[[2]], Line[{{x, y}, {x + a ∗ Cos[ϑ], y + a ∗ Sin[ϑ]}}]}]]
In[7] := Manipulate[Graphics[Join[{Black, Circle[]}, Ray0[y, a]],
           PlotRange−>{{−1 − a, 1.1}, {−1.1, a + 0.7}}],
      {{y, 0.6}, 0, 1}, {{a, 2}, 1, 10}]
```

Out[7] =

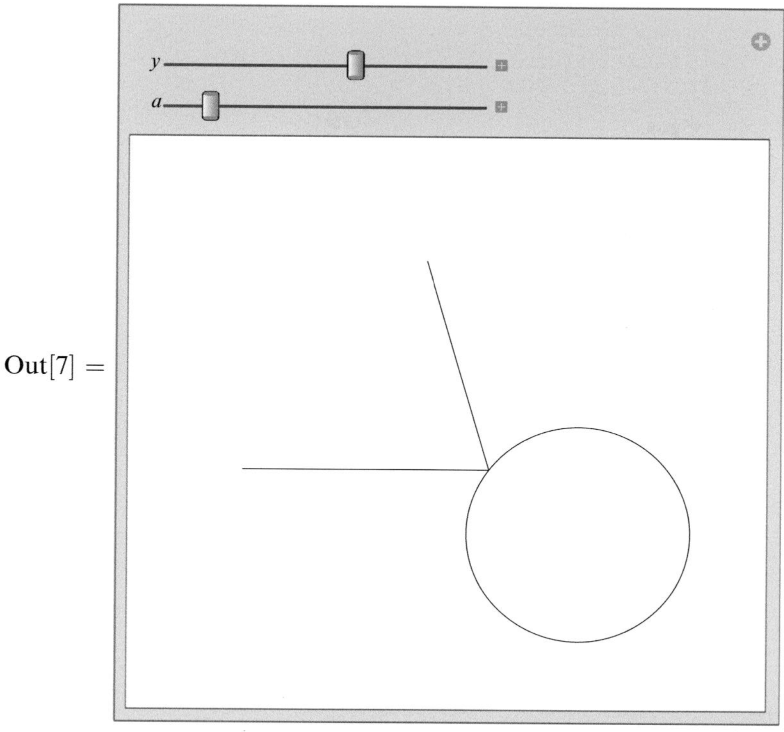


```
In[8] := Manipulate[
  Graphics[
    Join[{Black, Circle[]},
      With[{δ = (ρmax − ρmin)/M},
        If[δ > 0, Apply[Join, Table[Ray0[y, a], {y, ρmin + δ/2, ρmax, δ}]], {}]]],
    PlotRange−>{{−1 − a, 1.1}, {−1.1, a + 0.7}}],
  {{ρmin, 0}, 0, 1}, {{ρmax, 1}, 0, 1},
  {{M, 10}, Table[i, {i, 30}]}, {{a, 2}, 1, 10}]
```

Out[8] =

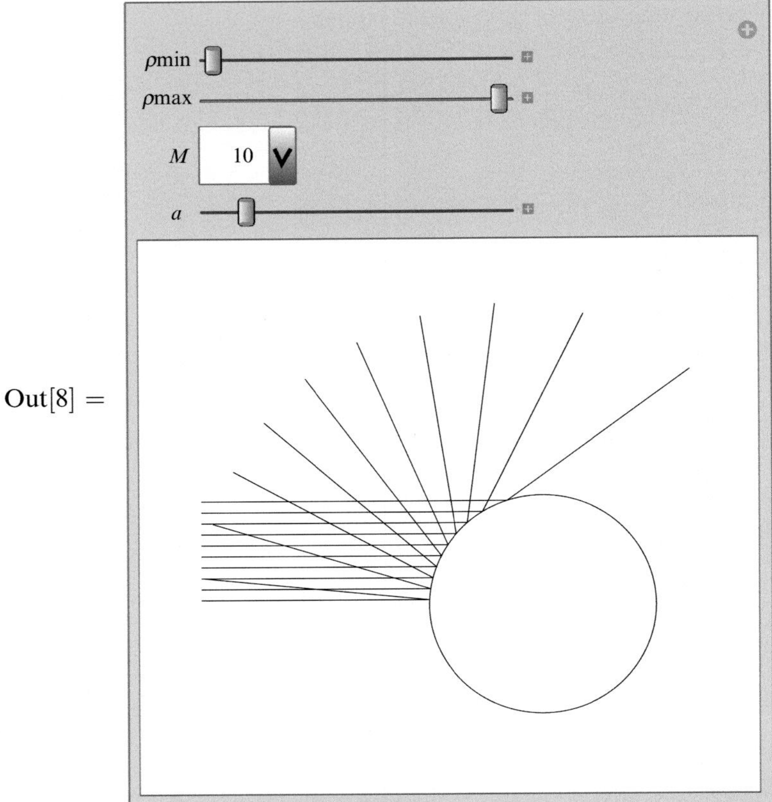

The scattering angle is

In[9] := $\vartheta 0 = \pi - 2 * \alpha$;

In[10] := Plot[Evaluate[$\vartheta 0$/. Snell], {ρ, 0, 1}]

Out[10] =

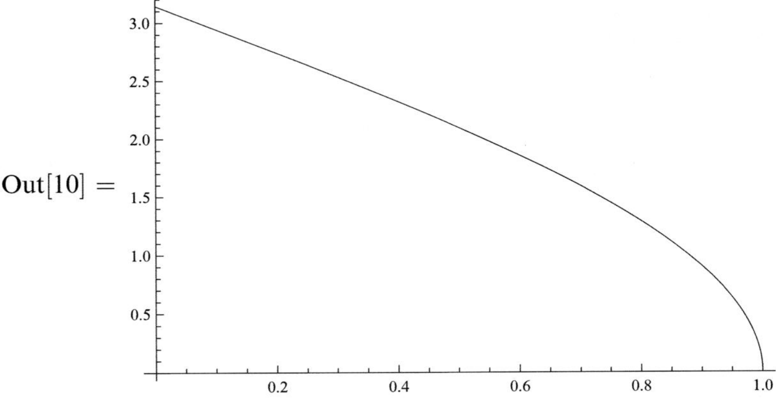

In order to calculate the differential cross section, we need to express ρ via the scattering angle ϑ:

In[11] := sol0 = Simplify[Solve[(ϑ0/. Snell) == ϑ, ρ], 0 < ϑ < π]

$\text{Out}[11] = \left\{\left\{\rho \to \text{Cos}\left[\frac{\vartheta}{2}\right]\right\}\right\}$

In[12] := ρ0 = ρ/. sol0[[1]]

$\text{Out}[12] = \text{Cos}\left[\frac{\vartheta}{2}\right]$

In[13] := Clear[sol0]

The angles α and β are

In[14] := Snell0 = {α−>(π − ϑ)/2, β−>ArcSin[Sin[α]/n]};

The area of the ring in the transverse plane corresponding to the scattering angles between ϑ and $\vartheta + \mathrm{d}\vartheta$, divided by $\mathrm{d}\Omega = 2\pi\,\text{Sin}[\vartheta]\,\mathrm{d}\vartheta$, is

In[15] := σ0 = Simplify[−D[ρ0^2, ϑ]/(2 * Sin[ϑ])]

$\text{Out}[15] = \frac{1}{4}$

Therefore, the cross sections for the two polarizations are

In[16] := σ0s[ϑ_] = Simplify[σ0 * Rs//. Snell0/. Water];
σ0p[ϑ_] = Simplify[σ0 * Rp//. Snell0/. Water];

In[17] := Plot[{σ0s[ϑ], σ0p[ϑ], (σ0s[ϑ] + σ0p[ϑ])/2}, {ϑ, 0, π},
PlotRange−>All, PlotStyle−>col]

Out[17] =

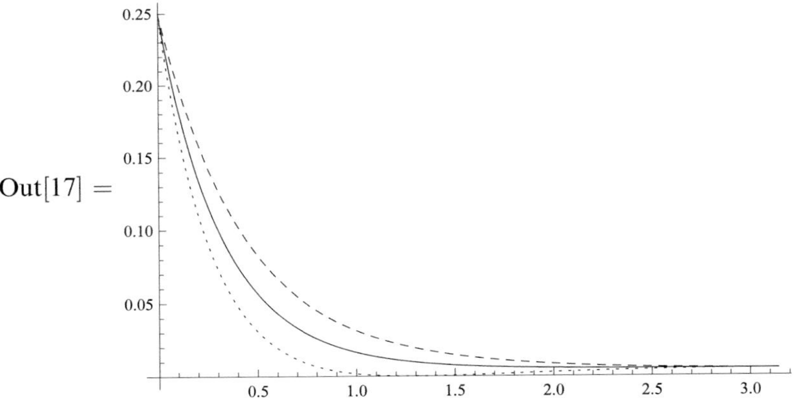

Note that there is a scattering angle ϑ at which the scattered light is completely polarized: its electric field is orthogonal to the scattering plane. This happens when α is equal to the Brewster angle αB

In[18] := αB = ArcTan[n]; αB/. Water

Out[18] = 0.927175

In this case

In[19] := βB = ArcTan[1/n];

so that $\alpha\text{B} + \beta\text{B} = \pi/2$ —the refracted ray is perpendicular to the reflected one:

```
In[20] := Graphics[{Line[{{−1,0},{1,0}}],Line[{{0,−1},{0,1}}],
    col[[1]],Line[{{0,0},{−1,1/n}}],
    col[[2]],Line[{{0,0},{1,1/n}}],
    col[[3]],Line[{{0,0},{1/n,−1}}],
    Black,Inset[Style["α",24],{−0.06,0.13}],Inset[Style["α",24],{0.06,0.13}],
    Inset[Style["β",24],{0.06,−0.13}]}]/.Water
```

Out[20] =

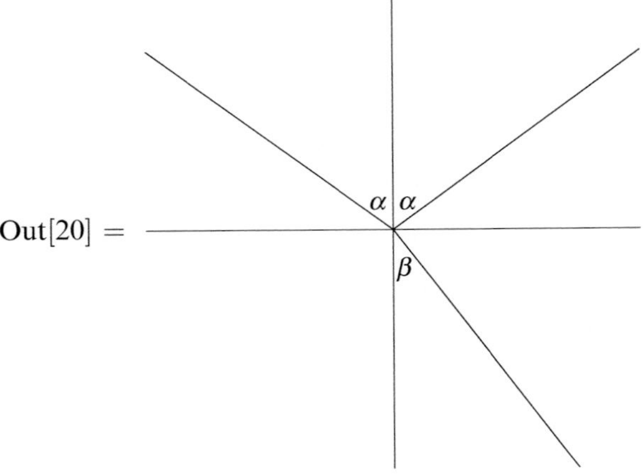

When the incoming light with the electric field in the scattering plane reaches water, electric dipoles in it oscillate along the direction perpendicular to the refracted ray; they don't radiate in the direction along this axis, i.e., don't produce the reflected ray: Rp = 0.

```
In[21] := Plot[σ0p[ϑ],{ϑ,0,π},PlotRange−>{0,0.01}]
```

Out[21] =

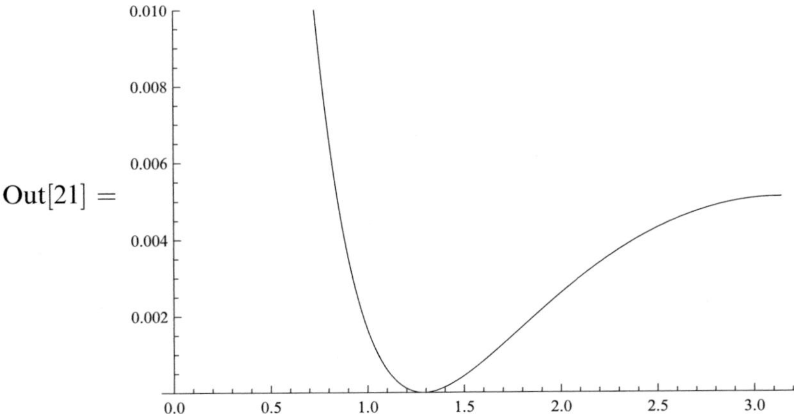


```
In[22] := ϑ0/.α−>αB/.Water
Out[22] = 1.28724
```

23.3 1 Ray Segment Inside the Drop

```
In[23] := Ray1[y_,a_] := With[{α = α/.Snell/.ρ->y,
        β = β/.Snell/.ρ->y/.Water,x = -Sqrt[1 - y^2]},
    With[{φ = π - α, ϑ = 2 * (β - α)},With[{ψ = φ + 2 * β - π},
        With[{x1 = Cos[ψ],y1 = Sin[ψ]},{col[[1]],Line[{{-1 - a,y},{x,y}}],
            col[[3]],Line[{{x,y},{x1,y1}}],
            col[[2]],Line[{{x1,y1},{x1 + a * Cos[ϑ],y1 + a * Sin[ϑ]}}]}]]]]
In[24] := Manipulate[Graphics[Join[{Black,Circle[]},Ray1[y,a]],
        PlotRange->{{-1 - a,1 + a},{-1.1,1.1}}],{{y,0.6},0,1},
    {{a,2},1,10}]
```

Out[24] =

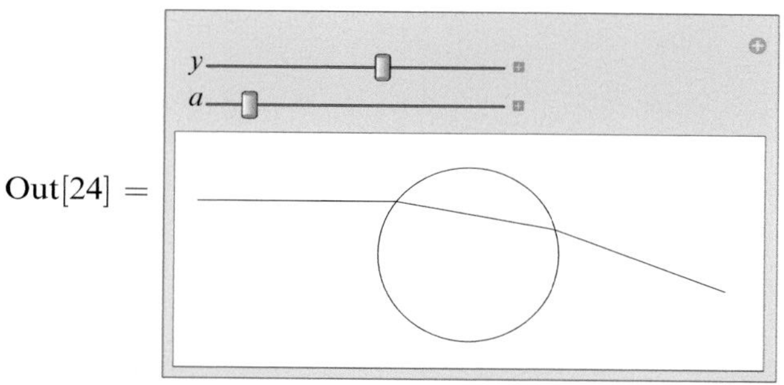

When entering water, the ray is deflected by $\alpha - \beta$ clockwise; when leaving water, it is deflected by the same angle again. The direction of the outgoing ray is

```
In[25] := ϑ1 = 2 * (β - α);
In[26] := Manipulate[
  Graphics[
    Join[{Black,Circle[]},
      With[{δ = (ρmax - ρmin)/M},
        If[δ > 0,Apply[Join,Table[Ray1[y,a],{y,ρmin + δ/2,ρmax,δ}]],{}]]],
    PlotRange->{{-1 - a,1 + a},{-1.1,1.1}}],
  {{ρmin,0},0,1},{{ρmax,1},0,1},
  {{M,10},Table[i,{i,30}]},{{a,2},1,10}]
```

Out[26] =

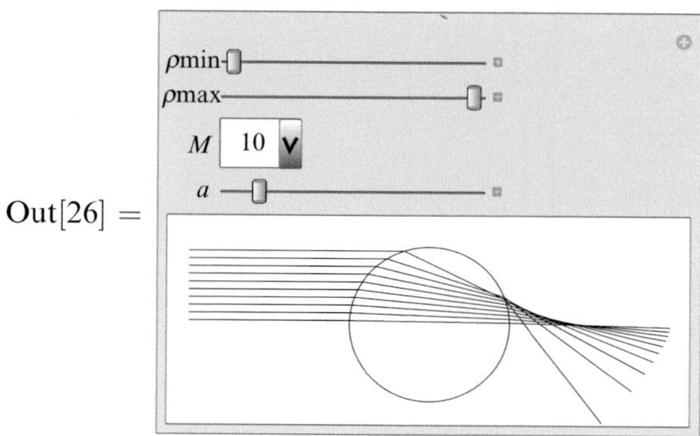

The scattering angle ϑ is obtained by reducing $\vartheta 1$ to the interval $[-\pi,\pi]$ and then taking Abs; in the present case, it is just $-\vartheta 1$.

In[27] := Plot[Evaluate[−ϑ1/. Snell/. Water], {ρ,0,1}]

Out[27] =

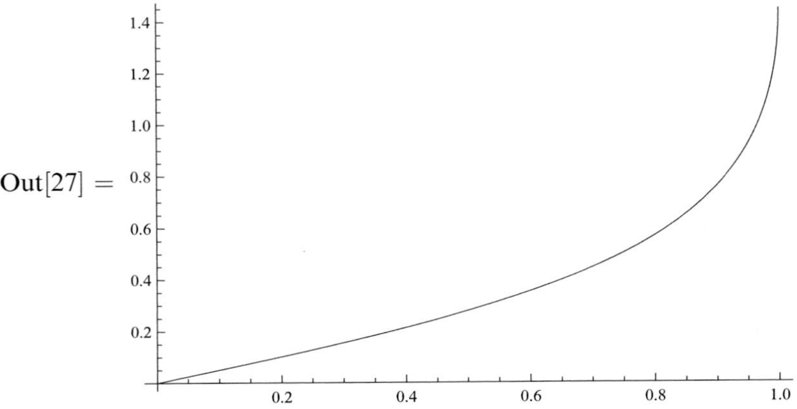

It varies from 0 to

In[28] := ϑ1m = Simplify[−ϑ1/. Snell/. ρ−>1]

Out[28] = $\pi - 2\,\text{ArcSin}\left[\frac{1}{n}\right]$

In[29] := ϑ1m/. Water

Out[29] = 1.4449

Now we have to solve the equation $\alpha - \beta = \vartheta/2$ for ρ.

In[30] := eq1 = (TrigExpand[Sin[α − β]]/. {Sin[α]−>ρ, Cos[α]−>Sqrt[1 − ρ^2], Sin[β]−>ρ/n, Cos[β]−>Sqrt[1 − (ρ/n)^2]}) == Sin[ϑ/2]

Out[30] = $-\frac{\rho\sqrt{1-\rho^2}}{n} + \rho\sqrt{1-\frac{\rho^2}{n^2}} == \text{Sin}\left[\frac{\vartheta}{2}\right]$

In[31] := sol1 = Solve[eq1, ρ]

Out[31] = $\left\{\left\{\rho \to -\frac{\sqrt{n^2 - n^2\text{Cos}[\vartheta]}}{\sqrt{2}\sqrt{1+n^2-2n\,\text{Cos}\left[\frac{\vartheta}{2}\right]}}\right\}, \left\{\rho \to \frac{\sqrt{n^2 - n^2\text{Cos}[\vartheta]}}{\sqrt{2}\sqrt{1+n^2-2n\,\text{Cos}\left[\frac{\vartheta}{2}\right]}}\right\}, \left\{\rho \to -\frac{\sqrt{n^2 - n^2\text{Cos}[\vartheta]}}{\sqrt{2}\sqrt{1+n^2+2n\,\text{Cos}\left[\frac{\vartheta}{2}\right]}}\right\}, \left\{\rho \to \frac{\sqrt{n^2 - n^2\text{Cos}[\vartheta]}}{\sqrt{2}\sqrt{1+n^2+2n\,\text{Cos}\left[\frac{\vartheta}{2}\right]}}\right\}\right\}$

We discard the negative solutions. The positive ones evaluated at ϑ1m are

In[32] := Simplify[ρ/. sol1[[{2,4}]]/. ϑ−>ϑ1m, n > 1]

Out[32] = $\left\{1, \sqrt{\frac{-1+n^2}{3+n^2}}\right\}$

So, the right solution is number 2:

```
In[33] := ρ1 = Simplify[Simplify[ρ/. sol1[[2]], {n > 1, 0 < ϑ < π}]/.
    Cos[ϑ]->1 - 2 * Sin[ϑ/2]^2, {n > 1, ϑ > 0, ϑ < π}]
```

$$\text{Out[33]} = \frac{n \,\text{Sin}\left[\frac{\vartheta}{2}\right]}{\sqrt{1+n^2-2n\,\text{Cos}\left[\frac{\vartheta}{2}\right]}}$$

```
In[34] := Clear[eq1, sol1]
```

The geometrical cross section for ϑ between ϑ and $\vartheta + \mathrm{d}\vartheta$, divided by $\mathrm{d}\Omega$, is

```
In[35] := σ1 = Simplify[D[ρ1^2, ϑ]/(4 * c2 * s2)/.
    {Cos[ϑ/2]->c2, Sin[ϑ/2]->s2}]
```

$$\text{Out[35]} = \frac{n^2\left(\text{c2} - 2\,\text{c2}^2 n + \text{c2}\, n^2 - n\,\text{s2}^2\right)}{4\,\text{c2}\left(1 - 2\,\text{c2}\, n + n^2\right)^2}$$

where $\text{c2} = \text{Cos}[\vartheta/2]$, $\text{s2} = \text{Sin}[\vartheta/2]$. The angles α and β are

```
In[36] := Snell1 = {α->ArcSin[ρ1], β->ArcSin[ρ1/n]};
```

The differential cross sections for the two polarizations are

```
In[37] := cs2 = {c2->Cos[ϑ/2], s2->Sin[ϑ/2]};
In[38] := σ1s[ϑ_] = Simplify[σ1 * (1 - Rs)^2/. Snell1/. cs2/. Water];
    σ1p[ϑ_] = Simplify[σ1 * (1 - Rp)^2/. Snell1/. cs2/. Water];
```

(the transmission coefficient is $T = 1 - R$, and transmission happens twice).

```
In[39] := Plot[{σ1s[ϑ], σ1p[ϑ], (σ1s[ϑ] + σ1p[ϑ])/2}, {ϑ, 0, ϑ1m/. Water},
    PlotRange->All, PlotStyle->col]
```

Out[39] =

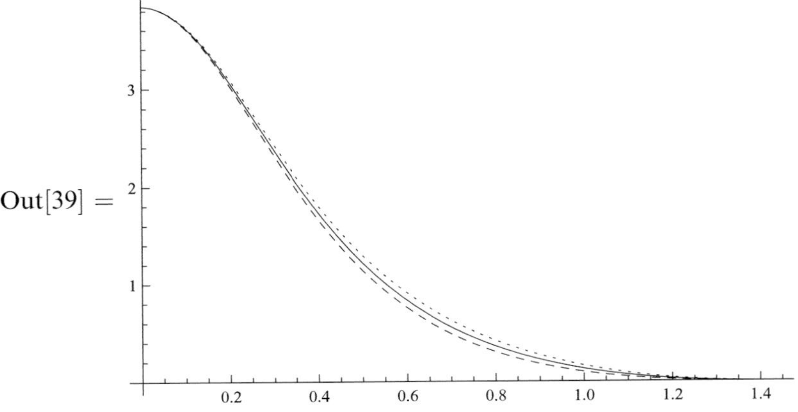

23.4 2 Ray Segments Inside the Drop

```
In[40] := Ray2[y_, a_] := With[{α = α/. Snell/. ρ->y,
    β = β/. Snell/. ρ->y/. Water, x = -Sqrt[1 - y^2]},
  With[{ϑ = 4 * β - 2 * α - π},
   Module[{φ = π - α, R = {col[[1]], Line[{{-1 - a, y}, {x, y}}]},
      x1 = x, y1 = y, x2, y2, ψ},
```

```
    ψ = φ + 2 * β − π; x2 = Cos[ψ]; y2 = Sin[ψ];
    R = Join[R, {col[[3]], Line[{{x1, y1}, {x2, y2}}]}];
    φ = ψ; x1 = x2; y1 = y2; ψ = φ + 2 * β − π; x2 = Cos[ψ]; y2 = Sin[ψ];
    R = Join[R, {col[[4]], Line[{{x1, y1}, {x2, y2}}]}];
    Join[R, {col[[2]], Line[{{x2, y2}, {x2 + a * Cos[ϑ], y2 + a * Sin[ϑ]}}]}]]]]
In[41] := Manipulate[Graphics[Join[{Black, Circle[]}, Ray2[y, a]],
      PlotRange−>{{−1 − a, 1.1}, {−1 − 0.7 * a, 1.1}}],
   {{y, 0.6}, 0, 1}, {{a, 2}, 1, 10}]
```

Out[41] =

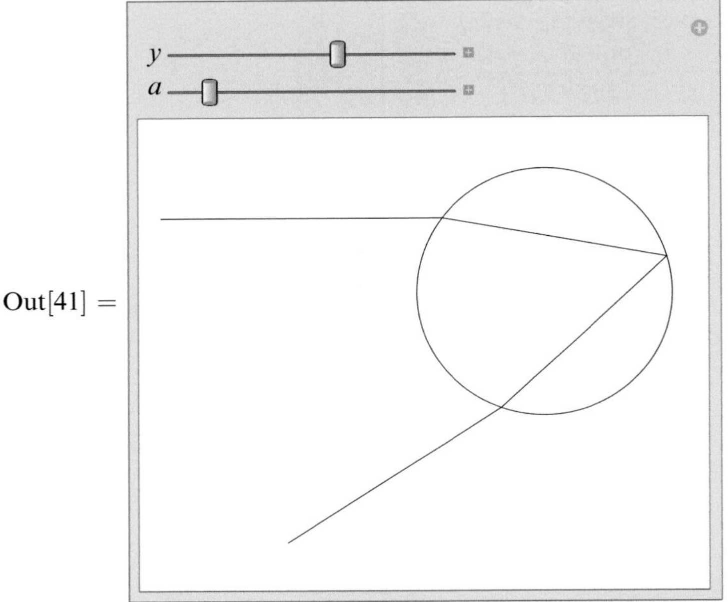

The second segment of the ray inside the drop is obtained from the first one by rotating by the angle $\pi - 2\beta$ clockwise; hence the outgoing ray is obtained from the one in the previous section by the same rotation:

```
In[42] := ϑ2 = 4 * β − 2 * α − π;
In[43] := Manipulate[
  Graphics[
   Join[{Black, Circle[]},
     With[{δ = (ρmax − ρmin)/M},
       If[δ > 0, Apply[Join, Table[Ray2[y, a], {y, ρmin + δ/2, ρmax, δ}]], {}]]],
   PlotRange−>{{−1 − a, 1.1}, {−1 − 0.7 * a, 1.1}}],
  {{ρmin, 0}, 0, 1}, {{ρmax, 1}, 0, 1},
  {{M, 10}, Table[i, {i, 30}]}, {{a, 2}, 1, 10}]
```

Out[43] =

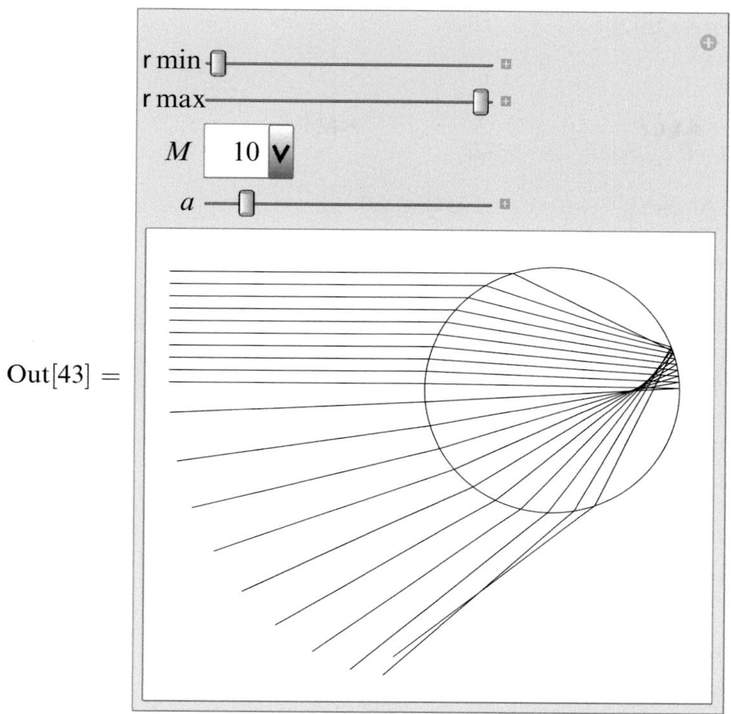

The real scattering angle is $\vartheta = -\vartheta 2$:
In[44] := Plot[$-\vartheta$2/. Snell/. Water, {ρ,0,1}]

Out[44] =

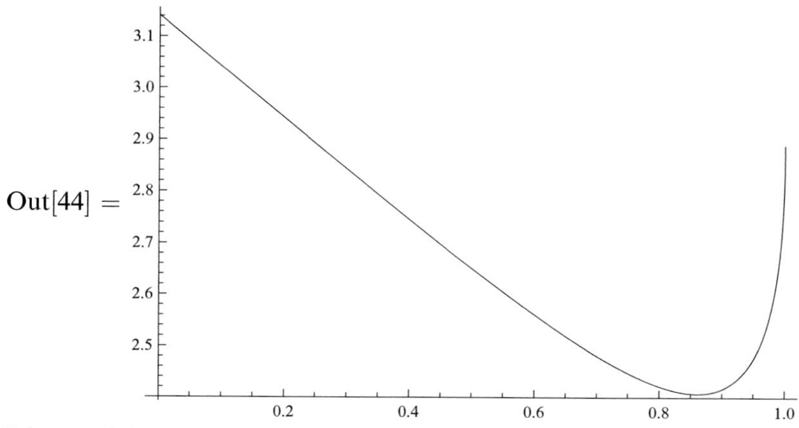

It has a minimum at
In[45] := s2r = Solve[D[ϑ2/. Snell, ρ] == 0, ρ]

Out[45] = $\left\{\left\{\rho \to -\frac{\sqrt{4-n^2}}{\sqrt{3}}\right\}, \left\{\rho \to \frac{\sqrt{4-n^2}}{\sqrt{3}}\right\}\right\}$

In[46] := ρ2r = ρ/.s2r[[2]]

$$\text{Out[46]} = \frac{\sqrt{4-n^2}}{\sqrt{3}}$$

equal to

In[47] := ϑ2r = −ϑ2/.Snell/.ρ−>ρ2r

$$\text{Out[47]} = \pi + 2\,\text{ArcSin}\left[\frac{\sqrt{4-n^2}}{\sqrt{3}}\right] - 4\,\text{ArcSin}\left[\frac{\sqrt{4-n^2}}{\sqrt{3}n}\right]$$

In[48] := {ρ2r, ϑ2r}/.Water

Out[48] = {0.860835, 2.40719}

Rays from a relatively wide ring around $\rho 2\text{r}$ have practically the same scattering angle $\vartheta 2\text{r}$. This contribution to the cross section tends to ∞ at this angle. When an observer sees a cloud of water drops illuminated by the sun, especially bright light rays arrive along the cone with angle $\pi - \vartheta 2\text{r}$. Usually, only a part of the circle is seen (the full circle can be sometimes observed from an airplane).

```
In[49] := With[{R = −Tan[ϑ2r/.Water]},Graphics3D[
  Join[{Opacity[0.1],Yellow,Cone[{{1,0,0},{0,0,0}},R]},
    Apply[Join,Table[
      Module[{x = 0.5 * (Random[] + 1),φ = 2 * π * Random[],y,z},
        y = R * x * Cos[φ]; z = R * x * Sin[φ];
        {Red,Line[{{0,y,z},{x,y,z}}],Blue,Line[{{x,y,z},{0,0,0}}]}],
      {n,50}]]],
  Boxed−>False,ViewPoint−>{−10,−30,0}]]
```

Out[49] =

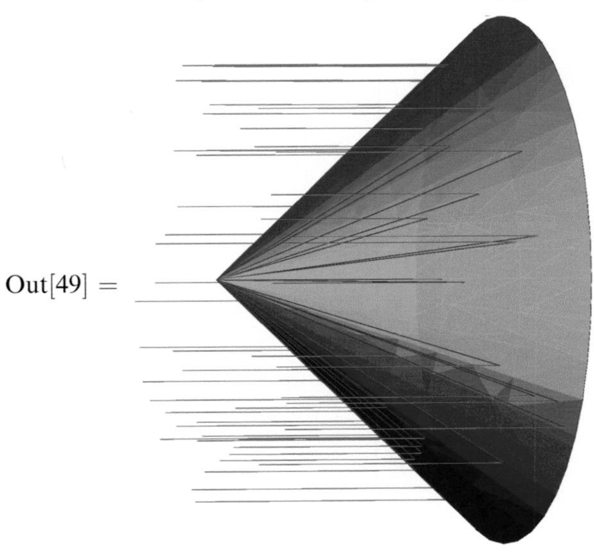

```
In[50] := Show[Import["rainbow.jpg"]]
```

Out[50] =

The scattering angle at $\rho = 1$ is

```
In[51] := ϑ2m = −ϑ2/.Snell/.ρ−>1
```

Out[51] $= 2\pi - 4\,\text{ArcSin}\left[\frac{1}{n}\right]$

```
In[52] := {ϑ2rw, ϑ2mw} = {ϑ2r, ϑ2m}/.Water;
```

Now we have to solve the equation $\alpha - 2\beta = (\vartheta - \pi)/2$ for ρ.

```
In[53] := eq2 = (TrigExpand[Sin[α − 2 ∗ β]]/.
      {Sin[α] → ρ, Cos[α] → Sqrt[1 − ρ^2],
         Sin[β] → ρ/n, Cos[β] → Sqrt[1 − (ρ/n)^2]}) == −Cos[ϑ/2]
```

Out[53] $= -\frac{\rho^3}{n^2} - \frac{2\rho\sqrt{1-\rho^2}\sqrt{1-\frac{\rho^2}{n^2}}}{n} + \rho\left(1 - \frac{\rho^2}{n^2}\right) == -\text{Cos}\left[\frac{\vartheta}{2}\right]$

```
In[54] := sol2 = Solve[eq2, ρ];
```

The solution number 3 is the smaller one; number 4 is the larger one (1 and 2 are negative).

```
In[55] := ρ2a = ρ/.sol2[[3]]; ρ2b = ρ/.sol2[[4]];
In[56] := Clear[eq2, sol2]
In[57] := p2a = ParametricPlot[{ϑ, ρ2a/.Water}, {ϑ, ϑ2r/.Water, π},
         PlotStyle → Blue];
      p2b = ParametricPlot[{ϑ, ρ2b/.Water}, {ϑ, ϑ2r/.Water, ϑ2m/.Water},
         PlotStyle → Red];
      Show[p2a, p2b, PlotRange → {{ϑ2r/.Water, π}, {0, 1}}]
```

Out[57] =

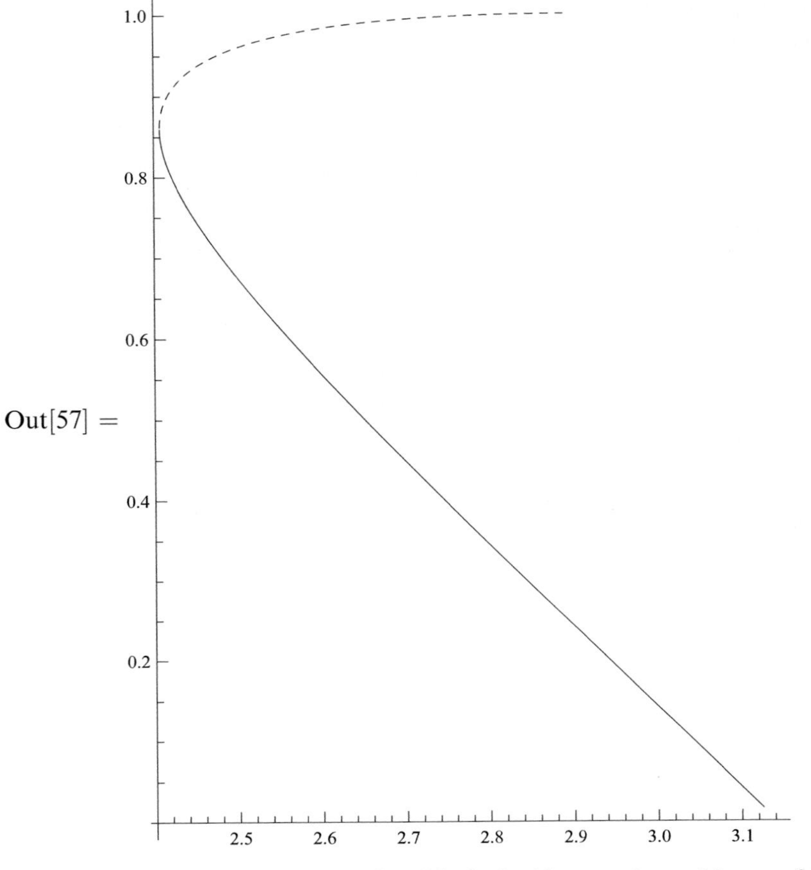

In order to scatter between ϑ and $\vartheta + \mathrm{d}\vartheta$, the incident ray has to hit one of the two rings, σ2a or (if $\vartheta < \vartheta$2m) σ2b.

```
In[58] := σ2a = −(D[ρ2a^2, ϑ]/(4 ∗ c2 ∗ s2)/. {Cos[ϑ/2] → c2, Sin[ϑ/2] → s2,
          Cos[ϑ] → c2^2 − s2^2, Sin[ϑ]−>2 ∗ c2 ∗ s2});
     σ2b = D[ρ2b^2, ϑ]/(4 ∗ c2 ∗ s2)/. {Cos[ϑ/2] → c2, Sin[ϑ/2] → s2,
          Cos[ϑ] → c2^2 − s2^2, Sin[ϑ]−>2 ∗ c2 ∗ s2};
```

The angles α and β in these two cases are

```
In[59] := Snell2a = {α−>ArcSin[ρ2a], β−>ArcSin[ρ2a/n]};
     Snell2b = {α−>ArcSin[ρ2b], β−>ArcSin[ρ2b/n]};
```

The differential cross sections for the two polarizations are

```
In[60] := σ2as[ϑ_] = σ2a ∗ (1 − Rs)^2 ∗ Rs/. Snell2a/. cs2/. Water;
     σ2ap[ϑ_] = σ2a ∗ (1 − Rp)^2 ∗ Rp/. Snell2a/. cs2/. Water;
     σ2bs[ϑ_] = σ2b ∗ (1 − Rs)^2 ∗ Rs/. Snell2b/. cs2/. Water;
     σ2bp[ϑ_] = σ2b ∗ (1 − Rp)^2 ∗ Rp/. Snell2b/. cs2/. Water;
In[61] := σ2s[ϑ_] := If[ϑ > ϑ2rw, σ2as[ϑ] + If[ϑ < ϑ2mw, σ2bs[ϑ], 0], 0];
     σ2p[ϑ_] := If[ϑ > ϑ2rw, σ2ap[ϑ] + If[ϑ < ϑ2mw, σ2bp[ϑ], 0], 0];
```

In[62] := Plot[{σ2s[ϑ], σ2p[ϑ], (σ2s[ϑ] + σ2p[ϑ])/2}, {ϑ, ϑ2rw + 0.001, π}, PlotRange−>{0, 1}, PlotStyle−>col]

Out[62] =

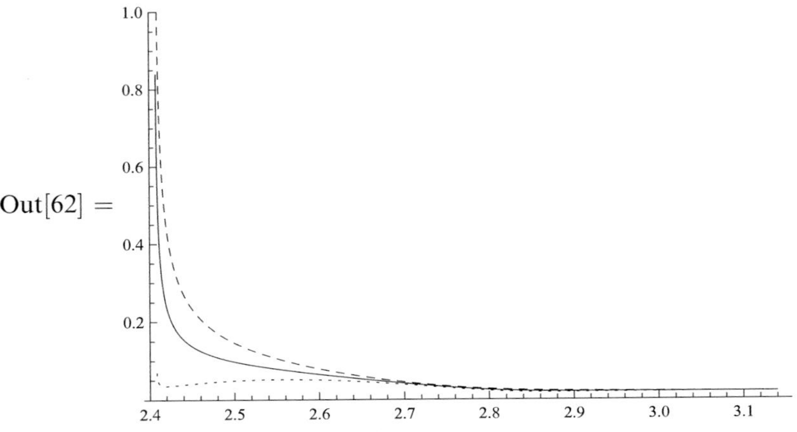

The considered contributions (0, 1, 2 ray segments inside the drop) to the cross sections with the s, p polarizations, as well as their sums, are

In[63] := {Plot[{σ0s[ϑ], σ1s[ϑ], σ2s[ϑ], σ0s[ϑ] + σ1s[ϑ] + σ2s[ϑ]}, {ϑ, 0, π}, PlotRange−>{0, 1}, PlotStyle−>col], Plot[{σ0p[ϑ], σ1p[ϑ], σ2p[ϑ], σ0p[ϑ] + σ1p[ϑ] + σ2p[ϑ]}, {ϑ, 0, π}, PlotRange−>{0, 1}, PlotStyle−>col]}

Out[63] =

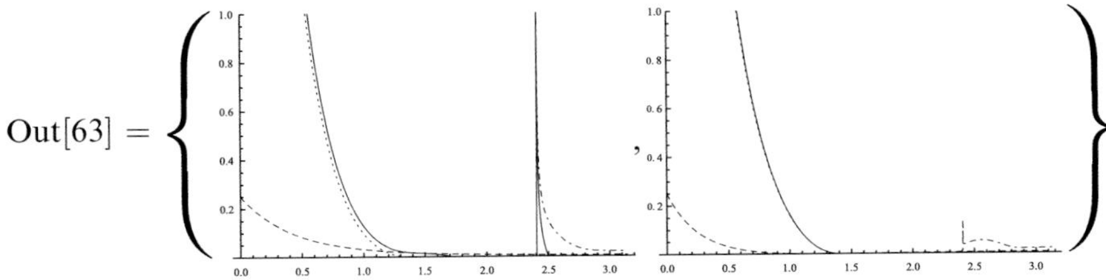

Of course, there are also higher contributions, not included here. The cross section is small for scattering angles below the rainbow peak; i.e., the sky just outside the rainbow is darker.

Near the rainbow peak, the last contribution (2 ray segments) is dominant; the ratio of the s and p polarizations is

In[64] := Rs2 = Simplify[TrigExpand[Rs/. Snell/. ρ−>ρ2r], n > 1];
Rp2 = Simplify[TrigExpand[Rp/. Snell/. ρ−>ρ2r], n > 1];
P2 = Simplify[(1 − Rs2)^2 ∗ Rs2/((1 − Rp2)^2 ∗ Rp2), n > 1]

$$\text{Out[64]} = \frac{(2+n^2)^6}{729n^4(-2+n^2)^2}$$

In[65] := P2/. Water

Out[65] = 25.3347

The rainbow light is highly linearly polarized, with the electric field orthogonal to the scattering plane, i.e., along the rainbow. The reason is that the incidence angle β of the ray which is about to reflect from the inner surface of the drop

```
In[66] := β/.Snell/.ρ−>ρ2r/.Water
Out[66] = 0.702055
```

is close to the Brewster angle

```
In[67] := βB/.Water
Out[67] = 0.643621
```

23.5 *L* Ray Segments Inside the Drop

Repeating the arguments from the previous sections, we obtain the direction of the outgoing ray $\vartheta\mathrm{L} = 2(\beta - \alpha) - (L-1)(\pi - 2\beta)$:

```
In[68] := ϑL = 2∗L∗β − 2∗α − (L − 1)∗π;
In[69] := Ray[L_,y_,a_] := With[{α = α/.Snell/.ρ−>y,
    β = β/.Snell/.ρ−>y/.Water},
  Module[{R = {col[[1]],Line[{{−1 − a,y},{−Sqrt[1 − y^2],y}}]},
      φ = π − α,φ1,ϑ = 2∗L∗β − 2∗α − (L − 1)∗π},
    Do[φ1 = φ + 2∗β − π;
      R = Join[R,{col[[m + 2]],Line[{{Cos[φ],Sin[φ]},{Cos[φ1],Sin[φ1]}}]}];
      φ = φ1,{m,L}];
    Join[R,{col[[2]],
      Line[{{Cos[φ],Sin[φ]},{Cos[φ] + a∗Cos[ϑ],Sin[φ] + a∗Sin[ϑ]}}]}]]]
In[70] := ϑ[L_,ρ_] = ϑL/.Snell/.Water;
In[71] := Manipulate[{
  Manipulate[
    Graphics[Join[{Black,Circle[]},Ray[L,y,a]],
      PlotRange−>{{−1 − a,1 + a},{−1 − a,1 + a}}],
    {{y,0.6},0,1},{{a,2},1,10}],
  Manipulate[
    Graphics[Join[{Black,Circle[]},
      With[{δ = (ρmax − ρmin)/M},
        If[δ > 0,Apply[Join,Table[Ray[L,y,a],{y,ρmin + δ/2,ρmax,δ}]],{}]]],
      PlotRange−>{{−1 − a,1 + a},{−1 − a,1 + a}}],
    {{ρmin,0},0,1},{{ρmax,1},0,1},
    {{M,10},Table[i,{i,30}]},{{a,2},1,10}],
  Plot[Abs[Mod[ϑ[L,ρ],2∗π,−π]],{ρ,0,1}]},
  {{L,3},Table[i,{i,0,4}]}]
```

0 1 2 3 4

y

a

Out[71] =

ρmin

ρmax

M 10

a

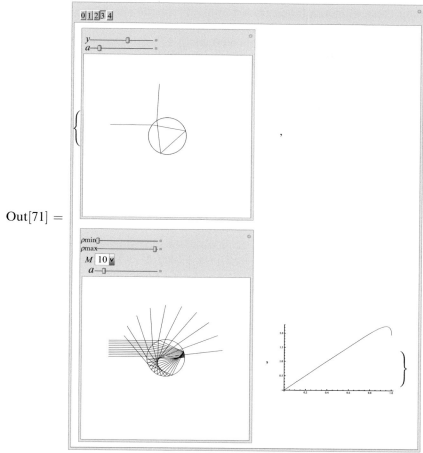

Scattering angles for $L = 0, 1, 2, 3, 4$ as functions of ρ are

In[72] := Plot[Evaluate[Table[Abs[Mod[ϑ[L,ρ], 2 * π, −π]], {L,0,4}]], {ρ,0,1}, PlotStyle−>col]

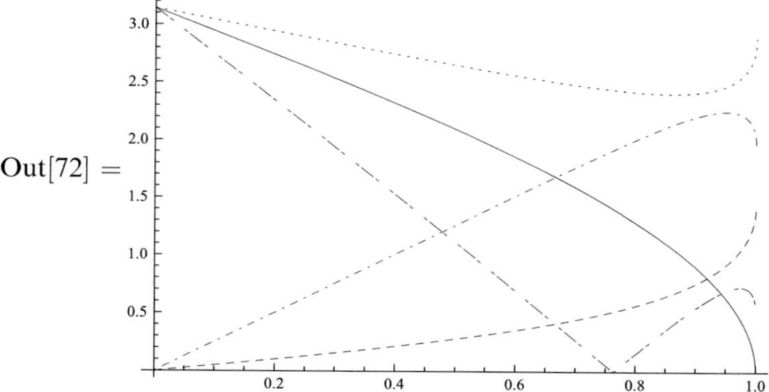

Naturally, those with even L start from π at $\rho = 0$; for odd L they start from 0. For $L \geqslant 2$ each one has a single extremum:

In[73] := sol = Solve[D[ϑL/.Snell, ρ] == 0, ρ]

$$\text{Out[73]} = \left\{\left\{\rho \to -\frac{\sqrt{L^2-n^2}}{\sqrt{-1+L^2}}\right\}, \left\{\rho \to \frac{\sqrt{L^2-n^2}}{\sqrt{-1+L^2}}\right\}\right\}$$

In[74] := ρL = ρ/.sol[[2]]

$$\text{Out[74]} = \frac{\sqrt{L^2-n^2}}{\sqrt{-1+L^2}}$$

In[75] := ϑr = ϑL/.Snell/.ρ->ρL

$$\text{Out[75]} = -(-1+L)\pi - 2\,\text{ArcSin}\left[\frac{\sqrt{L^2-n^2}}{\sqrt{-1+L^2}}\right] + 2L\,\text{ArcSin}\left[\frac{\sqrt{L^2-n^2}}{\sqrt{-1+L^2}n}\right]$$

These extrema produce rainbows at the angles (in degrees)

In[76] := Table[(π − Abs[Mod[ϑr/.Water, 2 ∗ π, −π]])/Degree, {L, 2, 4}]

Out[76] = {42.0781, 50.8908, 138.263}

The angles at which an observer sees the first and the second rainbow are

In[77] := {ϑr2, ϑr3} = {π + (ϑr/.L−>2), −π − (ϑr/.L−>3)}

$$\text{Out[77]} = \left\{-2\,\text{ArcSin}\left[\frac{\sqrt{4-n^2}}{\sqrt{3}}\right] + 4\,\text{ArcSin}\left[\frac{\sqrt{4-n^2}}{\sqrt{3}n}\right],\right.$$
$$\left.\pi + 2\,\text{ArcSin}\left[\frac{\sqrt{9-n^2}}{2\sqrt{2}}\right] - 6\,\text{ArcSin}\left[\frac{\sqrt{9-n^2}}{2\sqrt{2}n}\right]\right\}$$

The sky is somewhat darker between the first rainbow and the second one, because neither rays with $L = 2$ nor those with $L = 3$ come from these directions.

Until now, we discussed monochromatic light. Then each rainbow is just a bright arc of the same color. In fact, the refraction index of water n depends on the color (wavelength) of light (dispersion). It is larger for violet light than for red one. Therefore the positions of the maxima of intensity of the scattered light also depend on the color.

In[78] := Plot[{ϑr2/Degree, ϑr3/Degree}, {n, 1.325, 1.335}, PlotStyle−>col]

Out[78] =

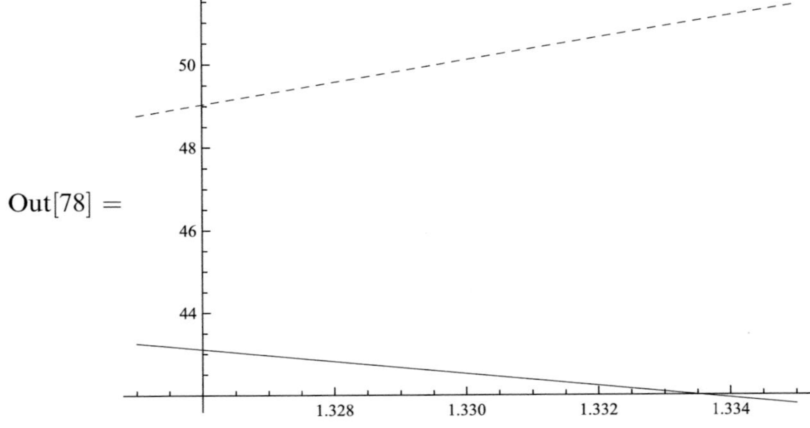

We see that the order of colors in the second rainbow is opposite to that in the first one; and the second rainbow is wider.

The ratios of s and p polarizations at the rainbow peaks are

In[79] := Rs0 = Simplify[TrigExpand[Rs/. Snell/. ρ−>ρL], {n > 1, L > 1}]

$$\text{Out[79]} = \frac{(-1+L)^2}{(1+L)^2}$$

In[80] := Rp0 = Simplify[TrigExpand[Rp/. Snell/. ρ−>ρL], {n > 1, L > 1}]

$$\text{Out[80]} = \frac{\left(L-n^2\right)^2}{\left(L+n^2\right)^2}$$

In[81] := P = Simplify[(1 − Rs0)^2 ∗ Rs0^(L − 1)/((1 − Rp0)^2 ∗ Rp0^(L − 1)), {n > 1, L > 1}]

$$\text{Out[81]} = \frac{\left(\frac{(-1+L)^2\left(L+n^2\right)^2}{(1+L)^2\left(L-n^2\right)^2}\right)^L \left(L^2-n^4\right)^2}{\left(-1+L^2\right)^2 n^4}$$

In[82] := Table[P/. Water, {L, 2, 4}]

Out[82] = {25.3347, 9.36736, 8.10737}

Higher rainbows are not so strongly polarized as the first one, because the incidence angle for the reflections inside the drop is not so close to the Brewster angle.

Chapter 24
Cyclohexane

24.1 Statement of the Problem

Cyclohexane molecule contains 6 C atoms connected cyclically by single chemical bonds; 2 H atoms are attached to each carbon one. Single bonds of a C atom have a fixed length and form fixed angles: if the C atom is put to the center of a tetrahedron, then its single bonds point to its vertices. The problem is: how many geometrical configurations (conformations, as chemists call them) of the cyclohexane molecule exist—one, several, or infinitely many (and if so, what is the dimensionality of this set).

Tetrahedron

What is the angle between single bonds? The unit vectors $a[1], \ldots\ a[4]$ (blue) are directed from the center of the tetrahedron (red) to its vertices.

```
In[1] := a[0] = {0,0,0}; a[1] = {0,0,1}; a[2] = {2 * Sqrt[2],0,−1}/3;
    a[3] = {−Sqrt[2],Sqrt[6],−1}/3; a[4] = {−Sqrt[2],−Sqrt[6],−1}/3;
In[2] := rc = 0.025; rs = 0.1;
In[3] := Graphics3D[{Blue,Cylinder[{a[0],a[1]},rc],Cylinder[{a[0],a[2]},rc],
        Cylinder[{a[0],a[3]},rc],Cylinder[{a[0],a[4]},rc],
        Red,Cylinder[{a[1],a[2]},rc],Cylinder[{a[1],a[3]},rc],
        Cylinder[{a[1],a[4]},rc],Cylinder[{a[2],a[3]},rc],
        Cylinder[{a[2],a[4]},rc],Cylinder[{a[3],a[4]},rc],
        Sphere[a[1],rs],Sphere[a[2],rs],Sphere[a[3],rs],Sphere[a[4],rs]},
    Boxed−>False, ViewPoint−>{10,10,4}]
```

A. Grozin, *Introduction to Mathematica® for Physicists*, Graduate Texts in Physics,
DOI 10.1007/978-3-319-00894-3_24,

Out[3] =

Let's check: the vectors are indeed unit and form equal angles with each other; hence all edges have equal lengths, so this is indeed a tetrahedron.

In[4] := MatrixForm[Table[a[i].a[j], {i, 1, 4}, {j, 1, 4}]]

Out[4]//MatrixForm =

$$\begin{pmatrix} 1 & -\frac{1}{3} & -\frac{1}{3} & -\frac{1}{3} \\ -\frac{1}{3} & 1 & -\frac{1}{3} & -\frac{1}{3} \\ -\frac{1}{3} & -\frac{1}{3} & 1 & -\frac{1}{3} \\ -\frac{1}{3} & -\frac{1}{3} & -\frac{1}{3} & 1 \end{pmatrix}$$

Cosine of the angle between these vectors is $-1/3$.

In[5] := Clear[a, rc, rs]

24.2 First Steps

Thus we have 6 vectors $a[1], \ldots, a[6]$ drawn from each C atom to the next one. They form a closed hexagon: $a[1] + \cdots + a[6] = 0$. Let the length of a single C–C bond be 1, then all the vectors are unit. Scalar products of neighboring vectors are $1/3$ (the sign has changed because now one of the vectors points in, not out). In order not to deal with the overall orientation of the molecule, let's consider invariant quantities—cosines of the angles $c[i, j] = a[i].a[j]$.

In[6] := c[i_, j_]/; $i > j$:= c[j, i]

In[7] := Do[c[i, i] = 1, {i, 1, 6}]

In[8] := Do[c[i, i + 1] = 1/3, {i, 1, 5}]; c[1, 6] = 1/3;

In[9] := MatrixForm[M = Array[c, {6,6}]]
Out[9]//MatrixForm =

$$\begin{pmatrix} 1 & \frac{1}{3} & c[1,3] & c[1,4] & c[1,5] & \frac{1}{3} \\ \frac{1}{3} & 1 & \frac{1}{3} & c[2,4] & c[2,5] & c[2,6] \\ c[1,3] & \frac{1}{3} & 1 & \frac{1}{3} & c[3,5] & c[3,6] \\ c[1,4] & c[2,4] & \frac{1}{3} & 1 & \frac{1}{3} & c[4,6] \\ c[1,5] & c[2,5] & c[3,5] & \frac{1}{3} & 1 & \frac{1}{3} \\ \frac{1}{3} & c[2,6] & c[3,6] & c[4,6] & \frac{1}{3} & 1 \end{pmatrix}$$

Linear Equations

Multiplying the vector equality $a[1]+\cdots+a[6]=0$ by each vector $a[i]$, we get 6 linear equations for $c[i,j]$.

In[10] := Eq = Table[Sum[$c[i,j]$, {j, 1, 6}] == 0, {i, 1, 6}]

$$\text{Out}[10] = \left\{\frac{5}{3}+c[1,3]+c[1,4]+c[1,5]==0, \frac{5}{3}+c[2,4]+c[2,5]+c[2,6]==0, \frac{5}{3}+c[1,3]+c[3,5]+c[3,6]==0, \frac{5}{3}+c[1,4]+c[2,4]+c[4,6]==0, \frac{5}{3}+c[1,5]+c[2,5]+c[3,5]==0, \frac{5}{3}+c[2,6]+c[3,6]+c[4,6]==0\right\}$$

Let's take $x=c[1,3]$, $y=c[3,5]$, $z=c[5,1]$ as independent variables and express the remaining ones via them.

In[11] := $c[1,3] = x$; $c[3,5] = y$; $c[1,5] = z$;
In[12] := s = Solve[Eq, {$c[1,4], c[2,4], c[2,5], c[2,6], c[3,6], c[4,6]$}][[1]]

$$\text{Out}[12] = \left\{c[1,4]\to -\frac{5}{3}-x-z, c[2,4]\to z, c[2,5]\to -\frac{5}{3}-y-z, c[2,6]\to y, c[3,6]\to -\frac{5}{3}-x-y, c[4,6]\to x\right\}$$

**In[13] := $c[1,4] = c[1,4]/.s$; $c[2,4] = c[2,4]/.s$; $c[2,5] = c[2,5]/.s$;
$c[2,6] = c[2,6]/.s$; $c[3,6] = c[3,6]/.s$; $c[4,6] = c[4,6]/.s$;**
In[14] := Clear[s]
In[15] := MatrixForm[$M = M$]
Out[15]//MatrixForm =

$$\begin{pmatrix} 1 & \frac{1}{3} & x & -\frac{5}{3}-x-z & z & \frac{1}{3} \\ \frac{1}{3} & 1 & \frac{1}{3} & z & -\frac{5}{3}-y-z & y \\ x & \frac{1}{3} & 1 & \frac{1}{3} & y & -\frac{5}{3}-x-y \\ -\frac{5}{3}-x-z & z & \frac{1}{3} & 1 & \frac{1}{3} & x \\ z & -\frac{5}{3}-y-z & y & \frac{1}{3} & 1 & \frac{1}{3} \\ \frac{1}{3} & y & -\frac{5}{3}-x-y & x & \frac{1}{3} & 1 \end{pmatrix}$$

24.3 Equations

Generating Combinations

Now let's recall that our vectors $a[1], \ldots, a[6]$ live in 3-dimensional space. Any 4 of them are linearly dependent. First we have to find a way to generate all 4-element lists made of the numbers from 1 to 6 in the increasing order. We shall accumulate them in the list L. The main work is done by the recursive function Gen. Its parameters: l—part of the list which has been already constructed; n—how many elements are to be added; a and b—boundaries of the interval from which numbers can be taken. If everything has been done ($n = 0$), the constructed list l is appended to the list of results L. Otherwise, we add each number i from the allowed interval to l and call Gen recursively to add $n-1$ numbers. The upper limit of the loop is determined by the requirement to have at least $n-1$ numbers in the interval from $i+1$ to b.

In[16] := L = {};
In[17] := Gen[l_, n_, a_, b_] := If[n ≤ 0, L = Append[L, l],
Do[Gen[Append[l, i], n − 1, i + 1, b], {i, a, b − n + 1}]]
In[18] := Gen[{}, 4, 1, 6]; L
Out[18] = {{1,2,3,4},{1,2,3,5},{1,2,3,6},{1,2,4,5},{1,2,4,6},{1,2,5,6},
{1,3,4,5},{1,3,4,6},{1,3,5,6},{1,4,5,6},{2,3,4,5},{2,3,4,6},
{2,3,5,6},{2,4,5,6},{3,4,5,6}}

It works. There are 15 4-combinations of 6 numbers.

Nonlinear Equations

For each set of 4 vectors, the determinant of the corresponding 4×4 submatrix of the matrix M (it is the square of the 4-dimensional volume spanned by these vectors) should be equal to 0.

In[19] := Eq = Table[Det[M[[l, l]]], {l, L}]

$$\text{Out}[19] = \left\{ -\frac{5}{3} - \frac{34x}{9} - \frac{23x^2}{9} - \frac{34z}{9} - \frac{20xz}{9} + \frac{2x^2z}{3} - \frac{23z^2}{9} + \frac{2xz^2}{3} + x^2z^2, \right.$$

$$-2 + \frac{2x}{9} + \frac{16x^2}{9} - \frac{40y}{9} + \frac{10xy}{9} + \frac{10x^2y}{3} - \frac{23y^2}{9} + \frac{2xy^2}{3} + x^2y^2 - \frac{40z}{9} +$$

$$\frac{10xz}{9} + \frac{10x^2z}{3} - \frac{32yz}{9} + \frac{10xyz}{3} + 2x^2yz - \frac{23z^2}{9} + \frac{2xz^2}{3} + x^2z^2,$$

$$-\frac{5}{3} - \frac{34x}{9} - \frac{23x^2}{9} - \frac{34y}{9} - \frac{20xy}{9} + \frac{2x^2y}{3} - \frac{23y^2}{9} + \frac{2xy^2}{3} + x^2y^2,$$

$$\frac{7}{3} + \frac{50x}{9} + \frac{16x^2}{9} + \frac{50y}{9} + \frac{98xy}{9} + \frac{10x^2y}{3} + \frac{16y^2}{9} + \frac{10xy^2}{3} + x^2y^2 + \frac{20z}{3} +$$

$$\frac{118xz}{9} + \frac{10x^2z}{3} + \frac{118yz}{9} + \frac{40xyz}{3} + 2x^2yz + \frac{10y^2z}{3} + 2xy^2z + 4z^2 +$$

$$\frac{20xz^2}{3}+x^2z^2+\frac{20yz^2}{3}+2xyz^2+y^2z^2,$$

$$-2-\frac{40x}{9}-\frac{23x^2}{9}+\frac{2y}{9}+\frac{10xy}{9}+\frac{2x^2y}{3}+\frac{16y^2}{9}+\frac{10xy^2}{3}+x^2y^2-\frac{40z}{9}-\frac{32xz}{9}+\frac{10yz}{9}+\frac{10xyz}{3}+\frac{10y^2z}{3}+2xy^2z-\frac{23z^2}{9}+\frac{2yz^2}{3}+y^2z^2,$$

$$-\frac{5}{3}-\frac{34y}{9}-\frac{23y^2}{9}-\frac{34z}{9}-\frac{20yz}{9}+\frac{2y^2z}{3}-\frac{23z^2}{9}+\frac{2yz^2}{3}+y^2z^2,$$

$$-2-\frac{40x}{9}-\frac{23x^2}{9}+\frac{2y}{9}+\frac{10xy}{9}+\frac{2x^2y}{3}+\frac{16y^2}{9}+\frac{10xy^2}{3}+x^2y^2-\frac{40z}{9}-\frac{32xz}{9}+\frac{10yz}{9}+\frac{10xyz}{3}+\frac{10y^2z}{3}+2xy^2z-\frac{23z^2}{9}+\frac{2yz^2}{3}+y^2z^2,$$

$$\frac{7}{3}+\frac{20x}{3}+4x^2+\frac{50y}{9}+\frac{118xy}{9}+\frac{20x^2y}{3}+\frac{16y^2}{9}+\frac{10xy^2}{3}+x^2y^2+\frac{50z}{9}+\frac{118xz}{9}+\frac{20x^2z}{3}+\frac{98yz}{9}+\frac{40xyz}{3}+2x^2yz+\frac{10y^2z}{3}+2xy^2z+\frac{16z^2}{9}+\frac{10xz^2}{3}+x^2z^2+\frac{10yz^2}{3}+2xyz^2+y^2z^2,$$

$$-2-\frac{40x}{9}-\frac{23x^2}{9}-\frac{40y}{9}-\frac{32xy}{9}-\frac{23y^2}{9}+\frac{2z}{9}+\frac{10xz}{9}+\frac{2x^2z}{3}+\frac{10yz}{9}+\frac{10xyz}{3}+\frac{2y^2z}{3}+\frac{16z^2}{9}+\frac{10xz^2}{3}+x^2z^2+\frac{10yz^2}{3}+2xyz^2+y^2z^2,$$

$$-\frac{5}{3}-\frac{34x}{9}-\frac{23x^2}{9}-\frac{34z}{9}-\frac{20xz}{9}+\frac{2x^2z}{3}-\frac{23z^2}{9}+\frac{2xz^2}{3}+x^2z^2,$$

$$-\frac{5}{3}-\frac{34y}{9}-\frac{23y^2}{9}-\frac{34z}{9}-\frac{20yz}{9}+\frac{2y^2z}{3}-\frac{23z^2}{9}+\frac{2yz^2}{3}+y^2z^2,$$

$$-2-\frac{40x}{9}-\frac{23x^2}{9}-\frac{40y}{9}-\frac{32xy}{9}-\frac{23y^2}{9}+\frac{2z}{9}+\frac{10xz}{9}+\frac{2x^2z}{3}+\frac{10yz}{9}+\frac{10xyz}{3}+\frac{2y^2z}{3}+\frac{16z^2}{9}+\frac{10xz^2}{3}+x^2z^2+\frac{10yz^2}{3}+2xyz^2+y^2z^2,$$

$$\frac{7}{3}+\frac{50x}{9}+\frac{16x^2}{9}+\frac{20y}{3}+\frac{118xy}{9}+\frac{10x^2y}{3}+4y^2+\frac{20xy^2}{3}+x^2y^2+\frac{50z}{9}+\frac{98xz}{9}+\frac{10x^2z}{3}+\frac{118yz}{9}+\frac{40xyz}{3}+2x^2yz+\frac{20y^2z}{3}+2xy^2z+\frac{16z^2}{9}+\frac{10xz^2}{3}+x^2z^2+\frac{10yz^2}{3}+2xyz^2+y^2z^2,$$

$$-2+\frac{2x}{9}+\frac{16x^2}{9}-\frac{40y}{9}+\frac{10xy}{9}+\frac{10x^2y}{3}-\frac{23y^2}{9}+\frac{2xy^2}{3}+x^2y^2-\frac{40z}{9}+\frac{10xz}{9}+\frac{10x^2z}{3}-\frac{32yz}{9}+\frac{10xyz}{3}+2x^2yz-\frac{23z^2}{9}+\frac{2xz^2}{3}+x^2z^2,$$

$$-\frac{5}{3}-\frac{34x}{9}-\frac{23x^2}{9}-\frac{34y}{9}-\frac{20xy}{9}+\frac{2x^2y}{3}-\frac{23y^2}{9}+\frac{2xy^2}{3}+x^2y^2\Big\}$$

These 15 polynomials of 3 variables must be equal to 0.

In[20] := Clear[*L*]

Gröbner Basis

Let's find a simpler set of polynomials having the same solution set—the Gröbner basis.

In[21] := GB = GroebnerBasis[Eq, {z, y, x}]

Out[21] = $\{-15-34x-23x^2-34y-20xy+6x^2y-23y^2+6xy^2+9x^2y^2,$
$-102-400x-284x^2+18x^4-69y-212xy+18x^2y+108x^3y+27x^4y-$
$69z-212xz+18x^2z+108x^3z+27x^4z,$
$18+54x+18x^2-6x^3+21y+41xy-9x^2y-9x^3y+21z+41xz-$
$9x^2z-9x^3z+20yz,$
$-15-34x-23x^2-34z-20xz+6x^2z-23z^2+6xz^2+9x^2z^2\}$

Do they factorize?

In[22] := GB = Map[Factor, GB]

Out[22] = $\{-15-34x-23x^2-34y-20xy+6x^2y-23y^2+6xy^2+9x^2y^2,$
$(3+x)(1+3x)\left(-34-20x+6x^2-23y+6xy+9x^2y-23z+6xz+9x^2z\right),$
$18+54x+18x^2-6x^3+21y+41xy-9x^2y-9x^3y+21z+41xz-9x^2z-$
$9x^3z+20yz,$
$-15-34x-23x^2-34z-20xz+6x^2z-23z^2+6xz^2+9x^2z^2\}$

In[23] := p1 = GB[[1]]; p2 = GB[[2]]/(3 + x)/(1 + 3 * x);
p3 = GB[[3]]; p4 = GB[[4]];

24.4 Projection onto the x, y Plane

Allowed Region

First let's find the projection of the solution set onto the x, y plane. What part of this plane are we interested in? First, x and y should lie between -1 and 1; second,

In[24] := c[3, 6]

Out[24] = $-\frac{5}{3}-x-y$

should also lie between -1 and 1.

In[25] := RegionPlot[$-1\le x\le 1$&& $-1\le y\le 1$&& $-1\le c[3,6]\le 1$,
{x, −1, 1}, {y, −1, 1}]

Out[25] =

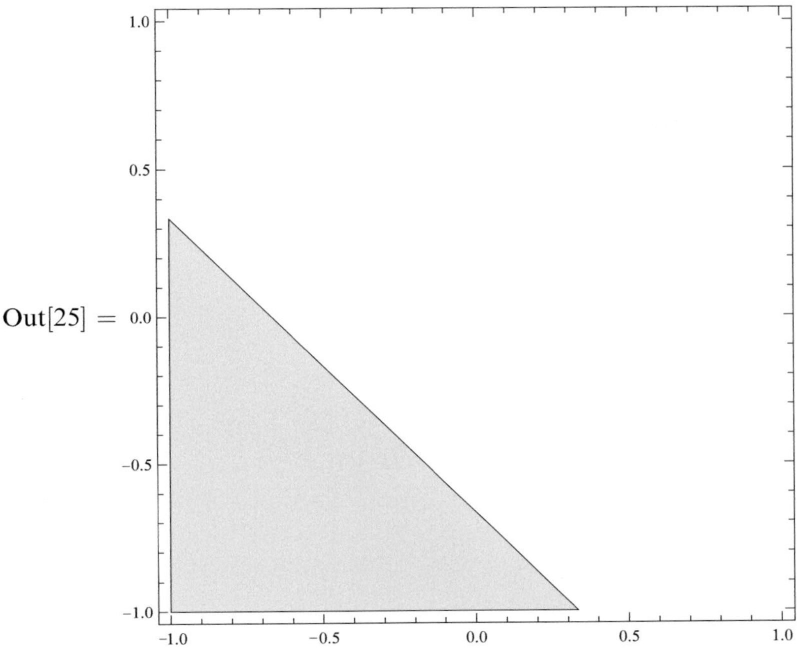

That is, the allowed region is the triangle with the vertices $(-1,-1)$, $(-1,1/3)$, and $(1/3,-1)$.

Solutions with $x = -1/3$

The second equation is satisfied at $x = -1/3$. What about the other ones?

In[26] := Eq = {p1,p3,p4}/.x−> − 1/3

Out[26] = $\left\{-\frac{56}{9}-\frac{80y}{3}-24y^2, \frac{20}{9}+\frac{20y}{3}+\frac{20z}{3}+20yz, -\frac{56}{9}-\frac{80z}{3}-24z^2\right\}$

In[27] := GB = GroebnerBasis[Eq, {z,y}]

Out[27] = $\{7+30y+27y^2, 1+3y+3z+9yz, 7+30z+27z^2\}$

In[28] := GB = Map[Factor,GB]

Out[28] = $\{(1+3y)(7+9y), (1+3y)(1+3z), (1+3z)(7+9z)\}$

So, we have found the solutions $x = y = z = -1/3$; $x = y = -1/3$, $z = -7/9$; and $x = z = -1/3$, $y = -7/9$.

In[29] := Eq/.{y−> − 1/3,z−> − 1/3}

Out[29] = {0,0,0}

In[30] := Eq/.{y−> − 1/3,z−> − 7/9}

Out[30] = {0,0,0}

In[31] := Eq/.{y−> −7/9,z−> −1/3}
Out[31] = {0,0,0}
It is clear from the symmetry argument that $y = z = -1/3$, $x = -7/9$ is also a solution.
In[32] := {p1,p2,p3,p4}/.{x−> −7/9,y−> −1/3,z−> −1/3}
Out[32] = {0,0,0,0}

Other Solutions

The first equation contains only x and y.
In[33] := p1
Out[33] = $-15 - 34x - 23x^2 - 34y - 20xy + 6x^2y - 23y^2 + 6xy^2 + 9x^2y^2$
In[34] := P1 = ContourPlot[p1 == 0, {x, −1, 0}, {y, −1, 0}]

Out[34] =

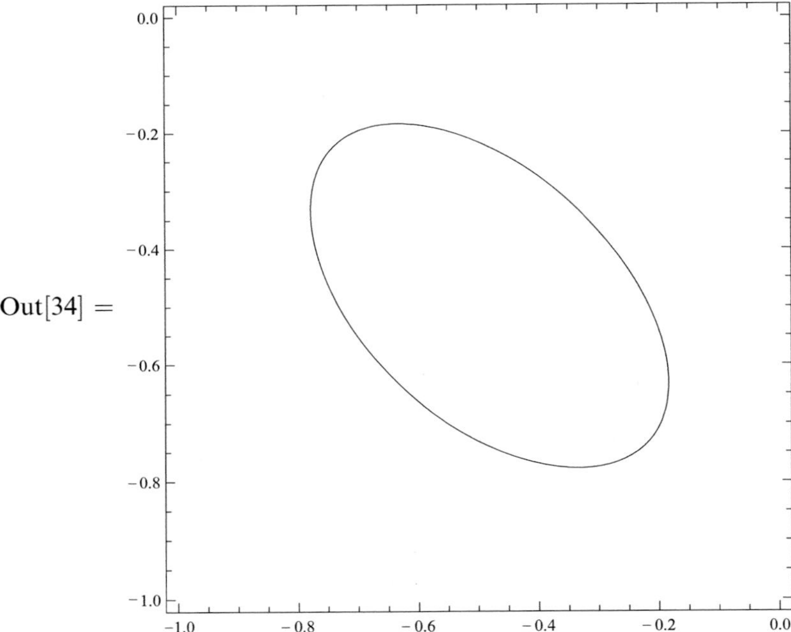

This equation is quadratic in y.
In[35] := Do[c[i] = Coefficient[p1, y, i], {i, 0, 2}]
Does $c[2]$ vanish somewhere?
In[36] := s = Solve[c[2] == 0, x]
Out[36] = $\left\{\left\{x \to \frac{1}{3}\left(-1 - 2\sqrt{6}\right)\right\}, \left\{x \to \frac{1}{3}\left(-1 + 2\sqrt{6}\right)\right\}\right\}$

In[37] := N[x/.s]
Out[37] = {−1.96633, 1.29966}
In our region $c[2] > 0$. The discriminant:
In[38] := d = Factor[c[1]^2 − 4 * c[0] * c[2]]
Out[38] = $32(-1+x)(7+9x)\left(1+6x+3x^2\right)$
In[39] := s = Solve[d == 0, x]
Out[39] = $\left\{\left\{x \to -\frac{7}{9}\right\}, \{x \to 1\}, \left\{x \to \frac{1}{3}\left(-3-\sqrt{6}\right)\right\}, \left\{x \to \frac{1}{3}\left(-3+\sqrt{6}\right)\right\}\right\}$
In[40] := N[x/.s]
Out[40] = {−0.777778, 1, −1.8165, −0.183503}
The discriminant is positive between $x = -7/9$ and
In[41] := xmax = x/.s[[4]]
Out[41] = $\frac{1}{3}\left(-3+\sqrt{6}\right)$
In[42] := Clear[s]
Two values of y (with pm = ±1) correspond to each x from this interval.
In[43] := y1 = (−c[1] + pm * Sqrt[d])/(2 * c[2])
Out[43] = $\left(34+20x-6x^2+4\sqrt{2}\text{pm}\sqrt{(-1+x)(7+9x)\left(1+6x+3x^2\right)}\right)/\left(2\left(-23+6x+9x^2\right)\right)$
In[44] := Clear[d]
Do the points we found earlier lie on this curve?
In[45] := {p1/.{x−> −1/3, y−> −1/3}, p1/.{x−> −1/3, y−> −7/9}, p1/.{x−> −7/9, y−> −1/3}}
Out[45] = {0, 0, 0}
Yes, they do. Let's denote these points A, B, C.
In[46] := pA = {−1/3, −1/3}; pB = {−1/3, −7/9}; pC = {−7/9, −1/3};
So, all solutions we are interested in project onto this curve in the x, y plane.

We shall need a few extra points on this curve. Let's find the second intersection with the diagonal $x = y$ (the first one is the point A).
In[47] := s = Solve[(p1/.y−>x) == 0, x]
Out[47] = $\left\{\{x \to -3\}, \left\{x \to -\frac{1}{3}\right\}, \left\{x \to \frac{1}{3}\left(3-2\sqrt{6}\right)\right\}, \left\{x \to \frac{1}{3}\left(3+2\sqrt{6}\right)\right\}\right\}$
In[48] := N[x/.s]
Out[48] = {−3., −0.333333, −0.632993, 2.63299}
In[49] := x0 = x/.s[[3]]
Out[49] = $\frac{1}{3}\left(3-2\sqrt{6}\right)$
Let's call this point D.
In[50] := pD = {x0, x0};
In[51] := Clear[s, x0]

Finally, let's introduce some additional point between A and B (rather arbitrarily; e.g., let it have $x = -1/4$) and call it E. Let its mirror image be the point F.

In[52] := x0 = −1/4;

In[53] := y0 = Simplify[y1/.{x−>x0, pm−>1}]

$$\text{Out[53]} = \frac{1}{383}\left(-229 - 10\sqrt{38}\right)$$

In[54] := pE = {x0, y0}; pF = {y0, x0};

In[55] := Clear[x0, y0]

This is shown in the plot.

```
In[56] := P2 = Graphics[{PointSize[Large],
    Red, Point[pA], Text[Style[A, Large], pA, {−1, −1}],
    Point[pB], Text[Style[B, Large], pB, {0, 1}],
    Point[pC], Text[Style[C, Large], pC, {1, 0}],
    Darker[Green], Point[pD], Text[Style[D, Large], pD, {1, 1}],
    Point[pE], Text[Style[E, Large], pE, {−1, 1}],
    Point[pF], Text[Style[F, Large], pF, {1, −1}]}];
In[57] := Show[P1, P2]
```

Out[57] =

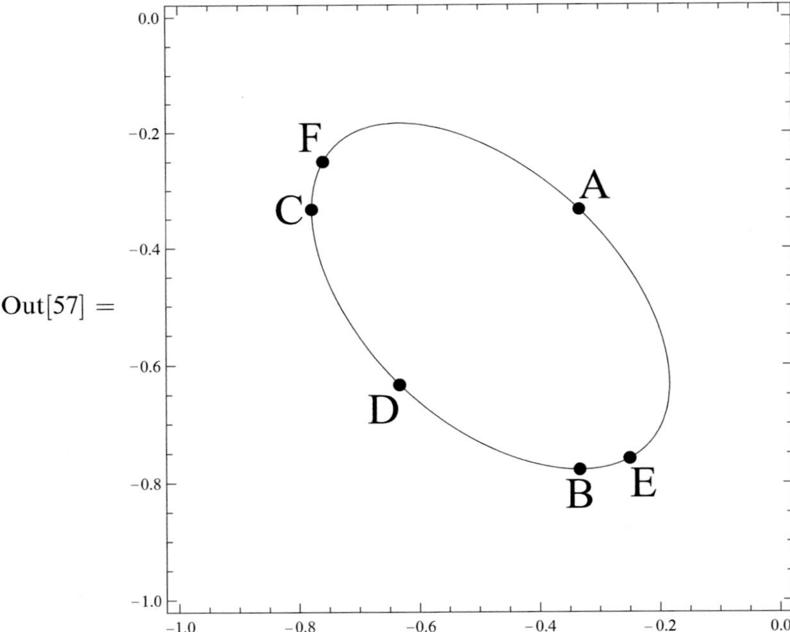

What's the reason for introducing the points D, E, F? Now it is easy to write down our curve parametrically. For $t \in [0, 1]$ let the point move from F to A, the motion along x being uniform.

```
In[58] := pFA[t_] := With[{xt = (1 − t) * pF[[1]] + t * pA[[1]]},
    {xt, y1/.{x−>xt, pm−> − 1}}]
In[59] := P1 = ParametricPlot[pFA[t], {t, 0, 1}, PlotRange−>{{−1, 0}, {−1, 0}}];
```

In[60] := Show[P1, P2]

Out[60] =

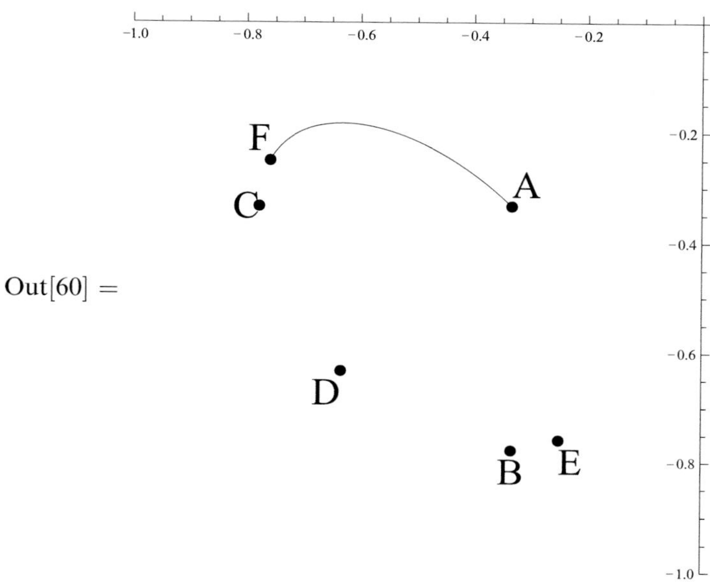

For $t \in [1,2]$ let the point move from A to E; this segment is mirror-symmetric to the previous one.

In[61] := pAE[t_] := With[{y0 = (2 − t) ∗ pA[[2]] + (t − 1) ∗ pE[[2]]},
{y1/.{x−>y0, pm−> − 1}, y0}]
In[62] := P1 = ParametricPlot[pAE[t], {t, 1, 2}, PlotRange−>{{−1, 0}, {−1, 0}}];
In[63] := Show[P1, P2]

Out[63] =

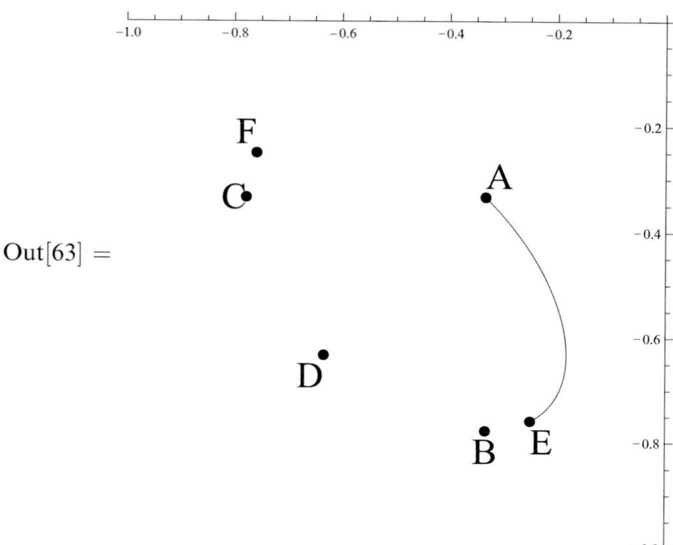

For $t \in [2,3]$ the point moves from E to D, the motion along x being uniform.

```
In[64] := pED[t_] := With[{x0 = (3 − t) ∗ pE[[1]] + (t − 2) ∗ pD[[1]]},
      {x0, y1/.{x−>x0, pm−>1}}]
In[65] := P1 = ParametricPlot[pED[t], {t, 2, 3}, PlotRange−>{{−1,0}, {−1,0}}];
In[66] := Show[P1, P2]
```

Out[66] =

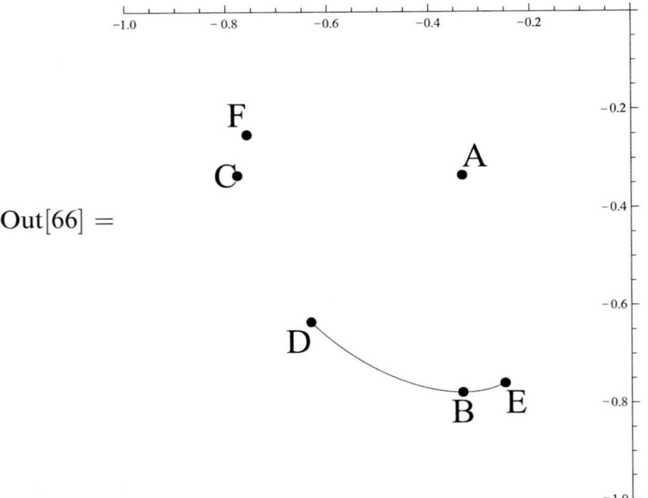

Finally, for $t \in [3,4]$ the point moves from D to F; this segment is mirror-symmetric to the previous one.

```
In[67] := pDF[t_] := With[{y0 = (4 − t) ∗ pD[[2]] + (t − 3) ∗ pF[[2]]},
      {y1/.{x−>y0, pm−>1}, y0}]
In[68] := P1 = ParametricPlot[pDF[t], {t, 3, 4}, PlotRange−>{{−1,0}, {−1,0}}];
In[69] := Show[P1, P2]
```

Out[69]=

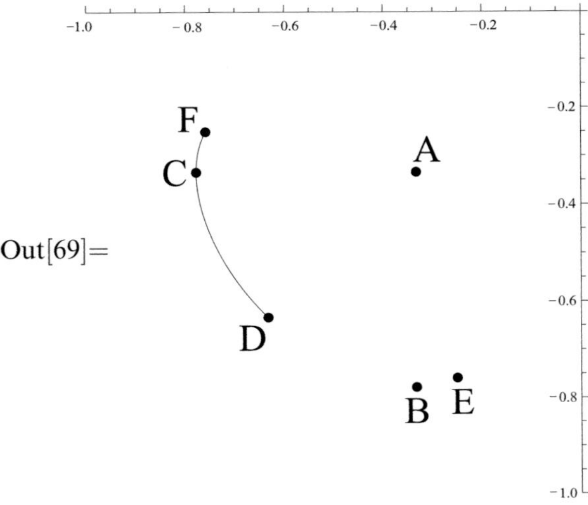

Later we shall join these segments and construct a parametric curve in the 3-dimensional space x, y, z.

24.5 Complete Analysis of the Solutions

How to find the value (or values) of z corresponding to some point x, y on our curve? It is easiest to use the second equation—it is linear in z.

In[70] := p2

Out[70] = $-34-20x+6x^2-23y+6xy+9x^2y-23z+6xz+9x^2z$

In[71] := Do[*c*[*i*] = Coefficient[p2, *z*, *i*], {*i*, 0, 1}]

Does $c[1]$ vanish somewhere in our region?

In[72] := *s* = Solve[*c*[1] == 0, *x*]

Out[72] = $\left\{\left\{x\to\frac{1}{3}\left(-1-2\sqrt{6}\right)\right\},\left\{x\to\frac{1}{3}\left(-1+2\sqrt{6}\right)\right\}\right\}$

In[73] := *N*[*x*/.*s*]

Out[73] = $\{-1.96633, 1.29966\}$

No, it does not.

In[74] := Clear[*s*]

So there is a single solution:

In[75] := z1 = −*c*[0]/*c*[1]

Out[75] = $\dfrac{34+20x-6x^2+23y-6xy-9x^2y}{-23+6x+9x^2}$

And what about the third and fourth equations?

In[76] := p3 = Numerator[Together[p3/.*z*−>z1]]

Out[76] = $-20\left(-15-34x-23x^2-34y-20xy+6x^2y-23y^2+6xy^2+9x^2y^2\right)$

In[77] := p4 = Numerator[Together[p4/.*z*−>z1]]

Out[77] = $-15-34x-23x^2-34y-20xy+6x^2y-23y^2+6xy^2+9x^2y^2$

In[78] := Cancel[p3/p1]

Out[78] = -20

In[79] := Cancel[p4/p1]

Out[79] = 1

They are satisfied automatically. What z corresponds to $x=y=-1/3$?

In[80] := z1/.{*x*−> −1/3, *y*−> −1/3}

Out[80] = $-\dfrac{7}{9}$

So, one of the solutions found earlier, namely $x=y=z=-1/3$, does not belong to our one-dimensional family of solutions. To summarize: we have found one isolated solution plus a one-dimensional family of solutions. In the parametric form:

In[81] := xyz[*t*_] := With[{xy = Which[*t* < 1, pFA[*t*], *t* < 2, pAE[*t*],
**   *t* < 3, pED[*t*], True, pDF[*t*]]},**
** {xy[[1]], xy[[2]], z1/.{*x*−>xy[[1]], *y*−>xy[[2]]}}]**

In[82] := P1 = ParametricPlot3D[xyz[*t*], {*t*, 0, 4},
** PlotRange−>{{−1,0},{−1,0},{−1,0}}, ViewPoint−>{10, 11, 12}];**

In[83] := p0 = {−1/3, −1/3, −1/3};

In[84] := P2 = Graphics3D[{Darker[Green], PointSize[Large], Point[p0]}];
In[85] := Show[P1, P2]

Out[85] =

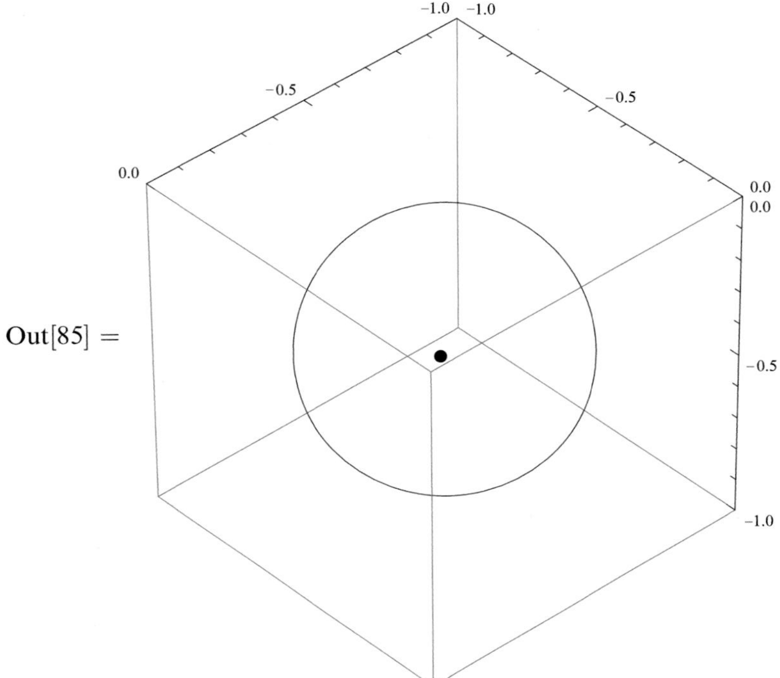

You can rotate this plot with your mouse to understand it better.

24.6 Shape of the Molecule

What does the cyclohexane molecule look like? Let's direct the x-axis along $a[1]$:
In[86] := *a*[1] = {1,0,0}
Out[86] = $\{1,0,0\}$
Let $a[2]$ lie in the x, y plane:
In[87] := *a*[2] = {1/3, 2 ∗ Sqrt[2]/3, 0}
Out[87] = $\left\{\frac{1}{3}, \frac{2\sqrt{2}}{3}, 0\right\}$
That is, the unit vector along y is a combination of $a[1]$ and $a[2]$:
In[88] := (3 ∗ *a*[2] − *a*[1])/(2 ∗ Sqrt[2])
Out[88] = $\{0,1,0\}$
The projections of $a[3]$ onto x and y are $c[1,3] = x$ and
In[89] := (3 ∗ *c*[2,3] − *c*[1,3])/(2 ∗ Sqrt[2])
Out[89] = $\frac{1-x}{2\sqrt{2}}$

The projection of $a[3]$ onto the z axis can be found from normalization:

In[90] := a[3] = {x, (1 − x)/(2 ∗ Sqrt[2]),
pm ∗ Sqrt[(1 − x) ∗ (7 + 9 ∗ x)]/(2 ∗ Sqrt[2])}

$$\text{Out[90]} = \left\{x, \frac{1-x}{2\sqrt{2}}, \frac{\text{pm}\sqrt{(1-x)(7+9x)}}{2\sqrt{2}}\right\}$$

where pm $= \pm 1$. Two molecule shapes correspond to a single set of values of x, y, z; they differ by the mirror reflection of the z coordinates. We shall discuss this matter in a moment.

In[91] := Table[Expand[a[i].a[3]]/.pm^2−>1, {i, 1, 3}]

$$\text{Out[91]} = \left\{x, \frac{1}{3}, 1\right\}$$

That is, the unit vector along the z axis is a combination of $a[1]$, $a[2]$, $a[3]$:

In[92] := Simplify[2 ∗ Sqrt[2]/Sqrt[(1 − x) ∗ (7 + 9 ∗ x)]∗
(a[3] − 3/8 ∗ (1 − x) ∗ a[2] + (1 − 9 ∗ x)/8 ∗ a[1])]

Out[92] = {0, 0, pm}

The rest is easy.

In[93] := Do[Print[a[i] = Simplify[{c[1, i], (3 ∗ c[2, i] − c[1, i])/(2 ∗ Sqrt[2]),
2 ∗ Sqrt[2] ∗ pm/Sqrt[(1 − x) ∗ (7 + 9 ∗ x)]∗
(c[3, i] − 3/8 ∗ (1 − x) ∗ c[2, i] + (1 − 9 ∗ x)/8 ∗ c[1, i])}]],
{i, 4, 6}]

$$\left\{-\frac{5}{3} - x - z, \frac{\frac{5}{3} + x + 4z}{2\sqrt{2}}, \frac{\text{pm}\left(1 + 9x^2 - 4z + 2x(7 + 6z)\right)}{2\sqrt{2}\sqrt{7 + 2x - 9x^2}}\right\}$$

$$\left\{z, -\frac{5 + 3y + 4z}{2\sqrt{2}}, -\frac{\text{pm}(-5 - 11y - 4z + x(5 + 3y + 12z))}{2\sqrt{2}\sqrt{7 + 2x - 9x^2}}\right\}$$

$$\left\{\frac{1}{3}, \frac{-1 + 9y}{6\sqrt{2}}, \frac{\text{pm}(-13 - 11y + x(-11 + 3y))}{2\sqrt{2}\sqrt{7 + 2x - 9x^2}}\right\}$$

Let's write a function which constructs the molecule for a given values of x, y, z and of the sign pm.

In[94] := rc = 0.1; rs = 0.25;

In[95] := Molecule[xyz_, s_] := Module[{r = {0, 0, 0}, r2, l = {Blue},
S = {x−>xyz[[1]], y−>xyz[[2]], z−>xyz[[3]], pm−>s}},
Do[r2 = r + (a[i]/.S); l = Append[l, Cylinder[{r, r2}, rc]]; r = r2, {i, 1, 6}];
r = {0, 0, 0}; l = Append[l, Red];
Do[r2 = r + (a[i]/.S); l = Append[l, Sphere[r, rs]]; r = r2, {i, 1, 6}];
Graphics3D[l]]

This is the isolated conformation of the cyclohexane molecule with $x = y = z = -1/3$. Use your mouse to understand it better.

In[96] := Show[Molecule[p0, 1], ViewPoint−>{15, −5, 5}, Boxed−>False]

Out[96] = 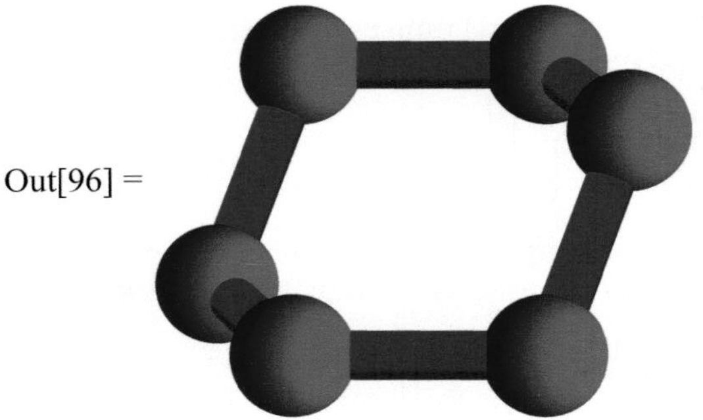

And this is the one-parameter family of conformations. To ensure smooth dependence on t, it is necessary to flip the sign pm when passing through the point C (where the expression under the radical sign vanishes). This happens at

```
In[97] := t0 = (621 - 8 * Sqrt[6])/159
```

Out[97] = $\frac{1}{159}\left(621 - 8\sqrt{6}\right)$

So, the molecule returns to its initial shape after we traverse the loop in the x, y, z space twice. You can see this conformation family especially clearly if you start animation.

```
In[98] := Manipulate[Show[
          Molecule[If[t > 4, xyz[t - 4], xyz[t]], If[t > t0&&t < t0 + 4, -1, +1]],
          PlotRange->{{-0.7, 1.7}, {-0.5, 1.9}, {-1.7, 1.7}},
          ViewPoint->{10, -10, 4}, Boxed->False],
     {t, 0, 8}]
```

Out[98] = 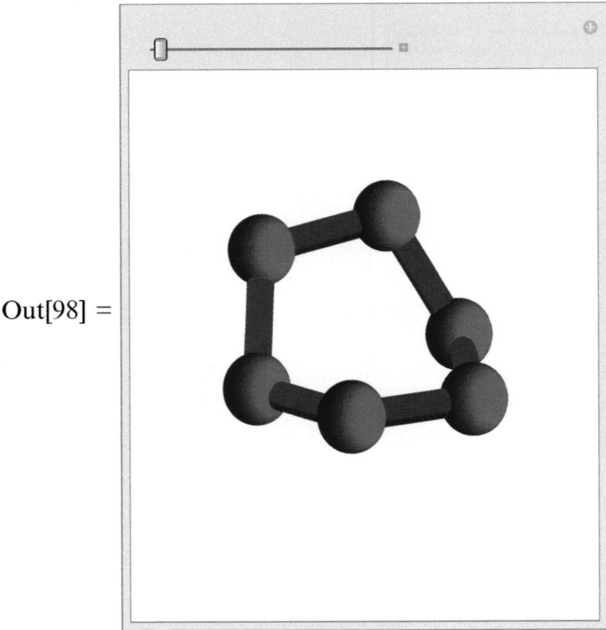

Chapter 25
Problems for Students

1. Write a procedure which returns the hydrogen wave function (in spherical coordinates, i. e., an expression containing r, θ, ϕ) for given quantum numbers n, l, m. Write a procedure to calculate the rate of the electric dipole transition [22] from the state n, l, m to the state n', l', m'.

2. Calculate Poisson brackets of the Hamiltonian, the angular momentum components, and the Runge–Lenz vector components [19] for a particle in the Coulomb field $U = -a/r$. Calculate commutators of the same quantities in quantum mechanics [18].

3. The hypergeometric function [23, 24, 27] is defined as the sum of the series

$$F(a,b,c,x) = \sum_{n=0}^{\infty} \frac{(a)_n (b)_n}{(c)_n} \frac{x^n}{n!},$$

where $(x)_n = x(x+1)\cdots(x+n-1)$ is the Pochhammer symbol. In many cases it can be expressed via simpler functions. Write a list of substitutions for simplifying hypergeometric functions. It is sufficient to consider only simplifications valid for an arbitrary x (not for specific values) where results are expressed via elementary functions. More general substitutions should be near the beginning of the list, then their particular cases can be eliminated.

4. Consider indefinite integrals of the form

$$\int A(x) \log B(x) \, \mathrm{d}x,$$

where $A(x)$ and $B(x)$ are rational functions of x. *Mathematica* can calculate such integrals, but often produces results in which some terms have imaginary parts in the region of x we are interested in. It is not easy to trace their cancellations. We'll suppose that $A(x)$ and $B(x)$ contain no parameters (except x), only numbers. We'll also suppose that *Mathematica* is able to find all roots of the denominator of $A(x)$, as well as of the numerator and the denominator of $B(x)$, and all these roots are real.

A. Grozin, *Introduction to Mathematica® for Physicists*, Graduate Texts in Physics,
DOI 10.1007/978-3-319-00894-3_25,

We are interested in a neighborhood of some point x_0; we want to get a result all terms of which are real near this point (if this is possible, of course). Implement the following obvious approach:

- Expand $A(x)$ into partial fractions with respect to x.
- Replace $\log B(x)$ by a combination of terms $\log(x-a_i)$ and $\log(a_i-x)$ (plus a constant) in such a way that they are all real near x_0.
- Multiply.
- Take integrals of $x^n \log(x-a)$ $(n \geq 0)$, $\log(x-a)/(x-b)^n$ $(n \geq 2)$ by parts to eliminate the logarithm. Don't use the *Mathematica* integrator—it can produce $\log(x-a)$ where $\log(a-x)$ is needed.
- We are left with the most difficult terms of the forms $\log(x-a)/(x-b)$ and $\log(a-x)/(x-b)$. By linear substitutions they reduce to 3 cases:

$$\int \frac{\log(y+1)}{y}\,dy = -\mathrm{Li}_2(-y)\,,$$
$$\int \frac{\log(y-1)}{y}\,dy = \log(y)\log(y-1)+\mathrm{Li}_2(1-y)\,,$$
$$\int \frac{\log(1-y)}{y}\,dy = -\mathrm{Li}_2(y)\,,$$

where y is positive near $x = x_0$ (the third formula is the definition of $\mathrm{Li}_2(y)$; the first one follows from it using the substitution $y \to -y$; the second one—using integration by parts).

The result must be real ($\log(x)$ is real at $x > 0$; $\mathrm{Li}_2(x)$—at $x < 1$). If this is impossible, print an error message.

5. Implement the algebra of Boolean expressions. They consist of the constants true and false, variables, the function not (one argument), and the functions and, or (an arbitrary number of arguments). The last two functions are commutative and associative. Take into account simplifications when one of the arguments is true or false; when two arguments coincide or equal to a and not$[a]$. Expressions should be reduced to the disjunctive normal form: "or" at the top level; its arguments can be "and"; their arguments can be "not" or variables.

6. Implement the algebra of quaternions.

7. Implement Dirac γ-matrix expressions, including trace calculations (in 4 dimensions [22] or in the general case of dimensional regularization, see, e.g., [24]). Pay no attention to efficiency.

8. Implement calculation of color factors of Feynman diagrams for the color group $SU(N_c)$ using the Cvitanović algorithm [27] (see also [24]).

9. Write a procedure to calculate two-loop massless propagator diagrams using integration by parts (see, e.g., [24]). Results should be linear combinations of the two basis integrals.

10. Hypergeometric functions whose argument is 1 and whose parameters contain a small parameters ε and tend to integers at $\varepsilon \to 0$ can be expanded in series in ε. The algorithm is described, e.g., in [24]; implement it.

11. Any polynomial over the field of complex numbers can be factorized into linear factors:

$$p(x) = \prod (x - a_i)^{d_i} ,$$

where a_i are its roots and d_i are their multiplicities (to simplify formulas, we have assumed that the leading coefficient is 1). Let's group factors with equal d_i:

$$p(x) = \prod p_i^{d_i} ,$$

where all d_i are distinct and the polynomials $p_i(x)$ have only simple zeros (are square-free). This square-free factorization can be obtained by a simple algorithm which uses only gcd (this is much simpler than the full factorization). Namely,

$$\gcd(p, p') = \prod p_i^{d_i - 1} .$$

Indeed, the polynomial $p(x)$ has zero of the order d_i at $x \to a_i$, and its derivative $p'(x)$ has zero of the order $d_i - 1$. Write a function to calculate square-free factorization using only gcd.

References

1. Buchberger, B., Collins, G.E., Loos, R. (ed.): Computer Algebra: Symbolic and Algebraic Computation, 2nd edn. Springer, Vienna (1983)
2. Davenport, J.H., Siret, Y., Tournier, E.: Computer Algebra: Systems and Algorithms for Algebraic Computation, 2nd edn. Academic Press, London (1993)
3. Geddes, K.O., Czapor, S.R., Labahn, G.: Algorithms for Computer Algebra. Kluwer Academic Publishers, Boston (1992)
4. von zur Gathen, J., Gerhard, J.: Modern Computer Algebra, 2nd edn. Cambridge University Press, Cambridge (2003)
5. Grozin, A.G.: Using REDUCE in High Energy Physics. Cambridge University Press, New York (1997); paperback edition (2005)
6. Wolfram, S.: The *Mathematica* Book, 5th edn. Wolfram Media, Champaign (2003)
7. Trott, M.: The *Mathematica* GuideBook for Symbolics. Springer Science+Business Media, Inc., New York (2006)
8. Trott, M.: The *Mathematica* GuideBook for Numerics. Springer Science+Business Media, Inc., New York (2006)
9. Trott, M.: The *Mathematica* GuideBook for Graphics. Springer, New York (2004)
10. Trott, M.: The *Mathematica* GuideBook for Programming. Springer, New York (2004)
11. Mangano, S.: *Mathematica* Cookbook, O'Reilly Media, Inc., Sebastopol, CA (2010)
12. Cox, D., Little, J., O'Shea, D.: Ideals, Varieties, and Algorithms, 3rd edn. Springer, New York (2007)
13. Arzhantsev, I.V.: Gröbner Bases and Systems of Algebraic Equations (in Russian), 3rd edn. MCCMO, Moscow (2003). http://www.mccme.ru/free-books/dubna/arjantsev.pdf
14. Kredel, H., Weispfenning, V.: J. Symbolic Computing dimension and independent sets for polynomial ideals **6**, 231 (1988)
15. Fateman, R.J.: A review of *Mathematica*. J. Symbolic Comput. **13**, 545 (1992). (http://www.cs.berkeley.edu/~fateman/papers/mma.pdf); http://www.cs.berkeley.edu/~fateman/papers/mma6rev.pdf
16. Bronstein, M.: Symbolic Integration I. Springer, Berlin (1997)
17. Davenport, J.H.: On the integration of algebraic functions. In: Lecture notes in computer science, vol. 102. Springer, New York (1981)
18. Landau, L.D., Lifshitz, E.M.: Quantum Mechanics: Non-relativistic Theory, 3rd edn. Butterworth-Heinemann, Oxford (1981)
19. Landau, L.D., Lifshitz, E.M.: Mechanics, 3rd edn. Butterworth-Heinemann, Oxford (1982)
20. Landau, L.D., Lifshitz, E.M.: The Classical Theory of Fields, 4th edn. Butterworth-Heinemann, Oxford (1980)
21. Landau, L.D., Lifshitz, E.M., Pitaevskii, L.P.: Electrodynamics of Continuous Media, 2nd edn. Butterworth-Heinemann, Oxford (1995)

A. Grozin, *Introduction to Mathematica® for Physicists*, Graduate Texts in Physics,
DOI 10.1007/978-3-319-00894-3,

22. Berestetskii, V.B., Lifshitz, E.M., Pitaevskii, L.P.: Quantum Electrodynamics, 2nd edn. Butterworth-Heinemann, Oxford (1982)
23. Prudnikov, A.P., Brychkov, Yu.A., Marichev, O.I.: Integrals and Series, vol. 3, Chapter 7. Gordon and Breach, New York (1990)
24. NIST Handbook of Mathematical Functions, ed. by F.W.J. Olver, D.W. Lozier, R.F. Boisvert, C.W. Clark, Cambridge University Press, Cambridge (2010). http://dlmf.nist.gov/
25. http://functions.wolfram.com/
26. Grozin, A.G.: Lectures on QED and QCD: Practical calculation and renormalization of one- and multi-loop Feynman diagrams. World Scientific (2007)
27. Cvitanović, P.: Group Theory. Princeton University Press, Princeton (2008). http://www.nbi.dk/GroupTheory/:

Index

A. Grozin, *Introduction to Mathematica® for Physicists*, Graduate Texts in Physics, DOI 10.1007/978-3-319-00894-3, © Springer International Publishing Switzerland 2014

Printed in the United States
By Bookmasters